Land Carbon Cycle Modeling

Carbon moves through the atmosphere, through the oceans, onto land, and into ecosystems. This cycling has a large effect on climate – changing geographic patterns of rainfall and the frequency of extreme weather – and is altered as the use of fossil fuels adds carbon to the cycle. The dynamics of this global carbon cycling are largely predicted over broad spatial scales and long periods of time by Earth system models. This book addresses the crucial question of how to assess, evaluate, and estimate the potential impact of the additional carbon to the land carbon cycle. The contributors describe a set of new approaches to land carbon cycle modeling for better exploring ecological questions regarding changes in carbon cycling; employing data assimilation techniques for model improvement; doing real- or near-time ecological forecasting for decision support; and combining newly available machine learning techniques with process-based models to improve prediction of the land carbon cycle under climate change. This new edition includes seven new chapters: machine learning and its applications to carbon cycle research (five chapters); principles underlying carbon dioxide removal from the atmosphere, contemporary active research and management issues (one chapter); and community infrastructure for ecological forecasting (one chapter).

Key Features

- Helps readers understand, implement, and criticize land carbon cycle models
- Offers a new theoretical framework to understand transient dynamics of the land carbon cycle
- Describes a suite of modeling skills – matrix approach to represent land carbon, nitrogen, and phosphorus cycles; data assimilation and machine learning to improve parameterization; and workflow systems to facilitate ecological forecasting
- Introduces a new set of techniques, such as semi-analytic spin-up (SASU), unified diagnostic system with a 1-3-5 scheme, traceability analysis, and benchmark analysis, and PROcess-guided machine learning and DAta-driven modeling (PRODA) for model evaluation and improvement
- Reorganized from the first edition with seven new chapters added
- Strives to balance theoretical considerations, technical details, and applications of ecosystem modeling for research, assessment, and crucial decision-making

Land Carbon Cycle Modeling

Matrix Approach, Data Assimilation, Ecological Forecasting, and Machine Learning

Second Edition

Edited by
Yiqi Luo and Benjamin Smith

CRC Press
Taylor & Francis Group
Boca Raton London New York

CRC Press is an imprint of the
Taylor & Francis Group, an **informa** business

Designed cover image: Benjamin Smith

Second edition published 2024
by CRC Press
2385 NW Executive Center Drive, Suite 320, Boca Raton FL 33431

and by CRC Press
4 Park Square, Milton Park, Abingdon, Oxon, OX14 4RN

CRC Press is an imprint of Taylor & Francis Group, LLC

First edition published by CRC Press 2022

Library of Congress Cataloging-in-Publication Data
Names: Luo, Yiqi, editor. | Smith, Benjamin, editor.
Title: Land carbon cycle modeling : matrix approach, data assimilation,
ecological forecasting, and machine learning / edited by Yiqi Luo and Benjamin Smith.
Description: Second edition. | Boca Raton : CRC Press, 2024. |
Includes bibliographical references and index.
Identifiers: LCCN 2023053960 (print) | LCCN 2023053961 (ebook) |
ISBN 9781032698496 (hbk) | ISBN 9781032711119 (pbk) | ISBN 9781032711126 (ebk)
Subjects: LCSH: Carbon. | Carbon cycle (Biogeochemistry)
Classification: LCC QE516.C37 L66 2024 (print) | LCC QE516.C37 (ebook) |
DDC 577/.144–dc23/eng/20240109
LC record available at https://lccn.loc.gov/2023053960
LC ebook record available at https://lccn.loc.gov/2023053961

ISBN: 978-1-032-69849-6 (hbk)
ISBN: 978-1-032-71111-9 (pbk)
ISBN: 978-1-032-71112-6 (ebk)

DOI: 10.1201/9781032711126

Typeset in Times LT Std
by Newgen Publishing UK

Contents

About the Editors ... ix
List of Contributors ... xi
Preface to the First Edition ... xiii
Yiqi Luo and Benjamin Smith
Preface to the Second Edition ... xv
Yiqi Luo and Benjamin Smith

UNIT ONE Fundamentals of Carbon Cycle Modeling

Chapter 1 Theoretical Foundation of the Land Carbon Cycle and Matrix Approach ... 3
Yiqi Luo

Chapter 2 Introduction to Modeling ... 11
Benjamin Smith

Chapter 3 Flow Diagrams and Balance Equations of Land Carbon Models ... 18
Yuanyuan Huang

Chapter 4 Practice 1: Carbon Flow Diagram and Carbon Balance Equations ... 24
Yuanyuan Huang

UNIT TWO Matrix Representation of Carbon Balance

Chapter 5 Developing Matrix Models for Land Carbon Models ... 29
Yuanyuan Huang

Chapter 6 Coupled Carbon-Nitrogen Matrix Models ... 34
Zheng Shi and Xingjie Lu

Chapter 7 Compartmental Dynamical Systems and Carbon Cycle Models ... 45
Carlos A. Sierra

Chapter 8 Practice 2: Matrix Representation of Carbon Balance Equations and Coding ... 51
Yuanyuan Huang

UNIT THREE Carbon Cycle Diagnostics for Uncertainty Analysis

Chapter 9 Unified Diagnostic System for Uncertainty Analysis ... 57
Yiqi Luo

Chapter 10 Matrix Phosphorus Model and Data Assimilation 63
Enqing Hou

Chapter 11 Principles Underlying Carbon Dioxide Removals from the Atmosphere 69
Yiqi Luo

Chapter 12 Practice 3: Diagnostic Variables in Matrix Models 74
Xingjie Lu

UNIT FOUR Semi-Analytic Spin-Up (SASU)

Chapter 13 Nonautonomous ODE System Solver and Stability Analysis 81
Ying Wang

Chapter 14 Semi-Analytic Spin-Up (SASU) of Coupled Carbon-Nitrogen Cycle Models 89
Xingjie Lu and Jianyang Xia

Chapter 15 Time Characteristics of Compartmental Systems 94
Carlos A. Sierra

Chapter 16 Practice 4: Efficiency and Convergence of Semi-Analytic Spin-Up (SASU) in TECO 98
Xingjie Lu

UNIT FIVE Traceability and Benchmark Analysis

Chapter 17 Overview of Traceability Analysis 105
Jianyang Xia

Chapter 18 Applications of the Transient Traceability Framework 112
Lifen Jiang

Chapter 19 Benchmark Analysis 121
Yiqi Luo and Forrest M. Hoffman

Chapter 20 Practice 5: Traceability Analysis for Evaluating Terrestrial Carbon Cycle Models 126
Jianyang Xia and Jian Zhou

UNIT SIX Introduction to Data Assimilation

Chapter 21 Data Assimilation: Introduction, Procedure, and Applications ... 133
Yiqi Luo

Chapter 22 Bayesian Statistics and Markov Chain Monte Carlo Method in Data Assimilation ... 140
Feng Tao

Chapter 23 Application of Data Assimilation to Soil Incubation Data ... 146
Junyi Liang and Jiang Jiang

Chapter 24 Practice 6: The Seven-Step Procedure for Data Assimilation ... 152
Xin Huang

UNIT SEVEN Data Assimilation with Field Measurements and Satellite Data

Chapter 25 Model-Data Integration at the SPRUCE Experiment ... 163
Daniel Ricciuto

Chapter 26 Application of Data Assimilation to a Peatland Methane Study ... 167
Shuang Ma

Chapter 27 Global Carbon Cycle Data Assimilation Using Earth Observation: The CARDAMOM Approach ... 173
Mathew Williams

Chapter 28 Practice 7: Data Assimilation at the SPRUCE Site ... 182
Shuang Ma

UNIT EIGHT Ecological Forecasting with EcoPAD

Chapter 29 Introduction to Ecological Forecasting ... 187
Yiqi Luo

Chapter 30 Ecological Platform for Assimilating Data (EcoPAD) for Ecological Forecasting ... 192
Yuanyuan Huang

Chapter 31 Community Cyberinfrastructure for Ecological Forecasting ... 199
Xin Huang and Lifen Jiang

Chapter 32 Practice 8: Ecological Forecasting at the SPRUCE Site ... 204
Jiang Jiang

UNIT NINE Machine Learning and its Applications to Carbon Cycle Research

Chapter 33 Introduction to Machine Learning and its Application to Carbon Cycle Research 211
Yuanyuan Huang

Chapter 34 Estimation of Terrestrial Gross Primary Productivity Using Long Short-Term Memory Network 217
Yao Zhang and Jinghao Qiu

Chapter 35 Machine Learning to Predict and Explain Complex Carbon Cycle Interactions 224
Julia K. Green

Chapter 36 Practice 9: Applications of Machine Learning to Predict Soil Organic Carbon Content 229
Feng Tao

UNIT TEN Process-based Machine Learning

Chapter 37 Introduction to Machine Learning and Neural Networks 237
Toby Dylan Hocking

Chapter 38 PROcess-Guided Deep Learning and DAta-Driven Modeling (PRODA) 244
Feng Tao and Yiqi Luo

Chapter 39 Hybrid Modeling in Earth System Science 253
Yu Zhou

Chapter 40 Practice 10: Deep Learning to Optimize Parameterization of CLM5 258
Feng Tao

Appendix 1: Matrix Algebra in Land Carbon Cycle Modeling 263
Ye Chen

Appendix 2: Introduction to Programming in Python 268
Xin Huang

Appendix 3: CarboTrain User Guide 275
Jian Zhou

Bibliography 285

Index 293

About the Editors

Yiqi Luo is Liberty Hyde Bailey Professor at Cornell University, USA. He obtained his PhD degree from the University of California, Davis in 1991 and did postdoctoral research at UCLA and Stanford University from 1991 to 1994, before he worked at the Desert Research Institute as Assistant and Associate Research Professor from 1994 to 1998, the University of Oklahoma as Associate, Full, and George Lynn Cross Professor from 1999 to 2017, and Northern Arizona University as Full and Regents Professor. Professor Luo has studied land carbon cycling using empirical and modeling approaches for more than 30 years. His research program has been focused on addressing two key issues: (1) how global change alters the structure and functions of terrestrial ecosystems, and (2) how terrestrial ecosystems regulate climate change. To address these issues, Dr. Luo's laboratory has conducted field global change experiments; developed terrestrial ecosystem models; synthesized extensive data sets using meta-analysis methods; integrated data and models with data assimilation techniques; and carried out theoretical and computational analysis. Particularly, his research team has recently developed a matrix approach to land carbon cycle modeling; applied data assimilation techniques to ecological research; and pioneered in ecological forecasting. Previously he has published three books, 36 book chapters, and close to 600 papers in peer-reviewed journals. He was a Highly Cited Researcher recognized by the Web of Science Group, Clarivate Analytics in 2018–2023. He was elected fellow of the American Association for the Advancement of Science (AAAS) in 2013; the American Geophysical Union (AGU) in 2016; and the Ecological Society of America (ESA) in 2018. This book, *Land Carbon Cycle Modeling: Matrix Approach, Data Assimilation, Ecological Forecasting, and Machine Learning 2e*, evolved from an international training course, *New Advances in Land Carbon Cycle Modeling*. The training course has been held six times from 2018 to 2023. The materials in the book have been partly or fully used by approximately 600 attendees of the training course.

Benjamin Smith is a Professor of Ecosystem Science, based in Sydney, Australia, where he is Research Director of Western Sydney University's Hawkesbury Institute for the Environment, a leading center for global change ecosystem research and innovation. Following undergraduate studies in biology at the University of Tasmania, Australia, Ben relocated to Dunedin, New Zealand, where he obtained his PhD from the University of Otago in 1996. Following postdoctoral posts in Sweden and Germany, he obtained tenure at Lund University in Sweden, transitioning to his current role at Western Sydney University in 2018. Benjamin Smith is known as a pioneer in the dynamic global vegetation modeling field. The original developer of the widely used LPJ-GUESS ecosystem model, he continues to lead a multi-institutional consortium of developers serving its international user community. As a Visiting Scientist with CSIRO Oceans & Atmosphere Flagship, he contributed to the implementation of vegetation demography, disturbance, and wildfire dynamics in the Australian Community Land Surface Model, CABLE. He is interested in the role of the biosphere in regional and global climate dynamics, using earth system models to examine feedback of ecosystems and land surface changes to the atmosphere and climate. He led the development of the first published regional earth system model, RCA-GUESS, and is active in the pan European consortium developing the global EC-EARTH ESM. An author of several influential papers in the global change modeling and assessment fields, Ben was recognized in the Clarivate Highly Cited Researcher list from 2019 to 2021.

Contributors

Ye Chen
Northern Arizona University, USA

Julia K. Green
University of Arizona, USA

Toby Dylan Hocking
Northern Arizona University, USA

Forrest Hoffman
Oak Ridge National Laboratory, USA

Enqing Hou
South China Botanical Garden, Chinese Academy of Sciences, China

Xin Huang
National Center for Atmospheric Research, USA

Yuanyuan Huang
Institute of Geographic Sciences and Natural Resources Research, Chinese Academy of Sciences, China

Jiang Jiang
Nanjing Forestry University, China

Lifen Jiang
Cornell University, USA

Junyi Liang
China Agricultural University, China

Yiqi Luo
Cornell University, USA

Xingjie Lu
Sun Yat-sen University, China

Shuang Ma
University of California, Los Angeles, USA

Jinghao Qiu
Peking University, China

Daniel Ricciuto
Oak Ridge National Laboratory, USA

Carlos Sierra
Max Plank Institute of Biogeochemistry, Germany

Zheng Shi
University of Oklahoma, USA

Benjamin Smith
Western Sydney University, Australia

Feng Tao
Cornell University, USA

Ying Wang
University of Oklahoma, USA

Mathew Williams
University of Edinburgh, UK

Jianyang Xia
East China Normal University, China

Yao Zhang
Peking University, China

Jian Zhou
Cornell University, USA

Yu Zhou
Cornell University, USA

Preface to the First Edition

Yiqi Luo
School of Integrative Plant Science, Cornell University, Ithaca, USA

Benjamin Smith
Hawkesbury Institute for the Environment, Western Sydney University, Richmond, Australia

The ecosystems of the vegetated land surface are critical to the health of the planet, its biodiversity, and people, providing benefits, so-called ecosystem services, on which human health, well-being and economic activity rely. The carbon cycle connects ecosystems on land to the atmosphere and the climate system. Rising temperatures, changing rainfall distributions, and more frequent extreme weather impact ecosystem services, often in a negative way. Rising atmospheric carbon dioxide concentrations – the main proximal cause of climate change – also affect ecosystems directly, enhancing photosynthesis and plant water-use efficiency in experimental settings, and almost certainly in nature. Due to a small imbalance between global photosynthesis, absorbing carbon dioxide from the atmosphere, and the return flux (mainly decomposition of litter and soil carbon) from the land to the atmosphere, a sizeable proportion of anthropogenic CO_2 emissions are reabsorbed by ecosystems – the land carbon sink. A key goal of international climate policy, to supplement fossil fuel reductions with negative emissions over a transitional period, largely relies on management interventions to preserve and enhance the land carbon sink. The land carbon cycle, then, is central both to mitigation (emissions reduction) and adaptation (management of climate impacts and risks) responses and policies. Developing effective measures requires predictions of how the system may be expected to respond under various scenarios. For this, of course, we need models.

While relevant to supporting climate science and policy, modeling is also gaining "market share" in environmental and ecological research generally. Several forces are involved in this trend. Ever increasing volumes of readily available data from observational and experimental networks are making it easier to parameterize and robustly validate models. Cheaper, more powerful computers, including cloud computing services, bring complex numerical algorithms and data assimilation methods within reach for many applications. New statistical and optimization methodologies have been developed and made accessible through convenient packages and toolboxes, useful not only to modelers but as platforms for collaboration between modelers and empirical scientists.

This book provides an overview of the current state of the land carbon cycle modeling field, exemplifying recent developments as described above. The book is built upon a summer training course, *New Advances in Land Carbon Cycle Modeling*, held annually since 2018 at Northern Arizona University. Over the four years the course has been offered, attendees from 32 countries in six continents have undertaken the training. The first training course in 2018 attracted about 40 participants, with a similar number in 2019. Due to the pandemic of COVID-19, the in-person course was replaced by an online version in 2020. Originally, we planned to have 25 attendees, but ended up with 85 participants from six continents. This grew further to 150 virtual attendees in 2021.

This book is mainly based on 31 lectures (including pre-training lectures) and ten practices prepared for the *New Advances in Land Carbon Cycle Modeling* training course in 2020 and 2021.

The book offers cutting-edge knowledge and techniques on carbon cycle science and modeling. We have designed ten training units in such a way that everyone can gain regardless of their prior background in modeling. The chapters range from theoretical foundation of land carbon cycling, traceability, and data assimilation to machine learning and ecological forecasting. Overall, four techniques are covered: the matrix approach to land carbon modeling; data assimilation for data-driven modeling; ecological forecasting; and combined machine learning with data assimilation to improve model prediction.

The organization of the book aligns with the training course, which has two blocks. The first block in units 1–5 is about the matrix approach to land carbon cycle modeling. The second block in units 6–10 is on data assimilation, ecological forecasting, and machine learning.

The matrix approach introduced in units 1–5 first describes a matrix equation. It is demonstrated that the matrix equation can unify land carbon cycle models; offer a new theoretical framework to guide carbon cycle research; help accelerate computational efficiency for spin-up; offer new analytics to diagnose model performance; and allow data assimilation of complex models. Five skills are covered in units 1–5, namely: drawing the carbon flow diagram and writing carbon balance equations of a model; developing matrix models from carbon balance equations and coding the matrix model; adding diagnostic variables to matrix models; adding semi-analytic spin-up (SASU) algorithms; and traceability analysis.

The matrix equation can be used to derive three diagnostic variables: carbon input, residence time, and carbon storage potential. The matrix equation can also be used to get an analytic solution of steady-state pool sizes, leading to SASU. The matrix equation is the foundation for traceability analysis. The chapters of units 1–5 explain these new skills.

Units 6–10 cover data assimilation, ecological forecasting, and machine learning. To realistically forecast ecological responses to climate change, three elements all need to be perfectly aligned: model structure, model parameterization, and the external forcing variables. The matrix approach discussed in units 1–5 is about process-based model structure. Data assimilation and machine learning in units 6–8 and 10 will help improve model parameterization. EcoPAD, a workflow system that is described in unit 9, will link real-time forcing to model forecasting.

Chapters in unit 6 describe the seven-step procedure of data assimilation. The seven steps are: defining a research objective; acquiring data sets; using one model; developing a cost function; minimizing mismatches between modeled and observed values with a global optimization method; estimating parameters; and predicting ecosystem responses. Chapters in units 7 and 8 are about applications of data assimilation to the SPRUCE field experiment and satellite observations, and evaluation of values of different data sets to constrain models and their predictions.

Chapters in unit 9 describe the Ecological Platform for Assimilating Data (EcoPAD) framework, which automatically ingests data into a model through a data assimilation system for ecological foresting.

Chapters in unit 10 introduce machine learning, a PROcess-guided deep learning combined with DAta-driven modeling (PRODA) approach, and its application to optimize parameterization of the CLM5 land surface model.

Most of the chapters are written in such a way as to be understandable by readers with minimal modeling background. Practices are targeted at a suitable level for such readers. A few chapters may require some mathematical background to be fully understood. The book offers three appendix chapters on, respectively: basic linear algebra; introductory programming with Python; and the Carbon Training (*CarboTrain*) package we use as a toolbox for the training course and the practice chapter in each unit. Depending on their prior knowledge, readers may choose to read these appendix chapters as optional supporting material to the chapters in units 1–10.

All the chapters are accompanied with pre-recorded lectures or practice instruction. These pre-recorded videos are available at https://www2.nau.edu/luo-lab/download/4th_training_course.php. (Please search for the videos using "ecolab Yiqi Luo" if the website is moved away from NAU.) If you plan to master skills described in the chapters, you may find it useful to read the book chapter; listen to the corresponding video; take the quiz at the end of each chapter after watching the video; and attempt the practice chapter at the end of each unit, following the pre-recorded instruction.

Please be aware that this book does not teach programming or how to code a model, nor does it teach how to do model development or modification. Readers with limited programming experience may, however, find the brief introduction to Python coding in appendix 2 useful.

The open access electronic version of this book has been made available thanks to financial contributions by Northern Arizona University, Lund University, and Oak Ridge National Laboratory. Finally, we wish to thank all 22 authors who have worked very hard for months to prepare the material for this book. We hope that you, the reader, find it useful and rewarding.

Yiqi Luo, Ben Smith, *September, 2021.*

Preface to the Second Edition

Yiqi Luo
School of Integrative Plant Sciences, Cornell University, Ithaca, USA

Benjamin Smith
Hawkesbury Institute for the Environment, Western Sydney University, Richmond, Australia

The Second Edition expands the coverage of this book to include newly developed modeling techniques within carbon cycle research, particularly within the realm of machine learning and artificial intelligence (AI). The second edition has one new unit (Unit 9) on machine learning applications within carbon cycle research, replacing a unit from the first edition on the value of data to constrain models and their predictions. The new unit has four chapters, respectively covering machine learning in carbon cycle research (Chapter 33), Long Short-Term Memory network (Chapter 34), machine learning to predict and explain complex carbon cycle interactions (Chapter 35), and a practice on applications of random forest machine learning to predict soil organic carbon content (Chapter 36).

In addition, this edition also adds new chapters on community cyberinfrastructure for ecological forecasting – added to the unit on ecological forecasting (Chapter 31) – and on hybrid modeling, added to the PRODA unit (Chapter 38). The new chapter *Principles Underlying Carbon Dioxide Removals from the Atmosphere* highlights carbon dioxide removal (CDR) activities as a growing application domain for carbon cycle modeling and data synthesis (Chapter 11). It replaces an earlier chapter on sensitivity analysis with the matrix model from the first edition of the book. There are also updates and minor corrections to several other chapters. Pre-recorded videos that are accompanied with all the chapters can be accessed via https://ecolab.cals.cornell.edu/?workshop. (Please search for the videos using "ecolab Yiqi Luo" if the website is moved away from Cornell University.)

We greatly appreciate the efforts of Dr. Lifen Jiang in supporting the editors in getting this edition over the line.

We have been heartened by the generally positive response to the first edition of this book, including from participants of the international training course, *New Advances in Land Carbon Cycle Modeling,* for which it constitutes a central resource. The feedback received has informed the revised scope and content of this current edition. We hope it continues to provide a useful and accessible resource that supports the growing interest in and use of quantitative approaches across carbon cycle science.

Yiqi Luo, Ben Smith, *October, 2023.*

Unit One

Fundamentals of Carbon Cycle Modeling

1 Theoretical Foundation of the Land Carbon Cycle and Matrix Approach

Yiqi Luo
Cornell University, Ithaca, USA

The land carbon cycle has been extensively studied yet its fundamental properties have not been fully understood. This chapter offers empirical evidence to demonstrate a general dynamic pattern that the land carbon cycle changes in a direction toward a moving attractor in response to global change. This general pattern is captured by a matrix equation. The relatively simple matrix equation can unify land carbon cycle models, accelerate computational efficiency for spin-up, diagnose model performance with new analytics, enable data assimilation with complex models to improve their predictive skills, and guide carbon cycle research with a new theoretical framework of dynamic disequilibrium.

CONVERGENCE OF THE LAND CARBON CYCLE

In the late 2010s, a deserted village, Houtouwan, on an island off the east coast of China was discovered to be completely overrun by vegetation approximately 20 years after dwellers left (Smithsonian Channel, The Abandoned Chinese Village that Nature Reclaimed) (Figure 1.1). What might happen to this place in another 20 years or even longer? It is likely that trees will gradually take over to form a coastal forest.

In fact, Mother Nature would overrun all urban places in the world if human disturbances were removed. Imagine that we could magically remove humans from any highly commercialized, heavily human-disturbed urban areas, such as Manhattan of New York City or Lujiazui of Shanghai, for 300 years: the place would soon be overrun by plants, animals, and microbes. Without any human activities, small trees would grow from cracks in concrete in five years, most of the high-rise buildings would collapse and forests would probably take over in 50 years. In 300 years, Manhattan would most likely be occupied by a deciduous forest similar to those in northeastern USA, and Lujiazui of Shanghai by some lowland forests.

Similarly, it has been repeatedly observed how vegetation takes over landscapes after natural disturbances occur. For example, following the 1988 Yellowstone fires – massive blazes that burned about 1.2 million acres in and around Yellowstone National Park – their size and severity led to a proclamation that Yellowstone had been destroyed. In fact, the burned landscape was retaken by thriving young lodgepole pine trees 30 years after the fires (Turner 2018). Secondary succession is an ecological term for this entirely natural process. Ecosystem succession has been extensively studied, mainly from the perspectives of species dynamics and community structures.

From a carbon cycle perspective, ecosystems converge toward some attractor states after anthropogenic and natural disturbances are removed. (Note that the states that ecosystems converge toward are moving attractors under global change.

FIGURE 1.1 Overrunning of the deserted village of Houtouwan by vegetation approximately 20 years after dwellers left the island that is situated off the east coast of China. Vegetation taking over after anthropogenic and natural disturbances are removed is caused by the internal processes of the carbon cycle that drive an ecosystem toward an attractor.

DOI: 10.1201/9781032711126-2

This point will be discussed later.) This convergence is done by "Mother Nature" and actually results from the internal processes of the land carbon cycle. What, then, are those internal processes? And, how can we mathematically represent them?

DONOR POOL-DOMINANT TRANSFER AND OTHER PROPERTIES THAT GOVERN THE LAND CARBON CYCLE

Before we answer those questions, let us briefly review the land carbon cycle. Carbon enters an ecosystem via photosynthesis. Photosynthetic products are partly allocated for autotrophic respiration and partly for growth of leaf, stem, and root. Once a plant or its parts die, they become litter and enter the litter pools. Litter decomposes, partly released to the atmosphere via heterotrophic respiration and partly incorporated into soil to become soil organic matter. Soil organic matter goes through decomposition and stabilization over and over again, driving soil carbon cycling. We will examine some of these processes to see what the best equation would be to represent them.

Let us first look at litterfall in which dead leaves fall from the tree canopy to the ground. To make it a subject of study, we define two new terms: donor pool and recipient pool. The donor pool donates litter whereas the recipient pool receives litter (Figure 1.2a). Litterfall is a rate process that moves carbon from the donor pool to the recipient pool. In this case, the rate of litter falling is proportional to the amount of litter in the donor pool while the amount of litter in the recipient pool has no effect on the rate of litterfall at all. Thus, the rate of litterfall is controlled by the donor pool.

The rate of litterfall, donated by $dX(t)/d(t)$, equals the donor pool size, $X(t)$, times a coefficient (k) as:

$$\frac{dX(t)}{dt} = kX(t) \tag{1.1}$$

This equation describes the *donor pool-dominated carbon transfer* (Figure 1.2a). Litterfall in the real world is also affected by wind and other environmental conditions over seasons. Modelers usually use an environmental scalar, $\xi(t)$,

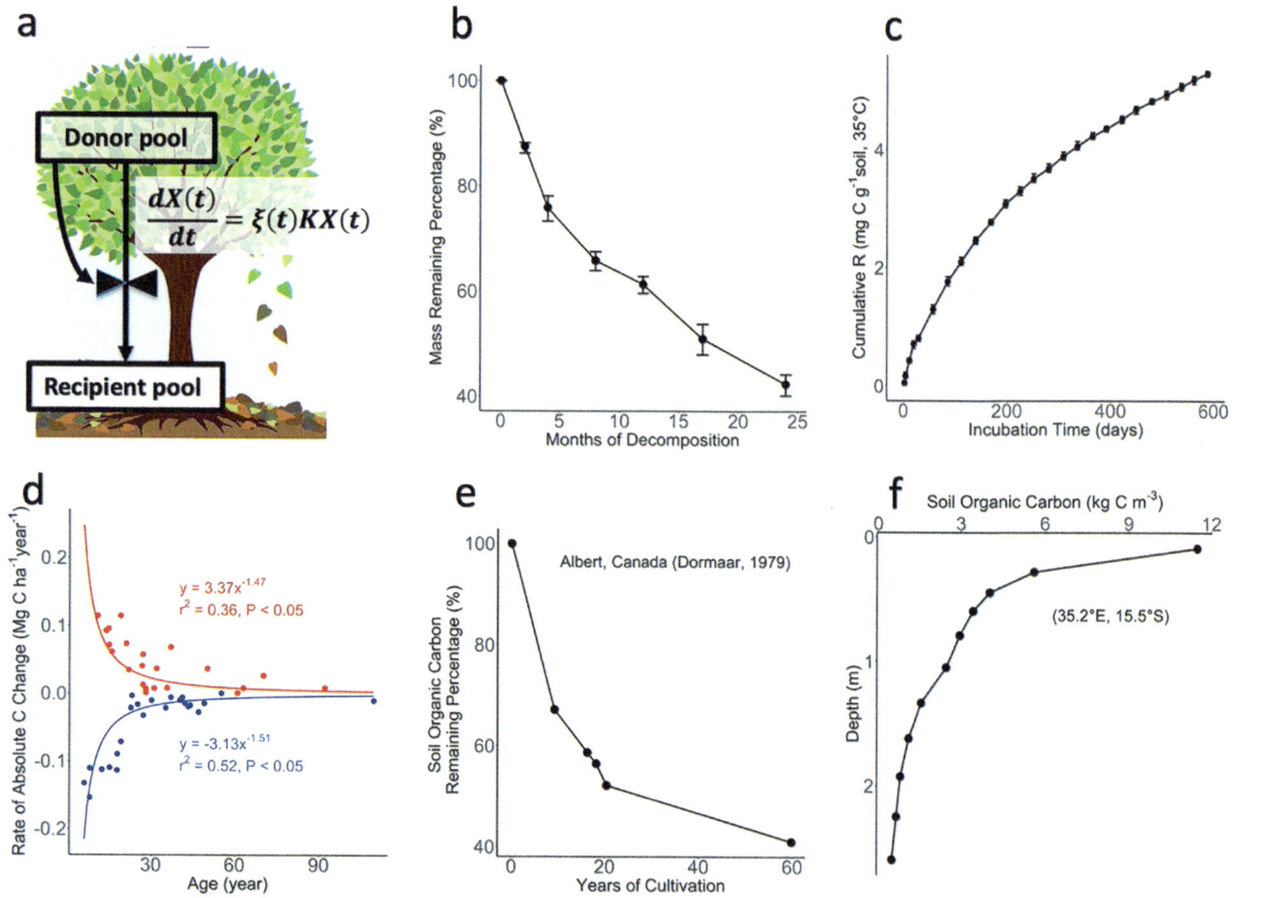

FIGURE 1.2 Macroscopic patterns of carbon transfer processes and distributions. The carbon transfer processes include (a) litterfall; (b) decomposition of ragweed (*Ambrosia psilostachya*) litter (Cheng et al. 2010); (c) soil organic carbon decomposition from incubation at 35 °C (replotting the data from Haddix et al. 2011); (d) rates of absolute carbon (C) change during the forest succession (Yang et al. 2011); (e) changes in soil organic matter after agricultural cultivation in Alberta, Canada (extracted data from Doomaar 1979); and (f) a vertical distribution of soil carbon with depth in Malawi, Africa, from a soil carbon database. The macroscopic patterns of almost all carbon transfer processes and distributions are governed by donor-pool dominated carbon transfer and can be described by the first-order decay function.

to account for the effects of phenology, wind, and other environmental factors on litterfall as:

$$\frac{dX(t)}{dt} = \xi(t)kX(t) \tag{1.2}$$

Once litter has fallen on the ground, it decomposes. Litter decomposition is usually studied with litterbags or wood logs on the ground or in air. Researchers initiate a study with a certain amount of litter in litterbags, place them in the field, then collect a subset of litterbags once every few weeks, find the dry weight, and calculate the dry weight remaining in comparison with the initial amount. A typical data set of litter decomposition shows that mass remaining becomes less and less as sampling time goes on (Figure 1.2b). This type of data set can usually be fitted by a first-order decay curve as in Equation 1.1. Thus, litter decomposition is often described by the same *donor pool-dominated carbon transfer* equation. In this case, k is often called the litter decay constant. Actually, k varies with litter type and location. Zhang et al. (2008) synthesized nearly 300 datasets from 70 studies all over the world. The study found that Equation 1.1 fits all data sets very well although the value of k greatly varies with litter types and environment. Cai et al. (2018) synthesized more than 1,600 data sets of straw decomposition for six types of crops. Their study fits a three-exponent equation to all the data sets. Thus, straw decomposition also follows the *donor pool-dominated carbon transfer*.

Another important process of the carbon cycle is soil organic carbon (SOC) decomposition. SOC decomposition is usually studied by soil incubation. That is, researchers collect soil samples from the field and put the samples in jars for a period of time. They then collect gas samples once in a while to measure the amount of carbon released from the soil sample through microbial respiration. Data are usually plotted either by measured CO_2 release or cumulative CO_2 released on the y-axis with time on the x-axis (Figure 1.2c). Almost all data follow a similar pattern. This pattern also can be well described by *donor pool-dominated transfer*.

Schädel et al. (2013) fitted data of SOC decomposition with one-, two-, and three-pool models. A one-pool model is developed to simulate SOC decomposition based on the assumption that all organic carbon compounds in the soil are homogeneous in term of decomposability. A two- or three-pool model assumes that soil organic compounds are heterogeneous and need different coefficients to represent decomposition of different cohorts (i.e., pools) of organic compounds. Schädel et al. (2013) found that two or three-pool models work well for SOC decomposition. Schädel et al. (2014) synthesized more than 120 data sets from permafrost regions. Xu et al. (2016) synthesized nearly 400 data sets from different locations all over the world. Both of the latter studies found that all their data sets (more than 500) can be well fitted by two or three-pool models. This suggests that the donor pool-dominated carbon transfer equation also works for SOC decomposition.

Yang, Luo, and Finzi (2011) synthesized more than 124 studies of soil carbon dynamics at different stages of forest succession. During the course of secondary succession, some forests gain carbon and some lose carbon (Figure 1.2d). In either case, carbon dynamics still can be described by the *donor-pool dominated carbon transfer*. Moreover, the soil organic matter remaining after long-term cultivation (Figure 1.2e) and the vertical distribution of soil organic carbon with depth (Figure 1.2f) both follow monotonic patterns, consistent with the donor-pool dominated carbon transfer.

So far, we have examined macroscopic patterns of key carbon transfer processes (e.g., litterfall, litter decomposition, and SOC decomposition) and distributions. These macroscopic patterns are very typical as almost ubiquitously revealed by thousands of field and laboratory studies. The macroscopic patterns can be well described by the donor pool-dominated carbon transfer equation.

The *donor pool-dominated carbon transfer*, then, is one of the four fundamental properties that govern dynamics of the land carbon cycle. The other three properties are (1) photosynthesis as the primary carbon influx pathway, (2) compartmentalization of carbon processes into plant, litter, and soil, and (3) the first-order kinetics of carbon transfer from the donor pool (Luo and Weng 2011) (Equation 1.1). Among the four properties, the *donor pool-dominated carbon transfer* is the most important property in determining the trajectory of the land carbon cycle (Luo, Keenan, and Smith 2015). If this property is altered in our model, the carbon cycle will not behave as we have observed in the real world.

The four properties fundamentally characterize the internal processes of the land carbon cycle. The internal processes drive the land carbon cycle to converge toward an attractor (Luo et al. 2017). This is the reason why places like Manhattan could become deciduous forests and Liujiazui of Shanghai could become lowland forests if human disturbances were removed. This convergence is applicable to almost any place on Earth.

Active research is going on to incorporate microbial processes and traits into carbon cycle models to account for the important role of microbes in catalyzing decomposition of soil organic matter (Chandel et al. 2023). Many of the microbial models were developed on purpose because the responsible researchers suspected that models based on first-order kinetics of carbon transfer from donor pools were too simple. However, these newly developed microbial models usually represent decomposition processes using nonlinear functions, such as Michaelis-Menten equations. Liang et al. (2018) tested different types of decomposition functions (e.g., first-order kinetics, Michaelis-Menten, and reverse Michaelis-Menten functions) with 84 data sets and found that the first-order kinetics function fit the observed macroscopic patterns of SOC decomposition better than the Michaelis-Menten or reverse Michaelis-Menten functions. It remains challenging to develop litter and SOC decomposition models that not only adequately represent our understanding of microbial processes but also are consistent with the macroscopic patterns observed from almost all experimental studies.

Next, we examine how well these four properties are represented in models.

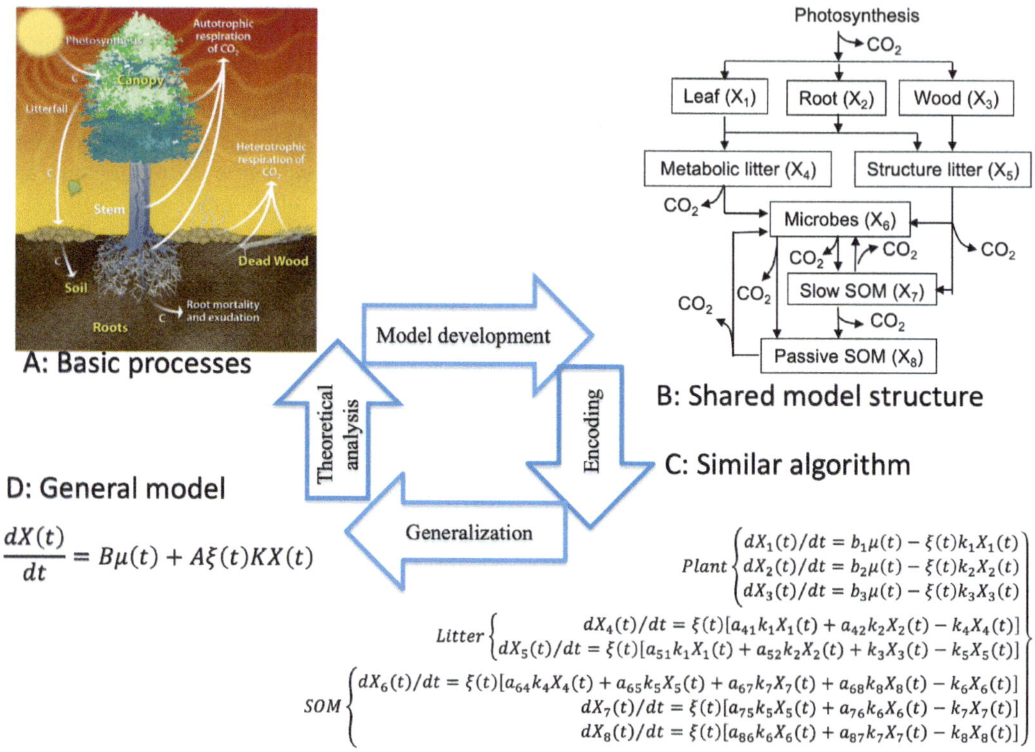

FIGURE 1.3 A generalized matrix model of the terrestrial carbon cycle. (A) The basic carbon cycle processes are represented by four fundamental properties for all terrestrial ecosystems. (B) The four properties have been incorporated into terrestrial carbon cycle models with a pool-and-flux structure. (C) The structure is typically encoded using a set of balance equations with carbon input into and output from each pool. (D) The balance equations of terrestrial carbon cycle models can be converted to a matrix equation. Thus, the matrix equation can be considered as a general system equation (or a dynamical equation) for the terrestrial carbon cycle.

THE MATRIX APPROACH TO MODEL REPRESENTATION OF THE LAND CARBON CYCLE

The four properties identified above are all well represented in models as long as the models use a so-called pool-and-flux or box-and-arrow structure. In this structure, we use pools to represent different carbon compartments and fluxes to represent carbon transfer among compartments or carbon input into and output out of the ecosystem. For this structure, we need one carbon balance equation to trace how much carbon gets into one pool and how much carbon leaves the pool. For example, the leaf pool, X_1, receives carbon from photosynthesis partitioned to leaves and loses carbon by senescence (Figure 1.3). Thus, we need one equation to calculate the amount of carbon resulting from photosynthesis and the amount of carbon lost to litterfall. The equation to describe the dynamics of carbon balance in the leaf pool over time can be represented by:

$$\frac{dX_1(t)}{dt} = b_1\mu(t) - \xi(t)k_1X_1(t) \tag{1.3}$$

where $\mu(t)$ represents the amount of carbon input from net primary production (NPP, photosynthesis minus autotrophic respiration), b_1 is the carbon partitioning from NPP to leaves, k_1 is the rate of senescence, and $\xi(t)$ is an environmental modifier. Thus, the change in the carbon pool size in the leaf pool $\left(\frac{dX_1(t)}{dt}\right)$ equals carbon input to the leaf pool $b_1\mu(t)$ minus carbon leaving the leaf pool $\xi(t)k_1X_1(t)$.

Extending this idea to all the eight pools of the Terrestrial Ecosystem (TECO) model, we have eight carbon balance equations to track carbon cycling in the ecosystem as:

$$\begin{cases} \left.\begin{array}{l} dX_1(t)/dt = b_1\mu(t) - \xi(t)k_1X_1(t) \\ dX_2(t)/dt = b_2\mu(t) - \xi(t)k_2X_2(t) \\ dX_3(t)/dt = b_3\mu(t) - \xi(t)k_3X_3(t) \end{array}\right\} \text{plane} \\ \left.\begin{array}{l} dX_4(t)/dt = \xi(t)\left[\alpha_{41}k_1X_1(t) + \alpha_{42}k_2X_2(t) - k_4X_4(t)\right] \\ dX_5(t)/dt = \xi(t)\left[\alpha_{51}k_1X_1(t) + \alpha_{52}k_2X_2(t) - k_3X_3(t) - k_5X_5(t)\right] \end{array}\right\} \\ \left.\begin{array}{l} dX_6(t)/dt = \xi(t)\left[\alpha_{64}k_4X_4(t) + \alpha_{65}k_5X_5(t) + \alpha_{6,}k_7X_7 + \alpha_{68}k_8X_8(t) - k_6X_6(t)\right] \\ dX_7(t)/dt = \xi(t)\left[\alpha_{75}k_5X_5(t) + \alpha_{76}k_6X_6(t) + k_7X_7(t)\right] \\ dX_8(t)/dt = \xi(t)\left[\alpha_{86}k_6X_6(t) + \alpha_{87}k_7X_7(t) + k_8X_8(t)\right] \end{array}\right\} \end{cases} \tag{1.4}$$

where X_i, $i = 1, 2, \ldots, 8$, is the amount of carbon, respectively, in three plant, two litter, and three SOC pools; b_1, b_2, and b_3 are plant partitioning coefficients to leaf, root, and wood; k_i, $i = 1, 2, \ldots, 8$, is the rate of carbon leaving the eight pools (i.e., process rate or exit rate); a_{ij}, $i = 1, 2, \ldots, 8$, $j = 1, 2, \ldots, 8$, is the transfer coefficient of carbon from pool j to pool i; and $\xi(t)$ is the environment modifier. The eight carbon balance equations in Equation 1.4 can be reorganized into a matrix form as:

$$\begin{pmatrix} dX_1(t)/dt \\ dX_2(t)/dt \\ dX_3(t)/dt \\ dX_4(t)/dt \\ dX_5(t)/dt \\ dX_6(t)/dt \\ dX_7(t)/dt \\ dX_8(t)/dt \end{pmatrix} = \begin{pmatrix} b_1 \\ b_2 \\ b_3 \\ 0 \\ 0 \\ 0 \\ 0 \\ 0 \end{pmatrix} \mu(t) + \xi(t) \begin{pmatrix} -1 & 0 & 0 & 0 & 0 & 0 & 0 & 0 \\ 0 & -1 & 0 & 0 & 0 & 0 & 0 & 0 \\ 0 & 0 & -1 & 0 & 0 & 0 & 0 & 0 \\ \alpha_{41} & \alpha_{42} & 0 & -1 & 0 & 0 & 0 & 0 \\ \alpha_{51} & \alpha_{52} & 1 & 0 & -1 & 0 & 0 & 0 \\ 0 & 0 & 0 & \alpha_{64} & \alpha_{65} & -1 & \alpha_{67} & \alpha_{68} \\ -1 & 0 & 0 & 0 & \alpha_{75} & \alpha_{76} & -1 & 0 \\ -1 & 0 & 0 & 0 & 0 & \alpha_{86} & \alpha_{87} & -1 \end{pmatrix} \times \begin{pmatrix} k_1 & & & & & & & \\ & k_2 & & & & & & \\ & & k_3 & & & & & \\ & & & k_4 & & & & \\ & & & & k_5 & & & \\ & & & & & k_6 & & \\ & & & & & & k_7 & \\ & & & & & & & k_8 \end{pmatrix} \begin{pmatrix} X_1(t) \\ X_2(t) \\ X_3(t) \\ X_4(t) \\ X_5(t) \\ X_6(t) \\ X_7(t) \\ X_8(t) \end{pmatrix} \quad 1.5$$

In this equation, a_{53} equals 1 as all the carbon from the wood pool goes to the structural litter pool. The above matrix equation can be succinctly expressed as:

$$\frac{dX(t)}{dt} = B\mu(t) + A\xi(t)KX(t) \quad 1.6$$

where X(t) is a vector of pool sizes, B is a vector of partitioning coefficients from carbon input to each of the pools, $\mu(t)$ is the carbon input rate, A is a matrix with −1 in the diagonal and transfer coefficients in the off-diagonal to quantify carbon movement along the pathways, K is a diagonal matrix of process rates (mortality for plant pools and decomposition coefficients for litter and soil pools) from donor pools, and $\xi(t)$ can be a scalar or a diagonal matrix of environmental modifiers to represent responses of the carbon cycle to changes in temperature, moisture, and oxygen. When $\xi(t)$ is a diagonal matrix, the environmental modifiers can be the same for all the pools or different for individual pools.

Equation 1.6 is generalizable to represent land carbon cycle models that follow first-order kinetics. It describes net carbon pool change, dX/dt, as a difference between carbon input $\mu(t)$, distributed to different plant pools via partitioning coefficients B, and carbon loss through the transformation matrices ($A\xi(t)K$) among individual pools $X(t)$.

As Equation 1.6 is generic to unify the land carbon cycle models that follow first-order kinetics, any single model is a special case of the generic equation. For example, the Community Land Model version 4.5 (CLM4.5) incorporates carbon transfer among seven pools per soil layer over 10 layers (Koven et al. 2013). The seven pools in each layer are metabolic litter, cellulose litter, lignin litter, coarse woody debris, fast soil organic matter, slow soil organic matter, and passive soil organic matter. Carbon vertical transfers mainly occur between one layer and the next layer down. Huang et al. (2018) converted the soil carbon module of CLM4.5 into one matrix equation as:

$$\frac{dX(t)}{dt} = B\mu(t) + A\xi(t)KX(t) + V(t)X(t) \quad 1.7$$

The matrix equation above has three items on the right side, i.e., carbon input and plant partitioning, carbon transfers and release, and vertical movement. The dimension of the matrix equation is 70 for 70 pools. Transfer matrix A, for example, is a 7×7 block matrix. Each element of the block matrix is a 10×10 matrix. The vertical movement matrix, $V(t)$, includes carbon transferring to the next layer up and the next layer down.

Equation 1.7 shares similar mathematic properties with Equation 1.6. Equation 1.7 has 70 dimensions whereas Equation 1.6 has eight dimensions. Equation 1.7 describes carbon transfers among 70 pools over 10 soil layers where Equation 1.6 describes carbon transfers among eight pools, including plant, litter, and soil pools without explicitly defined soil layers.

Lu et al. (2020) converted the CLM5 carbon and nitrogen cycles into four matrix equations, two for vegetation carbon and nitrogen cycles and two for soil carbon and nitrogen cycles (see Chapter 6). The vegetation carbon cycle in CLM5, for example, contains 18 pools, including six tissue pools: leaf, fine root, live stem, dead stem, live coarse root, and dead coarse root. Each tissue pool is accompanied by a storage pool and a transfer pool. A crop grain tissue pool, accompanied by a grain storage pool and a grain transfer pool, is added when the crop model is used. Vegetation carbon dynamics are controlled by phenology for the leaf onset and offset dates according to air temperature and soil water conditions. Harvest and fire remove part of plant carbon from ecosystems and part goes to litter pools according to the harvest rate. Natural death moves carbon from plant pools to litter pools at defined mortality rates. The fire module triggers the occurrence of occasional fire events based on the amount of the fuel (i.e., litter) and the soil moisture. When fire occurs, carbon in plant pools is partly released to the atmosphere and partly transferred to the litter pools based on their tissue quality and burned area. Thus, the vegetation carbon dynamics are represented by the following matrix equation:

$$\frac{dX_v}{dt} = B\mu(t) + \left(A_{phc}(t)K_{phc} + A_{gmc}(t)K_{gmc} + A_{fic}(t)K_{fic}X_v\right)(t) \quad 1.8$$

Here, X_v is an n-entry vector, representing vegetation carbon pool size. K is an $n \times n$ diagonal matrix. Its subscripts *phc*, *gmc*, and *fic* indicate carbon processes related to phenology, gap mortality (i.e., harvest from land use and natural mortality), and fire, respectively. The diagonal entries are the process (or exit) rates of all vegetation carbon pools due to phenology (K_{phc}), gap mortality (K_{gmc}), and fire (K_{fic}). Once converted, the matrix equations make CLM5 more modular, analytically clear, easily diagnosed, and computationally more efficient for spin-up.

Equation 1.8 similarly shares mathematical properties with Equation 1.6. Equation 1.8 uses three processes: phenology, mortality, and fire, to control carbon transfers among 18 vegetation pools whereas Equation 1.6 only uses the environmental scalar $\xi(t)$ to control carbon transfer.

Some other global models, such as CABLE, LPJ-GUESS and ORCHIDEE, have also been converted to matrix equations. Once a model is converted to a matrix equation, it appears quite simple; a first-order differential equation. In fact, the equation is not that simple as it represents a nonautonomous system, which is discussed below. With this unified matrix equation, we can explore the general properties of carbon models and the general behavior of the land carbon cycle. This is what we call the *matrix approach*.

Note that expression of a land carbon cycle model in a matrix form may not, in and of itself, represent much significant advance in carbon cycle research. The ability to express carbon dynamics in a matrix form was recognized several decades ago (e.g., Lasaga 1980, Sundquist 1985). The matrix models are merely used to represent a set of differential equations in a matrix form to describe carbon cycling among multiple pools of the earth system (Bolin and Eriksson 1958). What we propose is to take the matrix expression to unify land carbon cycle models, develop a theoretical framework to guide carbon research, pinpoint sources of model uncertainty (Chapters 17 and 18), accelerate spin-up of land carbon cycle models by tens of times (Chapters 14 and 16), and enable data assimilation (Chapter 38) (Luo et al. 2022).

The matrix representation can be adapted to accommodate nonlinear dynamics. For instance, Equation 1.6 can be modified for a nonlinear microbial model as:

$$\frac{dX(t)}{dt} = B(t)\mu(t) + A(t)\xi(t)K(t)X(t) \quad 1.9$$

Thus, carbon input U, plant carbon partitioning B, transfer coefficients A, and process rates K all are potentially functions of pool sizes X. Carlos Sierra's group from the Max Planck Institute for Biogeochemistry in Germany have converted many nonlinear microbial models into matrix equations and explored their general behavior (Sierra and Müller 2015, Metzler, Müller, and Sierra 2018).

THE PARADOX OF THE MATRIX EQUATION AND NONAUTONOMOUS SYSTEMS

While we argue that all the land carbon cycle models that follow first-order kinetics can be converted into a unified matrix form, the general equation is mathematically extremely simple (e.g., Equation 1.6). A paradox arises: how can such a simple equation represent the extremely complex phenomena of the carbon cycle observed in the real world?

To explore this paradox, we organized a workshop in 2012, sponsored by US National Institute for Mathematical and Biological Synthesis, NIMBioS. We invited 20 applied mathematicians and 20 ecologists to explore the paradox: why is the matrix equation (Equation 1.6) extremely simple whereas carbon cycle phenomena observed in the real world can be very complex? We presented this paradox to the workshop participants. In the first couple of days, it was very difficult to convince the group about this issue. Some ecologists said that ecosystems are complex and cannot be described by such a simple equation. They urged us to re-examine this mathematical expression. We told them that we have used thousands of data sets to verify the equation. The applied mathematicians told us that this equation is too simple to be interesting enough for them to do any study with. They suggested that we add a nonlinear term in the equation. We told them that if a nonlinear term is added to the equation, it no longer describes the land carbon cycle as revealed by data. We were stubborn and kept pointing out the paradox. After one and a half days, Dr. James Cushing, an applied mathematician from the University of Arizona, told us that the system we were studying is probably a nonautonomous system.

Nonautonomous systems have been a subject of mathematics research in recent decades. They comprise dynamical systems with input and parameters being time dependent. The matrix equation of the land carbon cycle as expressed by Equation 1.6, indeed, has its inputs and parameters being time dependent.

With this new insight, we organized a working group to study the nonautonomous system, sponsored once again by NIMBioS. The working group consisted of a few applied mathematicians and a few ecologists. One of the group members, Dr. Martin Rasmussen, a professor from Imperial College London, had coauthored two books on nonautonomous systems. He taught us a lot about how to study nonautonomous systems. We worked together over 3 years and had four meetings.

The working group ultimately made a few key findings. First, it found that nonlinear models of soil carbon decomposition (e.g., Equation 1.9) generate unrealistic responses to small perturbations and carbon input (Wang et al. 2014, 2016). A stability analysis and numerical simulations were conducted for two nonlinear microbial models (a two-pool model and a three-pool model) of soil carbon decomposition. Both models exhibit dampening oscillatory responses to small perturbations. In addition, the equilibrium pool sizes of litter or soil carbon are insensitive to carbon inputs in the nonlinear microbial models. This oscillatory behavior and insensitivity of soil carbon to carbon input exhibited by the nonlinear models have not been observed in the real world.

Second, we identified a mathematical foundation through proof of theorems on exponential stability to explain observed convergence of the land carbon cycle (Rasmussen et al. 2016). The land carbon cycle can be considered as a linear nonautonomous compartmental system described by a dynamical equation (i.e., Equation 1.6). The equation can be rewritten as:

$$\frac{dX(t)}{dt} = B\mu(t) + G(t)X(t) \quad 1.10$$

where

$$G(t) = A\xi(t)K$$

Matrix G is invertible. The entries g_{ij} (t) of G satisfy three conditions. First, there is always carbon to leave any individual pool as g_{ii} $(t) < 0$ for all i. Second, there is either no carbon flow pathway or some amount of carbon moving from pool j to pool i (i.e., g_{ij} $(t) \geq 0$ for all $i \neq j$). Third, carbon exiting pool i is either all going to other pools or some going to other pools and some being lost from the system via respiration (i.e., $\sum_{i=1}^{d} g_{ij}(t) \leq 0$ for all j). As the land carbon cycle satisfies the three conditions, the system is exponentially converging to a time-dependent attractor, which is very useful for our understanding of predictability of the land carbon cycle as discussed below.

The working group also found a mathematical representation for a transient dynamical equation of the land carbon cycle (Luo et al. 2017), which is discussed in the last section of this chapter.

PREDICTABILITY OF THE LAND CARBON CYCLE

In the very beginning of this chapter, we showed convergence of the land carbon cycle toward some attractor states by "Mother Nature". This "Mother Nature" is the exponential stability of the carbon cycle equation (Rasmussen et al. 2016). The certainty of any ecosystems converging to some attractor states after human and natural disturbances reflects the high predictability of the land carbon cycle (Figure 1.4).

Given one type of forcing, patterns of carbon cycle dynamics are usually highly predictable (Luo et al. 2015). For example, periodic climate forcing over seasons usually influences the carbon cycle system to generate periodicity in carbon fluxes. Similarly, one fire disturbance event generates a pulse release of carbon from land to the atmosphere followed by gradual recovery. The gradual recovery is well illustrated by vegetation overrunning the deserted village in China (Figure 1.1) and thriving young lodgepole pine trees recolonizing the burned landscape after the 1988 Yellowstone fires (Turner 2018) as described in the first section of this chapter.

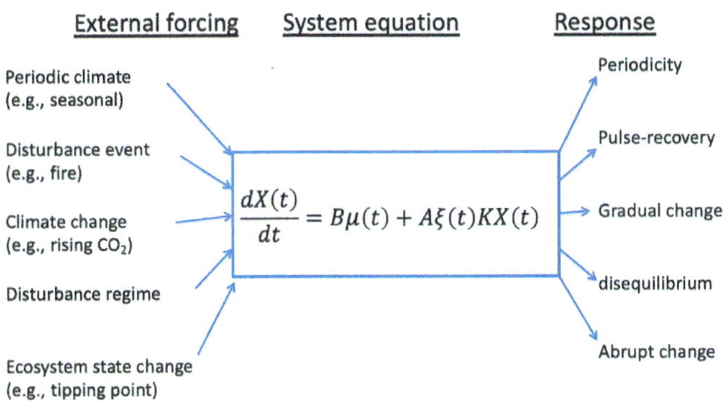

FIGURE 1.4 Predictability of the terrestrial carbon cycle. Responses of the carbon cycle to a given type of external forcing are highly predictable.

Global change factors, such as rising atmospheric CO_2 concentration and warming, usually generate gradual but directional changes in carbon cycle processes. Shifts in fire regimes usually lead to disequilibrium in the carbon cycle. And ecosystem state changes, due, for example, to land use change from forests to croplands, usually result in abrupt changes in ecosystem properties, such as photosynthetic capacity and carbon processes that depend on it. While many processes of the carbon cycle are highly predictable, precisely predicting carbon storage changes under climate change requires lots of data to constrain model parameterization (Luo and Schuur 2020).

DYNAMIC DISEQUILIBRIUM OF LAND CARBON CYCLE

It is well known that the carbon cycle dynamics at steady state can be described by two terms: carbon input and residence time. The product of these two terms determines the carbon storage capacity. However, directional climate change pushes the land carbon cycle out of steady state towards dynamic disequilibrium (Luo and Weng 2011). To quantify the dynamic disequilibrium, we need a third term, in addition to carbon input and residence time, to represent the transient dynamics of the land carbon cycle. The working group supported by NIMBioS worked very hard to find a mathematical expression of the transient dynamics. We derived a variety of mathematical formulations for the transient dynamics of the carbon cycle. One day we came up with the third term to describe the disequilibrium of the land carbon cycle. The original derivation was very complicated. After a few rounds of re-organization, we can now explain the mathematical derivation in a very simple way. Basically, we can multiply each term of Equation 1.6 by the inverse matrix of $A\xi(t)K$, $(A\xi(t)K)^{-1}$, and then re-organize the equation to get an equation of the form:

$$X(t) = \left(-A\xi(t)K\right)^{-1} B\mu(t) - \left(-A\xi(t)K\right)^{-1} X'(t) \quad 1.11$$

Then, the transient dynamics of the land carbon cycle can be expressed by:

$$X(t) = \tau_E \mu(t) - X_p(t) \quad 1.12$$

where τ_E is ecosystem residence time as defined by

$$\tau_E = \left(-A\xi(t)K\right)^{-1} B \quad 1.13$$

and X_p is carbon storage potential. It represents the disequilibrium term of carbon cycle as:

$$X_p(t) = \left(-A\xi(t)K\right)^{-1} X'(t) \tag{1.14}$$

The first part in the right side of Equation 1.12 is the carbon storage capacity $X_c(t)$ as:

$$X_c(t) = \tau_E \mu(t)$$

This transient equation is also the dynamical equation of the land carbon cycle. The term dynamical equation is often used in mathematics to describe how the state of a system changes over time. Although the formulation of the carbon dynamical equation is simple, it represents a nonautonomous system, which is influenced by five categories of external forcing variables (i.e., periodic climate, disturbance event, disturbance regime shift, climate change, and ecosystem state change) (Figure 1.4) (Luo et al. 2017). Those external forcing variables are superimposed on each other to influence carbon cycle dynamics, generating complex phenomena. That is the reason why we observe complex phenomena of carbon cycle dynamics in the real world. Behind the phenomena is a relatively simple dynamical equation to govern the trajectory of the carbon cycle.

In the following units of this book, we will show you how the matrix approach can unify land carbon cycle models, accelerate spin-up, offer new analytics for model diagnostics, and facilitate data assimilation to improve the fit of models to observations.

SUGGESTED READINGS

Luo Y, YY Huang, CA Sierra, JY Xia, A Ahlström, YZ Chen, O Hararuk, EQ Hou, LF Jiang, CJ Liao, XJ Lu, Z Shi, B Smith, F Tao, YP Wang. (2022) Matrix approach to land carbon cycle modeling. *Journal of Advances in modeling Earth System, 14* (7), DOI: 10.1029/2022MS003008

Luo YQ, ES Weng. (2011) Dynamic disequilibrium of the terrestrial carbon cycle under global change. *Trends in Ecology & Evolution, 26*, 96–104.

QUIZ

1. What are the four fundamental properties of internal land carbon cycle processes?
2. What are the five categories of external forcing to influence the carbon cycle?
3. Briefly describe the donor-pool dominated carbon transfer.
4. How do external forcing variables interact with internal processes to create the complex phenomena of the land carbon cycle?

2 Introduction to Modeling

Benjamin Smith
Western Sydney University, Richmond, Australia

This chapter introduces the concept of a model, and its role within modern research methodology and the scientific method. Some typical characteristics of the system dynamics models used in carbon cycle studies are described, alongside examples of different ways they can be applied in ecosystem and earth system research. With reference to in-depth discussion and examples throughout the book, we introduce the workflow of six steps you would typically follow when integrating a model within a robust research study design.

WHAT IS A MODEL?

The dictionary definition of a *model* is an *idealized, simplified, or down-sized representation* of something, the purpose of which is to *describe, explain, or depict* that something that the model represents, which we may refer to as its object. The model encodes information about certain, well-chosen aspects of the object, and its purpose is to convey that information to the viewer or user of the model in a clear, concise, and potentially useful way. Compared to the phenomenon or notion it represents, the model may constitute a simpler, more lucid, or even an exaggerated representation of those aspects it is designed to convey – and that is exactly the purpose (Figure 2.1). Models induce us to perceive – and may sometimes help us to understand – something significant about the object by discarding extraneous detail and focusing on the essential. Thus, models are designed to *simplify, explain, and communicate*, and these aspects are interdependent: a simple explanation of a complex idea makes it easier to understand and convey to others. Similarly, models are used in learning, research, and academic exchange as a way of simplifying, explaining or synthesizing, and communicating knowledge, data, and ideas, allowing us to put them into practice in useful ways.

MODELS IN RESEARCH

Models in a broad sense are fundamental to modern research methodology. In its simplest form, a model may be a mental construct describing a perceived pattern or relationship; for example, the simple observation that an increase in one variable tends to be consistently associated with an increase or decrease in another. The existence of patterns and relationships in nature has surely been noted by people throughout time. The fact that certain types of edible plants were often found growing in a particular physical situation, or associated with certain conspicuous species, would have been important practical knowledge in hunter-gatherer societies. From the beginnings of agriculture, farmers have taken note of environmental events that signaled propitious conditions for sowing and harvesting their crops, or that warned of hazards such as droughts, freezing conditions, or insect plagues. Philosophers of ancient Greece pioneered the documentation of patterns and relationships in nature in a systematic way, reasoning about their wider significance. In modern terminology, they could be said to be the first to employ modeling as part of the research process. The 'models' of this era were qualitative, mental models, describing and potentially suggesting explanations for observed patterns and relationships, and expressed in words, illustrations, or allegories.

Today, models, whether mental, conceptual, or mathematically formalized (Figure 2.2), feature in most scientific research. Oreskes (1994) stated: "Numerical models are a form of highly complex scientific hypothesis".

FIGURE 2.1 Models are common in everyday life, and serve a similar purpose as in learning, research, and academic interchange, namely to simplify complex information and facilitate communication and understanding.

DOI: 10.1201/9781032711126-3

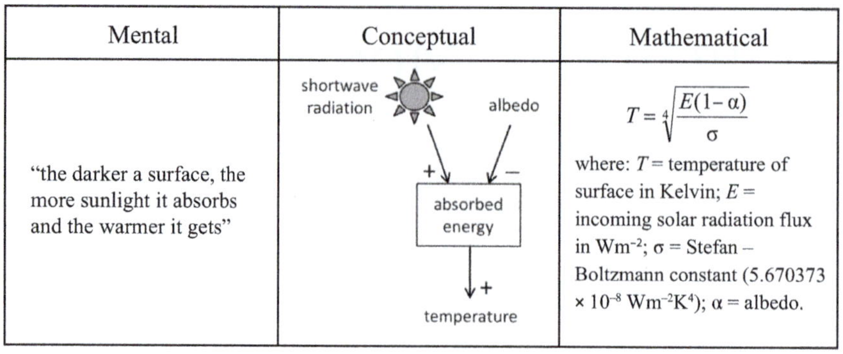

FIGURE 2.2 Three increasingly formal and precisely specified models that express a common hypothesis.

The hypothesis, of course, is an element or 'step' in the empirical scientific method, pioneered by Galileo in the 17th century. The core of this method is the experiment, by which observations from the real world are collected and examined in such a way as to shed light on the merit of a hypothesis. Often, several or many empirical studies addressing the same or related hypotheses are needed before scientists achieve consensus and the former hypothesis becomes an element of a consensus theory or 'settled science'. The hypothesis behind a given study is rarely entirely novel, but represents a step beyond the existing frontier of knowledge in the research area or field. We can think of the hypothesis as an 'informed guess' given what we already know thanks to the consensus knowledge and theory of the field, accrued over earlier studies and scientific discourse.

Similarly, the models we commonly use in research have the character that they bring together elements of established knowledge with ideas or informed guesses about things we do not yet know with certainty, or lack data to express in a precise, quantitative way. In this sense, a model provides a context or frame for integrating knowledge and observations to pose questions about the real world in the form of a testable hypothesis. A model is not necessarily 'true' or proven, but can stand as a formalized or explicit hypothesis of how the system under study works.

Many different kinds of models are applied within the environmental and earth sciences. This book focuses mainly on numerical simulation models, implemented as software algorithms and executed on a computer. A familiar class of simulation models are the numerical weather prediction (NWP) models used by weather service agencies (like the US National Weather Service) to produce daily and longer-term weather forecasts. These models depend for the (relative) accuracy of their predictions on knowledge of the physical processes that govern the dynamics of weather, and on extensive observations (such as measurements from weather stations, balloons, and satellites) both to 'train' the model, improving its fit with the measurements, and to fix the state or 'initial conditions'. Integrating observations in this way is called 'constraining' the model.

NWP models are an example of the class of models we refer to as being mechanistic or process-based. Many of the core equations of an NWP model express known physical relationships or laws governing energy balance and motion, applied to the three-dimensional atmosphere. Others, such as equations governing the formation and behavior of clouds, are parameterized, i.e., fitted to observations, but informed by physics. This combination of mechanistic (process) knowledge and empirical fitting is characteristic for most process-based models.

WAYS OF USING MODELS

Perhaps because weather prediction models and forecasts are so familiar from daily life, many people, including scientists who do not habitually work with numerical models as part of their methodological toolbox, tend to think of such models primarily as tools for prediction forward in time. Certainly, extrapolating beyond the range of observations (in time or space) can be one useful way to apply some types of models (though not all models are suitable for this). However, this is far from being the only way modeling can perform a useful role within research methodology. In fact, forward prediction and extrapolation, while interesting and useful in many science applications, is of little direct relevance to science's central endeavor to advance fundamental knowledge about the natural world. In general, the potential for new discoveries is greatest at the interface between modeling and observation, where information on real-world phenomena encoded in data meets the potential of modeling to explain or decode patterns in the data in terms of underlying mechanisms of cause and effect.

Box 2.1 shows some ways in which process-based models of the kind covered by this book can be applied within research on the land carbon cycle. In general, the potential for integration between modeling and empirical approaches increases as you go down the list. We will see examples of most of these modes of applying models in the course of this book.

SYSTEM DYNAMICS

Land carbon cycle models are at their core system dynamic models. Systems are fundamental to the organization and processes of human society and our daily lives. Think of a 'political system', 'educational system', or 'energy system' as examples. Fundamentally, systems are a mental framework

BOX 2.1 MODES OF MODEL APPLICATION

- project future changes and impacts, extrapolate beyond current data (e.g., Chapter 18 – C cycle transient responses to future climate change)
- scale-up findings from local studies to regional or global scale (e.g., Chapter 27 – model-data fusion of sub-continental GPP)
- characterize uncertainty, identify robust responses and relationships (e.g., Chapter 17 – tracing uncertainty sources in land carbon models)
- attribute observed relationships and patterns to underlying drivers and mechanisms (e.g., Chapter 35 – disentangling drivers of photosynthetic activity in the tropical Americas)
- synthesize the state of knowledge, identify gaps, generate hypotheses to guide empirical studies (e.g., Chapter 29 – ecological forecasting of field experiments)
- communicate scientific evidence to societal end-users and decision-makers (e.g., Chapter 11 – mitigation potential of alternative carbon dioxide removal solutions)

(increasing potential for integration with empirical studies ↓)

through which we may view, understand, or organize complex activities or ideas that involve or depend on linkages between different interacting elements or parts. The presence of interlinked elements, constituting a 'whole that is greater than the sum of its parts' is characteristic for a system. 'Systems thinking' can also be applied to nature and has had a powerful influence on the development of quantitative techniques in ecology and numerous other fields. General systems theory, formalized in the 1940s, depicts an entity under study as a network of interrelated elements whose properties and linkages influence the behavior of the system and its evolution over time. Systems theory provides a framework for formalizing a conceptual or hypothetical understanding of an object of study in terms of drivers, responding processes, and networks of interacting elements, unified through the consideration of flows – for example, energy, matter, or capital – across the network.

When we model the carbon cycle, the system in focus is an ecosystem, or more broadly the Earth system (constituting the Earth's interacting 'spheres' – the atmosphere, hydrosphere, biosphere, etc.). The origins of the ecosystem concept can be traced to the British botanist Arthur Tansley, writing in 1935, who argued for the need to consider organisms and their environment as part of a unified whole in order to understand patterns and changes in nature. Tansley noted that organisms are locked into constant interactions with their immediate physical environment, and that the interactions between individuals and species are mediated by the changes they impose on physical (and chemical) factors of the microenvironment through their growth and function. This notion that *interactions* between organisms and the air, soil, and water in which they occur govern pattern and process in nature is central to systems thinking in ecology.

TYPES OF LAND CARBON CYCLE MODELS

Today's biogeochemical models can be considered the end branches of a family tree having its roots in the work of the Odum brothers, Eugene and Howard, and their textbook, *Fundamentals of Ecology*, first published in 1953. They combined concepts from system dynamics theory with analogies from electrical engineering and thermodynamics to frame ecological systems as networks in which interactions result from energy flows between trophic species levels, linking organisms with their abiotic environment via biological analogues of concepts from electrical networks such as voltage, capacitance, and resistance. Modern biogeochemical models build on similar foundations to Howard Odum's depiction of energy flow for the Silver Springs aquatic ecosystem in Florida, USA (Odum 1971; Figure 2.3). Energy that enters the system as sunlight, powering the photosynthesis of producers (algae and aquatic plants), is used to drive the metabolism of the producers and the consumers that depend on them, up through the trophic chain. At each trophic step, energy is lost through respiration, returning heat to the environment. The Silver Springs ecosystem is here represented as a system of compartments (the trophic groups) that exchange energy with each other and with the external environment. Each compartment has a store of biochemical energy, linked to the abundance of organisms in that trophic group, and the size of this store changes over time as energy is gained and lost through processes like photosynthesis, herbivory, predation, and detritus production. All flows into, out of, and between compartments of the system are mirrored by changes in the sizes of the compartments themselves – the system is said to uphold *mass balance* (here expressed in units of energy). We can state that the Silver Springs ecosystem is here depicted as a *compartmental dynamic system*. Such a representation has some very useful mathematical properties for simulation and analysis of the system, as we shall see elsewhere in this book, and as discussed in detail in Chapter 7.

An alternative 'currency' for this model, instead of energy, would be the carbon that enters the system as CO_2 assimilated through photosynthesis, is transferred between trophic levels through herbivory, production of detritus, or predation, and is lost to the system as CO_2 released through autotrophic or heterotrophic respiration (in this aquatic system, there are also imports and exports due to streamflow and movement of organisms in and out of the study area). Similarly, models in use today typically adopt a carbon-based representation of the ecosystem and its network of pools and fluxes as their basic framework. An example of such a framework was presented

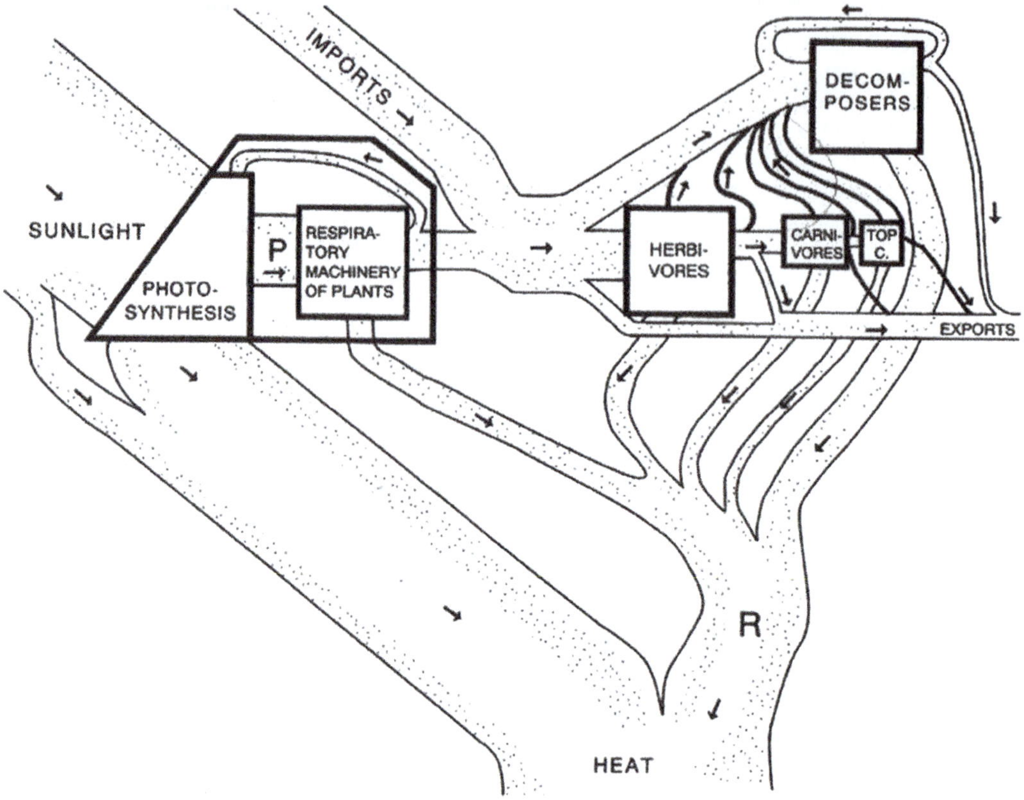

FIGURE 2.3 Energy flow diagram for the Silver Springs ecosystem in Florida.

Adapted by S. Maud from Odum, H.T. (1971) Environment, Power, and Society. Wiley-Interscience, New York.

for the TECO model in Chapter 1. Superimposed on this, many models incorporate representations of hydrological, nutrient, and energy cycles that interact with carbon cycle processes and states. In Chapter 6 we shall see how a nitrogen cycle may be added to the carbon-only version of TECO, to account for progressive nitrogen limitation under elevated atmospheric CO_2. Recent developments have likewise seen phosphorus cycle dynamics added to some widely used models, as discussed in Chapter 10.

Some of the main 'families' of modern land models incorporating representations of carbon processes are detailed in Table 2.1. These groupings and their defining characteristics as shown in the table are not clear-cut. On the contrary, each model family has evolved over time as investigators have enhanced and adapted existing tools for application to novel research questions and management problems. Each type of model was originally developed with a certain research goal in mind. For example, Soil-Vegetation-Atmosphere Transfer (SVAT) schemes, first developed in the 1980s, focus on land surface hydrology and energy balance, as important exchange processes that affect the dynamics of the atmosphere, driving weather and climate. SVAT models like the Simple Biosphere Model (SiB2; Sellers et al. 1996), which originally did not incorporate explicit carbon processes, were extended to incorporate a representation of canopy CO_2 exchange because an estimate of photosynthesis rate was required as a control on stomatal conductance and evapotranspiration, in turn affecting the partitioning of the vegetation-atmosphere energy flux into sensible and latent heat components. Modern land surface models (LSMs) coupled to earth system model (ESM) frameworks and used for global climate change simulations have evolved from SVAT models, progressively incorporating more processes and interactions, often adopted from schemes first developed for the other model families like forest growth models (used to simulate the growth and dynamics of trees and forest stands), terrestrial biogeochemistry models (adding belowground carbon and nutrient dynamics), and dynamic global vegetation models (DGVMs; adding competition between plant functional types or PFTs). In fact, current representatives of many of the model families shown in Table 2.1 have converged to incorporate many of the same algorithms, functionality, and underlying theory across model families. The frameworks contributing to global studies and assessments, such as annual updates of the global carbon budget (Sitch et al. 2015), and to account for biogeochemical and biophysical feedbacks to climate change in IPCC climate projections, are collectively termed terrestrial biosphere models (TBMs). Fisher et al. (2018) provides a useful account of the current status of, and research front for, TBMs incorporated within ESM frameworks.

MODELING WORKFLOW

When using models as part of a robust research study design, there are a number of steps to go through. Depending on the study and the prior work we are building on, by ourselves or others, we may leave out some steps, change the sequence, or iterate more than once over a given sequence of steps. However, most modelers would agree, in principle, that a study should feature the following steps, and that these should be systematically described in publications discussing the work, allowing others to fully understand and reproduce it.

Specify the Question or Hypothesis and Identify How Modeling Can Help

It is rather elementary that any scientific study begins with a question, and linked to this, a testable hypothesis or hypotheses. If the hypothesis can be expressed or captured in the form of a model, and there are data available to compare and contrast to the model results, modeling may be valid to consider as part of the study design.

Choose a Model

This step could also be labeled 'develop a model' or 'adapt a model'. Textbooks on modeling often advocate for building the model from first principles, starting from a conceptual diagram that captures the hypotheses of the study, to be reflected in the model. In an ideal world, this ensures the chosen model is targeted specifically to the question of the study, no more and no less. This principle is captured by the popular modeler saying "the model should be as simple as possible, but no simpler".

In practice, the carbon cycle literature contains only rare examples of studies using completely novel model structures and codes. Many current model frameworks have evolved over decades, involving many person-years of development, coding, evaluation, and application. Studies employing the model along the way turn up issues (scientific, as well as programming errors) focusing attention on needed revisions and improvements, but also increase confidence in the model where they demonstrate that it is able to reproduce patterns or relationships seen in observations, or in independent studies using alternative approaches. A typical ecosystem model is a sophisticated software application comprising hundreds to thousands of lines of computer code. Even though it could be argued that many models have grown overly complicated, to at least some degree this reflects the complexity of real ecosystems, the range of biotic and abiotic factors and interactions that govern their dynamics, and the steadily advancing research front in understanding and modeling natural systems.

Thus, in carbon cycle science, it is seldom efficient or realistic for an investigator to develop a new model completely from scratch, even if we accept that this might be the ideal. Rather, it is a question of choosing a model framework that fits the research question and study system in terms of criteria such as: model versus system complexity, assumptions of temporal and spatial scale, available evidence of model skill (e.g., past published studies on similar systems or questions), configurability of parameters, input files, and source code, and availability of code and documentation.

A common pitfall is to select a model that has convenient technical features but is poorly matched to the target study in terms of spatial and temporal scale. There are two essential aspects to this. One is that the same process can exhibit different sensitivity to its drivers depending on the scale of observation. For example, due to structural, functional, and compositional heterogeneity, photosynthesis measured at the leaf scale in a forest will show a different relationship to

TABLE 2.1
Model types used in contemporary land carbon cycle research

	Soil-vegetation-atmosphere transfer scheme (SVAT)	Forest growth model	Terrestrial biogeochemistry model	Dynamic global vegetation model (DGVM)	Biogeochemical land surface model (LSM)
Example	SiB2	3PG, 4C, SORTIE	CENTURY, TECO	IBIS, LPJ-GUESS, ED2	CABLE, ORCHIDEE, CLM4.5
Energy cycling/balance	✓	—	sometimes	sometimes	✓
Hydrological cycling/balance	✓	✓	✓	✓	✓
Canopy physiology/CO_2 exchange	✓	✓	✓	✓	✓
Plant C dynamics	—	✓	✓	✓	✓
Belowground C dynamics	—	sometimes	✓	✓	✓
Nutrient (N, P) dynamics	—	sometimes	✓	sometimes	sometimes
Plant functional types (PFTs)	static	static/dynamic	Static	dynamic	usually static
Stand dynamics	—	sometimes	—	sometimes	—
Typical application	local/global	local	local	regional/global	regional/global

drivers such as temperature, insolation, or CO_2 concentration compared to the canopy, stand, or landscape scale. Decomposition of soil organic matter will show a different apparent sensitivity to temperature (Q_{10}) in a chamber measurement over an hour, compared to an incubation experiment over a year, or a decades-long soil inventory dataset. This means that the structure and parameters of the most suitable model to study variations in photosynthesis or decomposition will differ depending on the temporal and spatial context of the system under study and the research question in focus. Second, different processes and entities of the ecosystem are important in controlling variations at different scales. For example, seasonal cycles of leaf production and shedding (phenology) are important for the productivity of nemoral forests at annual and longer time scales, but leaf area index could be prescribed (specified as a fixed value) when modeling the gas exchange of a nemoral tree over a diurnal cycle. Changes in the sizes of 'slow' and 'passive' soil organic matter (SOM) pools tend to dominate soil carbon dynamics on scales of centuries and millennia, but have no impact on variation in CO_2 flux in short-term chamber measurements. This implies that a single-pool SOM model may be suitable for the analysis of data from soil chambers, whereas a model used to analyze changes in global soil carbon under IPCC future climate projections needs to distinguish multiple SOM pools of different lability (this is further discussed in Chapter 23). Ensuring that the chosen model accounts for the processes and entities relevant to the scale and question of the study is a particularly important consideration in carbon cycle studies.

Verify that the Model Works

Modelers distinguish between 'verification' and 'validation'. The former seeks confirmation that the model implementation behaves as expected in a purely technical sense. When we run a numerical model we are 'solving' the model given the input data. Verification entails ensuring that the output data from the model matches the solution of the set of difference (or balance) equations the model (typically) encodes (see Chapter 3). In practice we seldom have the true (analytical) solution of these equations available to compare with the output of the model, but, based on our scientific knowledge of the system we are simulating, we can usually tell whether the output is reasonable or 'sane' given the forcing. For example, carbon pool sizes should not normally go negative, while fluxes such as GPP, autotrophic and heterotrophic respiration, or CH_4 emission, should be within a certain range.

Calibrate the Model

Some though not all models can be calibrated against different types of observational or measurement data. For a system dynamic model, calibration may involve tuning parameters of individual equations or processes based on measurements or estimates of those processes. For example, key parameters of a biochemical photosynthesis model may be calibrated based on gas exchange measurements of leaves, yielding a so-called A-c_i curve. When you choose an existing model framework for your study, this kind of process-level calibration will likely already have been done, though you may have good reason to perform your own calibration if you have relevant measurements from your system.

A greater challenge is performing calibration on the overall output of the model, where this emerges from interactions between different processes, drivers, and the evolving system state. For example, net biome production (NBP) in a global carbon cycle model may be the emergent balance between uptake (photosynthesis/GPP) and multiple release fluxes (autotrophic and heterotrophic respiration, emissions from wildfires) integrated across the land ecosystems of the world. The release fluxes in particular depend not only on current environmental drivers, but on the cumulative sizes of source pools such as SOM pools of different quality/lability in different climates. Because of the long response lags, or spin-up effect, in such a model, and due to spatial heterogeneity, errors in individual processes can rapidly cascade and accumulate to generate large bias in NBP, even if individual processes are well calibrated to measurement data, where such are available.

This book offers a number of the best available solutions to the considerable challenge of calibrating the emergent output of an ecosystem model, in all its complexity. In essence, these involve identifying the parameters across different process formulations of the model to which the model is most sensitive (with respect to a particular output variable, such as NBP), specifying the potential real-world range of each parameter, and searching the hyperdimensional space of this parameter set (within the realistic range) to find a combination of values that yields the best fit between the model output, given those parameter values, and observational data on the same variable. This approach, termed data assimilation or model-data fusion, is introduced in Chapter 21, and further elaborated and exemplified in subsequent chapters of Units 6–7.

Validate the Model

In the introduction to this chapter it was noted that all models, by definition, are a simplified depiction of the real thing they represent. This is not a failing or shortcoming but the very essence of a model. However, being simpler also implies that even a good model can never be expected to replicate the behavior of a real system exactly, given that some processes, interactions, and spatio-temporal details involved in the behavior of the real system are missing from the model. Thus, model error is inherent in the very concept and purpose of a model. Box (1979) stated: "all models are wrong, but some are useful".

The simple observation that a model can never be perfectly 'true' has led to the suggestion that a model cannot be validated, in the sense that validation implies confirmation of truth (Oreskes 1994). This argument is largely semantic however, and in practice we are of course very interested in knowing the extent to which the model behaves in a similar way to what we know, from observation and theory, of the real system under study. As Canham (2003) stated: "The process of evaluating model structure is clearly critical enough to warrant a specific term, and 'validation' appears to be the best candidate".

A robust validation strategy focuses not only on the emergent output of the model (e.g., NBP), but on the behavior of individual process representations and the assumptions they entail. This is because compensating errors and biases in different processes, or for example across the grid of a spatially-distributed model, can coincidentally produce the 'right result for the wrong reason', when focusing only on an emergent or integrated output such as NBP. This issue is formally termed *equifinality*, the potential to obtain similar results with different model structures and parameterizations (Luo et al. 2009). The Free-Air CO_2 Enrichment Model Data Synthesis (FACE-MDS) initiative has pioneered best practice in validation of complex ecosystem models using datasets from field experiments through an 'assumption-centered' approach (Medlyn et al. 2015). The alternative, perhaps more established and less labor-intensive, approach of *benchmarking* uses data on multiple output variables to simultaneously query the skill of the model in different dimensions, reflecting different processes and feedbacks (Chapter 19).

Design the Model Experiment

The model experiment, sometimes also termed the simulation protocol, tailors the configuration, forcing and output data from the model simulation so as to shed light on the research question of the study. As discussed in the introduction to this chapter, the model typically stands for, and incorporates, our hypothesis or hypotheses about the system and its responses to influences such as a shift in climate or a management intervention. Through the model experiment, we wish to probe whether, or under what assumptions and conditions, the model reproduces salient observations of the system, thereby testing our hypothesis as encapsulated by the model, and allowing us to draw inferences about the behavior of the system.

There are several elements to designing a robust model experiment, and some of these are challenging in the case of carbon cycle studies. One challenge concerns the initialization or *spin-up* of the model state. In general, the spin-up strives to attain a steady state for ecosystem carbon pools ahead of the main part of the model simulation, avoiding drift in the model output that could confound the response to a perturbation of the model drivers. Some challenges and solutions are presented in Chapter 13 and exemplified in Chapter 14. An alternative to the steady state assumption is discussed in Chapter 27.

The book contains many examples of model experiments tailored to different research questions and problems. Some examples were listed in Box 2.1.

SUMMARY

We have seen that the model, whether conceptual or formally specified, is an inherent component of research methodology and the scientific method, integrating the knowledge we have with hypotheses we may wish to pose about the system under study. Several different types of models are applied within carbon cycle research, but this book focuses on numerical simulation models encapsulating networks of plant and soil compartments and the processes that govern the flows of carbon and other elements across the network, influenced by environmental drivers. The workflow of six steps, further elaborated in other parts of the book, serves as a guide to how we may integrate modeling in the design of a robust scientific study.

SUGGESTED READING

Canham, C.D., Cole, J.J. & Lauenroth, W.K. 2003. *Models in Ecosystem Science*. Princeton University Press, Princeton, N.J.

QUIZ

1. "The model should be as simple as possible, but no simpler" – Discuss!
2. List four ways in which models and observations can be combined within ecosystem studies. What is the role of the model versus the observational data in each case?
3. The carbon cycle of an ecosystem can be depicted as a compartmental dynamic system. List four properties of such a system.
4. How is the spatial and temporal scale of a study relevant in selecting a suitable model?
5. Define these terms: calibration, verification, validation, benchmarking.

3 Flow Diagrams and Balance Equations of Land Carbon Models

Yuanyuan Huang
Institute of Geographic Sciences and Natural Resources Research,
Chinese Academy of Sciences, Beijing, China

Great oaks from little acorns grow. This chapter offers basic concepts and tools for building land carbon models. If we build them step-by-step, we can end up understanding complex land carbon models. The goal of this chapter is to understand carbon flow diagrams and balance equations, to be able to derive generic carbon balance equations from carbon flow diagrams, and *vice versa*.

CARBON FLOW DIAGRAM

Flow diagrams are not new to us. We use flow diagrams in different forms, complexities, and for different purposes. For example, we could draw a simple flow diagram to help us diagnose what might be the issue and what to do when a computer stops working (Figure 3.1a). Or you could find a complex flow diagram like the schematic of carbon, nitrogen, and phosphorus cycles considered in the model ORCHIDEE-CNP (Figure 3.1(b)), which illustrates sophisticated interactions among carbon, nitrogen, and phosphorus dynamics. A classic carbon flow diagram from IPCC reports is shown in Figure 3.1c. When it comes to carbon, we are generally interested in tracking the amount of carbon in space and time. In Figure 3.1c, you see different compartments or pools with different amount of carbon. You also see flows of carbon from one compartment to another.

Panel b is adapted from Sun et al., 2021 and Panel c from Ciais et al., 2014.

We use land carbon models to track temporal-spatial dynamics of carbon, to answer questions like: Where is carbon stored? How long does carbon stay in one place? When, how, where, and how much carbon is transferred? Depending on the scientific questions, we may also use land carbon models to track water, nutrients, and energy as they are parts of the critical environment or system that drives and interacts with carbon dynamics. Earth is such a system comprising numerous interacting processes. Here, we focus mainly on the biosphere and its carbon dynamics, but a lot of principles and methods introduced here are also applicable to water, nutrients, and energy. A flow diagram is very helpful in conceptualizing our issues, clarifying study boundaries, and disentangling complex interactions.

We use the terms stock (or storage, pool) and flow (or flux) very frequently in carbon studies. Suppose that you have a bank account. The total amount of money in your bank account is the stock. Every month, you deposit a certain amount of money, for example, from your salary. You also withdraw some money each month for your everyday expenses. This deposit and withdrawal are the flows into and out of your account. The same concept applies to carbon. We could take total carbon in the biosphere, for example, as our stock, and we track carbon flows into (deposits) and out of (withdrawals) the biosphere to understand the dynamics of carbon stock in the biosphere.

We need to know the boundaries of our system to be clear on what are our stocks, what are our flows and in which directions they are moving. Take terrestrial carbon flows as an example (Figure 3.2). When we take terrestrial plants as our study system, carbon in plants is the stock. Carbon that goes through photosynthesis, a process that turns carbon dioxide into sugars (carbohydrates, or organic carbon), is our incoming flow. Respiration involves using the sugars produced during photosynthesis plus oxygen to produce energy for plant growth. It is one of our outgoing carbon flows. Litterfall, in which the plant sheds leaves, fine roots, and branches as part of its phenological (seasonal) cycles, or under unfavorable conditions, is the second outgoing flow. If we take soil as our study system, soil carbon is the stock. Litterfall is the incoming carbon flow. Soil respiration is the outgoing carbon flow and carbon lost during runoff is another outgoing flow. In the field, soil respiration generally comprises two main components: the autotrophic respiration that takes place in plant roots and the heterotrophic respiration due to the breakdown of soil organic matter by microbes. If, instead, we take plant and soil together as our system, the carbon stock is the total carbon in plant and soil. The incoming carbon flux is the carbon flow through photosynthesis and the outgoing fluxes are plant respiration, soil respiration, and carbon lost during runoff. Litterfall is no longer an incoming or outgoing flow, but an internal flow linking the plant and soil carbon stocks.

CARBON BALANCE EQUATIONS

The principle for tracking carbon dynamics is the law of conservation of mass. Matter can neither be created nor destroyed, but chemical structure and physical form can change.

If we know the initial carbon pool size, and the values of carbon fluxes going into and out of the pool, we can track the dynamics of the carbon pool through time. This follows the law of conservation of mass. In terms of carbon, the change

DOI: 10.1201/9781032711126-4

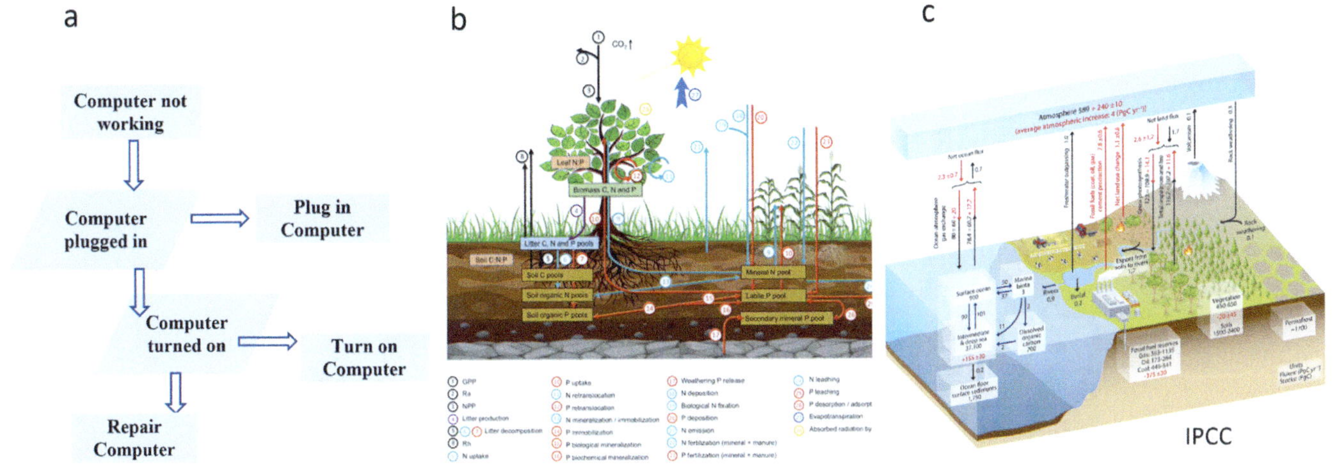

FIGURE 3.1 Examples of flow diagrams.

in its pool size is always equal to the mass of total incoming minus total outgoing flows. Suppose we have a carbon pool (I call it x here) that weighs 30,000 grams (gC) initially, i.e., at time, $t = 0$ (Figure 3.3). You could think about it as if you have \$30,000 in your bank account at the start of the year. Let's say the incoming flux is 2,000 gC year^{-1}. We call it the input rate I. In our monetary analogy, it corresponds to your annual salary. Let's say your income is relatively stable, then I does not change with time. For the outgoing flow, the donor pool-dominant transfer is a common example from the land carbon cycle. For example, if x represents the carbon contained in the biomass of a plant, respiration represents an outgoing flow that depends on the size of this pool: if we have a bigger plant, we would end up with more respiration. Back to our analogy: the higher your income, the more tax you pay. Suppose 20% of carbon in the plant pool is used up through respiration every year. This can be expressed as a turnover rate: a proportion of the pool that is used up and converted to an outgoing flux per unit time. Let's give it the symbol k. Here, $k = 0.2$ year^{-1}. So how much carbon will we have at the end of the year or at the beginning of the next year?

Here our time step is 1 year. When $t = 0$, we have $x = 30{,}000$ gC (initial pool size), $I = 2000$ gC year^{-1}, $k = 0.2$ year^{-1}.

When $t = 1$, our pool size $x\ (t = 1) = x\ (t = 0)$ + input – output = 30,000 + 2,000 × 1 – 0.2 × 30,000 × 1 = 26,000.

What if we want to know x when $t = 2$? It is the same logic. $x\ (t = 2) = x\ (t = 1)$ + input – output = 26,000 + 2,000 × 1 – 0.2 × 26,000 × 1 = 22,800. Similarly, we could derive the carbon stock in the third, fourth, fifth year, and so on. The mathematic equation for x could be summarized as:

$$x(t+1) = x(t) + I \times \Delta t - k \times x(t) \times \Delta t \qquad 3.1$$

which says x at the next time step equals x at the current time step, plus the change of x. The time step (1 year) is denoted by Δt. The change of x is expressed as the input flow minus the output flow. This is the basis for the Euler method, a numerical method we sometimes apply to solve the differential equation to be introduced below.

Equation 3.1 can be rewritten as,

$$\frac{x(t+1) - x(t)}{\Delta t} = I - kx(t) \qquad 3.2$$

When $\Delta t \to 0$, Equation 3.2 is the same as,

$$\frac{dx}{dt} = I - kx \qquad 3.3$$

Equation 3.2 is the difference version of the carbon balance equation, and Equation 3.3 is the differential version. Both equations embody the law of conservation of mass. Another way of putting this is, they uphold mass balance: the rate of change of a carbon pool equals the input rate minus the output rate.

If we have two carbon pools in our system, sometimes the outgoing flux of one pool might become the incoming flux of the other (Figure 3.4). For example, in our plant-soil system in Figure 3.2, the litterfall is an outgoing flow of the plant pool and an incoming flow into the soil pool. For the plant, the carbon balance equation can be written the same as in the above one-pool case. The carbon balance equation of the second (soil) pool also expresses the mass balance. Here the incoming rate is a fraction, say f_{21}, of the outgoing flux from pool x_1. The outgoing flux from pool x_1 is a constant annual proportion of the plant pool: k_1x_1. This outgoing flux comprises the fraction that goes into soil (e.g., through litterfall) and the fraction that does not enter soil (e.g., plant respiration). Combining these terms, we have the incoming flux of pool x_2 as $f_{21}k_1x_1$. The outgoing flux from x_2 is also pool-size dependent: k_2x_2. The carbon balance equations for this two-pool system are given by Equations 3.4 and 3.5:

$$\frac{dx_1}{dt} = I - k_1x_1 \qquad 3.4$$

$$\frac{dx_2}{dt} = f_{21}k_1x_1 - k_2x_2 \qquad 3.5$$

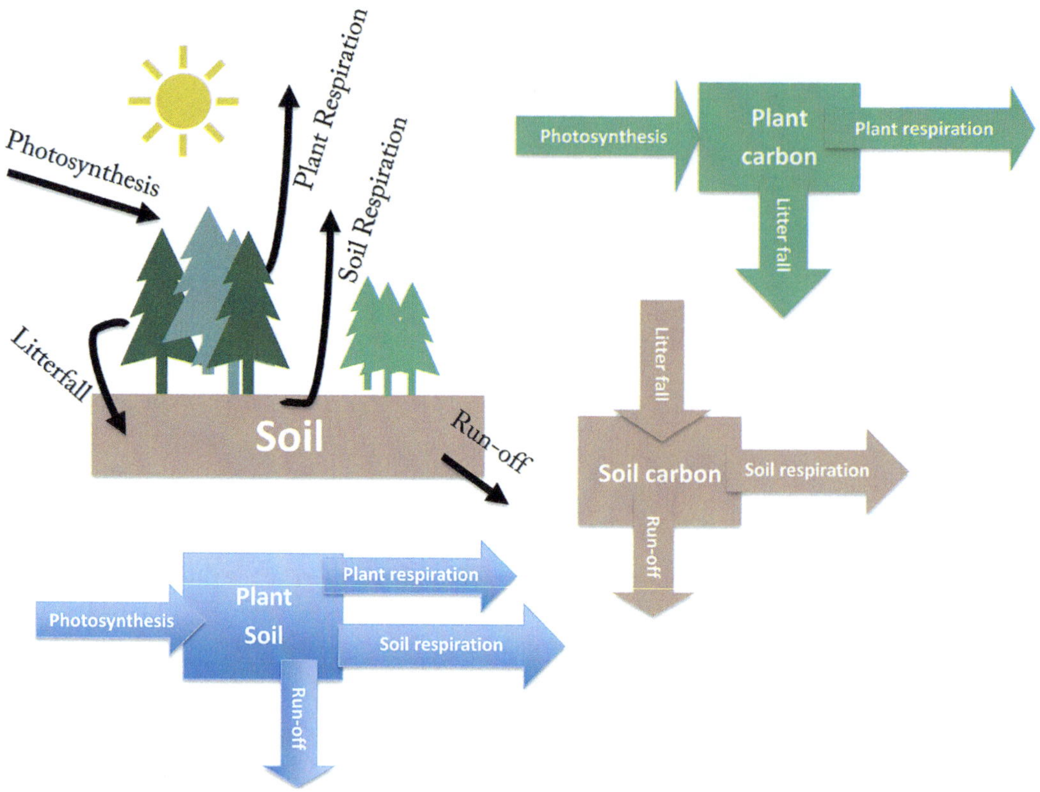

FIGURE 3.2 An idealized illustration of terrestrial processes of carbon flow.

FIGURE 3.3 A conceptual one-pool carbon model.

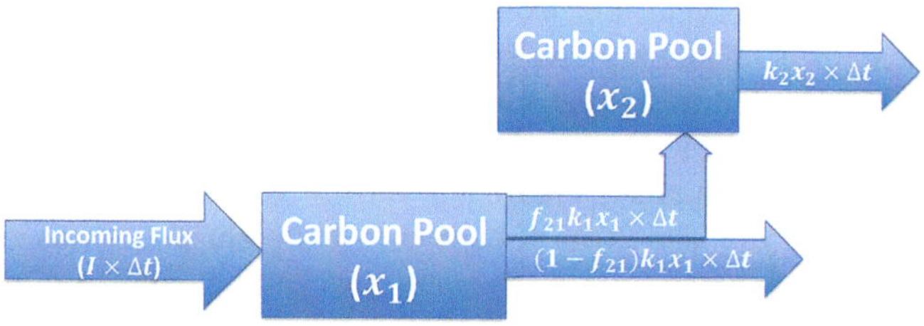

FIGURE 3.4 A conceptual two-pool carbon model.

We may extend this same logic to models with any number of pools. For each pool, we need to know what are its incoming and what are the outgoing fluxes. Figure 3.5 shows the carbon flow diagram for two well-known carbon models. The left one is the classic soil carbon model, CENTURY (simplified, Parton, et al., 1988). The right is the TECO model (Xu et al., 2006). Both models conceptualize our natural carbon cycles into different pools or 'compartments', represented by boxes in the diagram, and track the transfers of carbon among different pools, represented as arrows. The transfers could be parameterized as one constant turnover rate relative to the size of the donor (source) pool, or could be modeled by functions with multiple terms, parameters, and dependencies, capturing complex interactions with climate, soil, and other factors.

FROM FLOW DIAGRAM TO CARBON BALANCE EQUATIONS

We take the TECO model as an example to illustrate how to derive a system of carbon balance equations from the carbon flow diagram. In Figure 3.5, we see that TECO has seven carbon pools: two plant biomass carbon pools (foliage, woody biomass), two litter carbon pools (metabolic, structure), and three soil organic carbon pools (microbes, slow soil organic matter (SOM), and passive SOM). We use x_1 to x_7 to denote these pools. For each carbon pool, we need to specify the fluxes that enter and leave the pool. Carbon flux through photosynthesis is the external carbon input into our system of seven pools. We will denote this input rate by I. Photosynthetic carbon input is allocated into foliage and woody biomass, in relative proportions denoted by the allocation coefficients β_1 and β_2. The carbon balance equation for foliage pool (x_1) and woody pool go as Equations 3.6 and 3.7.

$$\frac{dx_1}{dt} = I\beta_1 - k_1 x_1 \tag{3.6}$$

$$\frac{dx_2}{dt} = I\beta_2 - k_2 x_2 \tag{3.7}$$

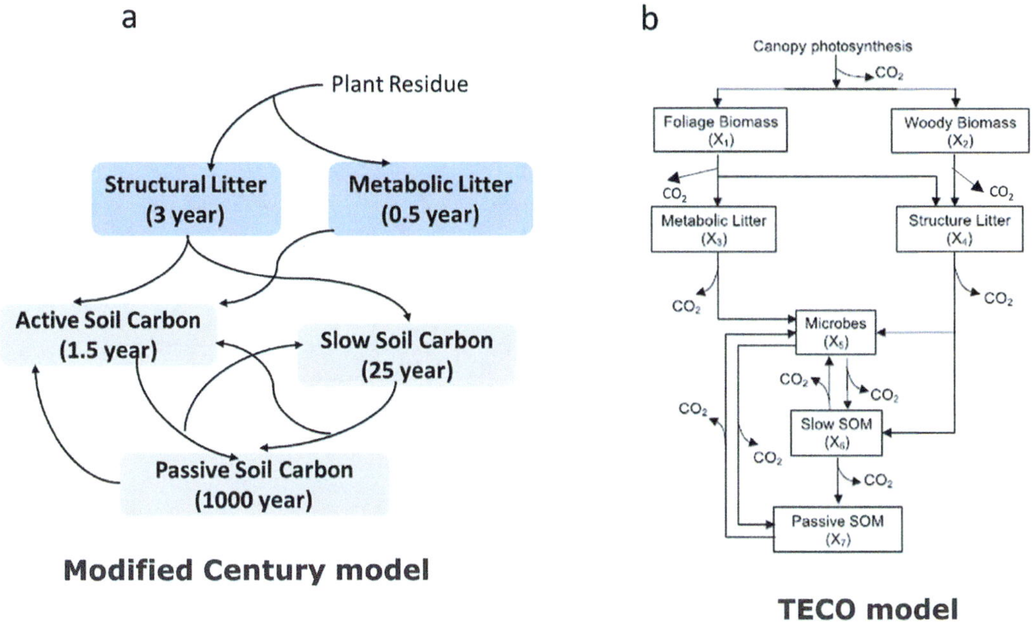

FIGURE 3.5 Flow diagrams of two carbon models: (a) CENTURY; (b) TECO.

a: adapted from Parton et al., 1988; b: adapted from Xu et al. 2006.

The structure litter pool (x_4) receives carbon input from pools x_1 and x_2. So the carbon balance equation of x_4 has two input fluxes. One is a fraction (f_{41}) of the outgoing carbon flux from x_1 (k_1x_1). The other is a fraction of the outgoing carbon flux from x_2. This incoming flux could be written as $f_{42}k_2x_2$. The outgoing flux for pool x_4 is dependent on the pool size and its turnover rate, k_4. We write this flux as k_4x_4. Equation 3.8 is the resulting carbon balance equation of the structure litter pool, x_4.

$$\frac{dx_4}{dt} = f_{41}k_1x_1 + f_{42}k_2x_2 - k_4x_4 \quad 3.8$$

The same procedure applies for the other carbon pools. We count how many arrows go into a pool and how many go out. One arrow normally corresponds to one flux. The boundary of the system in this example encompasses the seven organic matter pools. The atmosphere is not part of the system and we do not track further the fluxes that go into the atmosphere, i.e., respiration represented by arrows labelled 'CO_2' in Figure 3.5. That being said, how much carbon goes into the atmosphere as carbon dioxide is an important topic as it is closely linked to climate change. Carbon balance equations for x_3, x_5, x_6, x_7 are given below:

$$\frac{dx_3}{dt} = f_{31}k_1x_1 - k_3x_3 \quad 3.9$$

$$\frac{dx_5}{dt} = f_{53}k_3x_3 + f_{54}k_4x_4 + f_{56}k_6x_6 + f_{57}k_7x_7 - k_5x_5 \quad 3.10$$

$$\frac{dx_6}{dt} = f_{64}k_4x_4 + f_{65}k_5x_5 - k_6x_6 \quad 3.11$$

$$\frac{dx_7}{dt} = f_{75}k_5x_5 + f_{76}k_6x_6 - k_7x_7 \quad 3.12$$

A further example for multiple carbon pools is the ORCHIDEE model. Figure 3.6 shows the litter and soil organic carbon (SOC) component of ORCHIDEE. As a rule of thumb, around 50% of soil organic matter is SOC. Later versions of this model separate SOC into different layers according to soil depths. However, the version depicted in Figure 3.6 tracks seven pools – four litter pools and three SOC pools – with different layers lumped together. Therefore, we have seven carbon balance equations. The idea is the same as what we have already worked through for the TECO model. We track the arrows that go into and out of each pool. For example, for the active SOC pool, x_5, we have six arrows that go into this pool. Each arrow represents an incoming flux from a different pool, either litter or SOC. We will use f_{51}, for example, to represent the fraction of the carbon flux that goes from the first pool (aboveground metabolic litter) to the fifth pool (active SOC).

When we have a small number of carbon pools, it is not difficult to write down the carbon balance equations one-by-one. It would be very tedious to write down all the equations if we have many pools, for example, 100 for the version of ORCHIDEE that simulates SOC dynamics at different soil depths. This version, known as ORCHIDEE-MICT, tracks seven types of organic carbon (four litter compartments and three SOC compartments), similarly to the version depicted in Figure 3.6. ORCHIDEE-MICT has 32 soil layers (Huang, et al., 2018a). For each soil layer, the model has active, slow, and passive SOC pools. Such a vertical soil discretization is especially helpful and more realistic in modeling soil carbon

$$\frac{dx_1}{dt} = I_1 - k_1x_1$$

$$\frac{dx_2}{dt} = I_2 - k_2x_2$$

$$\frac{dx_3}{dt} = I_3 - k_3x_3$$

$$\frac{dx_4}{dt} = I_4 - k_4x_4$$

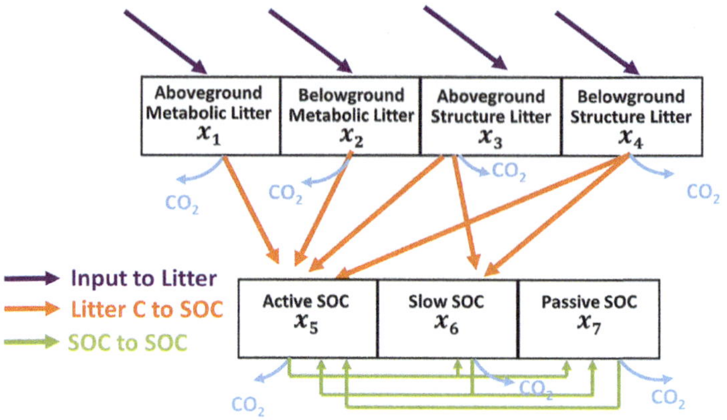

$$\frac{dx_5}{dt} = f_{51}k_1x_1 + f_{52}k_2x_2 + f_{53}k_3x_3 + f_{54}k_4x_4 + f_{56}k_6x_6 + f_{57}k_7x_7 - k_5x_5$$

$$\frac{dx_6}{dt} = f_{63}k_3x_3 + f_{64}k_4x_4 + f_{65}k_5x_5 - k_6x_6$$

$$\frac{dx_7}{dt} = f_{75}k_5x_5 + f_{76}k_6x_6 - k_7x_7$$

FIGURE 3.6 Flowchart and carbon balance equations of the ORCHIDEE model.

dynamics in permafrost regions. In total, the model has 32 × 3 + 4 = 100 pools. Instead of writing down the carbon balance equations one-by-one, we can put them together into one matrix equation, as shown in Figure 3.7. The matrix form of the carbon balance equations says the same thing: the rate of change in carbon pool size equals the input minus the output. Now it is not that obvious to spot the input fluxes and the output fluxes. They are folded into the matrices and depend on the sign of the elements in these matrices. The first item on the right side of the matrix equation in Figure 3.7 summarizes the external carbon input into the system. The second block of matrices summarizes the turnovers and transfers of fluxes among different organic carbon pools. The third block of matrices captures vertical processes between the adjacent soil layers, such as bioturbation, diffusion, and advection.

Variations in these matrices reflect structural or parametric differences among models. For example, the left side of Figure 3.8 shows the flow diagram of the litter and SOM component of the CLM4.5 model. CLM4.5 tracks coarse woody debris, metabolic litter, cellulose litter, lignin litter, fast, slow, and passive SOM in ten soil layers (Huang et al., 2018). Different from ORCHIDEE-MICT, CLM4.5 also tracks the vertical distribution of litter. The matrix equation here takes a similar general form. Elements in each matrix are different. The dimensions of the CLM4.5 matrices are 70×70 corresponding to 70 organic matter pools, while ORCHIDEE-MICT matrices are 100×100.

We will explore matrix representations in more detail in the following chapter. If you have understood the concepts introduced here, then big congratulations. These complex land carbon models are built upon these basics step by step.

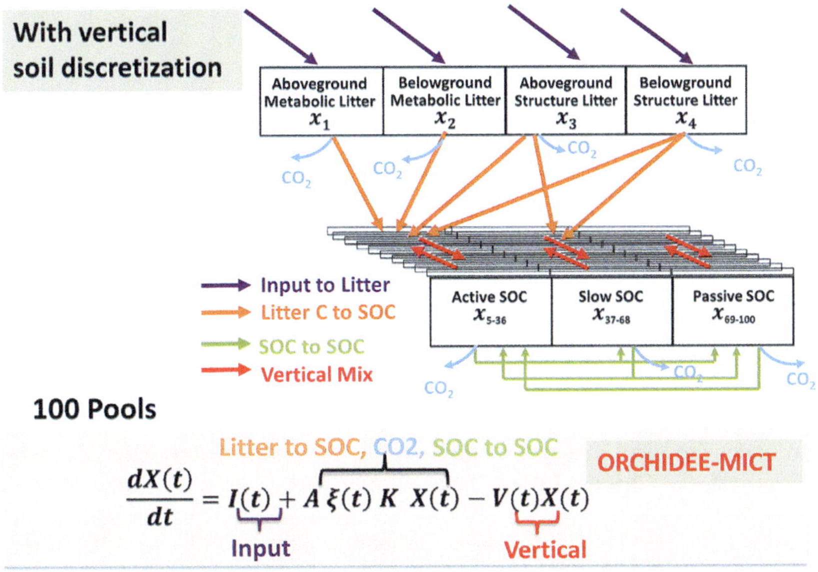

FIGURE 3.7 Flowchart and matrix-form carbon balance equation of the ORCHIDEE-MICT model with vertical soil discretization.

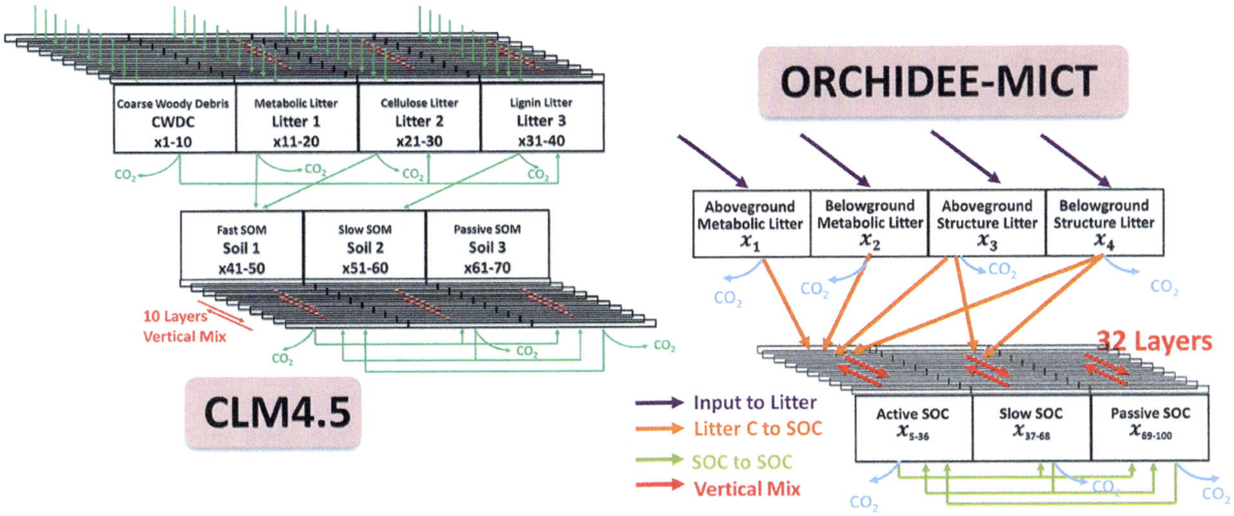

FIGURE 3.8 Flowcharts of the litter and soil organic carbon components of CLM4.5 and ORCHIDEE-MICT.

SUGGESTED READING

Luo YQ, Weng ES . (2011) Dynamic disequilibrium of the terrestrial carbon cycle under global change. *Trends in Ecology & Evolution*, *26*, 96–104.

QUIZ

1 What does a pool stand for?
2 What does a flux mean?
3 What is the principle of writing carbon balance equations?
4 Does burning of the organic matter by fire violate the carbon balance?

4 Practice 1

Carbon Flow Diagram and Carbon Balance Equations

Yuanyuan Huang
Institute of Geographic Sciences and Natural Resources Research, Chinese Academy of Sciences, Beijing, China

The exercises are designed to help you practice writing carbon balance equations from a carbon flow diagram and *vice versa*. The goal is to lead you to understand the concepts of flow, pool, and mass balance. Please refer to Chapter 3 if you have difficulties with the exercises.

INTRODUCTION

Land biogeochemical models all simulate carbon cycling through different pools in an ecosystem according to carbon balance equations, regardless of model structure. The balance equations are based on the law of conservation of mass which tells us that the change in carbon pool size is always equal to the net difference between the incoming fluxes to, and outgoing fluxes from, that pool. Different pools in the ecosystem are connected via carbon transfer between them. To track carbon flows among different carbon pools, scientists develop a carbon flow diagram that uses boxes to indicate pools and arrows to indicate fluxes. Thus, the carbon flow diagram is often called a box-arrow diagram. The models that track carbon flows among pools are often called pool-flux models. The matrix form of land carbon cycle models is built upon the carbon balance equations that are connected via carbon transfer among pools. To learn the matrix approach to land carbon cycle modeling, it is essential to understand the carbon flow diagram and carbon balance equation.

The first exercise of this unit is focused on writing carbon balance equations from a carbon flow diagram, whereas the second exercise is about drawing the carbon flow diagram based on carbon balance equations. Both drawing the carbon flow diagram from the carbon balance equations and writing the equations from the diagram assist us in understanding land carbon dynamics and developing matrix models.

EXERCISE 1 WRITING CARBON BALANCE EQUATIONS FOR THE CENTURY MODEL

Figure 4.1 shows a simplified carbon flow diagram for the CENTURY model. Plant residue is divided into structural and metabolic litter according to its lignin and nitrogen content. Structural and metabolic litters are decomposed by microbes. The resulting microbial products become the substrate for the formation of soil organic matter in three soil pools (i.e., active, slow, and passive). For the decomposition flux of the structural litter, a fraction (A) is incorporated into the slow soil organic carbon with a turnover time of 25 years. We assume some of this remaining fraction goes as microbial respiration, and the other part contributes to the active soil organic C. The flow diagram also tells us the transfer fluxes among different soil carbon pools.

Please write down the carbon balance equations based on the carbon flow diagram (Figure 4.1). Note that some details in the diagram such as L/N, 1–A, F (T), and CO_2 fluxes to the atmosphere differ from the original paper by Parton et al. (1988). You could neglect information such as (3y) in the box for structural C and other similar boxes when you do this exercise. You can read the paper by Parton et al. (1988) if you are interested in knowing more about this model.

To write the balance equations, you could use symbols as in Chapter 3 or develop your own symbol system. Fundamentally, you need to figure out the amount of carbon moving into one pool, the amount of carbon moving out of the pool, and changes in the amount of carbon in the pool (i.e., pool size change) in one unit of time. Here are a few tips. First, you could define state variables, such as x_1 to represent the pool of structural C, x_2 to represent the pool of metabolic C, etc. Then, the size change in the pool of structural C can be expressed by dx_1/dt. You could similarly write changes in carbon amounts in other pools. Second, you need to define rate variables to represent rates of carbon moving out of pools, such as k_1 for a rate of carbon moving out of the structural

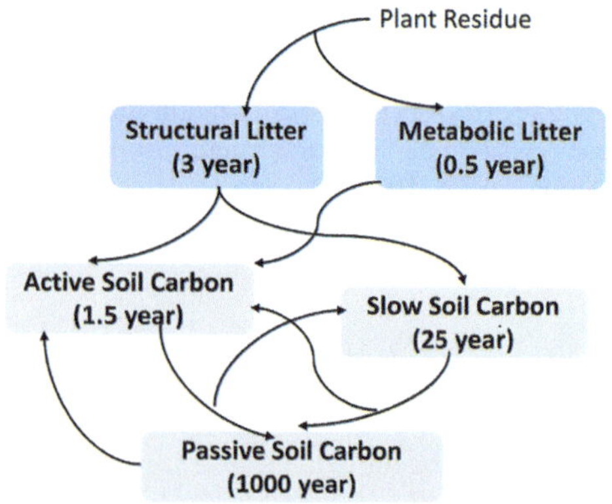

FIGURE 4.1 Flow diagram for the carbon flows in the CENTURY model (simplified).

Reproduced from Parton et al. (1988).

DOI: 10.1201/9781032711126-5

TABLE 4.1
Governing equations of RaSOM, pools and parameters, and their definitions

Pool	Description	Differential equation	
S	polymeric organic carbon	$\frac{dS}{dt} = I_S - F_S + \gamma_{B1}B + f_E\gamma_E$	4.1
	monomeric organic carbon	$\frac{dD}{dt} = I_D + F_S - F_D + \gamma_{B1}X + (1 - f_E)\gamma_E E$	4.2
X	reserve microbial biomass	$\frac{dX}{dt} = Y_X F_D - (\kappa - g + \gamma_{B1})X$	4.3
B	structural microbial biomass	$\frac{dB}{dt} = mX - (\gamma_{B1} + p_E)B$	4.4
E	extracellular enzymes	$\frac{dE}{dt} = p_E B - \gamma_E E$	4.5

where:
I_S: polymeric input flux (g C m^{-3} d^{-1})
I_D: monomeric input flux (g C m^{-3} d^{-1})
F_S: polymeric depolymerization flux (g C m^{-3} d^{-1})
F_D: monomeric uptake flux (g C m^{-3} d^{-1})
Y_X: yield coefficient for reserve biomass (unitless)
f_E: fraction of decayed extracellular enzymes contributing to the polymer pool (unitless)
γ_{B1}: microbial mortality rate (d^{-1})
γ_E: enzyme turnover rate (d^{-1})
κ: metabolic turnover rate (d^{-1})
g: growth rate (d^{-1})
p_E: enzyme production rate (d^{-1})
m: structural microbial biomass formation rate (d^{-1})

Modified from Tang and Riley (2015) by Rose Z. Abramoff.

C pool, k_2 for a rate of carbon moving out of the metabolic C pool, etc. With the rate of carbon moving out of a pool and a state variable of the pool, you might figure out how to calculate the amount of carbon leaving the pool. Then, you need to figure out the amount of carbon moving to a pool. The third step is to define the total amount of carbon input, let's say using a symbol I, from plant residue. Fourth, you need to figure out a fraction of the incoming carbon that enters a particular pool. For example, carbon input from plant residue partly goes to the structure C pool and partly goes to the metabolic pool. You need to define a symbol, b_1, as a fraction of incoming carbon going to the structural C pool and b_2 to the metabolic pool. Then the amount of carbon moving to the metabolic C pool is $b_2 \times I$. Similarly, you can figure out the amount of carbon moving to the structural C pool. However, it becomes much more complicated to figure out the amount of carbon that moves to the active soil C pool. The pool receives carbon from all the other four pools. Thus, it has four terms, which are: $f_{31} \times k_1x_1 + f_{32} \times k_2x_2 + f_{34} \times k_4x_4 + f_{35} \times k_5x_5$. Can you figure out how to get each of the terms? As a note, the symbol f_{31} describes the fraction of the carbon that leaves pool 1 and enters pool 3. If you can go over these steps successfully, you may obtain one carbon balance equation, $\frac{dx_4}{dt} = f_{41} \times k_1x_1 + f_{43} \times k_3x_3 - k_4x_4$, for the slow soil C pool. What are the carbon balance equations for the other pools?

EXERCISE 2

Drawing a carbon flow diagram for the ReSOM model based on its carbon balance equations

This exercise aims to develop your ability to draw a carbon flow diagram from carbon balance equations. ReSOM is a REaction-network-based model of Soil Organic Matter and microbes (Tang and Riley, 2013; Tang and Riley, 2015). The model simulates depolymerization of polymers, mineralization and sorption of monomers by microbes via extracellular enzymes and microbial assimilation, microbial death, and necromass decomposition.

Key equations that govern exchange of carbon among pools over time in the ReSOM model are listed in Table 4.1 together with definitions of symbols and parameters. All pools are in units of carbon mass per soil volume (g C m^{-3}).

To draw a carbon flow diagram of the ReSOM model from its carbon balance equations, you need to figure out the following items: (1) the number of pools and what they are; (2) carbon transfer between any of the two pools; (3) carbon input from outside of the system; and (4) carbon loss from the system. Here is a hint to figure out carbon transfer between two pools from a carbon balance equation: a positive sign before a flux (e.g., F_S in Equation 4.2) indicates carbon entering the pool, whereas a negative sign before a flux (e.g., F_D in Equation 4.2) indicates carbon leaving the pool.

Unit Two

Matrix Representation of Carbon Balance

5 Developing Matrix Models for Land Carbon Models

Yuanyuan Huang
Institute of Geographic Sciences and Natural Resources Research, Chinese Academy of Sciences, Beijing, China

The goal of this chapter is to understand how the carbon balance of a system could be represented in matrix form, to be able to write a matrix carbon model, and, for advanced readers, to think about potential applications of the matrix model for your own research.

WHAT IS THE MATRIX VERSION OF THE CARBON BALANCE EQUATION?

The matrix representation of the carbon balance of a multipool system was introduced in Chapters 1 and 3. Instead of writing down the carbon balance equations one-by-one for each pool, we can integrate all carbon balance equations within one matrix equation. This matrix equation captures the carbon balance for each carbon pool and its linkages to other carbon pools of the system via the flows (fluxes) between them. One thing to keep in mind is that the matrix version of the carbon model, though it uses a different notation, is mathematically identical to the original carbon model that has one carbon balance equation for each carbon pool. Benefits of the matrix version of carbon models include simplicity in model structure, high modularity in coding, clarity in diagnostics, and computational efficiency in spin-up. These properties are discussed in detail in the following Chapters: 6, 9, 14, 17, and 18.

The matrix version of many carbon models could be written as:

$$\frac{dX}{dt} = B * I + A * K * X \qquad 5.1$$

Each of these matrices could be time-dependent. For simplicity, we neglect the time dependence for now. The left-hand side of Equation 5.1 depicts the change rates of carbon pool sizes in our system. If we have, for example, seven pools, X here is a column vector with seven rows, one for each pool, which are the state variables of the system. The right-hand side of Equation 5.1 captures the incoming and outgoing terms, i.e., the fluxes entering or leaving the pools of the system. B is the allocation matrix, which functions to partition the external carbon input I into different carbon pools. In the plant-soil system of Chapter 3 (Figure 3.2), you could think of I as the photosynthetic carbon input, and B captures the fraction of plant-assimilated carbon that is allocated into different plant organs. K is the turnover rate matrix. A is the $n \times n$ dimensional transfer matrix, where n is the number of carbon pools. A captures the fraction of outgoing carbon flow from one pool that goes into a second pool. From the location and sign of elements in A, we could reconstruct the network of carbon transfers among different carbon pools.

The flow diagram and carbon balance equations of the TECO model were presented in Chapter 3 (Figure 3.5b, Equations 3.6–3.12). In the case of TECO, which has seven carbon pools, B and X are 7×1 vectors, while A and K are 7×7 matrices. K is a diagonal matrix with each diagonal element k corresponding to the turnover rate of a carbon pool. All other elements in K are set to zero. The diagonal elements of A are all -1. Non-zero elements of the off-diagonal parts of A correspond to the fraction of carbon fluxes transferred from the ith to the jth pool, where i is the column number and j the row number, starting at (1,1) from the top-left corner of the matrix. For example, f_{41} in the first column and fourth row tells us there are carbon fluxes transferred from the first to the fourth carbon pool, that is, from foliage biomass to structure litter. All transfer fluxes in the carbon flow diagram can be found in the A matrix. In the case of TECO, Equation 5.1 can be expanded as:

$$\begin{pmatrix} d_{x1}(t)/dt \\ d_{x2}(t)/dt \\ d_{x3}(t)/dt \\ d_{x4}(t)/dt \\ d_{x5}(t)/dt \\ d_{x6}(t)/dt \\ d_{x7}(t)/dt \end{pmatrix} = \begin{pmatrix} \beta_1 \\ \beta_2 \\ 0 \\ 0 \\ 0 \\ 0 \\ 0 \end{pmatrix} I + \begin{pmatrix} -1 & 0 & 0 & 0 & 0 & 0 & 0 \\ 0 & -1 & 0 & 0 & 0 & 0 & 0 \\ f_{31} & 0 & -1 & 0 & 0 & 0 & 0 \\ f_{41} & f_{42} & 0 & -1 & 0 & 0 & 0 \\ 0 & 0 & f_{53} & f_{54} & -1 & f_{56} & f_{57} \\ 0 & 0 & 0 & f_{64} & f_{65} & -1 & 0 \\ 0 & 0 & 0 & 0 & f_{75} & f_{76} & -1 \end{pmatrix}$$

$$\times \begin{pmatrix} k_1 & & & & & & \\ & k_2 & & & & & \\ & & k_3 & & & & \\ & & & k_4 & & & \\ & & & & k_5 & & \\ & & & & & k_6 & \\ & & & & & & k_7 \end{pmatrix} \begin{pmatrix} x_1 \\ x_2 \\ x_3 \\ x_4 \\ x_5 \\ x_6 \\ x_7 \end{pmatrix}$$

This matrix version of the carbon model is especially handy when we have many carbon pools. We will take the litter and soil organic carbon components of CLM 4.5 (70 pools) and ORCHIDEE-MICT (100 pools) as examples.

DOI: 10.1201/9781032711126-7

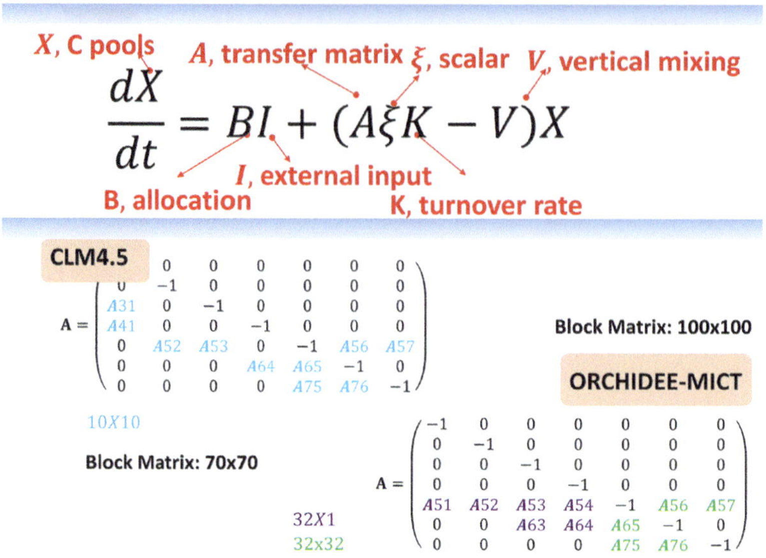

FIGURE 5.1 Matrix equation of CLM4.5 and ORCHIDEE-MICT.

The matrix equation for CLM4.5 and ORCHIDEE-MICT look similar to the matrix equation for TECO. In Figure 5.1 we have one more item: the *V* matrix, which captures the vertical transfers of carbon between adjacent soil layers. The *A* matrix is more complex than the earlier case. Both CLM4.5 and ORCHIDEE-MICT assume the fraction of carbon transferred between different organic carbon pools is the same for different soil depths. Here, *A* has the form of a block matrix, which means *A* is made up by smaller matrices, i.e., blocks. For CLM4.5, each block, for example, *A31*, is a 10 × 10 matrix, corresponding to 10 soil layers. The diagonal elements in *A31* are the same since the model assumes the fraction of transfers do not change with depth. For ORCHIDEE-MICT, it is a little more complex as litter pools are not separated by soil depth. So transfers of litter carbon fluxes to soil carbon fluxes (e.g., *A51*) are 32 × 1 vectors and zeros. For transfers among soil organic carbon pools, for example *A65*, the matrix has a dimension of 32 × 32, with each element representing the transfer from one of the 32 active soil organic carbon pools to one of the 32 slow soil organic carbon pools. ξ is the matrix of environmental scalars that quantify the deviation of the actual decomposition rate from the potential rate due to the non-optimal environmental conditions. We could add it to the TECO model also. Please check in Huang et al. (2018b) and Huang et al. (2018a) for more details if you are interested.

Based on these examples, we see that the matrix equations for different carbon models may take a similar form. But the structure of each matrix might be different. How each matrix is organized depends on the structure of the original model.

HOW TO DERIVE THE MATRIX EQUATION?

After we know what the matrix looks like, the next step is to derive the matrix model from the carbon balance equations.

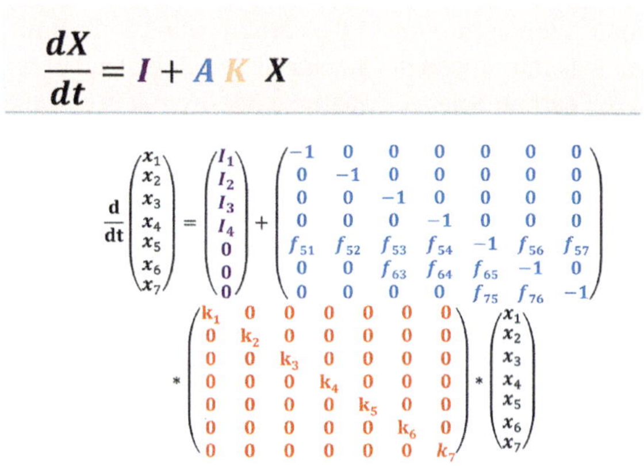

FIGURE 5.2 Matrix equation of the litter and soil organic carbon component of the ORCHIDEE model. The flow diagram and carbon balance equations are shown in Figure 3.6, Chapter 3.

Here we take the ORCHIDEE model in Figure 3.6 of Chapter 3 as an example. We assume we have I_1, I_2, I_3, and I_4 as inputs to the respective litter pools. Carbon from the litter pools is transferred into different soil organic carbon pools. Additionally, there are internal carbon transfers between different soil organic carbon pools. Figure 5.2 shows the matrix equation of this model. *X* and *I* are column vectors, and *A* and *K* are two-dimensional matrices. The dimensions of *X*, *I*, *A*, and *K* are determined by the number of carbon pools. *X* is easy to construct. We put the seven carbon pools x_1, x_2, ..., x_7 in a column. *I* is a column vector with I_1 to I_4 as the input rates for litter carbon pools and zero for soil organic carbon pools as these pools receive no external carbon inputs. *K* is a diagonal matrix with each diagonal element corresponding to the turnover rate of one of the carbon pools.

Mathematically, a $n \times n$ matrix (*A* matrix) multiplied by a $n \times 1$ vector (*X*) will produce a $n \times 1$ vector. The value of the

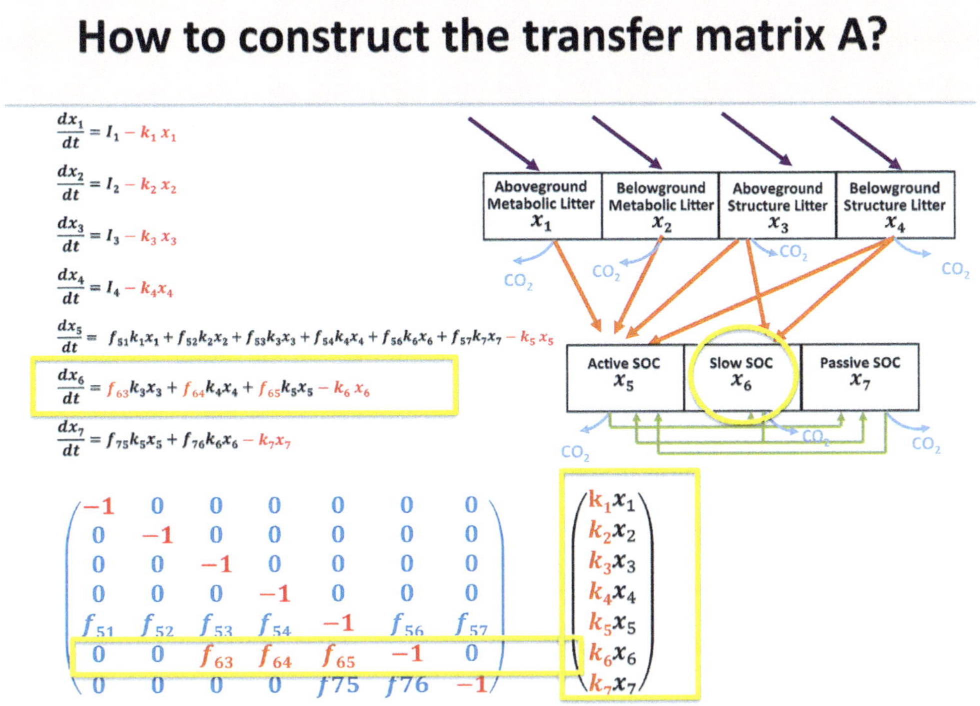

FIGURE 5.3 Illustration of the function of the transfer matrix A.

i^{th} row of this multiplication equals the sum of the products between elements from the ith row with the column vector. The multiplication of K and X will give us a column vector with elements k_1x_1, k_2x_2, ..., k_7x_7. For the multiplication of A by KX, we take the sixth row as an example (Figure 5.3). We have 0 multiply k_1x_1, 0 multiply k_2x_2, f_{63} multiply k_3x_3, f_{64} multiply k_4x_4, f_{65} multiply k_5x_5, and −1 multiply k_6x_6, which is exactly the right side of the carbon balance equation for the sixth carbon pool. The diagonal elements of A are all −1. There are many 0 elements in A. These 0s indicate there are no carbon transfers from the ith (column number) to j^{th} (row number) carbon pools.

If we look at each column of the A matrix, it summarizes all the fluxes that go out of the ith carbon pool (Figure 5.4). For example, after reading the fourth column of the A matrix, we know there is carbon transferred from the fourth to the fifth and sixth carbon pool if f_{54} and f_{64} are not equal to 0. This is exactly what we see in the carbon flow diagram (Figure 3.6 of Chapter 3). If we sum up all the off-diagonal elements in one column and the value is smaller than 1, for example, $f_{54} + f_{64} < 1$, this means there is carbon leaving the system which we do not track further. For example, if our model is focused on the storages of organic carbon in soil, we may not care about carbon that leaves the soil system in the form of CO_2 flux to the atmosphere.

Each row of the matrix A summarizes all the carbon fluxes that are transferred into the ith carbon pool. For example, if we look at the sixth row, we have f_{63}, f_{64}, f_{65} as non-zero, off-diagonal elements. This tells us there are fluxes from the third, fourth, and fifth carbon pools being transferred into this sixth carbon pool. The relevant transfers are highlighted in yellow in the carbon flow diagram of Figure 5.5.

We have seen that a single matrix equation can encompass the details of multiple processes simulated by a carbon model. Detailed information about the carbon dynamics is folded into different components of the matrices. If we expand the matrix calculation by row, we should get the exact set of carbon balance equations, one equation for each pool.

Next we are going to explore a little more on the matrix equation for models with vertical soil layers. For such a model, we require one more vertical transfer matrix, V (Figure 5.6). The X, I, and K matrices are generally constructed in the same way as in the earlier examples above. The transfer matrix A, with 100 × 100 dimension, is folded into the block matrix. Each of the smaller blocks in black, for example $A51$, is a depth-dependent vector, tracking the fraction of litter carbon transferred into SOC in different soil depths. The matrices in red are 32 × 32 diagonal matrices capturing transfers between different soil organic carbon categories in the same soil layer. Each item of the diagonal takes the same value, as ORCHIDEE-MICT assumes the transfer fractions are not depth-dependent.

The vertical transfer matrix can be represented by the block matrix shown in Figure 5.6. Most of its components are 0 except for the active, slow, and passive SOC pools. Each diagonal block is a tridiagonal matrix that describes vertical redistribution of corresponding carbon pools among different soil layers. As the vertical transfer rates are not differentiated among different types of carbon pools, $V55(t)$, $V66(t)$, and $V77(t)$ are identical. The subscript numbers indicate soil layers;

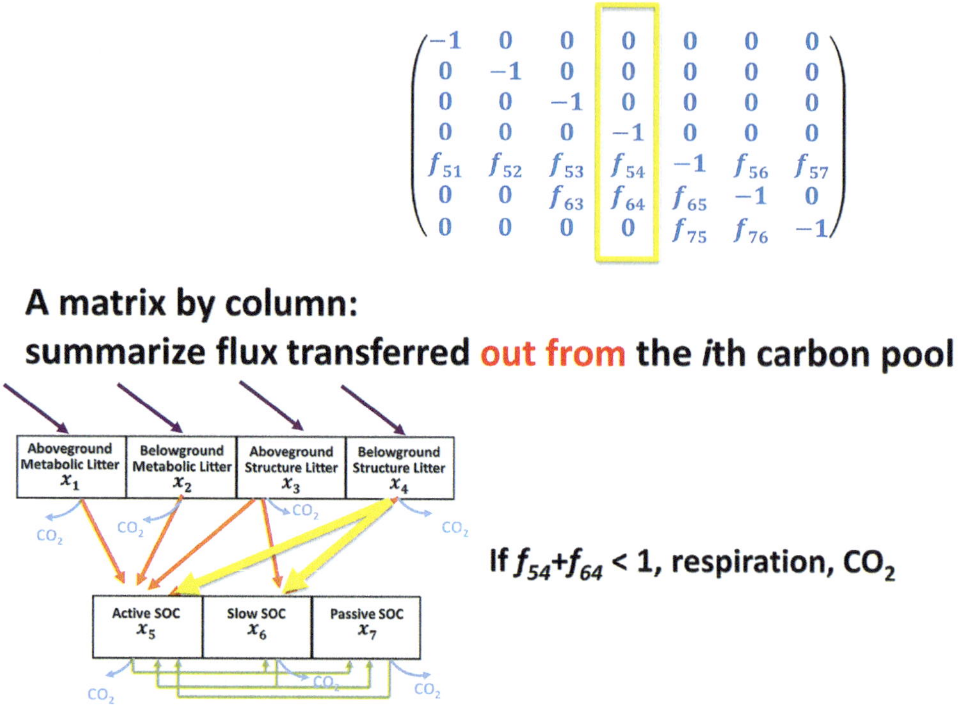

FIGURE 5.4 Illustration of the function of the column of the transfer matrix (*A* matrix).

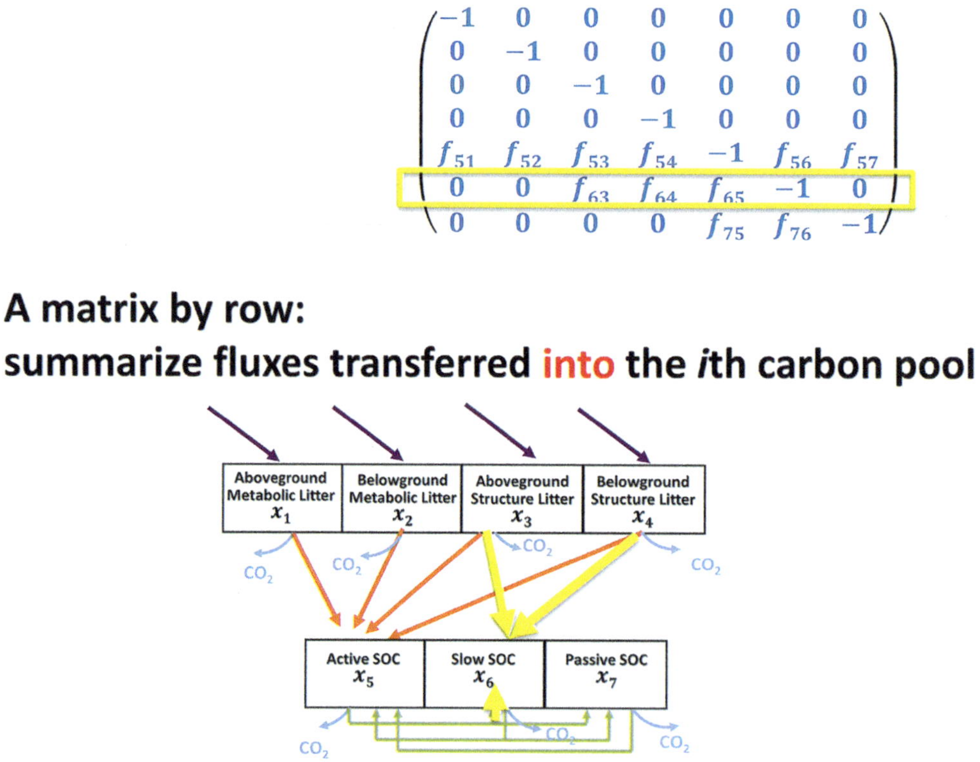

FIGURE 5.5 Illustration of the function of the row of the transfer matrix (*A* matrix).

h and g correspond respectively to the mixing rate between the current soil layer and the one above it, and the current soil layer and the one below it; z_i indicates the depth of soil layer i. Detailed information is available in Huang et al. (2018a).

In this chapter we have learned how to derive the matrix equation from the carbon balance equations. It would also be possible to derive the matrix equation directly from the carbon flow diagram. From the carbon flow diagram, we can

$$\frac{dX}{dt} = I + AK\,X - VX$$

$$V(t) = \begin{pmatrix} 0 & 0 & 0 & 0 & 0 & 0 & 0 \\ 0 & 0 & 0 & 0 & 0 & 0 & 0 \\ 0 & 0 & 0 & 0 & 0 & 0 & 0 \\ 0 & 0 & 0 & 0 & 0 & 0 & 0 \\ 0 & 0 & 0 & 0 & V55(t) & 0 & 0 \\ 0 & 0 & 0 & 0 & 0 & V66(t) & 0 \\ 0 & 0 & 0 & 0 & 0 & 0 & V77(t) \end{pmatrix}$$

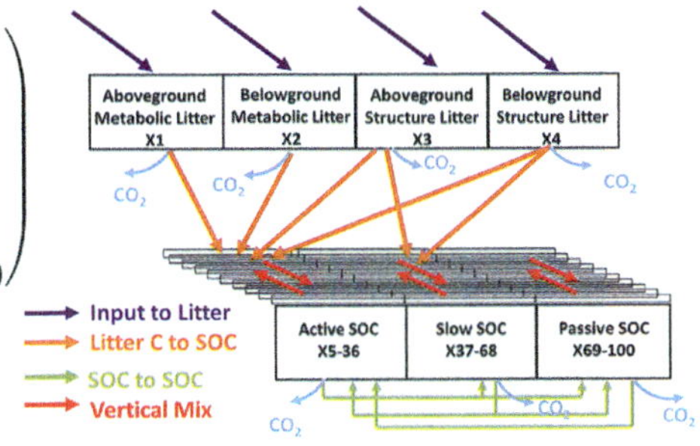

$$V66 = diag(z_1, z_2, \ldots, z_{32})^{-1} \begin{pmatrix} g_1 & -g_1 & 0 & 0 & \cdots & 0 & 0 & 0 \\ -h_2 & h_2+g_2 & -g_2 & 0 & \cdots & 0 & 0 & 0 \\ 0 & -h_3 & h_3+g_3 & -g_3 & \cdots & 0 & 0 & 0 \\ 0 & 0 & -h_4 & h_4+g_4 & \cdots & 0 & 0 & 0 \\ \vdots & \vdots & \vdots & \vdots & \ddots & \vdots & \vdots & \vdots \\ 0 & 0 & 0 & 0 & \cdots & h_{30}+g_{30} & -g_{30} & 0 \\ 0 & 0 & 0 & 0 & \cdots & -h_{31} & h_{31}+g_{31} & -g_{31} \\ 0 & 0 & 0 & 0 & \cdots & 0 & -h_{32} & h_{32} \end{pmatrix}$$

FIGURE 5.6 Illustration of the vertical transfer matrix *V* of the ORCHIDEE-MICT model.

determine the dimension of matrices (or vectors) from the number of carbon pools. The diagram also tells us the external carbon inputs, the incoming and outgoing fluxes of each pool, and the direction of these fluxes. These are basically what we need to construct the matrix equation.

SUGGESTED READINGS

Huang, Y. Y., Lu, X.J. et al. 2018. Matrix Approach to Land Carbon Cycle Modeling: A Case Study with the Community Land Model. *Global Change Biology 24*, 1394–1404. doi:10.1111/gcb.13948

Huang, Y. Y., Zhu, D. et al. 2018. Matrix-Based Sensitivity Assessment of Soil Organic Carbon Storage: A Case Study from the ORCHIDEE-MICT Model. *Journal of Advances in Modeling Earth Systems 10*, 1790–1808. doi:10.1029/2017ms001237

QUIZ

1 Is the sum of each row of the transfer matrix *A* always not bigger than 0?

2 Is the sum of each column of the transfer matrix *A* always not bigger than 0?

3 How would you add a new carbon pool into the matrix equation?

4 Why is a matrix equation exactly the same as a carbon balance equation in terms of describing the carbon cycle?

6 Coupled Carbon-Nitrogen Matrix Models

Zheng Shi
University of Oklahoma, Norman, USA

Xingjie Lu
Sun Yat-sen University, Guangzhou, China

Carbon sequestration in terrestrial ecosystems is strongly regulated by nitrogen processes. Many global land models now simulate the carbon and nitrogen interaction. The goal of this chapter is to understand how coupled carbon and nitrogen models at ecosystem and global scales may be represented in the matrix form. Basically, nitrogen transfers among all the organic nitrogen pools can be represented in one matrix equation that is equivalent to the carbon matrix equation. The carbon and nitrogen matrix equations are coupled through the C:N ratio. Mineral nitrogen dynamics, determined by nitrogen input, mineralization, plant uptake, and leaching, can also be described by one equation.

INTRODUCTION

Rising atmospheric carbon dioxide (CO_2) concentration tends to induce carbon (C) sequestration in terrestrial ecosystems. The conceptual framework of progressive nitrogen (N) limitation has predicted N limitation on future C sequestration in terrestrial ecosystems in response to rising atmospheric CO_2. The N limitation may become progressively stronger over time unless N fixation is stimulated and/or N losses are reduced, leading to increased N capital (Luo et al., 2004). In addition, the degree of N regulation on terrestrial C sequestration depends on changes in several C-N coupling parameters, such as the stoichiometric flexibility of C:N ratios of biomass compartments, changes in plant N uptake via soil exploration, and N redistribution from soil to vegetation. Encoding what we know about how C and N flow within ecosystems, and how these flows are coupled, can help to fully understand the strength of N regulation on C sequestration in terrestrial ecosystems.

In this chapter, we will show how C and N cycling are coupled in an ecosystem model and a global land model, and how the coupled processes can be represented in the matrix form.

MATRIX REPRESENTATION OF C-N COUPLING IN TERRESTRIAL ECOSYSTEM (TECO) MODEL

One of the coupled C and N models presented in this chapter is developed from the terrestrial ecosystem (TECO) model (Weng & Luo, 2011). TECO was first presented in Chapter 1, and its matrix representation introduced in Chapter 5. The coupled C-N version we will discuss here, called TECO-CN, has eight C and N pools in addition to one mineral N pool. C enters the ecosystem through canopy photosynthesis and is then allocated into foliage (X_1), wood biomass (X_2), and fine roots (X_3) (Figure 6.1). Similarly, N is absorbed by plants from mineral soil, and then partitioned among leaf (N_1), woody tissues (N_2), and fine roots (N_3). Detritus from plant biomass turnover is transferred to metabolic litter (X_4) and structure litter (X_5) pools, where it is decomposed by microbes (X_6). The structure litter (X_5) is partly respired while partly converted into fast (X_6) and slow soil organic matter (SOM, X_7). During these C cycling processes, N in plant detritus is also transferred among the same set of ecosystem pools (i.e., litter, fast, slow, and passive SOM). Mathematically, these C processes may be described by the following first-order ordinary differential equation:

$$\frac{dX}{dt} = B\mu + A\xi(t)KX(t) \qquad 6.1$$

where $X = (x_1, x_2, x_3, x_4, x_5, x_6, x_7, x_8)$, represents C pools in leaf, wood, fine roots, metabolic litter, structure litter, microbe, slow and passive SOM, respectively. $\xi(t)$ is a vector of environmental scalars accounting for temperature and moisture effects on all C decomposition. A describes C transformation among various ecosystem compartments, given as:

$$A = \begin{pmatrix} -1 & 0 & 0 & 0 & 0 & 0 & 0 & 0 \\ 0 & -1 & 0 & 0 & 0 & 0 & 0 & 0 \\ 0 & 0 & -1 & 0 & 0 & 0 & 0 & 0 \\ f_{4,1} & f_{4,2} & f_{4,3} & -1 & 0 & 0 & 0 & 0 \\ 1-f_{4,1} & 1-f_{4,2} & 1-f_{4,3} & 0 & -1 & 0 & 0 & 0 \\ 0 & 0 & 0 & f_{6,4} & f_{6,5} & -1 & f_{6,7} & f_{6,8} \\ 0 & 0 & 0 & 0 & f_{7,5} & f_{7,6} & -1 & 0 \\ 0 & 0 & 0 & 0 & 0 & f_{8,6} & f_{8,7} & -1 \end{pmatrix}$$

DOI: 10.1201/9781032711126-8

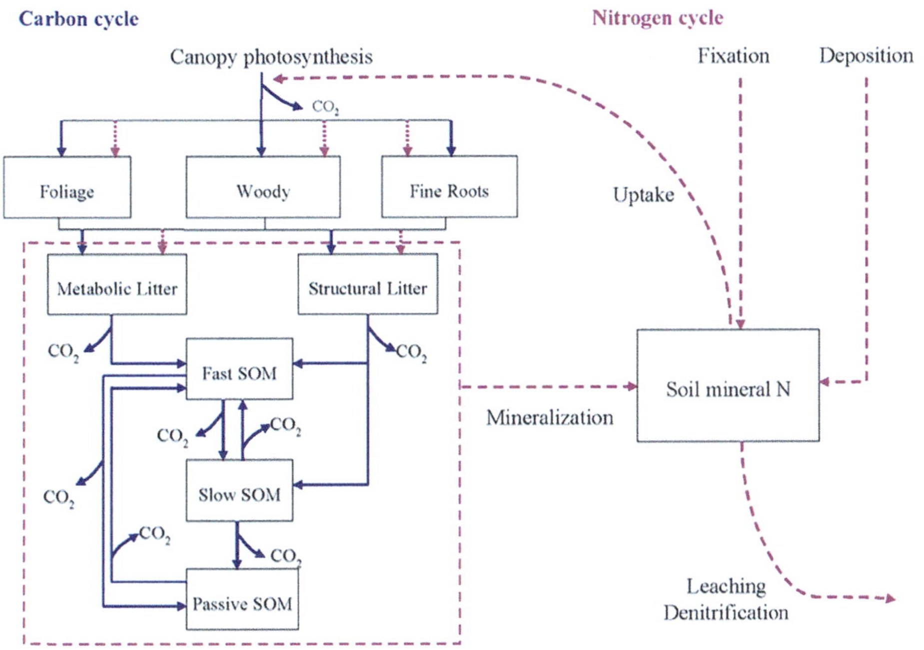

FIGURE 6.1 Carbon and nitrogen pools and pathways of carbon and nitrogen fluxes in the TECO-CN model. Blue arrows show carbon cycling processes, while pink arrows indicate nitrogen cycling processes. SOM = soil organic matter.

The non-zero elements (f_{ij}) in matrix A describe C transfer coefficients (i.e., the fractions of the C entering the ith pool from the jth pool), while the zero elements indicate no C flows between these two pools. K, which is an 8 × 8 diagonal matrix with diagonal entries given by vector $K = (k_1, k_2, k_3, k_4, k_5, k_6, k_7, k_8)$, is baseline turnover rate of carbon pools (i.e., the amounts of C per unit mass leaving each pool per day). I is the input and $B = (b_1, b_2, b_3, 0, 0, 0, 0, 0)$ is the allocations of input into ecosystem carbon pools.

The N processes can be described by:

$$\frac{dN}{dt} = A\xi(t)KR^{-1}X(t) + k_u N_{\min}(t)\Pi \qquad 6.2$$

where $N = (n_1, n_2, n_3, n_4, n_5, n_6, n_7, n_8)$, represents N pools in leaf, wood, fine roots, metabolic litter, structure litter, microbe, slow and passive SOM, respectively. R is an 8 × 8 diagonal matrix with diagonal entries given by vector $R = (r_1, r_2, r_3, r_4, r_5, r_6, r_7, r_8)$, representing C:N ratios in the eight organic N pools. $\Pi = (\pi_1, \pi_2, 1 - \pi_1 - \pi_2, 0, 0, 0, 0)$, is a vector of allocation coefficients expressing the proportion of plant N uptake that is allocated to leaf, wood, and fine root biomass pools. κ_u is the rate of plant N uptake per time step (year), expressed as a proportion of $N_{min}(t)$, the amount of plant available N (mineral N) in soil at time t. The dynamics of the mineral N pool are determined by balance between N input (i.e., N mineralization, biological fixation, and atmospheric deposition) and output through plant N uptake and N loss (i.e., leaching and gaseous N fluxes), which can be expressed by:

$$\frac{dN_{\min}}{dt}(t) = -(\kappa_u + \kappa_L)N_{\min}(t) + \xi(t)\varphi_1^* AKR^{-1}X(t) + F(t) \qquad 6.3$$

where κ_u and κ_L are rates of N uptake and loss, respectively. The second term on the right side of Equation 6.3 describes the amount of N released during mineralization. φ_1^* is the proportion for mineral N production. The C and N cycles are coupled through the parameter R which is C:N ratio. $F(t)$ represents N input through biological fixation and atmospheric deposition.

APPLICATION OF MATRIX REPRESENTATION OF C-N COUPLED MODEL

To illustrate the application of matrix forms of a C-N coupled model, we designed a case study to examine changes in C-N coupling parameters under CO_2 enrichment using a data assimilation approach. The case study uses measurements of C and N pools in various ecosystem compartments (i.e., foliage, woody tissues, fine roots, microbe, forest floor soil inorganic N, and mineral soil) and fluxes (i.e., litterfall, soil respiration and mineralization, and plant N uptake, N input from biological fixation and atmospheric deposition) obtained from the Duke Forest Free-Air CO_2 Enrichment (FACE) experiment in North Carolina, USA, during the period 1996–2005. Key parameters of TECO-CN (i.e., C:N ratio, N uptake, N allocation coefficient, N input, N loss, and the initial value of mineral soil N pool) were estimated through a Bayesian probabilistic inversion. The inversion was done separately for plots with ambient versus elevated CO_2 treatments, yielding one set of parameters for each (Shi et al., 2016).

Comparison of parameter distributions showed that plant N uptake, and C:N ratios in foliage, fine root, metabolic, and structural litter, were significantly higher under elevated than ambient CO_2, whereas CO_2 enrichment did not exert significant effects on C:N ratios in wood tissues and SOM. Moreover, elevated CO_2 led to decrease of C exit rates in foliage, woody biomass, structural litter, and passive SOM, indicating an increase of C residence time in these ecosystem compartments. By contrast, elevated CO_2 resulted in the increase of C exit rate in fine roots, demonstrating faster fine root turnover under CO_2 enrichment. In addition, C allocation to the foliage became smaller under elevated CO_2, while C allocation to fine roots tended to be larger under CO_2 enrichment.

The estimated parameters were then used for a forward analysis to examine ecosystem C and N dynamics under ambient and elevated CO_2 conditions at Duke Forest. Our results demonstrated that modeled N pools in foliage, woody tissues, fine roots, and forest floor closely matched with the corresponding measurements for both ambient and elevated CO_2 scenarios (Figure 6.2). However, TECO-CN could not capture the observed declining trend of microbial N content with time. In addition, the trained model did not simulate N dynamics in mineral soil well, partly due to the large variations in SOM measurements among different years.

MATRIX REPRESENTATION OF C-N COUPLING IN CLM5

The other C-N coupled model we will examine in this chapter is the Community Land Model version 5 (CLM5, Figure 6.3). To extend the ecosystem scale to global scale, CLM5 simulations represent various ecosystems on a grid covering the global land area. Like TECO-CN, the CLM5 biogeochemistry module includes carbon and nitrogen cycles for aboveground and belowground processes (Lawrence et al., 2019). The vegetation modules simulate the biogeochemical transfers among 18 carbon pools and 19 nitrogen pools, representing vegetation and soil compartments of a grid cell or tile. Tissue pools, storage pools, and transfer pools are included for each of leaf, fine root, live stem, dead stem, live course root, and dead course root compartments. In addition to C pools, N pools include one more retranslocation pool temporarily storing N from litterfall. The soil biogeochemistry module has 20 soil layers as a default setting. Each layer contains seven pools for organic C and organic N in each of metabolic litter, cellulose litter, lignin litter, coarse woody debris, fast soil organic matter, slow soil organic matter, and passive soil organic matter. Inorganic N pools, such as ammonium and nitrate, interact among each other, or with organic C and organic N pools, or with the environment through multiple nitrogen processes such as nitrification, denitrification, leaching, atmospheric N deposition, and biological N fixation. All these biogeochemical processes and pools in both vegetation and soil modules can be formulated as carbon or nitrogen balance equations.

We may reorganize the vegetation carbon and nitrogen balance equations of CLM5 into two matrix equations:

$$\frac{dC_{veg}}{dt} = B\mu_{Cin} + \big(A_{phc}(t)K_{phc} + A_{gmc}(t)K_{gmc} + A_{fic}(t)K_{fic}\big)C_{veg}(t) \tag{6.4}$$

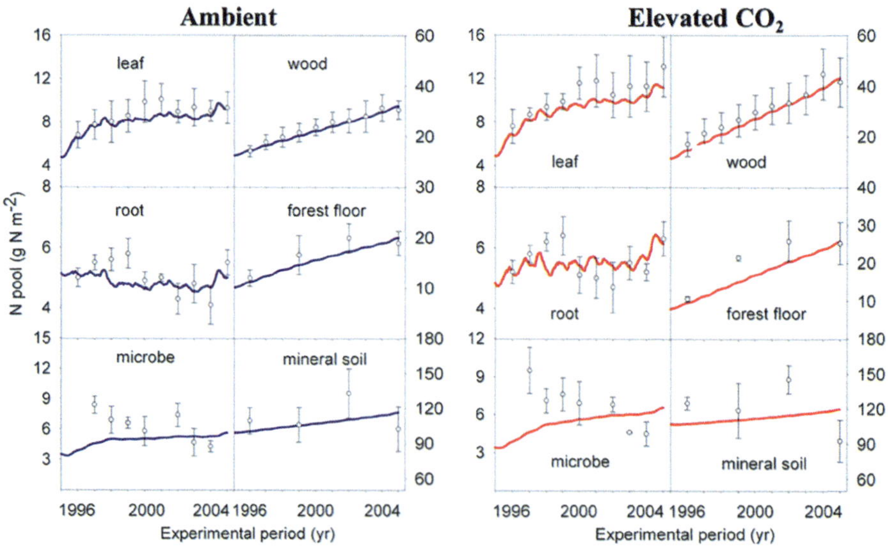

FIGURE 6.2 The comparisons of modeled vs. measured nitrogen pools in various ecosystem compartments under ambient CO_2 (blue lines) and elevated CO_2 (red lines).

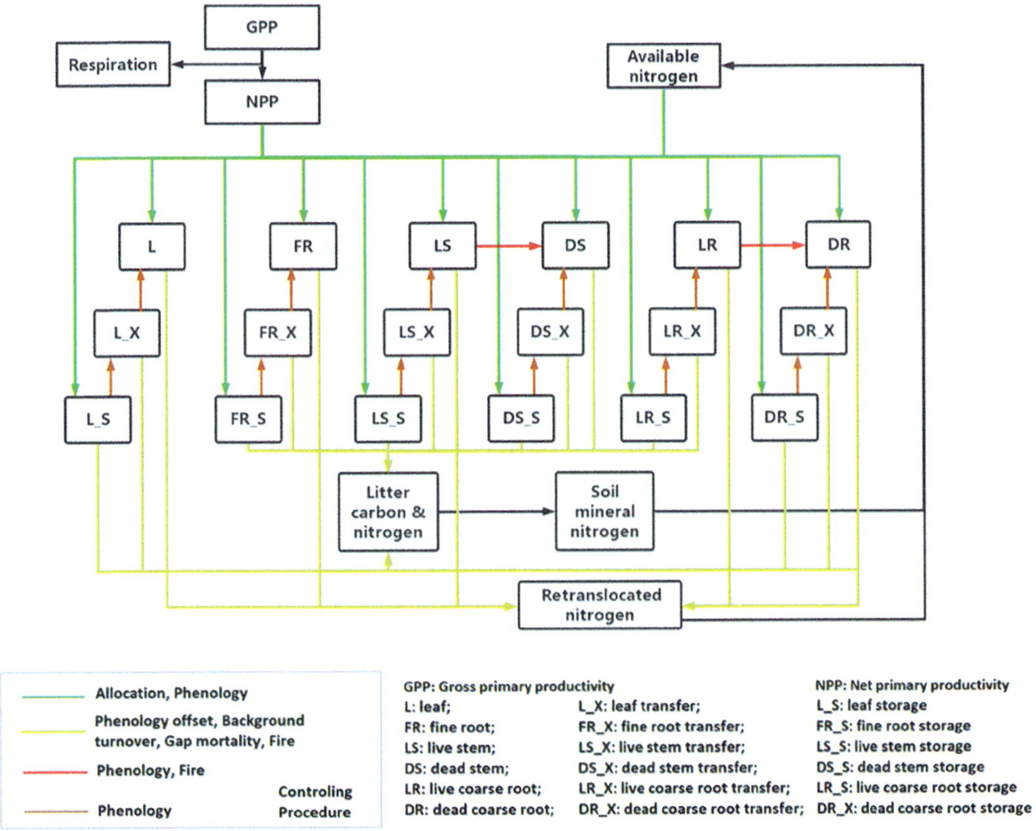

FIGURE 6.3 Carbon and nitrogen flow diagram of vegetation biogeochemical cycle in CLM5. GPP = gross primary production; NPP = net primary production.

$$\frac{dN_{veg}}{dt} = B\mu_{Nin} + \left(A_{phn}(t)K_{phn} + A_{fin}(t)K_{fin}\right)N_{veg}(t) \qquad 6.5$$

C_{veg} and N_{veg} are time-dependent state variables, which are vectors, each entry representing its respective vegetation pool size (gC m^{-2} and gN m^{-2}). I_{Cin} and I_{Nin} are scalars for C and N input, respectively. C input is the net primary productivity, which is the difference between gross primary productivity (i.e., plant photosynthesis) and autotrophic respiration. N input to the vegetation N cycle includes both biological N fixation and N uptake from roots. B_C and B_N are also vectors, representing allocation fractions of plant C and N input to individual pools. The K matrices are $n \times n$ diagonal matrices whose diagonal elements represent turnover rates (pool fraction per annual time step) due to different plant and vegetation processes: subscripts *ph*, *gm*, and *fi* indicate phenology, gap mortality (i.e., harvest from land use and natural mortality), and fire processes, respectively, as described by:

$$K_{ph} = \begin{pmatrix} k_{p1} & \cdots & 0 \\ \vdots & \ddots & \vdots \\ 0 & \cdots & k_{pn} \end{pmatrix}$$

$$K_{gm} = \begin{pmatrix} k_{n1} & \cdots & 0 \\ \vdots & \ddots & \vdots \\ 0 & \cdots & k_{nn} \end{pmatrix}$$

$$K_{fi} = \begin{pmatrix} k_{f1} & \cdots & 0 \\ \vdots & \ddots & \vdots \\ 0 & \cdots & k_{fn} \end{pmatrix}$$

The turnover rates in plant phenology matrix K indicate the leaf, root, live stem, and dead stem turnover due to phenology processes. The exit rates in gap mortality matrix K include harvest rates from land use plus natural mortality. The exit rates in fire matrix K represent the plant C loss rate due to fires.

A is a transfer coefficient matrix, representing C and N transfer among pools as specified in Equations 6.6–6.11 below for CLM5. Subscript c and n respectively denote carbon and nitrogen.

$$
A_{phc} = \begin{pmatrix}
-1 & 0 & a_{1,3} & 0 & 0 & 0 & 0 & 0 & 0 & 0 & 0 & 0 & 0 & 0 & 0 & 0 & 0 & 0 \\
0 & -1 & 0 & 0 & 0 & 0 & 0 & 0 & 0 & 0 & 0 & 0 & 0 & 0 & 0 & 0 & 0 & 0 \\
0 & a_{3,2} & -1 & 0 & 0 & 0 & 0 & 0 & 0 & 0 & 0 & 0 & 0 & 0 & 0 & 0 & 0 & 0 \\
0 & 0 & 0 & -1 & 0 & a_{4,6} & 0 & 0 & 0 & 0 & 0 & 0 & 0 & 0 & 0 & 0 & 0 & 0 \\
0 & 0 & 0 & 0 & -1 & 0 & 0 & 0 & 0 & 0 & 0 & 0 & 0 & 0 & 0 & 0 & 0 & 0 \\
0 & 0 & 0 & 0 & a_{6,5} & -1 & 0 & 0 & 0 & 0 & 0 & 0 & 0 & 0 & 0 & 0 & 0 & 0 \\
0 & 0 & 0 & 0 & 0 & 0 & -1 & 0 & a_{7,9} & 0 & 0 & 0 & 0 & 0 & 0 & 0 & 0 & 0 \\
0 & 0 & 0 & 0 & 0 & 0 & 0 & -1 & 0 & 0 & 0 & 0 & 0 & 0 & 0 & 0 & 0 & 0 \\
0 & 0 & 0 & 0 & 0 & 0 & 0 & a_{9,8} & -1 & 0 & 0 & 0 & 0 & 0 & 0 & 0 & 0 & 0 \\
0 & 0 & 0 & 0 & 0 & 0 & a_{10,7} & 0 & 0 & -1 & 0 & a_{10,12} & 0 & 0 & 0 & 0 & 0 & 0 \\
0 & 0 & 0 & 0 & 0 & 0 & 0 & 0 & 0 & 0 & -1 & 0 & 0 & 0 & 0 & 0 & 0 & 0 \\
0 & 0 & 0 & 0 & 0 & 0 & 0 & 0 & 0 & 0 & a_{12,11} & -1 & 0 & 0 & 0 & 0 & 0 & 0 \\
0 & 0 & 0 & 0 & 0 & 0 & 0 & 0 & 0 & 0 & 0 & 0 & -1 & 0 & a_{13,15} & 0 & 0 & 0 \\
0 & 0 & 0 & 0 & 0 & 0 & 0 & 0 & 0 & 0 & 0 & 0 & 0 & -1 & 0 & 0 & 0 & 0 \\
0 & 0 & 0 & 0 & 0 & 0 & 0 & 0 & 0 & 0 & 0 & 0 & 0 & a_{15,14} & -1 & 0 & 0 & 0 \\
0 & 0 & 0 & 0 & 0 & 0 & 0 & 0 & 0 & 0 & 0 & 0 & a_{16,13} & 0 & 0 & -1 & 0 & a_{16,18} \\
0 & 0 & 0 & 0 & 0 & 0 & 0 & 0 & 0 & 0 & 0 & 0 & 0 & 0 & 0 & 0 & -1 & 0 \\
0 & 0 & 0 & 0 & 0 & 0 & 0 & 0 & 0 & 0 & 0 & 0 & 0 & 0 & 0 & 0 & a_{18,17} & -1
\end{pmatrix}
$$

$$
A_{phn} = \begin{pmatrix}
-1 & 0 & a_{1,3} & 0 & 0 & 0 & 0 & 0 & 0 & 0 & 0 & 0 & 0 & 0 & 0 & 0 & 0 & 0 & a_{1,19} \\
0 & -1 & 0 & 0 & 0 & 0 & 0 & 0 & 0 & 0 & 0 & 0 & 0 & 0 & 0 & 0 & 0 & 0 & a_{2,19} \\
0 & a_{3,2} & -1 & 0 & 0 & 0 & 0 & 0 & 0 & 0 & 0 & 0 & 0 & 0 & 0 & 0 & 0 & 0 & a_{3,19} \\
0 & 0 & 0 & -1 & 0 & a_{4,6} & 0 & 0 & 0 & 0 & 0 & 0 & 0 & 0 & 0 & 0 & 0 & 0 & a_{4,19} \\
0 & 0 & 0 & 0 & -1 & 0 & 0 & 0 & 0 & 0 & 0 & 0 & 0 & 0 & 0 & 0 & 0 & 0 & a_{5,19} \\
0 & 0 & 0 & 0 & a_{6,5} & -1 & 0 & 0 & 0 & 0 & 0 & 0 & 0 & 0 & 0 & 0 & 0 & 0 & a_{6,19} \\
0 & 0 & 0 & 0 & 0 & 0 & -1 & 0 & a_{7,9} & 0 & 0 & 0 & 0 & 0 & 0 & 0 & 0 & 0 & a_{7,19} \\
0 & 0 & 0 & 0 & 0 & 0 & 0 & -1 & 0 & 0 & 0 & 0 & 0 & 0 & 0 & 0 & 0 & 0 & a_{8,19} \\
0 & 0 & 0 & 0 & 0 & 0 & 0 & a_{9,8} & -1 & 0 & 0 & 0 & 0 & 0 & 0 & 0 & 0 & 0 & a_{9,19} \\
0 & 0 & 0 & 0 & 0 & 0 & a_{10,7} & 0 & 0 & -1 & 0 & a_{10,12} & 0 & 0 & 0 & 0 & 0 & 0 & a_{10,19} \\
0 & 0 & 0 & 0 & 0 & 0 & 0 & 0 & 0 & 0 & -1 & 0 & 0 & 0 & 0 & 0 & 0 & 0 & a_{11,19} \\
0 & 0 & 0 & 0 & 0 & 0 & 0 & 0 & 0 & 0 & a_{12,11} & -1 & 0 & 0 & 0 & 0 & 0 & 0 & a_{12,19} \\
0 & 0 & 0 & 0 & 0 & 0 & 0 & 0 & 0 & 0 & 0 & 0 & -1 & 0 & a_{13,15} & 0 & 0 & 0 & a_{13,19} \\
0 & 0 & 0 & 0 & 0 & 0 & 0 & 0 & 0 & 0 & 0 & 0 & 0 & -1 & 0 & 0 & 0 & 0 & a_{14,19} \\
0 & 0 & 0 & 0 & 0 & 0 & 0 & 0 & 0 & 0 & 0 & 0 & 0 & a_{15,14} & -1 & 0 & 0 & 0 & a_{15,19} \\
0 & 0 & 0 & 0 & 0 & 0 & 0 & 0 & 0 & 0 & 0 & 0 & a_{16,13} & 0 & 0 & -1 & 0 & a_{16,18} & a_{16,19} \\
0 & 0 & 0 & 0 & 0 & 0 & 0 & 0 & 0 & 0 & 0 & 0 & 0 & 0 & 0 & 0 & -1 & 0 & a_{17,19} \\
0 & 0 & 0 & 0 & 0 & 0 & 0 & 0 & 0 & 0 & 0 & 0 & 0 & 0 & 0 & 0 & a_{18,17} & -1 & a_{18,19} \\
a_{19,1} & 0 & 0 & a_{19,4} & 0 & 0 & a_{19,7} & 0 & 0 & a_{19,10} & 0 & 0 & 0 & 0 & 0 & 0 & 0 & 0 & -1
\end{pmatrix}
$$

$$A_{gmc} = \begin{pmatrix}
-1 & 0 & 0 & 0 & 0 & 0 & 0 & 0 & 0 & 0 & 0 & 0 & 0 & 0 & 0 & 0 & 0 & 0 \\
0 & -1 & 0 & 0 & 0 & 0 & 0 & 0 & 0 & 0 & 0 & 0 & 0 & 0 & 0 & 0 & 0 & 0 \\
0 & 0 & -1 & 0 & 0 & 0 & 0 & 0 & 0 & 0 & 0 & 0 & 0 & 0 & 0 & 0 & 0 & 0 \\
0 & 0 & 0 & -1 & 0 & 0 & 0 & 0 & 0 & 0 & 0 & 0 & 0 & 0 & 0 & 0 & 0 & 0 \\
0 & 0 & 0 & 0 & -1 & 0 & 0 & 0 & 0 & 0 & 0 & 0 & 0 & 0 & 0 & 0 & 0 & 0 \\
0 & 0 & 0 & 0 & 0 & -1 & 0 & 0 & 0 & 0 & 0 & 0 & 0 & 0 & 0 & 0 & 0 & 0 \\
0 & 0 & 0 & 0 & 0 & 0 & -1 & 0 & 0 & 0 & 0 & 0 & 0 & 0 & 0 & 0 & 0 & 0 \\
0 & 0 & 0 & 0 & 0 & 0 & 0 & -1 & 0 & 0 & 0 & 0 & 0 & 0 & 0 & 0 & 0 & 0 \\
0 & 0 & 0 & 0 & 0 & 0 & 0 & 0 & -1 & 0 & 0 & 0 & 0 & 0 & 0 & 0 & 0 & 0 \\
0 & 0 & 0 & 0 & 0 & 0 & 0 & 0 & 0 & -1 & 0 & 0 & 0 & 0 & 0 & 0 & 0 & 0 \\
0 & 0 & 0 & 0 & 0 & 0 & 0 & 0 & 0 & 0 & -1 & 0 & 0 & 0 & 0 & 0 & 0 & 0 \\
0 & 0 & 0 & 0 & 0 & 0 & 0 & 0 & 0 & 0 & 0 & -1 & 0 & 0 & 0 & 0 & 0 & 0 \\
0 & 0 & 0 & 0 & 0 & 0 & 0 & 0 & 0 & 0 & 0 & 0 & -1 & 0 & 0 & 0 & 0 & 0 \\
0 & 0 & 0 & 0 & 0 & 0 & 0 & 0 & 0 & 0 & 0 & 0 & 0 & -1 & 0 & 0 & 0 & 0 \\
0 & 0 & 0 & 0 & 0 & 0 & 0 & 0 & 0 & 0 & 0 & 0 & 0 & 0 & -1 & 0 & 0 & 0 \\
0 & 0 & 0 & 0 & 0 & 0 & 0 & 0 & 0 & 0 & 0 & 0 & 0 & 0 & 0 & -1 & 0 & 0 \\
0 & 0 & 0 & 0 & 0 & 0 & 0 & 0 & 0 & 0 & 0 & 0 & 0 & 0 & 0 & 0 & -1 & 0 \\
0 & 0 & 0 & 0 & 0 & 0 & 0 & 0 & 0 & 0 & 0 & 0 & 0 & 0 & 0 & 0 & 0 & -1
\end{pmatrix}$$

$$A_{gmn} = \begin{pmatrix}
-1 & 0 & 0 & 0 & 0 & 0 & 0 & 0 & 0 & 0 & 0 & 0 & 0 & 0 & 0 & 0 & 0 & 0 & 0 \\
0 & -1 & 0 & 0 & 0 & 0 & 0 & 0 & 0 & 0 & 0 & 0 & 0 & 0 & 0 & 0 & 0 & 0 & 0 \\
0 & 0 & -1 & 0 & 0 & 0 & 0 & 0 & 0 & 0 & 0 & 0 & 0 & 0 & 0 & 0 & 0 & 0 & 0 \\
0 & 0 & 0 & -1 & 0 & 0 & 0 & 0 & 0 & 0 & 0 & 0 & 0 & 0 & 0 & 0 & 0 & 0 & 0 \\
0 & 0 & 0 & 0 & -1 & 0 & 0 & 0 & 0 & 0 & 0 & 0 & 0 & 0 & 0 & 0 & 0 & 0 & 0 \\
0 & 0 & 0 & 0 & 0 & -1 & 0 & 0 & 0 & 0 & 0 & 0 & 0 & 0 & 0 & 0 & 0 & 0 & 0 \\
0 & 0 & 0 & 0 & 0 & 0 & -1 & 0 & 0 & 0 & 0 & 0 & 0 & 0 & 0 & 0 & 0 & 0 & 0 \\
0 & 0 & 0 & 0 & 0 & 0 & 0 & -1 & 0 & 0 & 0 & 0 & 0 & 0 & 0 & 0 & 0 & 0 & 0 \\
0 & 0 & 0 & 0 & 0 & 0 & 0 & 0 & -1 & 0 & 0 & 0 & 0 & 0 & 0 & 0 & 0 & 0 & 0 \\
0 & 0 & 0 & 0 & 0 & 0 & 0 & 0 & 0 & -1 & 0 & 0 & 0 & 0 & 0 & 0 & 0 & 0 & 0 \\
0 & 0 & 0 & 0 & 0 & 0 & 0 & 0 & 0 & 0 & -1 & 0 & 0 & 0 & 0 & 0 & 0 & 0 & 0 \\
0 & 0 & 0 & 0 & 0 & 0 & 0 & 0 & 0 & 0 & 0 & -1 & 0 & 0 & 0 & 0 & 0 & 0 & 0 \\
0 & 0 & 0 & 0 & 0 & 0 & 0 & 0 & 0 & 0 & 0 & 0 & -1 & 0 & 0 & 0 & 0 & 0 & 0 \\
0 & 0 & 0 & 0 & 0 & 0 & 0 & 0 & 0 & 0 & 0 & 0 & 0 & -1 & 0 & 0 & 0 & 0 & 0 \\
0 & 0 & 0 & 0 & 0 & 0 & 0 & 0 & 0 & 0 & 0 & 0 & 0 & 0 & -1 & 0 & 0 & 0 & 0 \\
0 & 0 & 0 & 0 & 0 & 0 & 0 & 0 & 0 & 0 & 0 & 0 & 0 & 0 & 0 & -1 & 0 & 0 & 0 \\
0 & 0 & 0 & 0 & 0 & 0 & 0 & 0 & 0 & 0 & 0 & 0 & 0 & 0 & 0 & 0 & -1 & 0 & 0 \\
0 & 0 & 0 & 0 & 0 & 0 & 0 & 0 & 0 & 0 & 0 & 0 & 0 & 0 & 0 & 0 & 0 & -1 & 0 \\
0 & 0 & 0 & 0 & 0 & 0 & 0 & 0 & 0 & 0 & 0 & 0 & 0 & 0 & 0 & 0 & 0 & 0 & -1
\end{pmatrix}$$

$$A_{\text{fic}} = \begin{pmatrix}
-1 & 0 & 0 & 0 & 0 & 0 & 0 & 0 & 0 & 0 & 0 & 0 & 0 & 0 & 0 & 0 & 0 & 0 \\
0 & -1 & 0 & 0 & 0 & 0 & 0 & 0 & 0 & 0 & 0 & 0 & 0 & 0 & 0 & 0 & 0 & 0 \\
0 & 0 & -1 & 0 & 0 & 0 & 0 & 0 & 0 & 0 & 0 & 0 & 0 & 0 & 0 & 0 & 0 & 0 \\
0 & 0 & 0 & -1 & 0 & 0 & 0 & 0 & 0 & 0 & 0 & 0 & 0 & 0 & 0 & 0 & 0 & 0 \\
0 & 0 & 0 & 0 & -1 & 0 & 0 & 0 & 0 & 0 & 0 & 0 & 0 & 0 & 0 & 0 & 0 & 0 \\
0 & 0 & 0 & 0 & 0 & -1 & 0 & 0 & 0 & 0 & 0 & 0 & 0 & 0 & 0 & 0 & 0 & 0 \\
0 & 0 & 0 & 0 & 0 & 0 & -1 & 0 & 0 & 0 & 0 & 0 & 0 & 0 & 0 & 0 & 0 & 0 \\
0 & 0 & 0 & 0 & 0 & 0 & 0 & -1 & 0 & 0 & 0 & 0 & 0 & 0 & 0 & 0 & 0 & 0 \\
0 & 0 & 0 & 0 & 0 & 0 & 0 & 0 & -1 & 0 & 0 & 0 & 0 & 0 & 0 & 0 & 0 & 0 \\
0 & 0 & 0 & 0 & 0 & 0 & a_{10,7} & 0 & 0 & -1 & 0 & 0 & 0 & 0 & 0 & 0 & 0 & 0 \\
0 & 0 & 0 & 0 & 0 & 0 & 0 & 0 & 0 & 0 & -1 & 0 & 0 & 0 & 0 & 0 & 0 & 0 \\
0 & 0 & 0 & 0 & 0 & 0 & 0 & 0 & 0 & 0 & 0 & -1 & 0 & 0 & 0 & 0 & 0 & 0 \\
0 & 0 & 0 & 0 & 0 & 0 & 0 & 0 & 0 & 0 & 0 & 0 & -1 & 0 & 0 & 0 & 0 & 0 \\
0 & 0 & 0 & 0 & 0 & 0 & 0 & 0 & 0 & 0 & 0 & 0 & 0 & -1 & 0 & 0 & 0 & 0 \\
0 & 0 & 0 & 0 & 0 & 0 & 0 & 0 & 0 & 0 & 0 & 0 & 0 & 0 & -1 & 0 & 0 & 0 \\
0 & 0 & 0 & 0 & 0 & 0 & 0 & 0 & 0 & 0 & 0 & 0 & a_{16,13} & 0 & 0 & -1 & 0 & 0 \\
0 & 0 & 0 & 0 & 0 & 0 & 0 & 0 & 0 & 0 & 0 & 0 & 0 & 0 & 0 & 0 & -1 & 0 \\
0 & 0 & 0 & 0 & 0 & 0 & 0 & 0 & 0 & 0 & 0 & 0 & 0 & 0 & 0 & 0 & 0 & -1
\end{pmatrix}$$

$$A_{\text{fin}} = \begin{pmatrix}
-1 & 0 & 0 & 0 & 0 & 0 & 0 & 0 & 0 & 0 & 0 & 0 & 0 & 0 & 0 & 0 & 0 & 0 & 0 \\
0 & -1 & 0 & 0 & 0 & 0 & 0 & 0 & 0 & 0 & 0 & 0 & 0 & 0 & 0 & 0 & 0 & 0 & 0 \\
0 & 0 & -1 & 0 & 0 & 0 & 0 & 0 & 0 & 0 & 0 & 0 & 0 & 0 & 0 & 0 & 0 & 0 & 0 \\
0 & 0 & 0 & -1 & 0 & 0 & 0 & 0 & 0 & 0 & 0 & 0 & 0 & 0 & 0 & 0 & 0 & 0 & 0 \\
0 & 0 & 0 & 0 & -1 & 0 & 0 & 0 & 0 & 0 & 0 & 0 & 0 & 0 & 0 & 0 & 0 & 0 & 0 \\
0 & 0 & 0 & 0 & 0 & -1 & 0 & 0 & 0 & 0 & 0 & 0 & 0 & 0 & 0 & 0 & 0 & 0 & 0 \\
0 & 0 & 0 & 0 & 0 & 0 & -1 & 0 & 0 & 0 & 0 & 0 & 0 & 0 & 0 & 0 & 0 & 0 & 0 \\
0 & 0 & 0 & 0 & 0 & 0 & 0 & -1 & 0 & 0 & 0 & 0 & 0 & 0 & 0 & 0 & 0 & 0 & 0 \\
0 & 0 & 0 & 0 & 0 & 0 & 0 & 0 & -1 & 0 & 0 & 0 & 0 & 0 & 0 & 0 & 0 & 0 & 0 \\
0 & 0 & 0 & 0 & 0 & 0 & a_{10,7} & 0 & 0 & -1 & 0 & 0 & 0 & 0 & 0 & 0 & 0 & 0 & 0 \\
0 & 0 & 0 & 0 & 0 & 0 & 0 & 0 & 0 & 0 & -1 & 0 & 0 & 0 & 0 & 0 & 0 & 0 & 0 \\
0 & 0 & 0 & 0 & 0 & 0 & 0 & 0 & 0 & 0 & 0 & -1 & 0 & 0 & 0 & 0 & 0 & 0 & 0 \\
0 & 0 & 0 & 0 & 0 & 0 & 0 & 0 & 0 & 0 & 0 & 0 & -1 & 0 & 0 & 0 & 0 & 0 & 0 \\
0 & 0 & 0 & 0 & 0 & 0 & 0 & 0 & 0 & 0 & 0 & 0 & 0 & -1 & 0 & 0 & 0 & 0 & 0 \\
0 & 0 & 0 & 0 & 0 & 0 & 0 & 0 & 0 & 0 & 0 & 0 & 0 & 0 & -1 & 0 & 0 & 0 & 0 \\
0 & 0 & 0 & 0 & 0 & 0 & 0 & 0 & 0 & 0 & 0 & 0 & a_{16,13} & 0 & 0 & -1 & 0 & 0 & 0 \\
0 & 0 & 0 & 0 & 0 & 0 & 0 & 0 & 0 & 0 & 0 & 0 & 0 & 0 & 0 & 0 & -1 & 0 & 0 \\
0 & 0 & 0 & 0 & 0 & 0 & 0 & 0 & 0 & 0 & 0 & 0 & 0 & 0 & 0 & 0 & 0 & -1 & 0 \\
0 & 0 & 0 & 0 & 0 & 0 & 0 & 0 & 0 & 0 & 0 & 0 & 0 & 0 & 0 & 0 & 0 & 0 & -1
\end{pmatrix}$$

The off-diagonal entry, $a_{i,j}$, for matrix A represents a fraction of C or N leaving pool j that goes to pool i. The diagonal entries are set to −1 to represent that all the exiting C leaves pool j. The pool names referred by the subscript i or j can be found in Figure 6.3.

Interactions between vegetation C and N cycles in the original CLM5 are fully preserved in this matrix version of the CLM5 model. The original CLM5 has two modules to regulate C and N interactions for vegetation. First, the photosynthetic capacity, an important variable driving the carbon cycle, interacts with the nitrogen cycle in the LUNA module. LUNA optimizes N allocation to maximize the daily net photosynthetic carbon gain. Second, plant N uptake interacts with the carbon assimilation in the FUN module. FUN is based on the concept that N uptake requires the expenditure of energy in exchange for carbon. Both the modules are fully preserved to drive the C and N interactions in the matrix version of the model.

The implementation of soil C and N cycling extends the maximum soil layers from a fixed value of 10 in CLM4.5 (see Chapter 5) to a default value of 20 with flexibility to change in CLM5.

N balance equations in the original CLM5 are likewise amenable to a matrix representation. The soil organic C and N transfer among soil pools is formalized by the following matrix equations:

$$\frac{dC_{soil}}{dt} = \mu_{Csoil} + \big(A_{hc}\xi(t)K_h - V(t) - K_f(t)\big)C_{soil}(t) \tag{6.6}$$

$$\frac{dN_{soil}}{dt} = \mu_{Nsoil} + \big(A_{hn}\xi(t)K_h - V(t) - K_f(t)\big)N_{soil}(t) \tag{6.7}$$

As with vegetation, C_{soil} and N_{soil} are vectors of state variables, representing soil organic C and N pool sizes in gC m^{-3} and gN m^{-3}, respectively. I_{Csoil} and I_{Nsoil} are vectors representing plant litterfall into different litter C and N pools, respectively. A_{hc} and A_{hn} represent the horizontal transfers of C and N, respectively, which means transfers among pools within one soil layer. V stands for the rate of vertical mixing between the same types of pools across soil layers, which is the same for C and N. K_f is the rate of fire-induced litter loss, which is the same for C and N.

Matrix A_h, including both A_{hc} and A_{hn}, is a block matrix constituted by several matrices as:

$$A_{hc} \text{ or } A_{hn} = \begin{pmatrix} A_{11} & 0 & 0 & 0 & 0 & 0 & 0 \\ 0 & A_{22} & 0 & 0 & 0 & 0 & 0 \\ A_{31} & 0 & A_{33} & 0 & 0 & 0 & 0 \\ A_{41} & 0 & 0 & A_{44} & 0 & 0 & 0 \\ 0 & A_{52} & A_{53} & 0 & A_{55} & A_{11} & A_{11} \\ 0 & 0 & 0 & A_{64} & A_{65} & A_{66} & 0 \\ 0 & 0 & 0 & 0 & A_{75} & A_{76} & A_{77} \end{pmatrix}$$

Each of the matrices A_{ij} within the block matrix A_{hc} or A_{hn} is 20 × 20, corresponding to 20 soil layers. The non-zero, off-diagonal matrices A_{ij}, $i \neq j$, indicate C and N transfer among pools within one soil layer as:

$$A_{ij} = \begin{cases} \begin{bmatrix} -1 & \cdots & 0 \\ \vdots & \ddots & \vdots \\ 0 & \cdots & -1 \end{bmatrix} & i = j;\ \text{C and N cycles} \\[2ex] \begin{bmatrix} (1-r_{ij})T_{ij} & \cdots & 0 \\ \vdots & \ddots & \vdots \\ 0 & \cdots & (1-r_{ij})T_{ij} \end{bmatrix} & i \neq j;\ \text{C cycle}(A_{hc}) \\[2ex] \begin{bmatrix} (1-r_{ij})T_{ij}\dfrac{CN_{j,1}}{CN_{i,1}} & \cdots & 0 \\ \vdots & \ddots & \vdots \\ 0 & \cdots & (1-r_{ij})T_{ij}\dfrac{CN_{j,20}}{CN_{i,20}} \end{bmatrix} & i \neq j;\text{N cycle}(A_{hn}) \end{cases}$$

Each of the diagonal matrices A_{ii} is a negative identity matrix (i.e., a 20×20 matrix with diagonal elements equal to −1 and all other elements zero). The off-diagonal matrices A_{ij} have non-zero diagonal values, $(1 - r_{ij})T_{ij}$ for carbon and $\left(1-r_{ij}\right)T_{ij}\dfrac{CN_{j,k}}{CN_{i,k}}$ for nitrogen. A_{ij} represents transfer coefficients. r_{ij} is the respired fraction of C along the transfer pathway from pool j to i. T_{ij} represents a pathway fraction of C going to pool i from that leaving the jth pool due to decomposition. $CN_{j,k}$ represents the C:N ratio of pool j in layer k.

The diagonal matrices K_h and K_f indicate the turnover rates, respectively, due to decomposition (horizontal transfer) and fire, at different layers:

$$K_h = \begin{pmatrix} k_1 & \cdots & 0 \\ \vdots & \ddots & \vdots \\ 0 & \cdots & k_n \end{pmatrix}$$

$$K_f = \begin{pmatrix} k_{f1} & \cdots & 0 \\ \vdots & \ddots & \vdots \\ 0 & \cdots & k_{fn} \end{pmatrix}$$

Environmental scalars ξ are time-dependent variables and are the product of temperature scalar ξ_T, water scalar ξ_W, oxygen scalar ξ_O, depth scalar ξ_D, and nitrogen scalar ξ_N:

$$\xi(t) = \xi_T(t)\xi_W(t)\xi_O\xi_D\xi_N(t)$$

The vertical mixing coefficient matrix V is made up of 6 identical matrices v:

$$V(t) = \begin{pmatrix} v000000 \\ 0v00000 \\ 00v0000 \\ 0000000 \\ 0000v00 \\ 00000v0 \\ 000000v \end{pmatrix}$$

Note that vertical mixing of coarse woody debris (CWD) is not allowed in CLM5; therefore, the corresponding vertical mixing matrix is 0 for CWD. The matrix v is a tridiagonal matrix, indicating the vertical mixing only transfers between adjacent layers:

$$v = \mathrm{diag}\left(dz_1, dz_2, \ldots dz_{20}\right)^{-1}$$
$$\begin{pmatrix} g_1 & -g_1 & 0 & & & 0 & 0 & 0 \\ -h_2 & h_2+g_2 & -g_2 & \cdots & & 0 & 0 & 0 \\ 0 & -h_3 & h_3+g_3 & & & 0 & 0 & 0 \\ & \vdots & & \ddots & & & \vdots & \\ 0\ 0\ 0 & & & & h_{18}+g_{18} & -g_{18} & 0 \\ 0\ 0\ 0 & & & \cdots & -h_{19} & h_{19}+g_{19} & -g_{19} \\ 0\ 0\ 0 & & & & 0 & -h_{10} & h_{20} \end{pmatrix}$$

The subscripts represent the soil layer; g and h are vertical mixing rates related to upward and downward transfers.

As for the vegetation component of CLM5, interactions between C and N cycles in the soil from the original CLM5 model are fully preserved in the matrix version. In the original CLM5, nitrogen limits soil organic C decomposition. The N limitation is represented by the ratio between available mineral N and the total soil N demand. The soil N demand includes soil immobilization during soil decomposition and plant uptake demand. The dynamics of mineral N processes that are involved in C and N interactions can be represented by one equation as fully described in Shi et al. (2016). The equation on mineral N dynamics can be coupled with the matrix equations on organic C and N processes to analytically or semi-analytically explore their interactions. The matrix form recoding simplifies the original representation as a complex C and N transform network of a biogeochemistry model like CLM5. Traceable components of the C and N cycles can thereby be abstracted and help build up surrogate models with less computational cost than the original models. Additionally, more robust diagnostic capability is brought out by the matrix form.

GLOBAL VALIDATION OF THE CLM5 MATRIX MODEL FOR C AND N SIMULATIONS

The temporal dynamics of C and N storage simulated by the matrix modules was compared with the dynamics simulated by the original versions of these modules. The CLM5 matrix model and the original CLM5 use the same default initial size of vegetation and soil C and N storage without spin-up for the transition simulation. Modeled C and N storage in the total ecosystem, vegetation, and soil organic matter from the matrix model matched with those from the original CLM5 (Figure 6.4). It is worth mentioning that the matrix module is not an exact representation of the original model because different processes in the original model are updated stepwise, while all processes in the matrix model are updated simultaneously. However, our validation results show that the relative differences of soil C and N are less than 1% for most grid cells. Only 0.4% and 0.5% of the grid cells in vegetation C and N storage, respectively, diverged by more than 1% relative difference (Figure 6.4g–h). Only two of 1466 global grid cells in ecosystem C and N storage diverged by more than 1%.

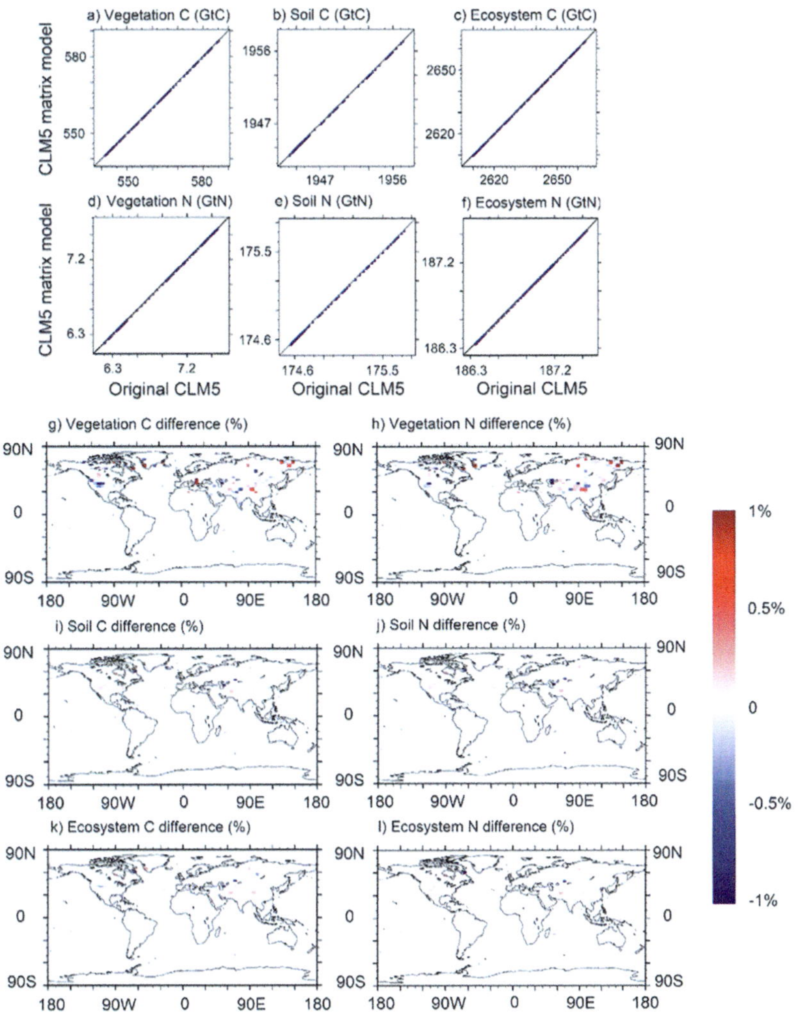

FIGURE 6.4 Comparison of simulated global carbon (C) and nitrogen (N) stocks from 1901 to 2014 between the CLM5 matrix model and the original CLM5. (a) Vegetation C storage, (b) soil C storage, (c) ecosystem C storage, (d) vegetation N storage, (e) soil N storage, and (f) ecosystem N storage are summed up over all land grid cells each year for comparison. Relative differences averaged over last four year (2011–2014) are calculated as: $\frac{X_{matrix} - X_{original}}{X_{original}} * 100$. X_{matrix} represents (g) vegetation C storage, (h) vegetation N storage, (i) soil C storage, (j) soil N storage, (k) ecosystem C storage, and (l) ecosystem N storage from the CLM5 matrix model. $X_{original}$ is the counterpart from the original CLM5.

Global annual means in all six C and N storages over 115 years perfectly lined up on the 1:1 line, signifying exact agreement (Figure 6.4a–f). The good agreements demonstrate that the matrix model can faithfully reproduce both temporal dynamics and spatial patterns of C and N states from the original model.

As mentioned before, minor divergence between the two model versions could be explained by the difference between the simultaneous nature of the calculations encapsulated by the matrix equations, versus the stepwise nature of the original model's algorithms. The matrix modules update C and N state variables only once within each time step, whereas the original modules update these state variables one after the other within each time step. This means that updates in a state variable calculated early in the sequence can affect the calculation for a subsequent state variable. For example, leaf turnover is updated by both phenology and fire in sequence in the original model. Thus, the turnover in leaf C from fire is proportional to the pool size after being updated by phenology in the original model, while the turnover in leaf C from fire is proportional to the pool size before phenology in the matrix model. As a consequence, simulated dynamics may slightly differ between the two model realizations. Nevertheless, the differences in modeled state variables due to different update methods of the two models were small enough not to generate notable differences as shown in global simulation of the C and N storage.

To summarize, we have shown in this chapter how matrix equations may be used to represent C-N coupled models for application at ecosystem and global scales. The new representation as a matrix equation for C-N coupled models has several advantages. The matrix form could help build up a surrogate model for parameter inversion and parameter sensitivity analysis, especially for understanding the role of different C-N interactions for N limitation of CO_2 fertilization. Previously, parameter inversions or parameter sensitivity analysis of terrestrial biogeochemical models like CLM5 have been prohibitively expensive computationally. The surrogate model in matrix form greatly saves the computational cost (Huang et al., 2018b; Tao et al., 2020). Another advantage of the matrix form is that it enables a strong diagnostic approach, the traceability analysis, which attributes the differences in simulated C and N storage into several traceable components (see Chapter 17). These traceable components convey critical ecological or biogeochemical meanings which can help us understand the key drivers of the spatial and temporal variability in terrestrial C and N cycles emerging from model simulations. For example, water scalars have been identified as the most significant traceable component to explain wide divergence between estimates of permafrost carbon storages driven by two reanalysis meteorological datasets, GSWP3 and CRUNCEP (Lu et al., 2020). The success in capturing the dynamics of biogeochemical cycling of a complex model such as CLM5 indicates the feasibility of implementing matrix form versions of other terrestrial biogeochemistry models.

SUGGESTED READINGS

Shi, Z., Yang, Y., Zhou, X., Weng, E., Finzi, A. C., & Luo, Y. (2016). Inverse analysis of coupled carbon–nitrogen cycles against multiple datasets at ambient and elevated CO2. *Journal of Plant Ecology*, *9* (*3*), 285–295.

Xingjie L., Zhenggang, D., Yuanyuan, H., David, L., Erik, K., Nathan C., Danica L., Negin, S., Edward, A., Schuur, G., Yiqi, L. (2020). Full implementation of matrix approach to biogeochemistry module of Community Land Model version 5 (CLM5). *Journal of Advances in Modeling Earth Systems.* In press. https://doi.org/10.1029/2020MS002105

QUIZ

1. What is the matrix form of ecosystem nitrogen cycling?
2. What are the differences between carbon and nitrogen cycles in the matrix form?
3. What are the key parameters to couple carbon and nitrogen cycles?
4. What drives the dynamic of the mineral nitrogen pool?

7 Compartmental Dynamical Systems and Carbon Cycle Models

Carlos A. Sierra
Max Planck Institute for Biogeochemistry, Jena, Germany

Models of the terrestrial carbon cycle are particular cases of compartmental dynamical systems, which are systems of differential equations that must conserve mass. This chapter introduces the main mathematical properties of compartmental dynamical systems and proposes a classification scheme that is useful for the analysis of carbon cycle models. This classification scheme distinguishes between models where carbon inputs and rates change over time or remain constant (nonautonomous versus autonomous models), and between models in which the amount of mass in compartments interact with mass in other compartments (nonlinearity). We show that simple concepts, such as steady state, may not be well defined for some groups of models, and present alternative concepts such as the pullback attractor for the analysis of models with no steady state. In addition, this chapter introduces the theoretical basis for the mathematical analysis of models written in matrix form.

INTRODUCTION

The matrix representation of models has emerged as a very general representation of ecosystem models, particularly models that track the movement of carbon, nitrogen, and other elements inside vegetation and soil pools (Mulholland and Keener 1974; Matis et al. 1979; Bolin 1981; Luo and Weng 2011; Xia et al. 2013; Luo et al. 2017). For soil organic matter models, some of the first representations in matrix form were the models of Bolker et al. (1998), Baisden and Amundson (2003), and Tuomi et al. (2009). For these authors, the matrix representation helped them organize the set of differential equations that resulted in their model in a more manageable and compact form. This is also the case in other fields of science such as biology or chemistry, where large sets of differential equations can be organized using this compact representation.

In fact, any model that represents the mass balance of a quantity such as atoms or molecules, can be represented in this form. Compared to other systems of differential equations, mass-balanced systems are special in the sense that all quantities are generally non-negative; i.e., the information that is fed into the model, and the predictions it produces can only exist inside the domain of the positive real numbers. Furthermore, the mass balance constraint leads to a special type of dynamical system known as a compartmental system.

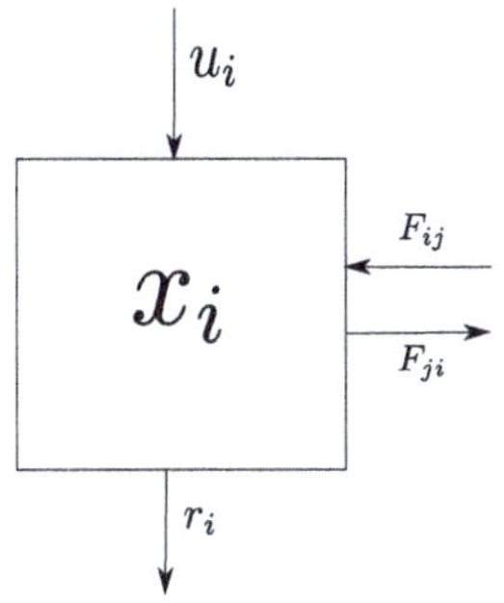

FIGURE 7.1 The mass balance of a single compartment.

We will now introduce the mathematical concept of compartmental systems, and will show that models written in compartmental form have a specific set of mathematical properties. These properties however, depend on the specific structure analyzed, mostly on the time dependence of the elements of the model and intrinsic nonlinearities.

DEFINITION OF COMPARTMENTAL SYSTEMS

We start by defining a *compartment* as an amount of material that is kinetically homogeneous and that follows the law of mass balance. The meaning of 'mass balance' is elaborated below. A compartmental system, therefore, is a set of compartments that exchange mass with each other and with the external environment. This implies that a compartmental system is an open system with an observer-defined boundary (Anderson 1983; Jacquez and Simon 1993).

Let's consider the mass stored in the compartment i, denoted by x_i, as the balance between (Figure 7.1):

- $u_i \geq 0$ inflow (uptake) from outside the system,
- $r_i \geq 0$ outflow (release) to outside the system,
- $F_{ji} \geq 0$ flow transfers from compartment i to compartment j,
- $F_{ij} \geq 0$ flow transfers from compartment j to compartment i.

The change in mass over time of this compartment, $\frac{dx_i}{dt} = \dot{x}_i$, must be balanced according to the equation:

$$\dot{x}_i = \sum_{j \neq i} \left(-F_{ji} + F_{ij} \right) + u_i - r_i,$$

DOI: 10.1201/9781032711126-9

where the constraints $F_{ij} \geq 0$, $u_i \geq 0$, and $r_i \geq 0$ must be met for all i, j, and t. The time dependence is omitted in the notation for simplicity, but all masses and flows may change over time.

An additional constraint for the system is that if the compartment is empty, no mass can flow out of it; i.e., if $x_i = 0$, then $r_i = 0$ and $F_{ji} = 0$ for all j, so that $\dot{x}_i \geq 0$.

If the flows F are continuously differentiable, i.e., they change smoothly over time without sudden jumps, we can define the flows as (Jacquez and Simon 1993):

$$F_{ji}(x) \equiv b_{ji}(x) \cdot x_i.$$

where b_{ji} is a coefficient of transfer from compartment i to compartment j. Then, we can write the mass balance equation for compartment i as:

$$\dot{x}_i = -\left(b_{0i} + \sum_{j \neq i} b_{ji}\right) x_i + \sum_{j \neq i} b_{ij} x_j + u_i.$$

The total outputs from compartment i can be expressed as $b_{ii} \equiv -\left(b_{0i} + \sum_{j \neq i} b_{ji}\right)$, then a general expression for each compartment satisfies the expression:

$$\dot{x}_i = \sum_j b_{ij} x_j + u_i.$$

A general expression for the entire system can be written as:

$$\dot{\boldsymbol{x}} = \mathbf{B}\boldsymbol{x} + \boldsymbol{u}, \qquad 7.1$$

where the elements may be time-dependent and the matrix **B** and vector $\boldsymbol{u}$ depend on the vector of states $\boldsymbol{x}$. Notice that in contrast to other chapters, we follow here a different notation and use **B** to denote a matrix. The system of Equation 7.1 is a *compartmental or reservoir system*, and the matrix **B** is called the compartmental matrix.

For any compartmental system, the compartmental matrix **B** has three properties:

- $b_{ii} \leq 0$ for all i, $t \geq 0$,
- $b_{ij} \geq 0$ for all $i \neq j$, $t \geq 0$,
- $\sum_{i=1}^{n} b_{ij} = \sum_{i \neq j} b_{ij} + b_{jj} = -z_j \leq 0$ for all j, $t \geq 0$.

In words, the compartmental matrix **B** must always meet the requirement that all its diagonal entries are non-positive, its off-diagonal entries non-negative, and the sum of all elements inside each column must be non-positive. This column sum represents the fraction of matter that is released from the system, and it is called the *fractional release coefficient* z_j because it can be used to compute the amount of material that is released to the external environment from each pool j. The total release from the system can be obtained with the expression:

$$\boldsymbol{r} = \boldsymbol{z} \circ \boldsymbol{x},$$

where $\boldsymbol{z}$ is the vector of fractional release coefficients and $\circ$ is the entry-wise product between the two vectors.

The property $-z_j \leq 0$ implies that **B** is a diagonally dominant matrix, which means that each element in the diagonal is greater than or equal to the column sum for this entry. Mathematically, **B** is diagonally dominant if $|b_{ii}| \geq \sum_{j \neq i} |b_{ij}|$, and strictly diagonally dominant if $|b_{ii}| \geq \sum_{j \neq i} |b_{ij}|$.

One important property of strictly diagonally dominant matrices is that they are invertible (Taussky 1949); i.e., there exists an inverse matrix $\mathbf{B}^{-1}$ such that $\mathbf{B} \cdot \mathbf{B}^{-1} = \mathbf{I}$, where **I** is the identity matrix. Compartmental systems that meet this property contain no traps (Jacquez and Simon 1993); i.e., all mass that enters the system eventually leaves from any of the output flows.

CLASSIFICATION OF COMPARTMENTAL SYSTEMS

In the derivation of the compartmental system (Equation 7.1), the explicit representation of time dependencies and nonlinearities was omitted. We will now introduce a classification scheme for compartmental systems based on these two properties, time dependencies (autonomy), and interaction among state variables (linearity). We call a model linear when the vector of inputs and the compartmental matrix are not dependent on the vector of states, and nonlinear otherwise. Similarly, we call a model autonomous when the mass inputs and the compartmental matrix are not explicitly time dependent, and nonautonomous otherwise (Table 7.1). This classification scheme leads to four distinct groups of compartmental systems, each with specific mathematical properties that we will explore in the following sections.

Autonomous Versus Nonautonomous Systems

In the autonomous case (Table 7.1), mass inputs and process rates in the system are constant. This implies that the external

TABLE 7.1
Classification of carbon cycle models according to their dependence on the vector of states (linearity), and on time (autonomy). Table cells are expressions for the differential equation describing $\dot{x}(t)$ that captures the change of mass contents with respect to time

x-dependence	Autonomous	Nonautonomous
Linear	$\boldsymbol{u} + \mathbf{B} \cdot \boldsymbol{x}(t)$	$\boldsymbol{u}(t) + \mathbf{B}(t) \cdot \boldsymbol{x}(t)$
Nonlinear	$\boldsymbol{u}(\boldsymbol{x}) + \mathbf{B}(\boldsymbol{x}) \cdot \boldsymbol{x}(t)$	$\boldsymbol{u}(\boldsymbol{x}, t) + \mathbf{B}(\boldsymbol{x}, t) \cdot \boldsymbol{x}(t)$

environment (e.g., solar radiation, air temperature, water content) are assumed constant. Although ecosystems are far from being surrounded by a constant environment, this assumption is sometimes useful to study basic properties of a system such as its long-term behavior.

However, it is important not to mix up concepts that belong to autonomous systems with concepts that do not apply for nonautonomous systems. For instance, an autonomous compartmental system generally converges to a steady state in the long term where the mass of each compartment does not change with time. In contrast, a nonautonomous system does not reach such a steady state because, by definition, the system is changing all the time. Therefore, it is wrong to talk about the steady state of a nonautonomous system (for additional details see Sierra et al. 2018).

Linear Versus Nonlinear Systems

In the linear case, the contents of compartments do not influence the rates at which mass flows into the system from the external environment, and do not influence the rates at which mass flows out of the compartments (Table 7.1). In other words, there are no feedbacks among compartment contents. However, nonlinear behavior can occur in ecosystems, for instance, when the amount of photosynthesis in the leaves depends on the amount of nonstructural carbohydrates or fine roots.

Nonlinear compartmental systems can show a very rich set of qualitative behaviors (Jacquez and Simon 1993; Anderson and Roller 1991), which for nonlinear autonomous systems range from sustained oscillations to catastrophic shifts to alternate states (Wang et al. 2014). In the nonlinear nonautonomous case, the time-dependent signals that affect the system introduce an even larger degree of complexity, which complicates the behavior of these systems further (Müller and Sierra 2017).

PROPERTIES AND LONG-TERM BEHAVIOR OF AUTONOMOUS COMPARTMENTAL SYSTEMS

Even though the assumption of a constant environment is unrealistic, autonomous models can be very useful in illustrating potential behavior of compartmental systems. In the following, we will present a few properties of autonomous systems that are useful for many applications, which include: long-term behavior of stocks and fluxes, behavior in the neighborhood of the steady state after a perturbation, the age structure of the compartments and the release flux, and the behavior of an impulsive tracer.

Linear Systems

We will consider first linear autonomous compartmental systems of the form

$$\dot{x}(t) = \boldsymbol{u} + \mathbf{B} \cdot \boldsymbol{x}(t), \qquad 7.2$$

with $\mathbf{B}$ invertible and some initial conditions at $t = 0$

$$\boldsymbol{x}(0) = \boldsymbol{x}_0.$$

One advantage of systems of the form of Equation 7.2 compared to the other systems in Table 7.1, is that it is possible to compute their analytical solution. The general solution of this model is given by:

$$\boldsymbol{x}(t) = e^{\mathbf{B}\cdot t}\boldsymbol{x}_0 + \left(\int_0^t e^{\mathbf{B}\cdot(t-\tau)} d\tau\right)\boldsymbol{u}, \qquad 7.3$$

where $e^{\mathbf{B}}$ is the matrix exponential.

Equation 7.3 shows that the solution of the system is composed of two terms. The first term accounts for the decomposition of the mass initially stored in the system at time zero. The second term accounts for decomposition of the inputs that entered the system until time t. At any given time, the mass stored in the system is the sum of both the remaining of the initial mass present at time $t = 0$ and all the un-decomposed mass that entered until time t.

The release of mass from the system is computed by multiplying the fractional release coefficients z_j by the amount of carbon stored in each pool as:

$$\boldsymbol{r}(t) = \left(z_j \cdot x_j(t)\right)_{j=1...n},$$

$$= \boldsymbol{z} \circ \boldsymbol{x}(t)$$

If the system runs for a very long time, it eventually reaches a point called the *steady state* where all inputs are equal to the outputs, and there are no changes in mass within the system. Technically, as $t \to +\infty$, $\boldsymbol{x}(t) \to \boldsymbol{x}^*$, where:

$$\boldsymbol{x}^* = -\mathbf{B}^{-1} \cdot \boldsymbol{u}, \qquad 7.4$$

and

$$\boldsymbol{r}^* = \left(z_j \cdot x_j^*\right)_{j...n}, \quad = \boldsymbol{z} \circ \boldsymbol{x}^*$$

Notice that the steady state does not depend on the initial conditions. It only depends on the compartmental matrix and the vector of external inputs, and represents the equilibrium point where the total amount of matter in the system and in the individual pools do not change, i.e., $\sum \dot{x} = 0$, and $\dot{x} = 0$, respectively.

Nonlinear Systems

In contrast to linear systems, nonlinear compartmental systems have no general explicit analytical solution. However, it is always possible to obtain a numerical solution of the system using any suitable numerical method (LeVeque 2007).

In most applications, we are interested in observing how the system evolves over time and eventually reaches a steady state. Therefore, it is of interest to find an equilibrium solution for the system:

$$\dot{x} = u(x) + \mathbf{B}(x) \cdot x, \quad 7.5$$

such that:

$$0 = u(x) + \mathbf{B}(x) \cdot x. \quad 7.6$$

However, it is not certain that a specific nonlinear system has an equilibrium solution, or in case there is one, that this equilibrium is unique. Anderson and Roller (1991) show special cases of nonlinear compartmental systems with constant inputs that have unique solutions, but these cases are too specific for our purposes here.

Certain combinations of parameter values and pool sizes may lead to the situation in which the matrix $\mathbf{B}(x)$ is not compartmental, and therefore the system may not be mass balanced. For this reason, it is useful to define a space in which a nonlinear system is well defined. Following Anderson and Roller (1991), we define $R_+^n := \{x \in R^n : x \geq 0\}$ as the set of all non-negative real numbers in an n-dimensional space. Since the mass in all compartments is always non-negative, the solutions of the system can only occupy this space. Now we define the space within R_+^n where all solutions of the system obey mass balance constraints as:

$$\Omega := \{x \in R_+^n : \mathbf{B}(x) \text{ is a compartmental matrix}\}.$$

The space Ω is the set of all possible states the system can take without violating mass balance. One important use of Ω is that it can be used to test whether a particular nonlinear model does not violate mass balance for any value of x and t.

For the case of constant inputs, i.e., u, Anderson and Roller (1991) propose an iteration strategy to find a steady-state solution for a nonlinear autonomous system. It consists of applying the formula:

$$x^{p+1} = -\mathbf{B}(x^p)^{-1} \cdot u, \; p = 0, 1, 2, \ldots,$$

until $x^{p+1} \approx x^p$. Notice that for this method to work, the compartmental matrix must be invertible. Also, the existence of one equilibrium point is not a guarantee that it is unique: other equilibria may exist as well. The choice of the starting $x^{p=0}$ may determine what equilibrium point the method will find.

Stability Analysis Near Equilibria

In many applications, it is of interest to study the behavior of a system as it approaches an equilibrium point, or the behavior of the system when it is slightly perturbed from this equilibrium. The study of these behaviors usually falls under the label *stability analysis*. Again, the stability analysis would differ depending on whether the autonomous system is linear or nonlinear.

Linear Systems

For linear autonomous compartmental systems (Equation 7.2), their long-term behavior can be studied by analyzing the eigenvalues and eigenvectors of the compartmental matrix **B**. It is well established that a compartmental matrix with constant coefficients has no eigenvalues with positive real part, which means that the mass inside the compartments never grows exponentially as long as inputs and rates are kept constant. This is ensured by the diagonally dominant property of the compartmental matrix.

In most applications, the eigenvalues of the linear autonomous compartmental matrix have a negative real part. In these cases, it is said that the compartmental system is asymptotically stable because all solutions converge in the long-term to the steady state of Equation 7.4. If the eigenvalues also contain a complex part, then the solution will approach the steady state through oscillations. If the eigenvalues contain no complex part, then the system approaches the steady state in the direction given by the eigenvector of the eigenvalue with the smallest absolute value of the real part.

A third possibility is that the compartmental matrix contains at least m eigenvalues with a real part equal to zero. In this case, it is said that the compartmental system contains m traps (Jacquez and Simon 1993). A trap is a compartment, or a set of connected compartments, where mass may flow in but cannot flow out. In this case, the system contains no equilibrium since **B** is not invertible and Equation 7.4 cannot be solved. The system therefore, will grow proportionally to the amount of mass entering the m traps.

Nonlinear Systems

For nonlinear systems, it is common to study the behavior of the system in the neighborhood of one or multiple equilibrium points. For compartmental systems, we are only interested in equilibria that reside in the space Ω, since they are the only ones that have a physical and biological interpretation.

We assume that the nonlinear autonomous system of Equation 7.5 has at least one equilibrium point in Ω, then we are interested in calculating the Jacobian matrix, defined as:

$$\mathbf{J}(x) = \frac{\partial(\mathbf{B}(x) \cdot x)}{\partial x},$$

at an equilibrium point $x = x* \in \Omega$. This Jacobian matrix tells us about the behavior of trajectories that are close to the steady state, which is a point in the phase plane. Then, the properties of the Jacobian matrix, particularly its eigenvalues, tell us about the stability of the system in the neighborhood of the equilibrium (Guckenheimer and Holmes 1983). It is possible to treat the nonlinear system as a linear system in the neighborhood of the equilibrium, and for this reason one can perform the same analysis of eigenvalues as in the linear case (Guckenheimer and Holmes 1983).

If there are eigenvalues with positive real part, trajectories are repelled away from the equilibrium point, which is considered *unstable* (Strogatz 1994). The existence of unstable equilibria is an indication of possible tipping points and alternative states for the system (Scheffer et al. 2001). However, it is often the case that the Jacobian matrix of a compartmental system is also a compartmental matrix, in which case the existence of unstable equilibria is excluded.

When this Jacobian matrix has a compartmental structure, the system is said to be *cooperative*, which means that if the mass of one compartment increases, the fluxes to other compartments also increase (Jacquez and Simon 1993). In this case, trajectories close to the equilibrium point are attracted to it, and in some particular cases this equilibrium may be unique (Jacquez and Simon 1993; Bastin and Guffens 2006). This particular case of a unique equilibrium point means that the system is *global asymptotically stable* or GAS (Müller and Sierra 2017).

PROPERTIES AND LONG-TERM BEHAVIOR OF NONAUTONOMOUS SYSTEMS

Nonautonomous compartmental systems behave in a completely different way to autonomous systems. Since the mass inputs and the rates change with time, it is not possible for them to converge to a fixed point in the state space. Also, the stability analysis tools for autonomous systems are of little use for nonautonomous systems. Methods to analyze nonautonomous systems are relatively new, and they are currently an active branch of mathematical research (Rasmussen 2007; Kloeden and Rasmussen 2011). Concepts from control engineering can also be very useful to study nonautonomous systems, particularly nonlinear ones (Sontag 1998). Again, we will split the concepts for linear versus nonlinear nonautonomous systems in the sections below.

Linear Systems

We will consider two cases for linear autonomous compartmental systems: (1) the case of time-dependent inputs and constant rates, and (2) the case of time-dependent inputs and rates.

The first case is given by a system of the form:

$$\dot{\boldsymbol{x}}(t) = \boldsymbol{u}(t) + \mathbf{B} \cdot \boldsymbol{x}(t),$$

with initial condition $\boldsymbol{x}(0) = \boldsymbol{x}_0$. If the vector-valued function $\boldsymbol{u}(t)$ is known, an analytical solution can be obtained as:

$$\boldsymbol{x}(t) = e^{\mathbf{B} \cdot t} \boldsymbol{x}_0 + \int_0^t e^{\mathbf{B} \cdot (t-\tau)} \cdot \boldsymbol{u}(\tau) \mathrm{d}\tau,$$

which is a general form for the linear autonomous solution of Equation 7.3. This analytical solution is only possible to compute because the rates in the compartmental matrix **B** are constant for all times, and therefore one can take advantage of the analytical properties of the matrix exponential.

For the second case, when both mass inputs and rates are time dependent, the system is expressed as:

$$\dot{\boldsymbol{x}}(t) = \boldsymbol{u}(t) + \mathbf{B}(t) \cdot \boldsymbol{x}(t), \qquad 7.7$$

for which an analytical solution cannot be computed. However, a semi-explicit solution for Equation 7.7 can be expressed in terms of the *state transition operator* $\Phi(t, t_0)$, which is a matrix whose product with the state vector at an initial time t_0 gives $\boldsymbol{x}(t)$ at a later time t. In other words, $\Phi(t, t_0) \cdot \boldsymbol{x}_0$ is the solution to the homogeneous equation $\dot{\boldsymbol{x}} = \mathbf{B}(t) \cdot \boldsymbol{x}$.

The semi-explicit solution of the linear nonautonomous system of Equation 7.7 can be expressed as:

$$\boldsymbol{x}(t, t_0, \boldsymbol{x}_0) = \Phi(t, t_0) \cdot \boldsymbol{x}_0 + \int_{t_0}^t \Phi(t, \tau) \cdot \boldsymbol{u}(\tau) \mathrm{d}\tau. \qquad 7.8$$

This solution explicitly depends on the initial conditions since for a nonautonomous system, where mass inputs and rates constantly change with time, the exact time and state when the system starts is of fundamental importance to compute a unique solution. In the autonomous case, solutions only depend on the time elapsed $t - t_0$, while in the nonautonomous case the solutions depend separately on the actual time t and the starting time t_0 (Kloeden and Rasmussen 2011).

Rasmussen et al. (2016) present a sufficient condition for the global exponential stability of the nonautonomous linear compartmental system. If the compartmental matrix **B** of the homogeneous system $\dot{\boldsymbol{x}} = \mathbf{B}(t) \cdot \boldsymbol{x}$ is strictly diagonally dominant for all t, then this system is exponentially stable. This means that there is a minimal rate at which the initial mass in the system decays. Now, for the inhomogeneous case (Equation 7.7), we can think of two solutions $s_1(t, t_1, \boldsymbol{x}_1)$ and $s_2(t, t_2, \boldsymbol{x}_2)$ that have different initial conditions. As a consequence of the exponential stability property, the two solutions are said to be *forward attracting*, i.e., they get close to each other as $t \to +\infty$.

Rasmussen et al. (2016) also showed that for linear nonautonomous compartmental systems that meet the sufficient condition for exponential stability, there exists a unique *pullback attracting solution* or *pullback attractor* which all solutions are attracted to. It is defined as:

$$v(t) := \int_{-\infty}^t \Phi(t, \tau) \cdot \boldsymbol{u}(\tau) \mathrm{d}\tau,$$

and can be interpreted as the solution on which the initial conditions have no influence (Kloeden and Rasmussen 2011). Therefore, the pullback attractor is the nonautonomous equivalent of the steady-state concept for autonomous systems (Carvalho et al. 2013).

A particular case is the linear nonautonomous system in which the mass inputs and the process rates are periodic. For example, this is the case of seasonal systems without noise in which the same periodic pattern for the mass inputs and for the process rates is repeated every year. More precisely, a periodic linear compartmental system is one in which $\boldsymbol{u}(t + T) = \boldsymbol{u}(t)$ and $\mathbf{B}(t + T) = \mathbf{B}(t)$ for a fixed period T and for all t. Mulholland and Keener (1974) showed that these types of systems have periodic solutions for which $\boldsymbol{x}(t + T) = \boldsymbol{x}(t)$. This periodic solution can be interpreted as a pullback attractor because it has no influence on the initial conditions.

Nonlinear Systems

Nonlinear nonautonomous compartmental systems are the most complex cases for their study and analysis. It is not possible in general to obtain analytical solutions, and, contrary to the autonomous case, it is not possible to study an equilibrium point for these systems because, by definition, compartment contents are always changing and they never reach a constant value.

As mass inputs and process rates change in a nonlinear nonautonomous compartmental system, it is possible that specific combinations of parameter values and compartment sizes lead the system outside the space Ω where mass balance consideration must be met. Therefore, it is always important to check that solutions for these systems are always inside this space; i.e., $\boldsymbol{x}(t, t_0, \boldsymbol{x}_0) \in \Omega$ for all t, where $\boldsymbol{x}(t, t_0, \boldsymbol{x}_0)$ is a solution trajectory of the nonlinear nonautonomous compartmental system of the form:

$$\dot{x}(t) = \boldsymbol{u}(\boldsymbol{x}(t), t) + \mathbf{B}(\boldsymbol{x}(t), t) \cdot \boldsymbol{x}(t). \quad 7.9$$

Concepts from control theory could be used to ensure that solutions are well behaved and inside Ω, and more importantly, within certain 'regions of stability' that solutions are attracted to (Müller and Sierra 2017; Kloeden and Rasmussen 2011).

Input-to-state stability (ISS) is a concept from the field of control theory that can be used to determine whether a nonlinear nonautonomous compartmental system meets stability properties. We say that a dynamical system is ISS if it is globally asymptotically stable in the absence of time-dependent perturbations, and if its trajectories are bounded by a function of the size of the input for all sufficiently large times (Sontag 1998; Müller and Sierra 2017). Therefore, we can expect the trajectories of an ISS system to remain within a certain region as long as the initial mass decays over time, and the mass inputs stay bounded within a certain limit.

We expect that for most applications, nonlinear nonautonomous compartmental systems meet the properties of ISS systems. However, mathematically showing that a system is ISS is not trivial, and this should be studied on a case-by-case basis (Sierra and Müller 2015; Müller and Sierra 2017).

FINAL REMARKS

The theory of compartmental dynamical systems offers a formal theoretical framework to express and analyze models of the carbon cycle and other biogeochemical elements that meet mass balance requirements. Using a matrix representation of carbon storage in ecosystem pools, it is possible to use the theory of compartmental dynamical systems to study important characteristics of models such as their long-term behavior, the presence of traps that retain carbon indefinitely in a model, and the response of ecosystem compartments to disturbances.

The representation of ecosystem models as compartmental systems is also useful to study system level properties of ecosystems (see Chapter 15). It is a useful mathematical representation that can relate ecosystem concepts to formal mathematical properties of dynamical systems.

SUGGESTED READING

General introductions to compartmental systems can be found in the monograph by Anderson (1983), and the comprehensive review of Jacquez and Simon (1993). More specific results about the application of compartmental systems to model the terrestrial carbon cycle can be found in the reference list and other chapters of this book.

QUIZ

1. According to the general classification of models with respect to their dynamical properties, what type of compartmental systems have a fixed-point steady-state?
2. Can linear compartmental systems show transitions through tipping points?
3. What is the analogue of a steady state for nonautonomous systems? Why?

8 Practice 2

Matrix Representation of Carbon Balance Equations and Coding

Yuanyuan Huang

Institute of Geographic Sciences and Natural Resources Research, Chinese Academy of Sciences, Beijing, China

This practice helps you to learn how to write a matrix equation from carbon balance equations using the CENTURY model as an example. You will also learn how to numerically solve the matrix equation through Python code using the CarboTrain package.

INTRODUCTION

In Chapter 5 we saw the benefits of representing a land carbon cycle model in a matrix form. This can be achieved by organizing the carbon balance equations of a model into one matrix equation. For many common models, the matrix version is mathematically identical to the original model without any changes in represented processes. Conversely, a matrix equation can be expanded row by row to give carbon balance equations for individual carbon pools. Exercise 1 focuses on how to derive the matrix version from the carbon balance equations of a carbon cycle model with multiple pools, the CENTURY model. Exercise 2 provides practice in coding and running the matrix model using Python through the CarboTrain package. Going through these exercises should give you some first-hand insight into the benefits of working with the matrix model, such as being easy to code and run model simulations.

EXERCISE 1 DERIVING THE MATRIX MODEL FROM CARBON BALANCE EQUATIONS

In this exercise we will develop a matrix form of the CENTURY model from its carbon balance equations. The carbon flow diagram for CENTURY was shown in Figure 4.1 of Chapter 4. The carbon cycle as depicted in the carbon flow diagram in Figure 4.1 can be represented by five carbon balance equations. If you performed Exercise 1 in Chapter 4, you have written down these carbon balance equations before. The equations are shown below:

$$\frac{dx_1}{dt} = I*\beta_1 - k_1 x_1$$

$$\frac{dx_2}{dt} = I*\beta_2 - k_2 x_2$$

$$\frac{dx_3}{dt} = f_{31}*k_1 x_1 + f_{32}*k_2 x_2 + f_{34}*k_4 x_4 + f_{35}*k_5 x_5 - k_3 x_3$$

$$\frac{dx_4}{dt} = f_{41}*k_1 x_1 + f_{43}*k_3 x_3 - k_4 x_4$$

$$\frac{dx_5}{dt} = f_{53}*k_3 x_3 + f_{54}*k_4 x_4 - k_5 x_5$$

where x_1, x_2, x_3, x_4, and x_5 correspond to structural carbon, metabolic carbon, active soil carbon, slow soil carbon, and passive soil carbon pools. I is the input rate from plant residue. β_1 and β_1 indicate the proportion of this carbon input that is allocated to structural and metabolic litter. fi_j are fractions of carbon transferred from pool j to pool i. k_1 to k_5 are rates of soil organic carbon decomposition.

To develop a matrix model from the carbon balance equations, we need to know the following items. The first item is the general form of the matrix equation. If you do not remember it, please go back to Chapter 5 to find the general form of the matrix equation. Second, you need to remember what the scalar, vectors, and matrices are in the general matrix equation. For example, is $\frac{dX}{dt}$ a scalar, vector, or matrix? What does it mean? Please list the scalar, vectors, and matrices in the general matrix equation. Third, you need to write all the scalar elements of these vectors and matrices. For example, $X = [x_1\ x_2\ x_3\ x_4\ x_5]^T$ indicates a vector of pools related to structural carbon, metabolic carbon, active soil carbon, slow soil carbon, and passive soil carbon. What are the other vectors and matrices of the matrix model?

EXERCISE 2 CODING AND RUNNING THE MATRIX MODEL

This exercise uses the package CarboTrain, which enables you to do exercises in this and other units with minimal background training in modeling and programming. Instructions for installing and working with CarboTrain are available in Appendix 3 of this book. The following steps will guide you through the coding and running of the matrix version of the TECO model using the CarboTrain package.

DOI: 10.1201/9781032711126-10

1 Run the default source code.
In the main window of CarboTrain, select Unit 2 and Exercise 2, specify the path of the output directory in which you wish to store your model result, click on Run Exercise. The notification "Task submitted!" will pop up. Click on OK. Wait until the notification "Finished" appears, then click on OK. Go to the output directory you specified. There are two files. One is result.png, which plots the change of seven carbon pools through time. A second file is output.csv, which saves the value of each carbon pool in each simulation year.

2 Understand the default source code.
In the CarboTrain main window, click on *Edit source code*. You will see the following source code (black text in your GUI). The code is based on the matrix model of TECO. The first part of the code loads packages and environment, and reads the output path you specified in the previous step.

The second part of the code, shown below, constructs the matrices and specifies our desired simulation length in years. The vector *iv_list* is the initial carbon pool size, *input_fluxes* specifies the input rate, *B* is the allocation vector, *A* is the transfer matrix, and *K* is the turnover rate matrix.

The third part of the code (Figure 8.3) calls the function *GeneralModel* to solve the carbon balance equations and generate the result.

```
# You specify values of Initial carbon pool size (iv_list), Input rate (input_fluxes),
# Allocation (B), Transfer (A), Turnover rate (K), and simulation length,
# GeneralModel solve the carbon balance equations and generate the result
mod = GeneralModel(times, B, A, K, iv_list, input_fluxes)

res = mod.get_x()
```

The fourth part of the code (Figure 8.4) generates the output figure (result.png) and saves results into the csv text file (output.csv).

3 Modifying the default source code.
To get a sense of how the code works and what controls the system dynamics, you could change some of the default values in the second part of the code. For example, if we change the number of simulation years (nyear) to 100, how does this affect the output from the run? If we make the passive soil carbon turnover faster by changing the default value of 0.00000154782 per day to 10^{-5} (1e-5 in Python) per day, what difference do you see in comparison with the default? By exploring and trying different settings and parameters you should learn to understand how different parts of the matrices control the system dynamics. Every time you modify the code through *Edit source code*, please make sure you save the code before you click on *Run Exercise*. It is good practice to change one place at a time and change the value back to the default value after you finish the practice.

4 Building a new carbon model (optional).
Suppose we have a system with leaf (pool 1), root (pool 2), wood (pool 3), metabolic litter (pool 4), structural litter (pool 5), fast soil organic matter (pool 6), and passive soil organic matter (pool 7). Carbon dynamics of the system are given by the matrix equation below. Based on the default TECO model source code, above, are you able to code this matrix model and check on carbon dynamics through time?

$$\frac{dX}{dt} = BI + AKX$$

$$B = \begin{bmatrix} \beta_1 \\ \beta_2 \\ 1-\beta_1-\beta_2 \\ 0 \\ 0 \\ 0 \\ 0 \end{bmatrix}$$

$$A = \begin{bmatrix} -1 & 0 & 0 & 0 & 0 & 0 & 0 \\ 0 & -1 & 0 & 0 & 0 & 0 & 0 \\ 0 & 0 & -1 & 0 & 0 & 0 & 0 \\ f_{41} & f_{42} & f_{43} & -1 & 0 & 0 & 0 \\ f_{51} & f_{52} & f_{53} & 0 & -1 & 0 & 0 \\ 0 & 0 & 0 & f_{64} & f_{65} & -1 & 0 \\ 0 & 0 & 0 & 0 & f_{75} & 0 & -1 \end{bmatrix}$$

$$K = \begin{bmatrix} k_1 & 0 & 0 & 0 & 0 & 0 & 0 \\ 0 & k_2 & 0 & 0 & 0 & 0 & 0 \\ 0 & 0 & k_3 & 0 & 0 & 0 & 0 \\ 0 & 0 & 0 & k_4 & 0 & 0 & 0 \\ 0 & 0 & 0 & 0 & k_5 & 0 & 0 \\ 0 & 0 & 0 & 0 & 0 & k_6 & 0 \\ 0 & 0 & 0 & 0 & 0 & 0 & k_7 \end{bmatrix}$$

We assume 40% of the net photosynthetic carbon is allocated to leaf, 35% to root, and the remaining to wood. $f_{41} = 0.7$; $f_{51} = 0.25$; $f_{42} = 0.65$; $f_{52} = 0.25$; $f_{43} = 0.15$; $f_{53} = 0.75$; $f_{64} = 0.3$; $f_{65} = 0.05$; $f_{75} = 0.04$; k_1 to k_7 are [0.0017, 0.002, 0.0001, 0.01,

```
from GeneralModel import GeneralModel
import numpy as np
import matplotlib.pyplot as plt
import sys

if __name__ == '__main__':

    output_folder = sys.argv[1]
```

FIGURE 8.1 First part of the source code of the matrix version of TECO.

0.001, 0.0001, 0.000001] per day. For other parameters, if not specified above (e.g., the input rate), we assume they take the same values as the default TECO model. Thus, you do not need to change these values.

Does your new matrix model work? Can you run it? What results do you get?

Click on *Open solutions* in CarboTrain to check if the source code of your new matrix model is similar to the code shown in Figure 8.5.

```
B = np.array([0.45, 0.55, 0, 0, 0, 0, 0]).reshape([7,1])   # allocation

f31 = 0.72; f41 = 0.28; f42 = 1; f53 = 0.45; f54 = 0.275; f64 = 0.275;
f65 = 0.296; f75 = 0.004; f56 = 0.42; f76 = 0.03; f57 = 0.45;

A = np.array([-1, 0, 0, 0, 0, 0, 0,
            0, -1, 0, 0, 0, 0, 0,
            f31, 0, -1, 0, 0, 0, 0,
            f41, f42, 0, -1, 0, 0, 0,
            0, 0, f53, f54, -1, f56, f57,
            0, 0, 0, f64, f65, -1, 0,
            0, 0, 0, 0, f75, f76, -1]).reshape([7,7])   # tranfer

#turnover rate per day of pools: foliage, wood, metabolic litter, structural
#litter, soil microbial,slow soil, passive soil
temp = [0.00176, 0.000100104, 0.021468, 0.000845, 0.008534, 8.976e-005, 0.00000154782]

K = np.zeros(49).reshape([7, 7])

for i in range(0, 7):
    K[i][i] = temp[i]

#Unit of turnover rate from day^-1 to second^-1
#1 day = 86400 seconds
K = np.multiply(K, 1/86400)

# Cinput_const
input_fluxes = 0.00002245 #

nyear = 10000   # number of simulation years

times = np.linspace(0, nyear*365*86400, num = nyear)

iv_list = [0,0,0,0,0,0,0]   # Initial carbon pool size
```

FIGURE 8.2 Second part of the source code of TECO.

```
# Plot carbon pools and save results into csv file
fig = plt.figure(6*2, figsize=(14, 7.68))
plt.subplots_adjust(left  = 0.1, right = 0.95, bottom = 0.10, top = 0.9, wspace =0.3, hspace =0.4)
x = list(range(1,nyear + 1, 1))

pool_names = ["foliage", "wood", "metabolic litter", "structural litter", "soil microbial", "slow soil", "passive soil"]

for i in range(1, 4):
    for j in range(1, 4):
        if ((i-1) * 3 + j) > 7 :
            break
        ax = plt.subplot(3, 3, (i-1) * 3 + j)
        ax.plot(x, res[(i-1) * 3 + j - 1,:])
        plt.xlabel("year", fontsize = 12)
        plt.ylabel(pool_names[(i-1) * 3 + (j-1)] + " pool ($g C m^{-2}$)", fontsize = 12)
plt.savefig(output_folder + "/result" + ".png", dpi = 500)
#plt.show()

print(res[:,nyear-1])   # print result of the last year

mod.write_output(output_folder + "./output.csv")
```

FIGURE 8.3 Call to the functions that solve the carbon balance equations and retrieve the result.

```
from GeneralModel import GeneralModel
import numpy as np
import matplotlib.pyplot as plt
import sys
if __name__ == "__main__":

    output_folder = sys.argv[1]

    B = np.array([0.4, 0.35, 0.25, 0, 0, 0, 0]).reshape([7,1])    # allocation

    f41 = 0.7; f51 = 0.25; f42 = 0.65; f52 = 0.25; f43 = 0.15; f53 = 0.75;
    f64 = 0.3; f65 = 0.05; f75 = 0.04;

    A = np.array([-1, 0, 0, 0, 0, 0, 0,
                  0, -1, 0, 0, 0, 0, 0,
                  0, 0, -1, 0, 0, 0, 0,
                  f41, f42, f43, -1, 0, 0, 0,
                  f51, f52, f53, 0, -1, 0, 0,
                  0, 0, 0, f64, f65, -1, 0,
                  0, 0, 0, 0, f75, 0, -1]).reshape([7,7])    # tranfer

    #turnover rate per day of pools:
    #leaf,root,wood, metabolic litter, structural litter, fast SOM, passive SOM
    temp = [0.0017, 0.002, 0.0001, 0.01, 0.001, 0.0001, 0.000001]

    K = np.zeros(49).reshape([7, 7])

    for i in range(0, 7):
        K[i][i] = temp[i]

    #Unit of turnover rate from day^-1 to second^-1
    #1 day = 86400 seconds
    K = np.multiply(K, 1/86400)

    # Cinput_const, assume to be constant
    input_fluxes = 0.00002245 #

    nyear = 10000

    times = np.linspace(0, nyear*365*86400, num = nyear)

    iv_list = [0,0,0,0,0,0,0]
```

FIGURE 8.4 Code segment that generates visual output and saves results to a text file.

```
mod = GeneralModel(times, B, A, K, iv_list, input_fluxes)

res = mod.get_x()

fig = plt.figure(6*2, figsize=(14, 7.68))
plt.subplots_adjust(left  = 0.1, right = 0.95, bottom = 0.10, top = 0.9, wspace =0.2, hspace =0)
x = list(range(1,nyear+1, 1))

pool_names = ["foliage", "root","wood", "metabolic litter", "structural litter", "fast soil", "passive soil"]

for i in range(1, 4):
    for j in range(1, 4):
        if ((i-1) * 3 + j) > 7 :
            break
        ax = plt.subplot(3, 3, (i-1) * 3 + j)
        ax.plot(x, res[(i-1) * 3 + j - 1,:])
        plt.xlabel("year", fontsize = 12)
        plt.ylabel(pool_names[(i-1) * 3 + (j-1)] + " pool ($g C m^{-2}$)", fontsize = 12)
plt.savefig(output_folder + "/result" + ".png", dpi = 500)
#plt.show()

print(res[:,nyear-1])

#mod.write_output("./output.csv")
```

FIGURE 8.5 Solution to Step 4.

Unit Three

Carbon Cycle Diagnostics for Uncertainty Analysis

9 Unified Diagnostic System for Uncertainty Analysis

Yiqi Luo
Cornell University, Ithaca, USA

All model intercomparison projects (MIPs) have shown large uncertainties in prediction of carbon sequestration among models and poor model-data matches. Although great efforts have been made, it is still difficult to identify causes of model uncertainty. This chapter offers a unified diagnostic system, which is also called a 1-3-5 scheme of diagnostics, for uncertainty analysis in carbon cycle modeling. The number 1 stands for one formula to unify the land carbon cycle models, the number 3 is for one three-dimensional (3D) space to evaluate all model outputs, and the number 5 is for five traceable components to pinpoint uncertainty sources via traceability analysis. Once the uncertainty sources are pinpointed, the model uncertainty can be shrunk to zero by standardizing the traceable components.

UNCERTAINTY IN LAND CARBON CYCLE MODELING

Hundreds of land models have been developed to predict the future state of ecosystems in an attempt to inform management practices for climate change mitigation. However, all model intercomparison projects (MIPs) have shown large uncertainties in prediction of carbon sequestration among models and poor model-data matches (Friedlingstein et al. 2006, 2014; Luo et al. 2015). For example, eleven earth system models (ESMs) participating in the Coupled Model Intercomparison Project Phase 5 (CMIP5) perform poorly in predicting the distribution of global land surface soil carbon (Todd-Brown et al. 2013). Similarly, the spread among 11 ESMs participating in the subsequent CMIP6 exercise has not changed significantly from CMIP5 results (Arora et al. 2020). The model predictions are an order of magnitude more uncertain over land than over ocean. Regionally, terrestrial ecosystem models did not accurately characterize a wide range of vegetation functional traits associated with net primary productivity (NPP) in the East Asian monsoon area (Cui et al. 2019).

Great efforts have been made in the past decades to identify causes of model uncertainty. For example, model development teams add more and more processes into land carbon models in the hope of representing ecosystem processes more realistically. This practice yields mixed results. Incorporation of the nitrogen cycle into more models has been suggested to reduce the spread of CMIP6, whereas different treatments of processes in permafrost regions result in divergent model predictions. In general, increasing details in process representations in models hinders our understanding of holistic system behavior (Sierra et al. 2018). Benchmark analysis has been used to evaluate model performance skills, quantify model-data mismatches, and identify processes that need to be improved (see Chapter 19). Data assimilation has been proposed to improve data-model consistency (see Chapter 38). None of these approaches, however, are sufficient to fully understand the causes of model uncertainty. Without identifying sources of uncertainty, it is extremely difficult to focus model improvement efforts to realistically predict future carbon dynamics.

This chapter introduces a 1-3-5 scheme as a unified diagnostic system for uncertainty analysis (Figure 9.1). This diagnostic system is based on the matrix approach to representation of land carbon cycle models in one (1) general equation without compromising any details of process representation. Land carbon cycle dynamics are defined by a three-dimensional (3D) space with axes of coordinates being carbon input, residence time, and carbon storage potential. Model uncertainty can be traced to five (5) variables, which are carbon input, plant partitioning, decomposition, carbon transfer, and environmental scalars. Thus, the 1-3-5 scheme provides an analytic framework to understand the structure of complex models, their dynamic behavior, and uncertainty.

ONE FORMULA TO REPRESENT LAND CARBON CYCLE MODELS

The unified diagnostic system offered by the matrix approach is based on one formula to unify land carbon cycle models. In

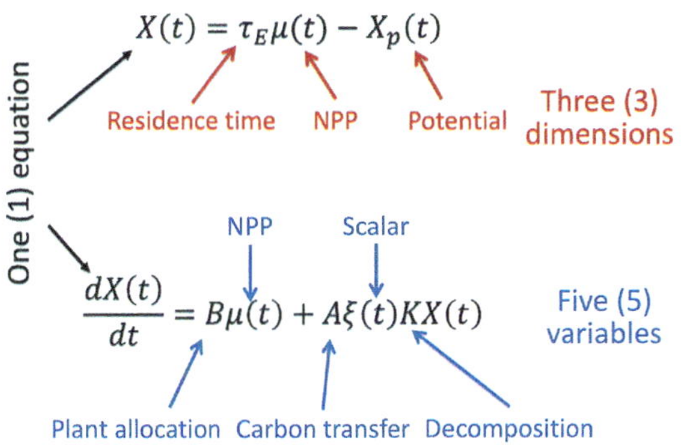

FIGURE 9.1 The unified diagnostic system with one (1) equation, three dimensions (3D), and five (5) variables.

DOI: 10.1201/9781032711126-12

Units 1 and 2 of this book, we have shown you that the matrix equation can unify land carbon cycle models.

At least 18 models have so far been converted to the matrix equations. These models are CLM3.5, CLM4, CLM4.5, CLM5, CABLE, LPJ-GUESS, IBIS, CASA, CENTURY, ORCHIDEE, TEM, TECO, DELAC2, ELM, GECO, FBDC, BEPS, and YASSO. In these models, we reorganize the original carbon balance equations into one matrix equation without changing any processes. For example, CLM5 uses hundreds of carbon balance equations to simulate carbon transfer among 18 plant pools of 17 vegetation types and 140 soil pools over 20 soil layers. These carbon balance equations are organized into one matrix equation for the 18 vegetation pools and one matrix equation for the 140 soil pools (Lu et al. 2020). The vegetation and soil matrix equations are connected through litterfall and can be integrated into one matrix equation.

Sierra and Müller (2015) reviewed all soil carbon models according to the principles underlying the models. The matrix equation we are concerned with in this book satisfies five principles: mass balance, substrate dependence, heterogeneity of decomposition rates, transformation of organic matter, and environmental effects (see Chapter 1). When decomposition rates or transfer rates are functions of substrate, we still can use a matrix equation to describe carbon dynamics. In this case, the matrix equation becomes a nonlinear model. Indeed, Carlos Sierra's group from Max Planck Institute of Biogeochemistry, Germany, uses the nonlinear matrix models to represent more than ten microbial models. Thus, we can understand models at different levels of complexity under one overarching theory. Please be mindful that nonlinear matrix models will have different properties from the linear ones (see Chapter 7).

With one general formula to represent all models, we can seek to understand the general behavior of the land carbon cycle and diagnose model uncertainty on a common ground. For example, the widely used model CLM5 includes 18 plant pools for each of the 17 vegetation types and 140 soil pools (i.e., 7 pools per layer over 20 layers) (Lu et al. 2020). Thus, CLM5 simulates carbon transfer among 158 pools (i.e., 140 for soil + 18 for plant) in one grid-cell if it is occupied by one vegetation type and 194 (i.e., 140 + 3 × 18) pools if one grid-cell is occupied by three vegetation types. As a consequence, the matrix equation has at least 158 elements in the pool vector. In contrast, the CABLE model has nine pools so that the pool vector has nine elements (Xia et al. 2012). Despite the different numbers of carbon pools treated by these two models, they share the fundamental structure represented in the matrix equation, a structure that is shared by all land carbon cycle models.

Besides the differences in pool numbers, each of the five components of the matrix equation (i.e., Equation 1.6 in Chapter 1) is represented differently either by fixed values, functions, or nested models (Luo et al. 2022). The differences in parameterization of the five components also contribute to different simulation results. For example, CABLE allocates 61% of NPP to roots, 23% to wood, and 16% to leaves (Figure 9.2a), whereas CLM3.5 allocates 43% of NPP to leaves, 16% to wood, and 41% to roots (Figure 9.2b) (Rafique

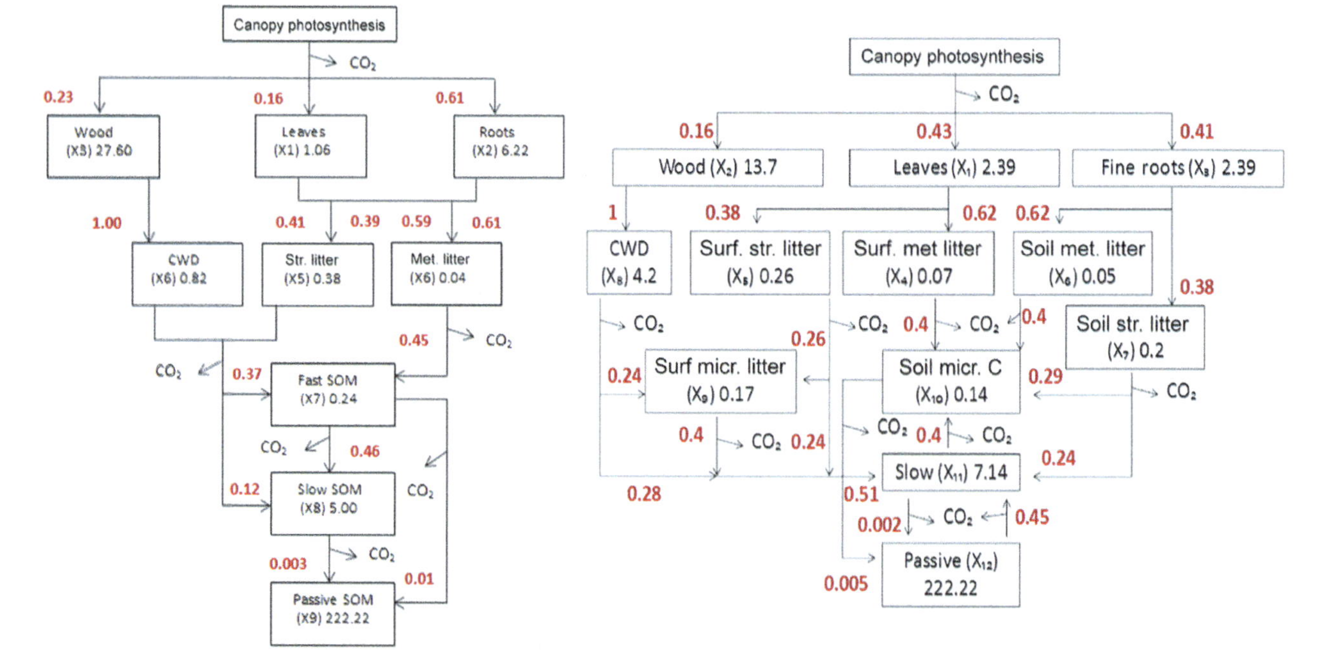

FIGURE 9.2 Model structure and parameterization of CABLE (a) and CLM3.5 (b). Carbon enters the system through photosynthesis and is partitioned among live plant pools. From live pools, carbon is transferred to litter pools, and then to soil carbon pools. Values in boxes show the pool residence times. Values outside the boxes in red show the partitioning and transfer coefficients. Abbreviations are CWD for coarse woody debris, Str. for structural litter, Met. for metabolic litter, Surf. for surface litter, micr. for microbial biomass, and SOM for soil organic matter.

Adopted from Rafique et al. 2016.

et al. 2016). Similarly, a large difference exists in carbon transfers from live plants to litter and soil. CABLE transfers dead tissues into three litter pools (including coarse woody debris, CWD) after senescence, whereas CLM3.5 transfers dead plant tissues to six litter pools (including CWD) after mortality. CLM3.5 and CABLE also differ in representing decomposition (i.e., *K* matrix in Figure 9.1). While both models can be represented by one matrix formula, CLM3.5 realizes each of the five components differently from CABLE, leading to different model projections of carbon dynamics. Despite these differences in model structure and parameterization, all models can be represented by the same matrix equation.

A THREE-DIMENSIONAL (3D) SPACE TO DESCRIBE MODEL OUTPUTS

Now let us explore what the 3D space means for evaluating model outputs. Chapter 1 explains that we need three variables: carbon input, residence time, and carbon storage potential to describe the transient dynamics of the land carbon cycle. The three variables become three dimensions to form a 3D space to which model outputs can be mapped, no matter how differently models are executed, for evaluation on a common ground. The mathematical foundation for using the 3D space to evaluate the transient dynamics of the carbon cycle is presented in detail in a paper by Luo et al. (2017). Here is a description of each of the three dimensions.

The first dimension of transient dynamics is carbon input through primary production. The primary production, being either gross primary production (GPP) or net primary production (NPP), quantifies the amount of carbon that enters an ecosystem to go through a variety of processes of the land carbon cycle before being released back to the atmosphere. Of the three dimensions, carbon input via GPP or NPP has been studied most.

The second dimension of transient dynamics is carbon residence time. Residence time is approximated as a mean value (i.e., mean residence time, MRT) or turnover time by dividing pool by flux. The approximation is valid when the carbon cycle is at equilibrium but yields substantial deviation from residence times for a multiple compartmental system when the carbon cycle is not at equilibrium (Lu et al. 2018b). Residence time or transit time is explained in detail in Chapter 15 and can be estimated from data assimilation for individual ecosystems. Overall, the residence time quantifies the duration of carbon staying in an ecosystem before being released back to the atmosphere.

The terms NPP and residence time together quantify equilibrium carbon storage capacity. Figure 9.3 shows simulated carbon storage capacity by three land models, the Beijing Climate Center (BCC) model, the Canada (CAN) model, and the Community Earth System Model Biogeochemical module (CESM GBC). In response to climate change, all three models simulate increases in NPP but decreases in residence time. The CESM BGC simulates the smallest NPP and residence time, leading to the lowest carbon storage capacity. The BCC model simulates the largest

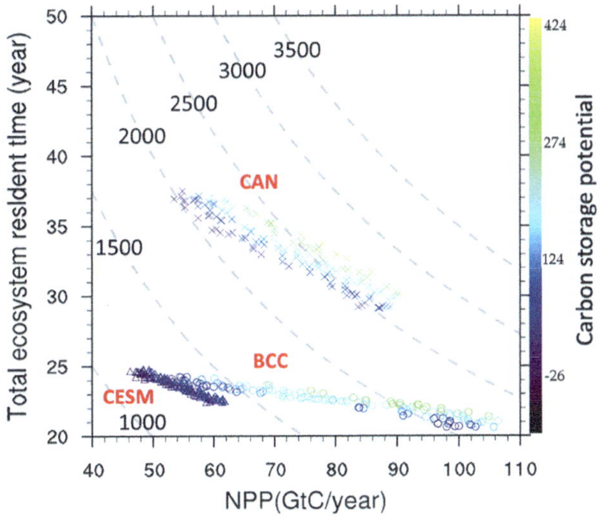

FIGURE 9.3 The 3D model output space (NPP in x-axis, carbon residence time in y-axis, and carbon storage potential in color), for three models from CMIP5. The points represent the global annual values of carbon storage for the three variables. The contours represent the carbon storage capacity. Colors show the values of the carbon storage potential.

increase in NPP, leading to a large increase in carbon storage capacity. The Canada model simulates the highest residence time and thus estimates the highest carbon storage capacity.

However, when models are used to simulate responses of ecosystems to climate changes, the carbon cycle is no longer at equilibrium but in dynamic disequilibrium. Therefore, we need the third term, the carbon storage potential, to describe disequilibrium of the land carbon cycle. The CESM GBC model has the smallest carbon storage potential whereas the CAN and BCC models have high storage potential (Figure 9.3).

We use one more example to explain the three dimensions (3D) of land carbon cycle dynamics, particularly the third dimension. The third dimension, carbon storage potential, is a relatively new concept and worth more explanation.

A free-air CO_2-enrichment (FACE) experiment was conducted in Duke Forest in North Carolina, USA. The FACE experiment started in the mid-1990s and ended in the late 2000s. The CO_2 concentration in the treatment rings was elevated by 200 parts per million above the ambient CO_2 concentration.

To illustrate the concept of 3D space of carbon dynamics in the FACE experiment, let us make a few assumptions. Let us assume that NPP is 1000 g C m^{-2} yr^{-1} and residence time is 40 years at ambient CO_2 concentration. Then, we have the steady-state carbon pool size in the Duke Forest as 40 kg C m^{-2} before the FACE treatment. We assume that elevated CO_2 treatments increase NPP by 40% but have no effect on residence time. The steady-state pool size is 56 kg C m^{-2} at elevated CO_2 treatments. For the sake of simplicity, we also assume no seasonal and diurnal changes in forcing variables and carbon processes.

In this example, we have two equilibrium states of carbon storage (i.e., carbon storage capacity), one at ambient [CO_2],

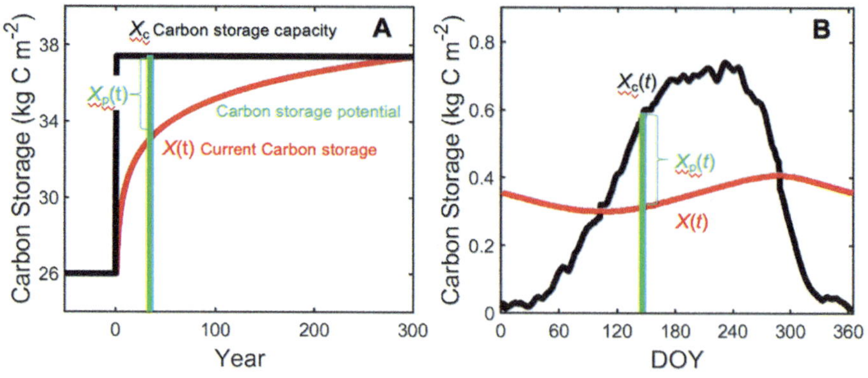

FIGURE 9.4 Carbon storage dynamics as determined by carbon storage capacity and potential. Panel A presents an ideal case in which carbon storage capacity (X_c) is a constant while carbon storage potential (X_p (t)) and carbon storage itself ($X(t)$) vary with time. In this case, the capacity is assumed to abruptly increase by 40%, mainly due to instantaneous increase in carbon input as in an elevated CO_2 experiment (Luo & Reynolds, 1999). Consequently, the potential immediately increases and then gradually declines as $X(t)$ increases toward the equilibrium (i.e., the carbon storage capacity at elevated CO_2 treatment). Panel B illustrates time-dependent $X(t)$, its capacity and potential in a nonautonomous system over day of year (DOY). Seasonal change in the capacity is due to change in carbon input, which is low in winter and high in summer. The capacity is a moving target that $X(t)$ chases. The rate of chasing is proportional to the storage potential.

equaling 40 kg C m^{-2} and the other at the elevated [CO_2], denoted as X_E, equaling 56 kg C m^{-2} (Figure 9.4A). At the very beginning of the FACE experiment, we need to determine in which direction and how fast the current carbon storage $X(t)$ would change. Generally speaking, we expect that the carbon storage changes toward the equilibrium state at elevated [CO_2] (X_E) but the rate of the change is fast in the first few years and slow in later years. Let us denote the difference between the equilibrium carbon storage and current storage to be the carbon storage potential X_p (t). The rate of carbon storage change is proportional to the potential X_p (t).

The above example illustrates three key findings about carbon storage dynamics. First, carbon storage is always moving toward the carbon storage capacity. Second, the capacity is the ultimate attractor which current carbon storage chases (or changes toward). Third, the rate of the carbon storage change is proportional to the carbon storage potential.

The Duke FACE example assumes equilibrium carbon storage capacity. However, the carbon storage capacity, which equals NPP times residence time, is changing over time in the real world as both NPP and residence time vary with time. This is the reason why the carbon cycle in the terrestrial ecosystem is mathematically considered a nonautonomous system.

We use simulated seasonal change of fine root biomass in Harvard Forest (Luo et al. 2017) as an example to illustrate the nonautonomous system of carbon cycle dynamics. The carbon storage capacity of the fine root pool is a theoretical quantity, namely NPP times residence time. As NPP is low in winter and high in summer, the root carbon storage capacity as indicated by the black line is low in winter and high in summer (see Figure 9.4B). The red line indicates the current carbon storage in fine roots, which is higher than the black line in the winter and lower than the black line in the summer. The red line is always moving toward the black line. That is a mathematical explanation for why fine root amount declines in fall and winter but increases in spring and summer.

For a nonautonomous system in which both NPP and residence time are changing with time, the carbon storage capacity still controls the direction of carbon storage change. Likewise, the carbon storage potential still determines the rate of carbon storage change. Therefore, carbon input, residence time, and carbon storage potential form a 3D space within which all model outputs can be evaluated and compared.

Zhou et al. (2018) have applied the 3D space to evaluate model performance from three model intercomparison projects (MIPs): CMIP5, TRENDY, and MsTMIP. The 3D space is defined by NPP as x-axis, residence time as y-axis, and color for carbon storage potential to place outputs from 25 models in the same space to evaluate their performance (see Chapter 18 for details). The three variables can also be plotted to indicate current carbon storage in relatively smooth lines, the carbon storage potential in shaded areas, and the carbon storage capacity in the zig-zag lines over a time course. This is the first time we are able to evaluate all model outputs from different MIPs in a simple, 3D space. See Chapter 18 for more applications of the 3D space to model evaluation.

FIVE TRACEABLE COMPONENTS FOR TRACEABILITY ANALYSIS

Now, let us examine the five traceable components of carbon cycle dynamics, which have been effectively used in traceability analysis.

The matrix equation describes land carbon dynamics via carbon input (e.g., NPP), allocation of carbon input to different plant parts, carbon process rates via litterfall or decomposition of organic matter, carbon transfer among pools, and environmental scalar. They consist of five components of the matrix equation. Mathematically, the five components are largely independent from each other in most models (Xia et al. 2012) although they may interact in the real world. Moreover, each component can be further traced to its subcomponents as far as individual carbon cycle processes and their parameter values. This mathematical property enables us to trace sources of model uncertainty to individual processes and parameters via traceability analysis (see Chapter 17).

This traceability analysis was first developed by Xia et al. (2013) for the equilibrium carbon storage capacity and expanded by Jiang et al. (2017) and Zhou et al. (2018) to the transient dynamics of the land carbon cycle. The traceability analysis can trace the model uncertainty in simulated carbon storage hierarchically along the traceable pathways down to vegetation traits, climate forcing, and soil attributes. Chapter 17 shows an authentic traceability analysis that requires matrix models to generate all the traceable components so that it enables uncertainty analysis to be analytically transparent. Chapter 18 shows post-MIP traceability analysis that does not require matrix models but is applied to any model output. The post-MIP traceability analysis can attribute uncertainty among models to variations in NPP, residence time, and carbon storage potential. The variations in NPP, residence time, and carbon storage potential can be further traced to environmental scalars and model parameters.

For example, the Australian CABLE model predicts lower carbon storage capacity than CLM3.5 due to lower NPP. But CABLE has higher residence time than CLM3.5 (Rafique et al. 2016). The higher residence time in CABLE is mainly caused by lower decomposition coefficients, leading to higher baseline residence time. The environmental scalars among the two models are similar. The example shows that the traceability framework can help trace sources of model uncertainty down to individual processes or parameter values.

SHRINKING MODEL UNCERTAINTY TO ZERO

Once the uncertainty sources are pinpointed via traceability analysis, the model uncertainty can be shrunken to zero by standardizing the traceable components. Hou et al. (2023) converted the carbon cycle module of eight land models (i.e., TEM, CENTURY4, DALEC2, TECO, FBDC, CASA, CLM4.5, and ORCHIDEE) into eight matrix models. The eight models differ greatly in complexity, with the number of carbon pools ranging from 2 to 101. Using the same gross primary production (GPP) and environmental variables (e.g., soil temperature and water content) to drive all the models, the eight matrix models simulate ecosystem carbon dynamics very differently. A traceability analysis indicates that the model uncertainty is mainly due to inter-model difference in baseline carbon residence time (Figure 9.5).

Once the uncertainty sources were identified, Hou et al. (2023) standardized parameters in the matrix models to investigate their contributions to the uncertainty. Standardizing baseline residence time (τ_b) and environmental scalar (ξ) reduced most of the across-model variations in the net ecosystem production and carbon storage. In comparison, standardizing parameters related to plant carbon use efficiency

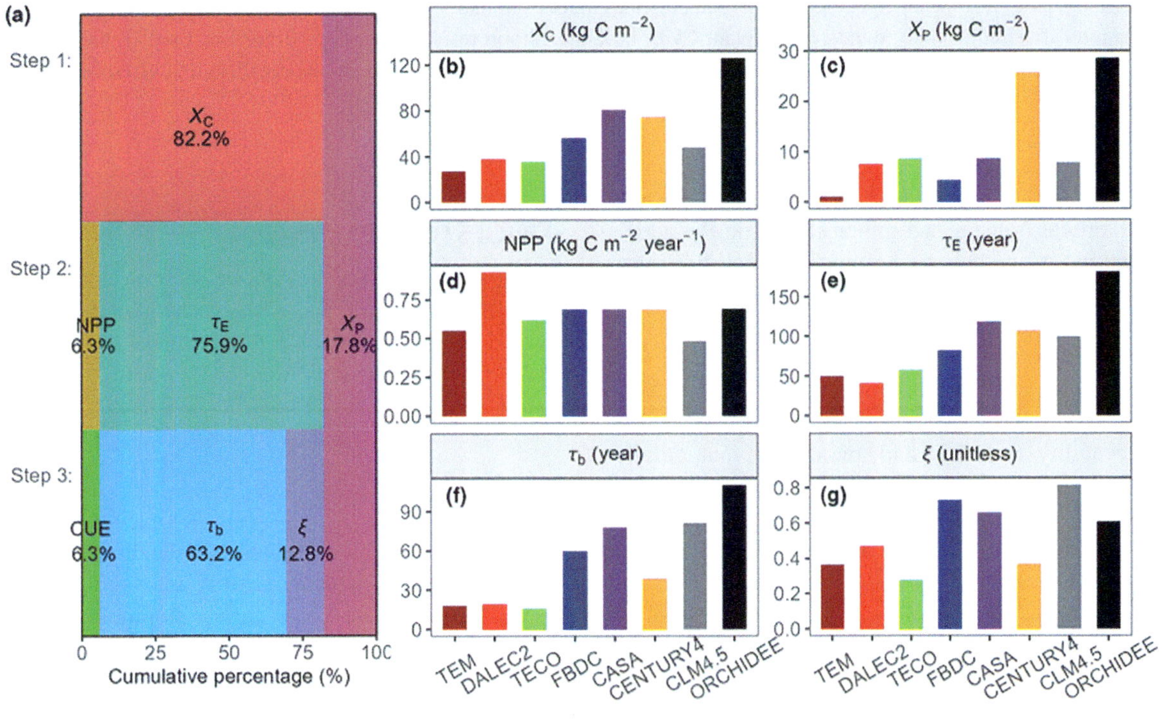

FIGURE 9.5 Authentic traceability analysis of simulated ecosystem carbon dynamics. (a) Model uncertainty (i.e., across-model spread) in ecosystem carbon storage is traced into model components by three steps. First, the spread is attributed to inter-model variations in (b) carbon storage capacity (X_C) and (c) carbon storage potential (X_P). Second, the variation in X_C is attributed to inter-model variations in (d) net primary production (NPP) and (e) ecosystem carbon residence time (τ_E). Third, the variation in τ_E is attributed to inter-model variations in (f) baseline carbon residence time (τ_b) and (g) environmental scalars (ξ); the variation in NPP is attributed wholly to inter-model variation in plant carbon use efficiency (CUE) as the same GPP drove the simulation of all the models. Adopted from Hou et al. 2023.

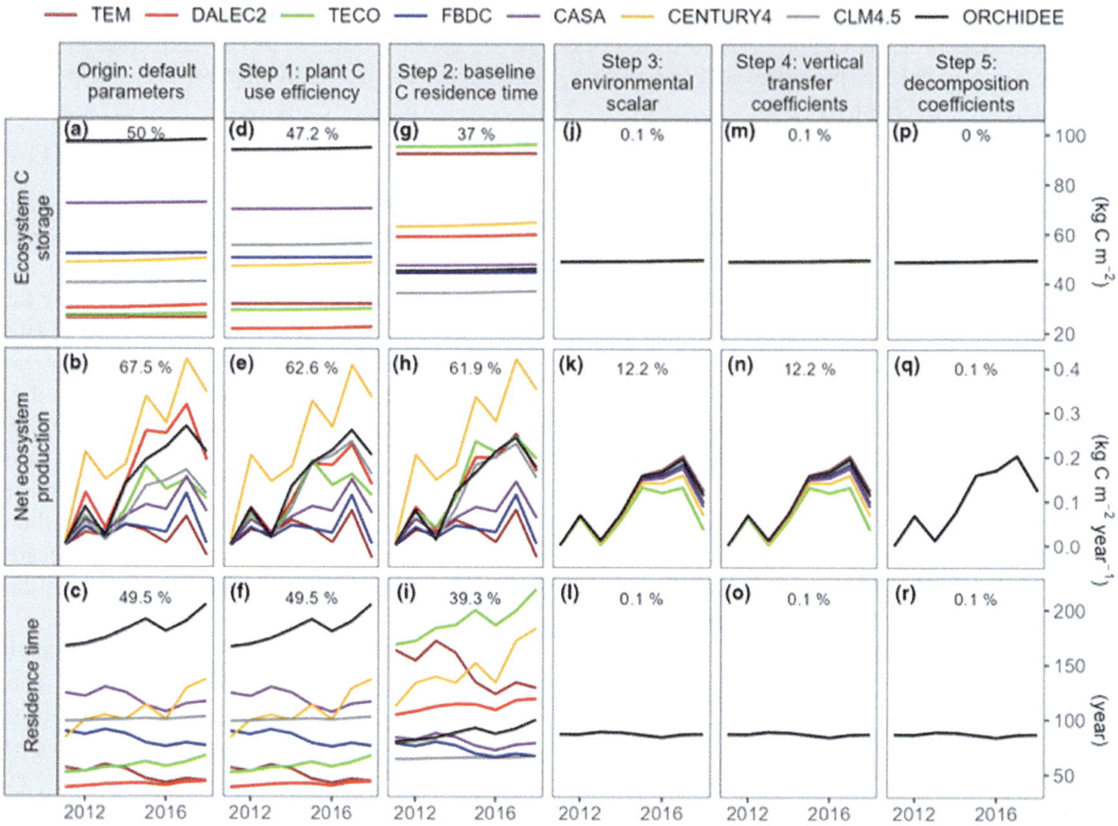

FIGURE 9.6 Model uncertainty (i.e., across-model spread) in ecosystem carbon dynamics are shrunken after standardizing parameter values. Simulated ecosystem carbon storage, net ecosystem production, and carbon residence time in the (a–c) original model run and (d–r) model runs after Steps 1–5, respectively. Origin model run indicates model simulations with the default parameter values. Steps 1–5 indicate model simulations after sequentially standardizing plant CUE, baseline carbon residence time, environmental scalar, and vertical transfer coefficients, and homogenizing decomposition coefficients. Percentage in each subpanel indicates the coefficient of variance of time-averaged simulations.

Adopted from Hou et al. (2023).

(CUE) and vertical transfer had minor effects on the across-model variations. Nonetheless, the model-model differences were shrunk to zero (i.e., identical predictions) among the eight models once the environmental scalar functions were adjusted and values of parameters *B*, *A*, and *K* were standardized to the same outputs for each of the terms (Hou et al., 2023) (Figure 9.6).

The traceability analysis and parameter manipulation demonstrate that the matrix approach makes model uncertainty a tractable issue: through simple mathematical adjustments in the matrix equation terms, we can analytically and precisely track the model uncertainty down to its sources. This is a step forward for understanding the sources of uncertainty in model intercomparison projects (MIPs).

Unit 3 of this book shows you how we can add diagnostic variables in matrix models so that we can use the 1-3-5 scheme for uncertainty analysis and traceability analysis. Unit 5 will specifically show you how to do traceability analysis.

SUGGESTED READING

Hou E, S Ma, YY Huang, Y Zhou, HS Kim, E López-Blanco, L Jiang, J Xia, F Tao, C Williams, M Williams, D Ricciuto, PJ Hanson, YQ Luo. 2023. Across-model spread and shrinking in predicting peatland carbon dynamics under global change. *Global Change Biology*. https://doi.org/10.1111/gcb.16643

QUIZ

1 What is the 1-3-5 scheme of the diagnostics system?
2 Briefly describe traceability analysis.
3 What does the term "carbon storage potential" mean?
4 A nonautonomous system is (choosing one)
  a a system which does not conserve mass balance.
  b a system with its properties changing with time.
  c a system with constant pool sizes.
  d a system with its pool sizes changing with time.

10 Matrix Phosphorus Model and Data Assimilation

Enqing Hou
South China Botanical Garden, Chinese Academy of Sciences, Guangzhou, China

Soil phosphorus supply regulates terrestrial carbon dynamics. To improve our understanding of this regulation, we need to improve our understanding of soil phosphorus dynamics first. This chapter first briefly reviews research progress in modeling soil phosphorus dynamics, with a focus on the diversity of representations among models. The chapter then demonstrates how to construct a soil phosphorus model and how to transfer it to a matrix form. Finally, an example study is presented to show how assimilating data into the matrix model can improve our understanding of soil phosphorus dynamics and availability. Overall, this chapter demonstrates that the matrix approach and data assimilation are very useful techniques to study terrestrial nutrient dynamics.

INTRODUCTION

Phosphorus (P) is a key element of macromolecules such as deoxyribonucleic acid (DNA), ribonucleic acid (RNA), and adenosine triphosphate (ATP), of which the former two are carriers of genetic information and the last is the carrier of energy during biochemical reactions. Given the important roles of P in life, plants and other organisms need P for growth and reproduction. In natural ecosystems, plants mainly obtain P through uptake from the soil. Since soil P supply is not always sufficient to meet plant P demand, P limits plant production and its response to elevated atmospheric carbon dioxide (CO_2) in many terrestrial ecosystems. Moreover, P can regulate ecosystem carbon storage and cycling by affecting soil microbial activity. Therefore, improving our understanding of terrestrial P dynamics and P–C interactions is crucial to realistically simulate the land C cycle as well as its responses to future global change.

The application of data assimilation to quantify terrestrial C dynamics is well established (e.g., Luo et al. 2016). However, data assimilation to inform a soil P model has rarely been tried (Hou et al. 2019), despite suitable soil P measurements being available in the literature (Hou et al. 2018). The data assimilation approach is potentially complementary to other currently available techniques (e.g., isotope dilution technique) in quantifying soil P dynamics. For example, it can use multiple sources of existing observations (e.g., soil P pool size and plant P uptake), simultaneously quantify the rates of all major soil P processes, and provide information about the uncertainties of the parameters related to soil P. Moreover, it may be particularly useful to quantify the dynamics of slow-cycling soil P pools (e.g., soil occluded P), which can hardly be achieved by any currently available experimental technique.

This chapter will first briefly review research progress in modeling soil P dynamics, with a focus on the diversity of representations among models. The chapter will then propose how a matrix framework and data assimilation can potentially improve our modeling of soil P dynamics. After that, an example study is presented to show how assimilating data into a matrix model can help to improve our understanding of soil P dynamics and availability.

A BRIEF OVERVIEW OF SOIL P DYNAMICS MODELS

Soil P dynamics is a key component of terrestrial P dynamics. To account for soil P supply as a determining factor for plant growth, land models are increasingly being extended to incorporate soil P dynamics. The first well-known land model to include soil P dynamics was CENTURY, published in the 1980s (Parton et al. 1988). In the last decade, because of growing interest in P cycle regulation of the land C cycle, over ten land models have been extended by incorporating soil P dynamics.

Current soil P models share some common features. Most are constructed based on the well-known soil P pools including soil labile P (readily available to plants), organic P (in organic forms), secondary mineral P (associated with secondary minerals such as iron and aluminum oxides), primary mineral P (apatite P), and occluded P (associated with soil clay and minerals and not directly available to plants). The dynamics of P as it accumulates and flows among these pools are usually expressed by a set of equations, based on empirical understanding of soil P dynamics. Despite a generally good understanding of soil P processes, most soil P models are not well parameterized, calibrated, or validated, because of limited long-term observations of soil P dynamics. In current modeling practice, soil P models are typically not fully spun up to a steady state before formal simulations, even though spin-up to steady state is normally regarded as essential for land modeling. High computational cost associated with the initialization of slow-cycling soil P pools such as soil occluded P, as well as the unidirectional change in soil primary mineral P with time, make a full spin-up impractical for most current models.

Current P models differ both in structure and parameters. Regarding structure, some models include both water-soluble P and labile P pools. The former can be directly available to plants and the latter may not be directly available to plants but exchangeable with the water-soluble P pool. However,

DOI: 10.1201/9781032711126-13

other models do not include a soil water-soluble P pool and assume that soil labile P is directly available to plants. Both empirical studies and theory have suggested that soil occluded P pool may deplete in some conditions (e.g., an anaerobic environment) and accumulate in some other conditions (e.g., an aerobic environment). However, most land models that incorporate soil P dynamics assume that soil occluded P accumulates all the time. Moreover, model structure with respect to soil organic P differs substantially among models, because soil organic P dynamics are coupled with soil organic C dynamics via soil C:P ratios in models, while models have very different structures for soil organic C dynamics (e.g., different numbers of soil organic C pools and vertically resolved vs. unresolved schemes).

Parameter values scaling soil P dynamics also differ among models. The literature provides many observations of soil P pools but few observations of soil P fluxes, especially fluxes from slow-cycling soil P pools (e.g., secondary mineral P and occluded P). Given these limitations, most soil P dynamics models are poorly parameterized and thus suffer from large uncertainties when they are used to predict future changes. Moreover, regulation by environmental factors (e.g., soil temperature and moisture content) of soil P dynamics and enzyme-mediated soil P mineralization are not well represented in current soil P models. In summary, the modeling of soil P dynamics is still in its infancy.

MATRIX APPROACH TO SOIL P MODELING AND DATA ASSIMILATION

The matrix approach and data assimilation can help advance soil P modeling in several ways. First, a matrix representation of the soil P system makes it easier to depict, understand, and compare models than the traditional approach using many difference equations. In the standard approach, change per time step in each soil P pool is represented by one equation with multiple state variables and parameters. Different soil P pools are mathematically linked by cross-terms in multiple equations, but such linkages may not be apparent to experimentalists who may have the data to validate the model but may not have any modeling experience. A matrix representation of soil P dynamics transfers the equations for different soil P pools into a unified form, no matter how many equations or what type of model, as long as a model is structured to describe transfers among multiple soil P pools. The matrix representation of soil P dynamics facilitates our understanding of model structure (e.g., the number of pools and the linkages among pools) and enables a direct comparison of structure among models.

A matrix representation of soil P dynamics also has the advantage of enabling a fast model spin-up and data assimilation that may be difficult or impossible with a traditional P model due to the prohibitive amount of computational power required. Soil P pools besides primary mineral P are generally in equilibrium in natural terrestrial ecosystems. Therefore, equilibrium soil P pool sizes are usually needed before a model simulation or data assimilation procedure. The turnover of soil occluded P in the field is typically slow (hundreds to tens of thousands of years), potentially even slower than the turnover of soil passive organic C (hundreds to thousands of years) which constitutes a bottleneck in spin-up of carbon-only ecosystem models. Therefore, it is a computational challenge to spin up a soil P model in the traditional way (e.g., repeat forcing many times). If soil P dynamics are represented in a unified matrix form, the semi-analytical spin-up (SASU) method introduced in Chapter 14 can be used to perform model spin-up, with the pool size of soil primary mineral P set to be constant. Application of SASU to soil P modeling may accelerate model spin-up by one or more orders of magnitude and may make it computationally feasible to assimilate soil P measurements from the field into models.

Data assimilation enables a quantitative and predictive understanding of soil P dynamics and availability. The understanding is fundamental to an accurate management of soil P availability, which can further meet the societal needs of efficient use of P fertilizer in croplands, prohibiting eutrophication of water bodies, and developing strategies to alleviate P constraints on terrestrial C sequestration. To this end, an example is given below to show how matrix model and data assimilation approaches can be used in soil P studies to gain insights into soil P dynamics and availability.

AN EXAMPLE OF APPLYING A MATRIX MODEL AND DATA ASSIMILATION TO SOIL P

Data Selection and Description

The example assimilates datasets reported in Guo et al. (2000) into a soil P dynamics model. Guo et al. (2000) reported consistent (i.e., using the same fractionation procedure) and repeated (seven or eight times) measurements of P fractions of eight soils that represent four of the twelve major USDA soil types. These datasets are selected because most soil P fractions changed substantially during the study period, which potentially offers a constraint on modeled soil P dynamics. Additionally, the soil samples are representative of common soil types, promoting the scope of applicability of the model. A less desirable feature of these datasets is that they are derived from experiments performed in the relatively artificial conditions of a greenhouse. One likely consequence is that the P pools in these experiments might have higher turnover rates than in the field.

In a nutshell, Guo et al. (2000) used crops to remove labile P from the eight soils to trigger changes in P fractions in these soils over a total of 14 cropping periods. After every two croppings, they sampled small amounts of the soils to determine soil P fractions using a modified Hedley P fractionation procedure. They fractionated soil P into several soil P fractions, corresponding to P pools with different turnover times and plant availability. Further details of the experimental design, cropping, sampling, and preparation of the soils, and determination of the P fractions and physicochemical properties of the soils, are available in Guo et al. (2000).

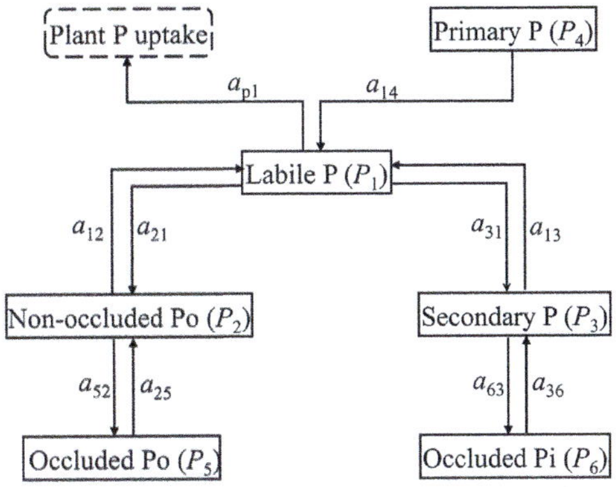

FIGURE 10.1 Schematic representation of the soil P model. Primary P indicates primary mineral P; secondary P indicates secondary mineral P; *Pi* indicates inorganic P; *Po* indicates organic P. An '*a*' on an arrow indicates the coefficient of soil P transformation: plant immobilization (a_{p1}), weathering (a_{14}), microbial immobilization (a_{21}), mineralization (a_{12}), sorption/precipitation (a_{31}), desorption/dissolution (a_{13}), and solid-phase transformations (a_{25}, a_{52}, a_{36}, and a_{63}).

Derived from Hou et al. (2019).

Construction of the P Matrix Model

Our example follows the study of Hou et al. 2019. Here we construct a soil P dynamics model that aligns with the datasets acquired by Guo et al. (2000). Our model groups the measured soil P fractions into six ecologically meaningful soil P pools, and we track the dynamics of these pools in the model. The six pools are labile P (P_1 in Figure 10.1), non-occluded *Po* (P_2 in Figure 10.1), secondary mineral P (P_3 in Figure 10.1), primary mineral P (P_4 in Figure 10.1), occluded *Po* (P_5 in Figure 10.1), and occluded *Pi* (P_6 in Figure 10.1). The labile P is inorganic P that is readily available to plants. The non-occluded *Po* is organic P that is sorbed to soil particles or secondary minerals (e.g., aluminum and iron oxides) and that can be mineralized by enzymes. The secondary mineral P is inorganic P that is sorbed to secondary minerals, which is not readily available to plants but is exchangeable with the labile P. The primary mineral P is P that is associated with primary minerals and exists mainly as apatite P. The occluded *Po* is organic P that is stabilized by soil minerals and that is presumably not mineralizable by enzymes unless dissolved. The occluded *Pi* is inorganic P that is occluded by soil minerals or aggregates and turns over very slowly. In Guo et al. (2000), the occluded *Po* and occluded *Pi* were not separated but determined together as residual P (P not extracted by the chemical reagents used). A proportion of residual P in organic forms (OP_o) is introduced here to represent the amounts of occluded *Po* (calculated as residual P × OP_o) and occluded *Pi* (calculated as residual P × (1–OP_o)). We consider all major soil P transformations in our model, as detailed in Figure 10.1. We calculate plant P uptake during a specific period as the sum of the decreases in $P_{1\text{–}6}$ during the period. For instance, we calculate plant P uptake after cropping 14 as the difference between the sum of $P_{1\text{–}6}$ at cropping 0 and the same sum after cropping 14. P leaching is not considered in our model, because soil moisture content was maintained near soil available water capacity during the experiment. Atmospheric P deposition is not considered in our model either because the experiment was performed in a greenhouse. Moreover, litterfall is disregarded, because the plants were young (≤45 days of growth) and unlikely to produce any litterfall. Matlab code and the eight datasets of soil P fractions are freely accessible via Hou et al. (2019).

Reflecting the structure depicted in Figure 10.1, we represent soil P dynamics in the model by the following set of balance equations.

$$\begin{cases} \dfrac{dP_1(t)}{dt} = a_{14}P_4(t) + a_{12}P_2(t) + a_{13}P_3(t) - k_1P_1(t) \\ \dfrac{dP_2(t)}{dt} = a_{21}P_1(t) + a_{25}P_5(t) - k_2P_2(t) \\ \dfrac{dP_3(t)}{dt} = a_{31}P_1(t) + a_{36}P_6(t) - k_3P_3(t) \\ \dfrac{dP_4(t)}{dt} = -k_4P_4(t) \\ \dfrac{dP_5(t)}{dt} = a_{52}P_2(t) - k_5P_5(t) \\ \dfrac{dP_6(t)}{dt} = a_{63}P_3(t) - k_6P_6(t) \end{cases} \tag{10.1}$$

Note that we do not list plant P uptake in Equation 10.1, because we treat it as a soil P flux. We do not use a_{p1} in Equation 10.1, because we can estimate it directly from a_{21} and a_{31} by the following equation:

$$a_{p1} = 1 - a_{21} - a_{31} \tag{10.2}$$

We can summarize the set of balance Equations 10.1 by the following first-order matrix equation:

$$\frac{dP(t)}{dt} = AKP(t) \tag{10.3}$$

where *A*, *K*, and *P(t)* are the matrices given by

$$A = \begin{pmatrix} -1 & a_{12} & a_{13} & 1 & 0 & 0 \\ a_{21} & -1 & 0 & 0 & 1 & 0 \\ a_{31} & 0 & -1 & 0 & 0 & 1 \\ 0 & 0 & 0 & -1 & 0 & 0 \\ 0 & 1-a_{12} & 0 & 0 & -1 & 0 \\ 0 & 0 & 1-a_{13} & 0 & 0 & -1 \end{pmatrix}$$

$$K = \text{diag}(k) = \begin{pmatrix} k_1 & 0 & 0 & 0 & 0 & 0 \\ 0 & k_2 & 0 & 0 & 0 & 0 \\ 0 & 0 & k_3 & 0 & 0 & 0 \\ 0 & 0 & 0 & k_4 & 0 & 0 \\ 0 & 0 & 0 & 0 & k_5 & 0 \\ 0 & 0 & 0 & 0 & 0 & k_6 \end{pmatrix}$$

$$P(t) = \begin{pmatrix} P_1(t) \\ P_2(t) \\ P_3(t) \\ P_4(t) \\ P_5(t) \\ P_6(t) \end{pmatrix}$$

Matrix A gives the transfers of P between the individual P pools, as described by the arrows in Figure 10.1. The elements (a_{ij}) are the P transfer coefficients, representing the fraction of P entering the ith (row) pool from the jth (column) pool. a_{52} is calculated as $1 - a_{12}$; a_{63} is calculated as $1 - a_{13}$; a_{14}, a_{25}, and a_{36} are fixed at 1.0. K is a 6 × 6 diagonal matrix representing the release rates of six soil P pools (units: g P g^{-1} P d^{-1}; for convenience, g g^{-1} d^{-1} was used in the following), i.e., the amount of P leaving each of the soil P pools per day. $P(t)$ describes the sizes of the soil P pools at time t.

Model Validation and Data Assimilation

We first validate the matrix P model with the eight datasets of soil P pool measurements in Guo et al. (2000). We use the same parameter values for simulating P pools of all the eight soils. In general, the soil P model can simulate temporal changes in soil P pools reasonably well, with a better performance for labile P, secondary mineral P, and occluded P than for non-occluded *Po* and primary mineral P.

We then use a data assimilation approach to estimate the values of parameters describing soil P dynamics. The approach, known as the Metropolis-Hastings algorithm, is described in more detail in Unit 6, where it is applied to the case of terrestrial C dynamics in Chapter 22. We run data assimilation formally for five replicates and 500,000 times for each soil to examine the convergence of the parameters. We test the convergence of the sampling chains by the Gelman-Rubin (*G-R*) diagnostic method to ensure that the within-run variation is roughly equal to the between-run variation. The Gelman-Rubin method is described in detail in Chapter 22.

After data assimilation, the soil P model can simulate well temporal changes in P pools of all the eight soils. Results on two soils are shown in Figure 10.2. Relationships between the measured and modeled soil P pools were mostly significant ($P < 0.05$), with R^2 values generally larger for labile P (mean 0.90), secondary mineral P (0.82), and occluded P (0.87) than for non-occluded *Po* (0.43) and primary mineral P (0.50). The relatively poor simulation of non-occluded *Po* was probably due to its relatively large measurement errors and dynamic nature.

NEW KNOWLEDGE EMERGING FROM DATA ASSIMILATION WITH THE MATRIX MODEL

Soil P Dynamics Quantified by Data Assimilation

Maximum likelihood estimates and uncertainty of the turnover rates of all soil P pools were estimated, including those of soil occluded *Pi* and occluded *Po*, which are rarely quantified in measurements. Estimated parameter values for a slightly weathered soil and a strongly weathered soil are shown in Table 10.1. Turnover rates of soil inorganic P pools generally decreased in the following order: labile P > secondary mineral P > occluded *Pi* (Table 10.1). The turnover rate of soil non-occluded *Po* was faster than that of soil occluded *Po* (Table 10.1).

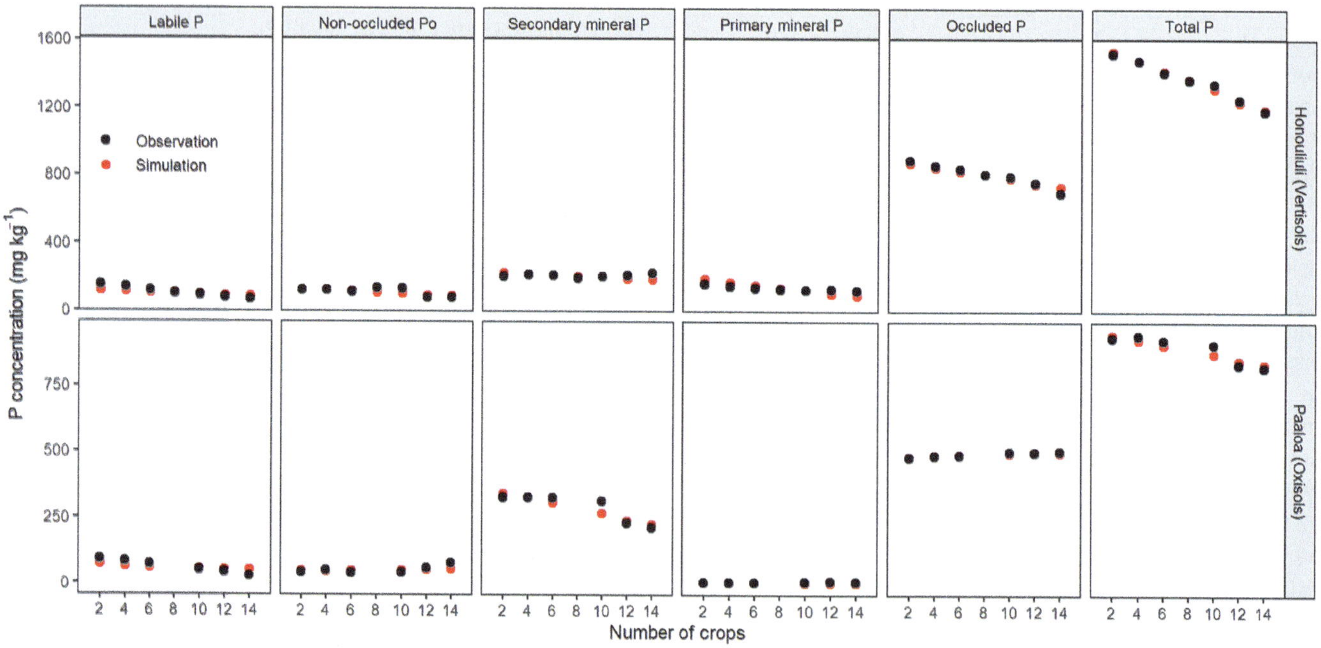

FIGURE 10.2 Observed vs. simulated temporal changes in P pools of two soils. The two soils are Honouliuli and Paaloa, which are typical Vertisols (slightly weathered) and Oxisols (strongly weathered), respectively.

Derived from Hou et al. (2019).

The turnover rates of soil P pools (Table 10.1) were generally comparable with the values reported in previous studies using isotopic and spectroscopic measurements in the laboratory. However, the turnover rates of soil P pools here could be higher than those in the field, because the datasets used for data assimilation were derived from an experiment in a greenhouse environment, where plant P uptake and soil P depletion could be faster than in the field.

Rates of transformations among all major P pools were also estimated (Table 10.1). These estimates can convey deep insights into soil P dynamics and soil P bioavailability. For example, the proportions of labile P flowing to secondary mineral P (mean 0.35) and non-occluded *Po* (0.52) were on average larger than that flowing to plants (0.13) (Table 10.1). This result suggests that soil secondary minerals and microbes were stronger competitors of soil labile P than plants. Both turnover rate of labile P and transfer coefficients related to labile P differed among soils (Table 10.1), suggesting that labile P dynamics varied among soils. This result suggests that not only the amount but also the dynamics of soil labile P control soil P availability.

The estimated model parameters from data assimilation provide information about the datasets and the model in two other aspects. Firstly, posterior distributions of parameters can tell how well model parameters were constrained by the datasets. Secondly, relationships among model parameters can reflect relationships defined by the model structure, correlations between the soil P pools, errors, or any combination of all three.

Soil P Dynamics in Relation to Other Ecosystem Properties

Soil P dynamics were revealed to differ between the lightly and the strongly weathered soils (Table 10.1). Turnover rate of secondary mineral P was approximately four times higher in the lightly weathered soil (0.044 g g^{-1} d^{-1}) than in the highly weathered soils (0.012 g g^{-1} d^{-1}) (Table 10.1). The proportion of labile P flowing to plants was higher in the slightly weathered soil (0.16) than in the strongly weathered soils (0.09); the opposite was true for the proportion of labile P flowing to secondary mineral P (slightly: 0.33; strongly: 0.38) (Table 10.1).

The proportion of labile P flowing to plants increased with soil pH; correspondingly, the proportion of labile P flowing to secondary mineral P decreased with soil pH (Figure 10.3c). Relationships of soil organic C concentration with model parameters were generally opposite to those of soil pH (Figure 10.3). These results suggest the regulation of plant and soil microbial competition for labile P by soil pH and

TABLE 10.1
Physicochemical properties and maximum likelihood estimates of model parameters describing *P* dynamics of two soils

Parameter	Unit	Honouliuli	Paaloa
Soil order		Vertisols	Oxisols
Weathered extent		Slightly	Strongly
Total P concentration at the start of experiment	g kg^{-1}	1840	596
0.5 M $NaHCO_3$ extractable P concentration	mg kg^{-1}	26.3	1.1
pH in water		7.26	5.05
Organic *C* concentration	g kg^{-1}	16.6	40
Exchange cation concentration	$cmol_c$ kg^{-1}	30.6	5.5
Acid ammonium oxalate extracted Fe concentration	g kg^{-1}	3.4	7.48
Acid ammonium oxalate extracted Al concentration	g kg^{-1}	1.43	2.98
Sand content	g kg^{-1}	56.5	53.8
Silt content	g kg^{-1}	363.6	205.4
Clay content	g kg^{-1}	580	740.8
Turnover rate of labile P (k_1)	g g^{-1} d^{-1}	0.04	0.048
Turnover rate of non-occluded *Po* (k_2)	g g^{-1} d^{-1}	0.022	0.073
Turnover rate of secondary mineral P (k_3)	g g^{-1} d^{-1}	0.044	0.012
Turnover rate of primary mineral P (k_4)	g g^{-1} d^{-1}	0.00193	0.00015
Turnover rate of occluded *Po* (k_5)	g g^{-1} d^{-1}	0.0077	0.0067
Turnover rate of occluded *Pi* (k_6)	g g^{-1} d^{-1}	0.0062	0.009
Coef. of transfer from labile P to non-occluded *Po* (a_{21})	Unitless	0.51	0.53
Coef. of transfer from labile P to secondary mineral *P* (a_{31})	Unitless	0.33	0.38
Coef. of transfer from labile P to plant (a_{p1})	Unitless	0.16	0.09
Coef. of transfer from non-occluded *Po* to labile P (a_{12})	Unitless	0.97	0.23
Coef. of transfer from non-occluded *Po* to occluded *Po* (a_{52})	Unitless	0.03	0.77
Coef. of transfer from secondary mineral P to labile P (a_{13})	Unitless	0.23	0.72
Coef. of transfer from secondary mineral P to occluded *Pi* (a_{63})	Unitless	0.77	0.28
Proportion of occluded *P* in organic form (OP_o)	Unitless	0.28	0.02

Derived from Hou et al. (2019).

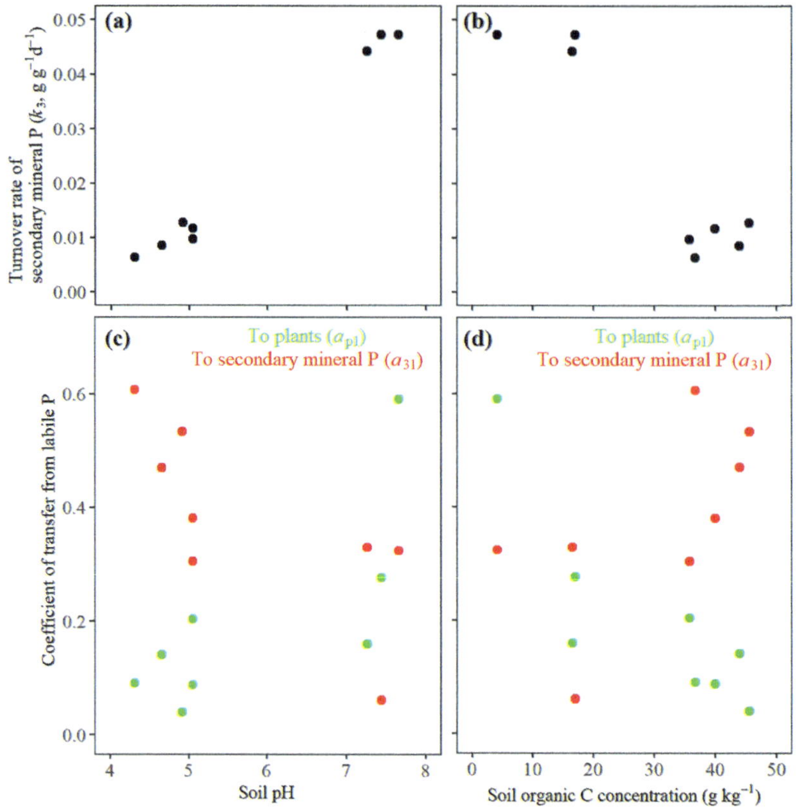

FIGURE 10.3 Soil P dynamics in relation to soil pH and organic C concentration. Turnover rate of secondary mineral P vs. (a) soil pH ($R^2 = 0.99$, $P < 0.001$) and (b) soil organic C concentration ($R^2 = 0.87$, $P = 0.001$). (c) Coefficients of transfer from labile P to plants (green; $R^2 = 0.53$, $P = 0.042$) and secondary mineral P (red; $R^2 = 0.71$, $P = 0.009$) vs. soil pH. (d) Coefficients of transfer from labile P to plants (green; $R^2 = 0.56$, $P = 0.034$) and secondary mineral P (red; $R^2 = 0.38$, $P = 0.106$) vs. soil organic C concentration.

Derived from Hou et al. (2019).

organic C concentration. The proportion of labile P flowing to secondary mineral P decreased with increasing soil pH. This was probably because of the decrease in soil P sorption capacity with increasing soil pH. This mechanism also explains the increase in turnover rate of secondary mineral P with increasing soil pH (Figure 10.3a). The increase in the proportion of labile P flowing to secondary mineral P with increasing soil organic C concentration (Figure 10.3d) may be attributable to the sorption of P to organic-metal complexations in soils. This mechanism additionally explains the decrease in turnover rate of secondary mineral P with increasing soil organic C concentration (Figure 10.3b).

SUMMARY

Despite increasing research efforts, there are still some uncertainties in the structure of soil P models and much larger uncertainties in the parameter values of such models. These uncertainties can be potentially reduced by the adoption of matrix and data assimilation approaches in soil P modeling. The matrix representation can make it easier to depict, understand, and compare soil P models compared to the traditional representation as a set of balance equations. Matrix approaches can also reduce the computational costs of models by enabling semi-analytical spin-up and make it computationally more feasible to assimilate soil P observations into models. An example study showed how to build a soil P model in matrix form, and how assimilating soil P observations into the matrix model can expose insights into soil P dynamics and availability. Overall, this chapter shows that assimilating soil P observations into a matrix model of soil P dynamics can improve our understanding of soil P dynamics and availability.

SUGGESTED READING

Hou, E., X. Lu, L. Jiang, D. Wen and Y. Luo (2019). Quantifying soil phosphorus dynamics: a data assimilation approach. *Journal of Geophysical Research: Biogeosciences 124*: 2159–2173.

QUIZ

1. Why is a unified matrix equation preferred over traditional balance equations in soil P studies?
2. Was turnover rate of soil labile P constant across soils?
3. Which model parameter is soil labile P pool size most sensitive to?
4. What purpose was data assimilation used for in the example study?

11 Principles Underlying Carbon Dioxide Removals from the Atmosphere

Yiqi Luo
Cornell University, Ithaca, USA

It is essential to reach net zero greenhouse gases (GHG) emissions. Various methods have been proposed to remove carbon dioxide from the atmosphere. Carbon dioxide removal (CDR) is complemental to dramatic reduction in anthropogenic emissions of GHG from fossil fuel burning, other industrial processes, and land management practices. This chapter introduces basic concepts of CDR, briefly describes land-based CDR, and outlines biogeochemical principles for evaluation and design of CDR strategies.

INTRODUCTION

The Earth's surface temperature has increased about 1.2°C since the industrial revolution in the late 1800s. It is higher than at any time in the past 100,000 years. The temperature was the highest on record in the last decade (2011–2020) and has been higher in each of the last four decades than any previous decade since 1850. Climate warming is a key facet of climate change, which also features intense droughts, water scarcity, severe fires, rising sea levels, flooding, melting polar ice, catastrophic storms, and other weather extremes. Climate change affects many aspects of humanity, including the ability to grow food, our health, housing, and safety. The increase in the Earth surface temperature primarily results from fossil fuel burning, deforestation, and other agricultural and industrial practices. These human activities result in emissions and accumulation of greenhouse gases (GHG), mainly carbon dioxide (CO_2) and methane (CH_4), in the atmosphere to trap the long-wave radiative heat, leading to climate warming (IPCC, 2021). To bend the curve of climate warming, it is essential to reduce GHG emissions and remove carbon dioxide from the atmosphere. Carbon dioxide removal (CDR) is a general term for a number of practices and technologies that transfer CO_2 from the atmosphere to long-term sinks in land, ocean, and/or geological reservoirs, slowing climate warming.

This chapter aims to introduce the concepts of CDR and nature-based climate solutions, describe various strategies of CDR, and explain carbon cycle principles for evaluating and designing CDR strategies.

CARBON DIOXIDE REMOVAL (CDR)

CDR is a process through which human activities and management cause carbon dioxide to be removed from the atmosphere, and durably stored as carbon in geological, terrestrial, or ocean reservoirs, or in products used in commercial or industrial processes, such as building materials or powerfuels. The effect of CDR is to contribute to negative carbon emissions. Policy frameworks such as the Paris Agreement generally recognize that CDR is required to achieve net zero CO_2 and GHG emissions in individual nations and around the globe in the time frame necessary to avoid dangerous climate change. CDR can counterbalance emissions that are technically difficult to eliminate in some agricultural and industrial practices. It is complementary to immediate and deep emissions reductions specified by nationally determined contributions (NDCs) (IPCC, 2023).

NDCs are national plans, including policies and measures, that governments aim to implement in response to climate change to achieve the global targets of reducing global GHG emissions and limit anthropogenic climate warming to well below 2°C, preferably 1.5°C, relative to the pre-industrial level by 2100 (Figure 11.1). NDCs encompass actions to reach net zero emissions, increase adaptation to the harmful effects of climate change, and adjust financial flows for supporting GHG emissions reduction. Reducing emissions and CDR are the two primary tools available to governments to achieve net zero emissions. Emissions reduction can occur through improved resource-use efficiency, sustainable development, and substituting fossil energy with renewable alternatives. In the energy supply sector, for example, CO_2 emissions are reduced through cutting down energy use and generating electricity from low-carbon sources. Power plants fired by fossil fuels (e.g., coal and natural gas) are gradually phased out while wind and solar power is increasingly used to generate electricity for transportation, heating buildings, operating industrial facilities, and other applications.

CDR is essential as a complement to emissions reduction to reach net zero GHG emissions around 2050 and keep climate warming by 2100 within the 1.5°C target. The required amount of CO_2 to be removed from the atmosphere depends on scenarios of socio-economic challenges to mitigation and adaptation as explored under illustrative model pathways. Under a low energy demand scenario with numerous optimistic assumptions on technology advancement and lifestyle changes, a small amount of CDR (~3 Gt CO_2 yr^{-1}) is enough to limit climate warming to within 1.5°C by 2100 (IPCC, 2019). Under a scenario of fossil fuel intensive economic development with a high energy demand, the annual negative emissions via CDR are required to be as great as 20 Gt CO_2 yr^{-1}.

DOI: 10.1201/9781032711126-14

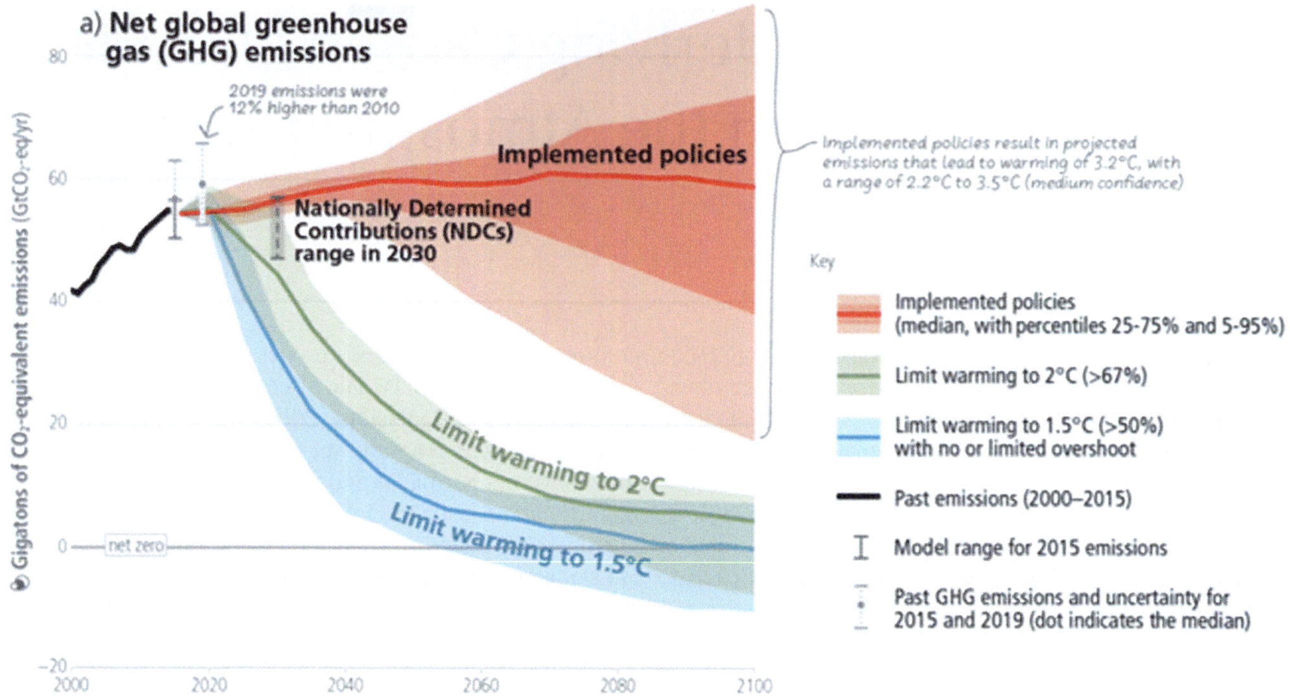

FIGURE 11.1 Global emissions pathways consistent with implemented policies and mitigation strategies. Colored ranges denote the 5th to 95th percentile across the global modeled pathways falling within a given category. The red ranges depict emissions pathways assuming policies that were implemented by the end of 2020. Ranges of modeled pathways that limit warming to 1.5°C (>50%) with no or limited overshoot are shown in light blue and pathways that limit warming to 2°C (>67%) are shown in green. Global emission pathways that would limit warming to 1.5°C (>50%) with no or limited overshoot and also reach net zero GHG in the second half of the century do so between 2070–2075.

Adopted from IPCC, 2023.

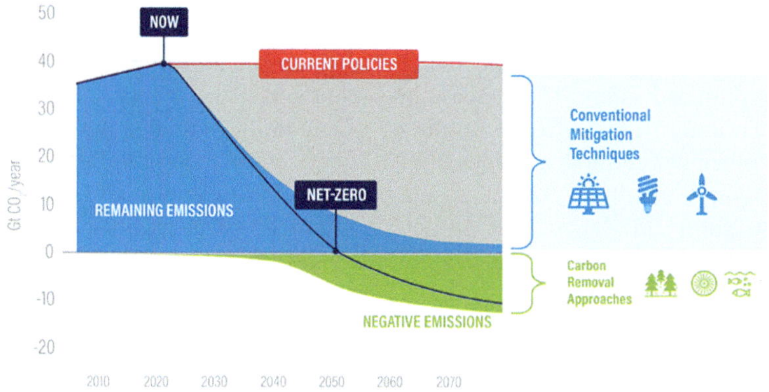

FIGURE 11.2 The role of carbon removal in bringing emissions to net zero by mid-century consistent with limiting global warming to 1.5°C above pre-industrial levels. Faster and/or deeper emissions reductions could reduce the role for carbon removal; slower and/or weaker emissions reductions would increase the need for carbon removal. Estimates, including both natural and technological carbon removal approaches, range from 5 to 16 billion metric tons per year globally by 2050. Gt CO_2/y = billions of metric tons of carbon dioxide per year. Based on IPCC (2019).

Adopted from World Resources Institute.

Presently CDR removes only a small fraction of global emissions. Most CDR strategies remain largely unproven to date and raise substantial concerns about adverse side-effects on environmental and social sustainability. Nonetheless, the existing CDR methods have the potential to remove up to 10 Gt CO_2 yr^{-1}. According to various modeling projections for the middle-of-the-road scenarios, CDR needs to remove 6 Gt CO_2 yr^{-1} in 2050, which is equivalent to about 13% of the current global emissions, to reach net zero (Figure 11.2).

Carbon dioxide removal includes nature-based climate solutions and technology-driven methods. The nature-based climate solutions emphasize the sustainable management and use of natural features and processes for reducing GHG emissions and storing carbon in land and ocean ecosystems.

Such solutions are inspired and supported by nature, which are cost-effective, simultaneously provide environmental, social, and economic benefits, and help build resilience. Examples include practices of improved forest management to help forest owners increase the carbon stored in their forested lands, reducing fertilizer use in agriculture to decrease emissions of nitrous oxide (a potent greenhouse gas), and restoring coastal wetlands to sequester carbon in submerged sediment. In addition to carbon sequestration, restoring coastal wetlands, such as mangroves, can help control coastal erosion resulting from waves and wind, provide nursery zones for marine life for sustaining fisheries, and mitigate impacts of sea level rise on coastlines.

The technology-based methods include direct air capture with carbon sequestration, biochar carbon removal, enhanced rock weathering, and bioenergy with carbon capture and storage (BECCS). Direct air capture (DAC) requires technology that uses chemical or physical processes to extract carbon dioxide directly from the ambient air. The storage of the extracted CO_2 in safe long-term storage often requires geological techniques to accomplish. Enhanced rock weathering requires engineering techniques to crash rocks to powder to increase reaction surface.

To meaningfully support climate mitigation, CDR must lead to (1) enhancements of carbon uptake from the atmosphere and/or GHG emissions reductions that are additional to a baseline scenario, (2) net cooling, taking into account side effects on water and energy cycling which could offset or even outweigh the gains in carbon uptake or emissions reductions, (3) the permanence of climate benefits, and (4) avoiding leakage of mitigation benefits, for instance by reducing carbon influx or ecosystem residence time in the area to which management is applied (Novick et al. 2022).

LAND-BASED CDR STRATEGIES

Land-based CDR methods include afforestation, reforestation, agricultural practices that sequester carbon in soils (carbon farming), wetland and coastal restoration, and bioenergy with carbon capture and storage (BECCS). The cost-effective, land-based CDR methods are estimated to have the potential to deliver 8–13.8 Gt $CO_2eq\ yr^{-1}$ between 2020 and 2050 (Roe et al. 2021). Cost-effective mitigation potentials represent a more realistic and actionable target grounded in public willingness to pay for climate mitigation, and therefore are more relevant in policy-making than technical potentials. The technical potentials measure the possible removal of CO_2 with available technology, regardless of the cost.

Cost-effective potential is high in the sector of forests and other ecosystems and moderate in the sector of agriculture (Roe et al. 2021). In the sectoral estimates, CDR potential due to restoration in forests and other ecosystems, and

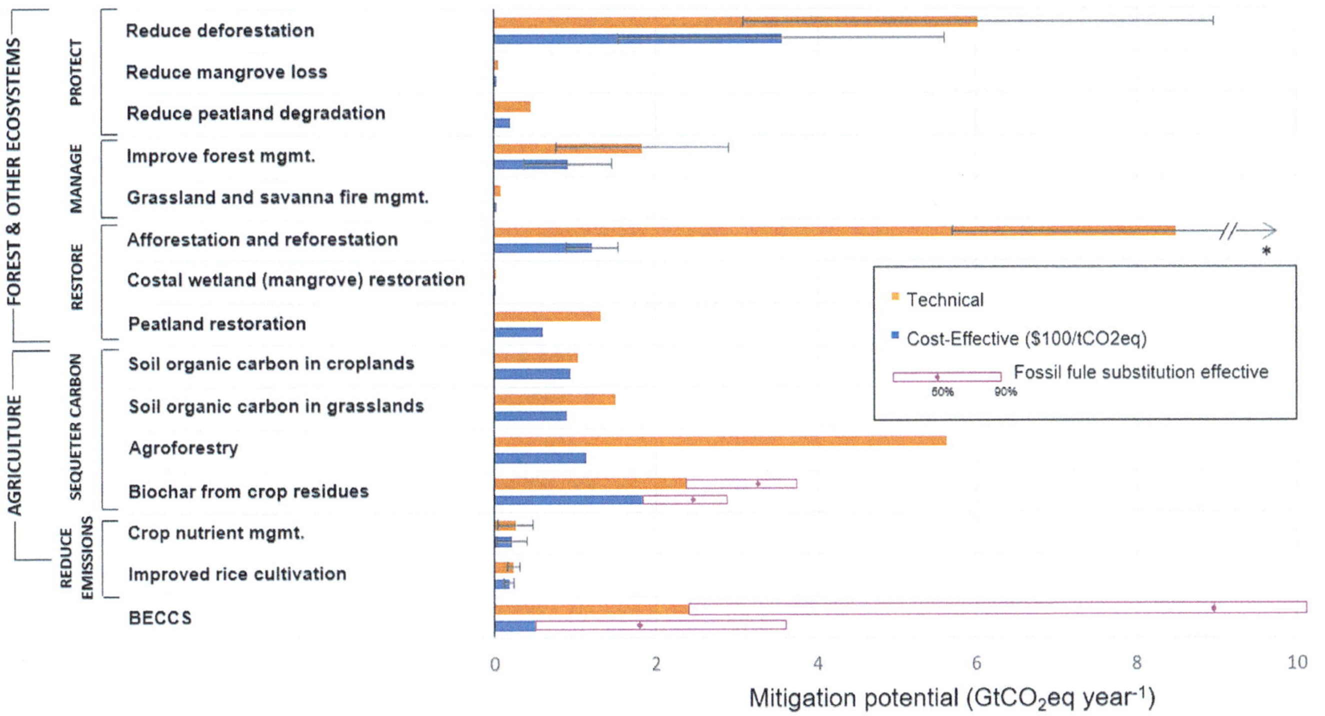

FIGURE 11.3 Climate mitigation potentials for 20 land-based measures in 2020–2050. Technical and cost-effective (≤$100/t$CO_2$eq) mitigation potentials are provided for each measure using a sectoral approach. The 20 measures are grouped into four systems-level mitigation categories, and seven management-level categories. For measures with more than one dataset, the bar graph represents the mean estimate, and the error bars represent the min and max potential range. Global mitigation potentials of substituting fossil fuels were estimated for BECCS, biochar, and manure management, shown in pink outline bars, illustrating the median and 90th percentile values.

Adopted from Roe et al. 2021.

measures of carbon sequestration in agriculture (excluding bioenergy and carbon capture and storage) is 20.3 ± 3.0 Gt CO_2 yr^{-1} for technical and 6.6 ± 0.3 Gt CO_2 yr^{-1} for cost-effective (Figure 11.3). The sectoral estimates have large CDR potentials from agriculture – agroforestry, biochar, and soil carbon sequestration (4.8 $GtCO_2$ yr^{-1} up to \$100/tCO_2) (Figure 11.3). Estimated mitigation from bioenergy and carbon capture and storage is 2.5 Gt CO_2eq yr^{-1} and 0.5 Gt CO_2eq yr^{-1}, respectively, with technical and cost-effective potential.

Land-based climate mitigation measures are also known as Agriculture, Forestry and other Land Uses (AFOLU) mitigation or nature-based climate solutions, which benefit human well-being and biodiversity. Land-based measures reduce GHG emissions and enhance carbon removals through supply-side interventions in forests and other ecosystems (i.e., protection, management, and restoration), agriculture (i.e., reducing emissions and enhancing carbon sequestration), and bioenergy (i.e., reducing fossil fuel emissions and sequestering carbon). Almost all countries have included AFOLU measures in their Nationally Determined Commitments (NDCs) under the Paris Agreement, either as specifically listed actions or by including the land sector in their broader GHG reduction targets. Globally, AFOLU-related NDC actions make up about 25% of planned GHG reductions with most focus on reducing deforestation.

PRINCIPLES FOR EVALUATING AND DESIGNING CDR STRATEGIES

The CDR strategies shown in Figure 11.3 involve either increasing carbon influx into long-term residence pools or lengthening times (or durability or permanence) in high-influx pools, or both. For example, producing biochar from crop residues and transferring it to the cropland soil can accelerate the influx of carbon into long-term carbon storage pools of the soil. Peatland restoration aims to lengthen carbon residence time of peat carbon by recreating anaerobic environments to depress peat decomposition, relative to reference ecosystems before restoration. Protection of existing ecosystems from deforestation, mangrove loss, and peatland degradation can be an effective strategy to avoid carbon loss from pools with long residence times.

The amount of land carbon storage, then, is determined by two parameters: the carbon influx and residence time (Luo and Weng 2011, Chapter 1). Consequently, either substantially increasing carbon influx into long residence pools or lengthening residence times of carbon in high-influx pools, or both, is a general principle that can be used to evaluate existing CDR strategies and guide the design of new ones that can effectively remove CO_2 from the atmosphere. This principle satisfies some of the default criteria of CDR, including additionality, permanence, and avoiding leakage. Increasing carbon influxes or lengthening residence time will result in additionality in carbon storage. Permanence is measured by carbon residence time. Avoiding leakage is demonstrated by quantifying carbon influx or residence time, and showing there is no decrease relative to reference conditions in the region of study.

Many of the current CDR techniques, such as afforestation and reforestation, coastal wetland restoration, and peatland restoration, mainly lengthen carbon residence time in comparison with their respective reference ecosystems. For example, afforestation removes carbon dioxide from the atmosphere by converting a grassland or cropland into a forested area. While the established forest may have similar carbon input (i.e., net primary production, NPP) as the previous grassland, input carbon is partially allocated to wood biomass, which has a longer residence time by decades or centuries compared to grass or crop biomass. When afforestation establishes a forest on an area of bare land, carbon input increases in the first several years, but then flatlines as the seedlings grow into trees and the forest canopy closes. From this point onwards, the carbon dioxide removal mainly results from allocation of input carbon to wood biomass that has a long residence time. Similarly, restoration of wetlands, peatlands, and forest lands (i.e., reforestation) all results in carbon dioxide removal mainly by shifting the allocation of carbon towards the longer residence time pools of the vegetation and/or soil. While biochar has longer residence time than its precursor biomass, a substantial fraction of carbon is immediately released to the atmosphere from biomass during pyrolysis. Thus, the increase in residence time by biochar must be discounted for the shortened residence time of the released carbon.

Among these existing CDR techniques, most of them alter carbon residence time more than carbon input. It appears that there are more options for managing carbon residence time than carbon input. Carbon residence time can change from a few months or years to thousands of years (Table 11.1). For example, plant materials, once submerged under water, can be preserved for thousands of years. In comparison, plant litter in upland ecosystems is usually decomposed within months or years. Once wood materials are buried in soil vaults, the wood carbon can have a residence time of thousands of years.

We can use this insight to design future CDR techniques using methods that can substantially lengthen carbon residence time. One example of manipulating residence time is through wood vault, a method of constructing a wood storage facility, to bury woody mass on a mega-ton scale in underground anaerobic environments and, thus, prevent wood from decay (Zeng and Hausmann 2022). The buried wood is expected to be preserved

TABLE 11.1
The potential to increase carbon residence time for various types of CDR

Type of CDR	Increase in carbon residence time
Afforestation and reforestation	50–300 years
Biochar	>100 years
Wood products, such as furniture	20–100 years
Wood vault	300–5000 years
Litter under anaerobic conditions	300–5000 years

for thousands of years. A wood vault unit that occupies 1 hectare of surface land, with 10 m effective depth, can store up to 100,000 m^3 of wood and sequester 0.1 $MtCO_2$. It takes 10,000 wood vault units of such a size to store 1 Gt CO_2 y^{-1}. The cost is estimated at $10–$50 per ton CO_2, which can be considered cost-effective. The land surface of a fully closed wood vault still can be used for recreation, agriculture, or a solar farm. The high durability, verifiability, and low cost of wood mass preservation in wood vaults makes it an attractive option for CDR.

SUGGESTED READING

Ning Zeng and Henry Hausmann, 2022. Wood Vault: Remove atmospheric CO_2 with trees, store wood for carbon sequestration for now and as biomass, bioenergy and carbon reserve for the future. *Carbon Balance and Management*, 17:2. https://doi.org/10.1186/s13021-022-00202-0

QUIZ

1. Carbon dioxide removal (CDR) is a human intervention to remove carbon dioxide from the atmosphere.
   a. True
   b. False
2. A principle to evaluate effectiveness of a CDR strategy is to examine if it:
   a. increases carbon influxes into long residence time pools.
   b. lengthens carbon residence time in high-influx pools.
   c. both increases carbon influxes and lengthens residence time.
   d. all the above.
3. Net zero carbon dioxide emissions means that emitted carbon dioxide from fossil fuel burning, deforestation, and other processes are fully counterbalanced by negative emissions via various CDR strategies.
   a. True
   b. False
4. Carbon dioxide removal includes the following methods (choose as many as fit).
   a. Reforestation by planting trees in degraded forest lands.
   b. Agricultural practices to restore soil fertility.
   c. Using chemical fertilizers to increase crop yields.
   d. Reclamation of wetlands to grow rice crop.

12 Practice 3

Diagnostic Variables in Matrix Models

Xingjie Lu
Sun Yat-sen University, Guangzhou, China

This practice helps you understand diagnostic variables in biogeochemistry matrix models. The key diagnostic variables include carbon storage capacity, storage potential, residence time, and input. We use the matrix version of the TECO model to demonstrate how to incorporate diagnostic variables from carbon balance equations. We verify the theory that carbon storage capacity represents an attractor of carbon storage dynamic. Meanwhile, we understand how changes in carbon turnover rate and input allocation fraction affect the carbon residence time and therefore the carbon storage capacity. Carbon residence time and carbon input are two essential diagnostics to characterize the steady state carbon storage in land carbon cycle modeling.

MOTIVATION OF THE UNCERTAINTY DIAGNOSTICS

Land carbon cycle models are frequently used to estimate biosphere-atmosphere feedback. However, different models often provide quite divergent predictions of land carbon uptake and storage as revealed via model intercomparison projects. Great efforts have been made to understand differences among models. Even so, we still have difficulties in identifying causes of model differences.

To improve understanding of why models behave differently, we need effective diagnostic tools. This is one of the motivations for developing the matrix approach. The traceability analysis with the matrix approach offers a framework to understand each traceable component first and then assemble them together to analyze the carbon storage dynamics.

The matrix approach enables decomposition of modeled carbon storage to a few traceable components according to the mathematical properties of the matrix equation. In this way, we can conduct traceability analysis to understand the causes of model uncertainty. Traceability analysis is discussed in detail in Unit 5 of this book.

THE MATHEMATICAL FOUNDATION OF THE DIAGNOSTICS OF LAND CARBON CYCLE MODELS

Previous chapters have demonstrated that the carbon storage dynamics in land carbon cycle models can be formulated in a matrix form (Luo et al. 2017; Chapter 1):

$$\frac{dX(t)}{dt} = B\mu(t) + G(t)X(t) \tag{12.1}$$

where $X(t)$, as a vector, represents carbon storage in multiple pools, $u(t)$ as a scalar represents the amount of carbon input from net primary production (NPP, i.e., photosynthesis minus autotrophic respiration), B, as a vector, is the carbon partitioning from NPP to multiple pools, and G, as a matrix, indicates the carbon transfer network among pools.

Equation 12.1 can be reformulated to a *diagnostic format*:

$$X(t) = X_c(t) - X_p(t) \tag{12.2}$$

where X_c is carbon storage capacity and X_p is carbon storage potential. The carbon storage capacity can be further decomposed into ecosystem residence time (τ_E) and carbon input (u):

$$X_c(t) = \tau_E u(t) \tag{12.3}$$

where ecosystem residence time is defined by:

$$\tau_E = -G^{-1}B \tag{12.4}$$

The carbon storage potential is the product of the inverse of matrix G and the carbon storage growth rate:

$$X_p(t) = -G^{-1}\frac{dX(t)}{dt} \tag{12.5}$$

In this practice, we are going to focus on the carbon storage capacity (X_c), carbon storage potential (X_p), carbon residence time (τ_E), and carbon input (u), which are the most important diagnostics in the matrix approach.

Equations 12.3–12.5 show that three terms: carbon storage capacity (X_c), carbon storage potential (X_p), and carbon residence time (τ_E), are related to the matrix G. Therefore, one of the most critical steps to calculate the above diagnostics is to find the matrix G, which equals $A(t)\xi(t)K$.

Exercise 1 and ***Exercise 2*** are to identify diagnostic variables from the TECO and CLM5 matrix models.

DOI: 10.1201/9781032711126-15

Exercise 1

The matrix equation of the TECO model can be written into one matrix equation:

$$\frac{dX(t)}{dt} = B\mu(t) + A(t)\xi(t)KX(t) \qquad 12.6$$

Here, μ is the C input scalar (gC m^{-2} s^{-1}), $\boldsymbol{B}$ represents the C allocation fraction vector (unitless), $\boldsymbol{A}$ is the partitioning coefficient matrix (unitless), $\boldsymbol{K}$ is a diagonal turnover rate matrix (s^{-1}), and $\boldsymbol{X}$ is C pool size vector (gC m^{-2}).

Ex 1.1 Find the **matrix G** of Equation 12.6. This matrix was defined in Equation 12.1: ***G*** = …

Ex 1.2 Write down the **diagnostic form** of Equation 12.6: ***X(t)*** = …

Ex 1.3 Identify the residence time, C input, C storage capacity and C storage potential from the diagnostic form.

Residence time: $\tau_E = \ldots$
C input: $\mu = \ldots$
C storage capacity: $X_c = \ldots$
C storage potential: $X_p = \ldots$

Exercise 2 (Optional)

The matrix equation of the CLM5 vegetation carbon cycle model can be written in the form of one matrix equation:

$$\frac{dX(t)}{dt} = Bu(t) + A_{ph}(t)K_{ph}(t)X(t) + A_{gm}K_{gm}X(t) + A_{fi}K_{fi}(t)X(t) \qquad 12.7$$

where μ is the C input scalar, (gC m^{-2} s^{-1}), $\boldsymbol{B}$ represents the C allocation fraction vector (unitless), $\boldsymbol{A}$ is the partitioning coefficient matrix (unitless), $\boldsymbol{K}$ is a diagonal turnover rate matrix (s^{-1}), and $\boldsymbol{X}$ is C pool size vector (gC m^{-2}). The subscripts, ***ph***, ***gm***, and ***fi*** represent the partitioning coefficient and turnover rate related to phenology, gap mortality, and fire processes.

Ex 2.1 Find the **matrix *G*** of Equation 12.7: ***G*** = …

Ex 2.2 Write down the **diagnostic form** of Equation 12.7: ***X(t)*** = …

Ex 2.3 Identify residence time, C input, C storage capacity, and C storage potential from the diagnostic form.

Residence time: $\tau_E = \ldots$
C input: $\mu = \ldots$
C storage capacity: $X_c = \ldots$
C storage potential: $X_p = \ldots$

CARBON STORAGE CAPACITY AND CARBON STORAGE POTENTIAL

The carbon storage capacity and carbon storage potential are two important diagnostics. The carbon storage capacity represents the maximal amount of carbon that a land ecosystem can store. The carbon storage potential represents the difference between carbon storage capacity and current carbon storage. The two diagnostics are the first tier variables to define land carbon dynamics for any modeling study. The carbon storage capacity is the attractor and the sign of the carbon storage potential represents the direction. That is, the carbon storage will ultimately reach carbon storage capacity regardless of initial values, carbon inputs, and other model parameters. **Exercise 3** will allow us to verify this notion.

Exercise 3

Run the TECO matrix model (Ex 3.1), then change the initial value (Ex 3.2) and input (Ex 3.3). Learn whether carbon storage changes to approach the carbon storage capacity. Learn how carbon storage potential changes.

Ex 3.1. Follow instructions to run the TECO matrix model in CarboTrain:

a Select **Unit 3**
b Select **Exercise 3**
c Select **Default**
d Select **Set Output Folder**
e **Open Source Code**
f Read lines 10–44, get familiar with parameters, input, and initial value. The file test_p3.py can be edited at this step.
g **Run Exercise**
h Check results in Output Folder. Time-dependent variation in carbon input, each pool size, residence time, total ecosystem carbon storage, total ecosystem carbon storage capacity, and carbon storage potential appear in the output file output.xls. Figures are in results.png.

Questions: Does carbon storage change towards carbon storage capacity? How does carbon storage potential change?

Ex 3.2. Try different initial values

a Repeat Ex 3.1, but select "**Change initial pool size I**" instead of "**Default**". **Open source code** and change the initial value to (1000.0 20000.0 50.0 3000.0 200.0 15000.0 40000.0) at line 44 of `test_p3.py`. **Run Exercise** and check the total ecosystem carbon storage and total ecosystem carbon storage capacity in `results.png`.

b Repeat Ex 3.1, but select "**Change initial pool size II**" instead of "**Default**". **Open source code** and change the initial value to (280.0 5000.0 15.0 800.0 50.0 4000.0 10000.0) at line 44 of `test_p3.py`. **Run Exercise** and check the total ecosystem carbon storage and total ecosystem carbon storage capacity in `results.png`.

Questions: If the initial pool size changes, does carbon storage change towards carbon storage capacity? Does the carbon storage capacity depend on the initial value? How does carbon storage potential change?

Ex 3.3. Try different C inputs

a Repeat Ex 3.1, but select "**Change carbon input I**" instead of "**Default**". **Open source code** and change the C input to 0.00001123 at line 38 of `test_p3.py`. **Run Exercise** and check the total ecosystem carbon storage and total ecosystem carbon storage capacity in `results.png`.

b Repeat Ex 3.1, but select "**Change carbon input II**" instead of "**Default**". **Open source code** and change the C input to 0.00004490 at line 38 of `test_p3.py`. **Run Exercise** and check the total ecosystem carbon storage and total ecosystem carbon storage capacity in `results.png`.

Questions: If the carbon storage capacity changes, does carbon storage still change towards carbon storage capacity? Does the carbon storage capacity depend on the C input? How does carbon storage potential change? Would the carbon storage always equal the carbon storage capacity at steady-state? What does zero carbon storage potential stand for?

These questions are critical to the conceptual foundation of the carbon storage capacity and carbon storage potential. Results from Exercise 3 guide you to understand why carbon storage capacity and carbon storage potential are important diagnostics for carbon cycle modeling and how the definitions of the two variables reflect the inherent property of land carbon cycle.

RESIDENCE TIME AND CARBON INPUT

Now let us further explore the carbon storage capacity, which is described by two additional traceable components: residence time and carbon input. Climate change causes changes in carbon cycling, usually via changes in residence time and/or carbon input. For example, rising atmospheric CO_2 concentration usually enhances carbon input (μ). Rising air temperature usually increases soil carbon decomposition and thus turnover rate of soil pools (K). When a forest is converted to cropland under land use change, the carbon allocation to woody tissue is usually set to zero, leading to changes in the carbon allocation fraction (B). Thus, changes in ecosystem carbon storage in response to various climate change factors can be explained by diagnostics related to residence time and carbon input in the matrix approach. ***Exercise 4*** will let us explore how changes in parameter values related to residence time or carbon input result in changes in the carbon storage capacity.

EXERCISE 4

Run the TECO matrix model, and make the following changes in parameters or carbon input (Ex 4.1–4.3). Observe how parameters influence the carbon storage capacity.

Ex 4.1 Try a different turnover rate.

a Repeat Ex 3.1, but Select "**Exercise 4**", and "**Low foliage turnover**". **Open source code** and change the foliage turnover rate to 8.8×10^{-4}, which is the first element of "temp" at line 26. **Run Exercise** and check the total ecosystem carbon storage capacity, residence time, and carbon input at steady state in `output.xls` and `results.png`.

b Repeat Ex 3.1, but Select "**Exercise 4**", and "**Low passive soil turnover**". **Open source code** and change the passive soil turnover rate to 7.739×10^{-7}, which is the last element of "temp" at line 26. **Run Exercise** and check the total ecosystem carbon storage capacity, residence time, and carbon input at steady state in `output.xls` and `results.png`.

Questions: Is the carbon storage capacity in Ex 4.1a or Ex 4.1b lower or higher than that in the default model (Ex 3.1)? Can the lower or higher carbon storage capacity be attributed to residence time or carbon input? How do changes in turnover of foliage or passive soil carbon impact residence time or carbon input? Which impact is stronger? Is it the impact of lower foliage turnover rate or the impact of lower passive soil turnover rate? Why?

Ex 4.2 Try different allocation fractions.

a Repeat Ex 3.1, but Select "**Exercise 4**", and "**High allocation to foliage**". **Open source code** and change the allocation fraction to (0.55, 0.45…), which are the first two elements of "B" at line 10. **Run Exercise** and check the total ecosystem carbon storage capacity, residence time, and carbon input at steady state in `output.xls` and `results.png`.

b Repeat Ex 3.1, but Select "**Exercise 4**", and "**High allocation to wood**". **Open source code** and change the allocation fraction to (0.2, 0.8…), which are the first two elements of "B" at line 10. **Run Exercise** and check the total ecosystem carbon storage capacity, residence time, and carbon input at steady state in `output.xls` and `results.png`.

Questions: Is the carbon storage capacity in Ex 4.2a or Ex 4.2b lower or higher than that in the default model (Ex 3.1)? Can the lower or higher carbon storage capacity be attributed to residence time or carbon input? How do changes in allocation fraction impact residence time or carbon input? Why?

Ex 4.3 Try multiple changes of parameters in different C input and turnover rate.

a Repeat Ex 3.1, but Select "**Exercise 4**", and "**Multiple Changes I**". **Open source code** and change the C input to 1.123×10^{-5} at line 26 and 38, and change the slow and passive soil turnover rate to 2.99×10^{-5} and 5.159×10^{-7}, which are the last two elements of "temp" at line 25. **Run Exercise** and check the total ecosystem carbon storage capacity, residence time, and carbon input at steady state in `output.xls` and `results.png`.

b Repeat Ex 3.1, but Select "**Exercise 4**", and "**Multiple Changes II**". **Open source code** and change the C input to 4.49 × 10^{-5} at line 26 and 38, and change the slow and passive soil turnover rate to 2.69 × 10^{-4} and 4.643 × 10^{-6}, which are the last two elements of "temp" at line 25. **Run Exercise** and check the total ecosystem carbon storage capacity, residence time, and carbon input at steady state in `output.xls` and `results.png`.

Questions: What are the common features or differences in carbon storage dynamics and carbon storage capacity between Ex 4.3a and Ex 4.3b? Can you see how residence time and carbon input cause differences in carbon storage capacity? What causes these differences in residence time or carbon input?

Understanding how carbon storage capacity changes with carbon residence time and carbon input changes is essential in diagnostics for carbon cycle modeling. This will be further explored in Unit 5 on traceability analysis.

Diagnostic capability is one of the most important benefits offered by the matrix approach. This practice only provides you with simple cases to understand the terrestrial ecosystem carbon cycle. To understand the land carbon cycle at the earth system scale, we need to analyze wider ranges of spatial and temporal variations from multiple models in response to rising atmospheric CO_2 concentration and climate warming (Lu et al. 2018).

SUGGESTED READINGS

Luo, Y. Q., Shi, Z., Lu, X. J., Xia, J. Y., Liang, J. Y., Jiang, J., et al. (2017). Transient dynamics of terrestrial carbon storage: Mathematical foundation and its applications. *Biogeosciences*, *14*(*1*), 145–161

Lu, X., Du, Z., Huang, Y., Lawrence, D., Kluzek, E., Collier, N., Lombardozzi, D., Sobhani, N., Schuur, E., Luo, Y., (2020). Full implementation of matrix approach to biogeochemistry module of Community Land Model version 5 (CLM5). *Journal of Advances in Modeling Earth Systems*, *12*: e2020MS002105.

Unit Four

Semi-Analytic Spin-Up (SASU)

13 Nonautonomous ODE System Solver and Stability Analysis

Ying Wang
University of Oklahoma, Norman, USA

The matrix equation model for the terrestrial carbon dynamics is a system of nonautonomous ordinary differential equations, ODEs, which naturally inherits the mathematical difficulties in the solution process and stability studies of its equilibria. This chapter introduces the mathematical properties of this matrix equation model. In particular, we will study: (1) the analytical solution of the matrix equation model, examining a three-pool terrestrial carbon dynamics representation to demonstrate the solution process; and (2) the stability analysis of the matrix equation model.

INTRODUCTION

We consider the following system of nonautonomous ordinary differential equations (ODEs):

$$X' = \xi(t)ACX + Bu(t) \tag{13.1}$$

where $X(t)$ is a vector of carbon pool sizes; X_0 is a vector of initial values of the carbon pools; $\xi(t)$ is an environmental scalar representing effects of temperature and moisture on the carbon transfer among pools; A and C are carbon transfer coefficients between plant, litter, and soil pools; $u(t)$ is the photosynthetically fixed carbon and usually estimated by canopy photosynthetic models; and B is a vector of partitioning coefficients of the photosynthetically fixed carbon to plant pools. For background to this equation, see Chapter 1.

Our goal is to develop an analytical solution of Equation 13.1, and to understand and predict how the equilibrium state stability is impacted by various environmental scalar functions $\xi(t)$, transfer matrices A and C, and photosynthetic input $u(t)$.

ANALYTICAL SOLUTION

In mathematical language, governing Equation 13.1 is a system of non-homogeneous nonautonomous ODEs, and the derivation of its solution is complicated. Therefore, instead of deriving the solution directly, we will start with studying the analytical solution of a scalar non-homogeneous nonautonomous ODE, and then extend it to the system situation.

FIRST ORDER NON-HOMOGENEOUS SCALAR EQUATION

A first-order linear non-homogeneous scalar equation has the following general form:

$$x' + p(t)x = q(t) \tag{13.2}$$

A standard way to solve (13.2) is the integrating factor method, which proceeds as follows. Let:

$$P(t) = \int_{t_0}^{t} p(s)\,ds,$$

then the integrating factor is given by:

$$h(t) = e^{P(t)}. \tag{13.3}$$

After multiplying the integrating factor (13.3) to both sides of (13.2), we get:

$$e^{P(t)}\left(x' + p(t)x\right) = e^{P(t)}q(t)$$

$$e^{P(t)}x' + e^{P(t)}p(t)x = e^{P(t)}q(t) \tag{13.4}$$

Integration by parts allows us to combine the left-hand side of (13.4) into one derivative:

$$\left(e^{P(t)}x\right)' = e^{P(t)}q(t).$$

Therefore, integrating the entire equation gives us the analytical solution formula for the first-order linear non-homogeneous scalar equation (13.2):

$$e^{P(t)}x = \int_{t_0}^{t} e^{P(s)}q(s)\,ds + C$$

$$x(t) = e^{-P(t)}\left(\int_{t_0}^{t} e^{P(s)}q(s)\,ds + C\right) \tag{13.5}$$

where:

$$C = x(t_0)$$

Let us take a look at the following straightforward example to get ourselves familiar with this solution procedure:

$$x' - x = e^{2t}. \tag{13.6}$$

DOI: 10.1201/9781032711126-17

The integrating factor in this case is:

$$h(t) = e^{P(t)} = e^{\int_{t_0}^{t} -1ds} = e^{-t+t_0}. \quad 13.7$$

Now, we multiply the integrating factor (13.7) to the ODE (13.6):

$$\left(e^{-t}x\right)' = e^{t},$$

then integrate this equation, we get:

$$e^{-t}x = e^{t} + C,$$

therefore:

$$x = e^{2t} + Ce^{t}$$

where:

$$C = x(0) - 1$$

One-Pool Model

Now, we consider a one-pool model, that is, the vector X in (13.1) becomes a scalar, then (13.1) becomes a first-order linear non-homogeneous scalar equation (13.2) with

$$p(t) = -\xi(t)AC, \quad q(t) = Bu(t).$$

where $X(t)$, A, C, and B are all scalars, instead of matrices. We can apply the solution formulae (13.5) to solve the one-pool terrestrial carbon cycle system model (13.1), we have that:

$$X(t) = e^{-P(t)}\left(\int_{t_0}^{t} e^{P(s)}Bu(s)ds + C\right)$$

where:

$$P(t) = \int_{t_0}^{t} -\xi(s)ACds, \quad C = X(t_0)$$

Homogeneous Nonautonomous ODEs System

Before moving on to study the analytical solution for the general n-pool model (13.1), we first prepare ourselves with the solution to an initial value problem of a homogeneous nonautonomous ODEs system in the following form:

$$\begin{cases} X' = A(t)X \\ X(t_0) = e_i \end{cases} \quad 13.8$$

where:

$$e_i = \Big(\underbrace{0,\cdots,0}_{i-1 \text{ copies of } 0}, 1, \underbrace{0,\cdots,0}_{n-i \text{ copies of } 0}\Big)^T$$

is an n-vector with only the ith entry equal to 1, and all the other entries equal to 0. The term "homogeneous" means that every term in Equation 13.8 includes X, and the term "nonautonomous" means that X' depends on the independent variable t explicitly, i.e., the right-hand side of (13.8) contains the term $A(t)$ which is a function of t. Note that system (13.8) is one system with n different initial conditions. If we solve (13.8) for $i = 1, 2, \cdots, n$, we will get solutions $X^{(1)}$, $X^{(2)}$, $\cdots X^{(n)}$ corresponding to the n different initial conditions e_i's ($i = 1, 2, \cdots, n$), and this set of solutions forms a fundamental matrix:

$$\Phi(t) = \left[X^{(1)}X^{(2)}\cdots X^{(n)}\right] \quad 13.9$$

of the homogeneous nonautonomous ODEs System:

$$X' = A(t)X. \quad 13.10$$

Non-homogeneous Nonautonomous ODEs System

Now we consider the following nonhomogeneous nonautonomous ODEs system:

$$X' = A(t)X + g(t). \quad 13.11$$

The difference between Equation 13.11 and Equation 13.10 is that (13.11) includes a non-homogeneous term $g(t)$, i.e., a term which does not include X. That is why Equation 13.11 is called a non-homogeneous nonautonomous ODEs system. To look for the solution of Equation 13.11, we will need to employ the fundamental matrix $\Phi(t)$ given in Equation 13.9 of the corresponding homogeneous system (13.10). Assume that the solution of Equation 13.11 has the following form:

$$X = \Phi(t)Q(t) \quad 13.12$$

where:

$$Q(t) = \begin{pmatrix} q_1(t) \\ \vdots \\ q_n(t) \end{pmatrix}$$

is a n-vector, i.e., the solution is a linear combination of the basis of the solution space of the corresponding homogeneous system (13.10). Then we plug the solution formula (13.12) into Equation 13.11, we will get:

$$\Phi'(t)Q + \Phi(t)Q' = A\Phi Q + g. \quad 13.13$$

Since $\Phi(t)$ is a fundamental matrix of Equation 13.10, that means that every column of $\Phi(t)$ is a solution of Equation 13.10, therefore we have that the fundamental matrix $\Phi(t)$ also satisfies Equation 13.10, i.e.:

$$\Phi'(t) = A\Phi(t). \qquad 13.14$$

Equation 13.14 allows us to cancel the $\Phi'(t)Q$ and $A\Phi(t)Q$ terms on the left and right hand sides of Equation 13.13, then Equation 13.13 becomes:

$$\Phi(t)Q' = g,$$

Equivalently:

$$Q' = \Phi^{-1}(t)g(t)$$

because matrix Φ as a fundamental matrix, is clearly invertible. To solve for Q, we integrate this equation, and get:

$$Q = \int_{t_0}^{t} \Phi^{-1}(s)g(s)ds + C$$

Therefore, the solution to the non-homogeneous nonautonomous ODEs system (13.11) is:

$$X(t) = \Phi(t)Q = \Phi(t)C + \Phi(t)\int_{t_0}^{t} \Phi^{-1}(s)g(s)ds \qquad 13.15$$

where:

$$C = \Phi^{-1}(t_0)X(t_0)$$

N-Pool Model

The n-pool model:

$$X' = \xi(t)ACX + Bu(t)$$

$$X(0) = X_0 \qquad 13.16$$

is in principle a non-homogeneous nonautonomous ODEs system (13.11) with:

$$A(t) = \xi(t)AC$$

$$g(t) = Bu(t).$$

Therefore, we can apply the solution formulae (13.15) to solve (13.16):

$$X(t) = \Phi(t)C + \Phi(t)\int_{t_0}^{t} \Phi^{-1}(s)g(s)ds = \underbrace{e^{AC\int_0^t \xi(\sigma)d\sigma}}_{(a)} X_0 + \left(\underbrace{\int_0^t e^{AC\int_s^t \xi(\sigma)d\sigma} u(s)ds}_{(d)} \right) B \qquad 13.17$$

Notice that, in this case, the fundamental matrix of the corresponding homogeneous system:

$$X' = \xi(t)ACX$$

is:

$$\Phi(t) = e^{AC\int_0^t \xi(\sigma)d\sigma}. \qquad 13.18$$

The calculation of the matrix exponential $e^{AC\int_0^t \xi(\sigma)d\sigma}$ is rather complicated. In principle, a matrix exponential e^A is defined in the same way as the scalar exponential:

$$e^x = \sum_{n=0}^{\infty} \frac{x^n}{n!}$$

to be:

$$e^A = \sum_{n=0}^{\infty} \frac{A^n}{n!}.$$

If the matrix:

$$A = diag(a_i)_{i=1}^{n}$$

is a diagonal matrix with $\{a_i\}_{i=1}^{n}$ as the diagonal entries, then,

$$e^A = diag(e^{a_i})_{i=1}^{n}.$$

If the matrix A is diagonalizable by an invertible matrix P (for example, the eigenvector matrix of A), i.e.,

$$A = P\Lambda P^{-1}$$

where Λ is a diagonal matrix, then the matrix exponential becomes

$$e^A = Pe^{\Lambda}P^{-1}.$$

From this formulation, we see that the essential step to calculate a matrix exponential is to diagonalize that matrix. Now, we apply this essential step to derive the details in the solution formulae (13.17). In order to calculate terms (a) and (d) in (13.17), we first introduce an auxiliary function:

$$\phi(t) = e^{ACt}. \qquad 13.19$$

To calculate this matrix exponential, we first find the eigenvalues of the matrix product AC:

$$\lambda_i, \; i = 1, \cdots, n$$

and a complete set of orthonormal eigenvectors of AC:

$$v_i, \; i = 1, \cdots, n$$

and define the eigenvector matrix:

$$V = (v_1, \cdots, v_n)$$

so that the matrix product AC can be diagonalized by the eigenvector matrix V, i.e.:

$$AC = Vdiag(\lambda_i)_{i=1}^{n} V^{-1},$$

which also gives that:

$$ACt = Vdiag(\lambda_i t)_{i=1}^{n} V^{-1},$$

then $\phi(t)$ given in (13.19) can be calculated by:

$$\phi(t) = e^{ACt} = Vdiag(e^{\lambda_i t})_{i=1}^{n} V^{-1} \qquad 13.20$$

Since the fundamental matrix $\Phi(t)$ given in (13.18) can be written as:

$$\Phi(t) = \phi(f(t)), \text{ where } f(t) = \int_0^t \xi(\sigma)d\sigma$$

this matrix exponential is therefore given as:

$$\begin{aligned}\Phi(t) &= e^{AC\int_0^t \xi(\sigma)d\sigma} = \phi(f(t)) \\ &= Vdiag(e^{\lambda_i f(t)})_{i=1}^{n} V^{-1} \\ &= Vdiag\left(e^{\lambda_i \int_0^t \xi(\sigma)d\sigma}\right)_{i=1}^{n} V^{-1}.\end{aligned}$$

hence, terms (a) and (d) in (13.17) are calculated as following:

$$(a) = e^{AC\int_0^t \xi(\sigma)d\sigma} X_0 = Vdiag(e^{\lambda_i f(t)})_{i=1}^{n} V^{-1} X_0 \qquad 13.21$$

and:

$$\begin{aligned}(\mathrm{d}) &= \left(\int_0^t e^{AC\int_s^t \xi(\sigma)d\sigma} u(s)ds\right)B \\ &= \int_0^t V\left(diag\left(e^{\lambda_i \int_s^t \xi(\sigma)d\sigma}\right)_{i=1}^{n}\right)V^{-1}u(s)dsB \\ &= V\int_0^t \left(diag\left(e^{\lambda_i \int_s^t \xi(\sigma)d\sigma}\right)_{i=1}^{n}\right)u(s)dsV^{-1}B \\ &= Vdiag\left(\int_0^t e^{\lambda_i \int_s^t \xi(\sigma)d\sigma} u(s)ds\right)_{i=1}^{n} V^{-1}B\end{aligned} \qquad 13.22$$

This finishes the calculation of the solution (13.17) to the n-pool model (13.16). However, we want readers to note that the calculation of the eigenvalues λ_i's and eigenvectors v_i's, which ultimately form the eigenvector matrix V of the carbon transfer coefficients matrix product AC, is far from trivial in most of the practical cases.

Mathematica Calculation for the Analytical Solution of a Three-Pool Model

We consider a three-pool model with:

$$A = \begin{pmatrix} -1 & 0 & 0 \\ 0 & -1 & 0 \\ 0 & 1 & -1 \end{pmatrix} \quad C = \begin{pmatrix} c_1 & 0 & 0 \\ 0 & c_2 & 0 \\ 0 & 0 & c_3 \end{pmatrix} \quad B = \begin{pmatrix} b \\ 0 \\ 0 \end{pmatrix}$$

$$\xi(t) = k_1 e^{b_1 t} \qquad u(t) = k_2 e^{b_2 t}, \qquad 13.23$$

and we use the mathematical software Mathematica to carry out the following calculations. The matrix product AC is:

$$AC = \begin{pmatrix} -c_1 & 0 & 0 \\ c_1 & -c_2 & 0 \\ 0 & c_2 & -c_3 \end{pmatrix}$$

The eigenvalues of AC are:

$$\lambda_1 = -c_1,\ \lambda_2 = -c_2,\ \lambda_3 = -c_3$$

and the eigenvector matrix of AC is:

$$V = \begin{pmatrix} \frac{(c_1 - c_2)(c_1 - c_3)}{c_1 c_2} & 0 & 0 \\ \frac{-c_1 + c_3}{c_2} & -1 + \frac{c_3}{c_2} & 0 \\ 1 & 1 & 1 \end{pmatrix}$$

Let the initial value X_0 be:

$$X_0 = \begin{pmatrix} x_1 \\ x_2 \\ x_3 \end{pmatrix}$$

then term (a) given in (13.21) has three components:

$$a_1 = e^{-\frac{c_1(-1+e^{b_1 t})k_1}{b_1}} x_1$$

$$a_2 = \frac{e^{\frac{(c_1+c_2)(-1+e^{b_1 t})k_1}{b_1}}\left(-c_1 e^{\frac{c_2(-1+e^{b_1 t})k_1}{b_1}} x_1 + e^{\frac{c_1(-1+e^{b_1 t})k_1}{b_1}}\left(-c_2 x_2 + c_1(x_1+x_2)\right)\right)}{c_1 - c_2}$$

and a_3 has a much longer expression, and is omitted here. To facilitate the calculation of term (*d*) given in (13.22), we assume that $b_2 = b_1 = b$, then term (*d*) given in (13.22) has three components:

$$d_1 = \frac{b\left(1 - e^{-\frac{c_1(-1+e^{b_1 t})k_1}{b_1}}\right)k_2}{c_1 k_1}$$

$$d_2 = \frac{b\left(c_1 - c_1 e^{-\frac{c_2(-1+e^{b_1 t})k_1}{b_1}} + c_2\left(-1 + e^{-\frac{c_1(-1+e^{b_1 t})k_1}{b_1}}\right)\right)k_2}{(c_1 - c_2)c_2 k_1}$$

$$d_3 = \frac{bc_1c_2^2\left(-\frac{(-1+\frac{c_3}{c_2})\left(-1+e^{-\frac{c_1(-1+e^{b_1 t})k_1}{b_1}}\right)}{c_1} - \frac{(c_1-c_3)\left(-1+e^{\frac{c_2(-1+e^{b_1 t})k_1}{b_1}}\right)}{c_2^2} - \frac{(1-\frac{c_1}{c_2})\left(-1+e^{-\frac{c_3(-1+e^{b_1 t})k_1}{b_1}}\right)}{c_3}\right)k_2}{(c_1-c_2)(c_1-c_3)(c_2-c_3)k_1}$$

Even for a seemingly simple three-pool model, the computation of the analytical solution of this model is already very complicated. For models with more pools, the calculation could be much more challenging for many cases.

STABILITY

Instantaneous Steady State

The instantaneous steady-state of (13.1):

$$X' = \xi(t)ACX + Bu(t) \qquad 13.24$$

is defined as:

$$X_{ss}(t) = -\xi^{-1}(t)u(t)C^{-1}A^{-1}B(t) \qquad 13.25$$

The instantaneous steady state provides the $X(t)$ with zero rate of change at time t.

Instantaneous Steady State for a Three-Pool Model

Let us re-visit the three-pool model (13.23) given in the previous section. A simple calculation shows that the inverse of the carbon transfer coefficients matrices AC is:

$$\text{Inverse}(AC) = \begin{pmatrix} -1/c_1 & 0 & 0 \\ -1/c_2 & -1/c_2 & 0 \\ -1/c_3 & -1/c_3 & -1/c_3 \end{pmatrix}$$

and therefore the instantaneous steady state solution is:

$$X_{ss}(t) = -\xi^{-1}(t)u(t)\text{Inverse}(AC)B(t) = \begin{pmatrix} \frac{bk_2}{c_1k_1} \\ \frac{bk_2}{c_2k_1} \\ \frac{bk_2}{c_3k_1} \end{pmatrix}$$

The Mathematica computational results show that all the eigenvalues $(\lambda_i)_{i=1}^{n}$ of the matrix AC are strictly negative, therefore the instantaneous steady state solution (13.25) is asymptotically stable. We numerically investigate an example to verify that the solution $X(t)$ does converge to the instantaneous steady-state solution $X_{ss}(t)$ as $t \to \infty$. To measure the convergence, we need to study the following function:

$$\begin{aligned} \text{distance}(t) &= X(t) - X_{ss}(t) \\ &= (a) + (d) - X_{ss}(t) \end{aligned} \qquad 13.26$$

where (*a*), (*d*), and $X_{ss}(t)$ are given in (13.21, 13.22, 13.25).

In this numerical investigation, we choose $c_1 = 1/10$, $c_2 = 3/10$, $c_3 = 2/10$, $b = 1$, $k_1 = 1$, $k_2 = 1$, $b_1 = 1$, and the initial value $X_0 = (5/10, 2/10, 3/10)$ in the three-pool model (13.23). Figure 13.1 shows the three components of the solution $X(t)$, the instantaneous steady-state solution $X_{ss}(t)$, and the distance given in (13.26).

An interesting question to ask is how the parameters affect the rate of convergence to the instantaneous steady state solution. This question will be left to the readers to explore.

Global Attractor

As the name suggests, a global attractor attracts all the solutions regardless of the initial profiles. We will introduce the definition of the global attractor, how to numerically estimate the global attractor, and a mathematical study of the convergence of the solutions to the global attractor as $t \to \infty$. In the end of this section, we will also give the condition when the global attractor and the instantaneous steady-state solution given in (13.25) converge to each other.

Consider a non-homogeneous nonautonomous system in the general form:

$$X'(t) = A(t)X(t) + b(t) \qquad 13.27$$

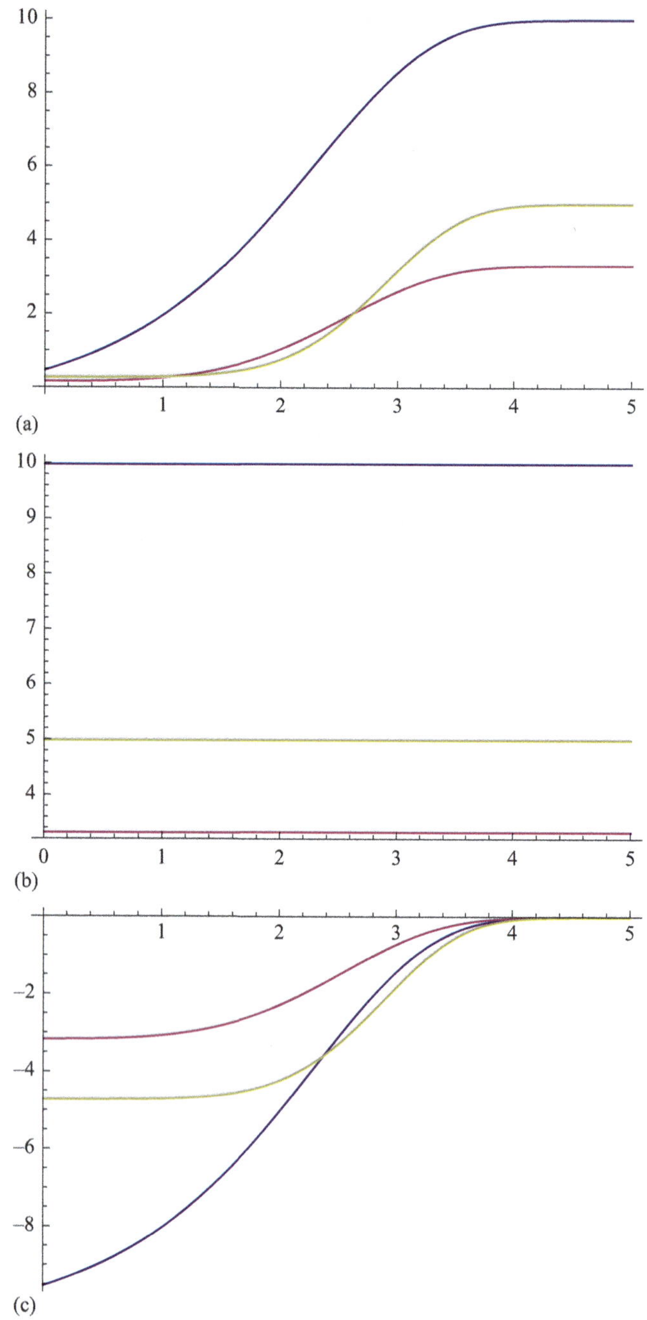

FIGURE 13.1 The three panels show (a) the solution $X(t)$, (b) the equilibrium solution $X_{ss}(t)$, and (c) the distance versus time, for $0 \le t \le 5$. The blue, red, and green curves represent the first, second, and third components, respectively.

where $A(t)$ has the block matrix form:

$$A(t) = \begin{pmatrix} A_{11}(t) & 0 \\ A_{21}(t) & A_{22}(t) \end{pmatrix}$$

$$A_{11}(t) \in R^{d_1 \times d_1}, 0 \in R^{d_1 \times d_2}, A_{21}(t) \in R^{d_2 \times d_1},$$
$$A_{22}(t) \in R^{d_2 \times d_2}, \text{and } d_1 + d_2 = d.$$

Assume that:

1. $A_{11}(t)$ is a lower triangular matrix
2. $\|b(t)\| \le B$ for some $B > 0$ (i.e., $b(t)$ is bounded)
3. $\exists \delta < 0$, such that
   - $a_{ii}(t) \le \delta$, for $i = 1, \cdots, d_1$
   - $a_{ii}(t) < 0$, for $i = d_1 + 1, \cdots, d$ (i.e., $A_{22}(t)$ has negative diagonal entries)
   - $|a_{ii}(t)| + \delta \ge \sum_{j \ne i, j > d_1} |a_{ij}(t)|$, for $i = d_1 + 1, \cdots, d$, (i.e., $A_{22}(t)$ is diagonally dominant)

Then:

$$\| \Phi(t,s) \| \le Ke^{\delta(t-s)} \quad 13.28$$

for some $K > 0$ and $t > s$, where $\Phi(t, s)$ is the fundamental matrix of:

$$X'(t) = A(t)X(t)$$

i.e., the solution of the general nonautonomous equation (13.27) can be written as:

$$X(t) = \Phi(t,s)X(s) + \int_s^t \Phi(t,\tau)b(\tau)d\tau$$

Let:

$$\mu(t) = \int_{-\infty}^{t} \Phi(t,s)b(s)ds \quad 13.29$$

then $\mu(t)$ is the unique global attractor of the nonautonomous system (13.27).

Notice that based on (13.28, 13.29) and the assumption on boundedness of $b(t)$,

$$\begin{aligned} \| \mu(t) \| &\le \int_{t-T}^{t} \| \Phi(t,s)b(s) \| ds \\ &+ \int_{-\infty}^{t-\tau} \| \Phi(t,s)b(s) \| ds \\ &\le \int_0^T \| \Phi(t,t-\sigma)b(t-\sigma) \| d\sigma \\ &+ \int_{-\infty}^{t-\tau} Ke^{\delta(t-s)} \| b(s) \| ds \\ &\le \int_0^T \| \Phi(t,t-\sigma)b(t-\sigma) \| d\sigma \\ &+ KBe^{\delta T}/|\delta| \\ &\xrightarrow{T\to\infty} \int_0^T \| \Phi(t,t-\sigma)b(t-\sigma) \| d\sigma \end{aligned} \quad 13.30$$

Equation 13.30 gives a way to numerically estimate the global attractor $\mu(t)$.

All the solutions (regardless of the initial conditions) will converge to $\mu(t)$ as $t \to \infty$. This is because:

$$X(t)-\mu(t) = \Phi(t,s)\big(X(s)-\mu(s)\big)$$

Therefore:

$$\| X(t)-\mu(t) \| \le Ke^{\delta(t-s)} \| \big(X(s)-\mu(s)\big) \| \overset{t\to\infty}{\longrightarrow} 0, \text{ since } \delta < 0$$

The Global Attractor of the N-Pool Model

Consider:

$$X' = \xi(t)ACX + Bu(t), \tag{13.31}$$

its global attractor is:

$$\begin{aligned}
\mu(t) &= \int_{-\infty}^{t} \Phi(t,s)Bu(s)ds \\
&= \int_{-\infty}^{t} e^{AC\int_s^t \xi(\sigma)d\sigma}(-1)\xi(s)ACX_{ss}(s)ds \\
&= \left[e^{AC\int_s^t \xi(\sigma)d\sigma} X_{ss}(s) \right]_{s=-\infty}^{s=-t} - \int_{-\infty}^{t} e^{AC\int_s^t \xi(\sigma)d\sigma} X'_{ss}(s)ds \\
&= X_{ss}(t) - \int_{-\infty}^{t} e^{AC\int_s^t \xi(\sigma)d\sigma} X'_{ss}(s)ds
\end{aligned}$$

Therefore:

$$\begin{aligned}
\mu(t) - X_{ss}(t) &= -\int_{-\infty}^{t} e^{AC\int_s^t \xi(\sigma)d\sigma} X'_{ss}(s)ds \\
&= -\int_{-\infty}^{T} e^{AC\int_s^t \xi(\sigma)d\sigma} X'_{ss}(s)ds \\
&\quad - \int_{T}^{t} e^{AC\int_s^t \xi(\sigma)d\sigma} X'_{ss}(s)ds
\end{aligned}$$

GENERAL STABILITY STATEMENTS

CASE 1 CONSIDER A HOMOGENEOUS SYSTEM OF ODES WITH CONSTANT COEFFICIENT MATRIX:

$$X' = AX.$$

Let λ_j, $j = 1, \cdots, n$ counting multiplicity be the eigenvalues of matrix A, and let $\tilde{X}$ be the equilibrium state.

- if $\mathrm{Re}\lambda_j < 0$ for all $j = 1, \cdots, n$, then $\tilde{X}$ is asymptotically stable.
- if $\mathrm{Re}\lambda_j \le 0$ for all $j = 1, \cdots, n$, and the eigenvalues with real part equals to zero are simple, then $\tilde{X}$ is stable.
- if there exists λ_j, such that $\mathrm{Re}\lambda_j = 0$, but is not simple, then $\tilde{X}$ is unstable.
- if there exists λ_j, such that $\mathrm{Re}\lambda_j > 0$, then $\tilde{X}$ is unstable.

CASE 2 CONSIDER A HOMOGENEOUS SYSTEM OF NONAUTONOMOUS ODES:

$$X' = AX + B(t)$$

where $B(X)$ is a continuous function. Let λ_j, $j = 1, \cdots, n$ counting multiplicity be the eigenvalues of matrix A, and let $\tilde{X}$ be the equilibrium state.

- if $\mathrm{Re}\lambda_j < 0$ for all $j = 1, \cdots, n$, and $\lim_{t\to\infty}\|B(t)\| = 0$, then $\tilde{X}$ is asymptotically stable.
- if $\mathrm{Re}\lambda_j \le 0$ for all $j = 1, \cdots, n$, and the eigenvalues with real part equals to zero are simple, and $\int_{t_0}^{\infty}\|B(t)\| < \infty$ then $\tilde{X}$ is stable.
- if there exists λ_j, such that $\mathrm{Re}\lambda_j > 0$, and $\lim_{t\to\infty}\|B(t)\| = 0$, then $\tilde{X}$ is unstable.

CASE 3 CONSIDER THE FOLLOWING GENERAL SYSTEM:

$$X' = AX + B(t)X + f(t, X)$$

where $B(t)$ is a continuous function. Let X_0 be an equilibrium solution. If:

$$\lim_{t\to\infty} \| B(t) \| = 0$$

and $f(t)$ is continuous in t, and has continuous partial derivative in x:

$$\lim_{x\to 0} \frac{\| f(t,x) \|}{x} = 0 \text{ uniformly in t}$$

then:

a If $\text{Re}\lambda_j < 0$, where λ_j are the eigenvalues of A, then $X = X_0$ is asymptotically stable.
b If there exists an eigenvalue λ_j, such that $Re\lambda_j > 0$, then $X = X_0$ is unstable.

SUGGESTED READING

C. H. Edwards, D. E. Penney, and D. T. Calvis. 2014. *Differential Equations and Boundary Value Problems (5/e)*, Pearson. ISBN 9780321796981

QUIZ

Find the general solution of the following ordinary differential equations and system of ODEs.

$$y' + y = 2, y(0) = 0$$

$$xy' + 2y = 3x, y(1) = 5$$

$$2xy' - 3y = 9x^3$$

$$\begin{pmatrix} x' \\ y' \end{pmatrix} = \begin{pmatrix} 0 & 8 \\ -2 & 0 \end{pmatrix} \begin{pmatrix} x \\ y \end{pmatrix}$$

$$\begin{pmatrix} x' \\ y' \end{pmatrix} = \begin{pmatrix} 0 & -1 \\ 10 & -7 \end{pmatrix} \begin{pmatrix} x \\ y \end{pmatrix}$$

$$\begin{pmatrix} x' \\ y' \end{pmatrix} = \begin{pmatrix} -3 & 2 \\ -3 & 4 \end{pmatrix} \begin{pmatrix} x \\ y \end{pmatrix}$$

$$\begin{pmatrix} x' \\ y' \end{pmatrix} = \begin{pmatrix} 1 & 2 \\ 2 & 1 \end{pmatrix} \begin{pmatrix} x \\ y \end{pmatrix} + \begin{pmatrix} 3 \\ -2 \end{pmatrix}$$

$$\begin{pmatrix} x' \\ y' \end{pmatrix} = \begin{pmatrix} 9 & 1 \\ -8 & -2 \end{pmatrix} \begin{pmatrix} x \\ y \end{pmatrix} + \begin{pmatrix} 2 \\ t \end{pmatrix} e^t$$

$$\begin{pmatrix} x' \\ y' \end{pmatrix} = \begin{pmatrix} 1 & -5 \\ 1 & -1 \end{pmatrix} \begin{pmatrix} x \\ y \end{pmatrix} + \begin{pmatrix} 2 \sin t \\ -3 \cos t \end{pmatrix}$$

Find the critical point(s) of the following systems, and determine whether it is asymptotically stable, stable, or unstable.

$$\begin{pmatrix} x' \\ y' \end{pmatrix} = \begin{pmatrix} -2 & 1 \\ 1 & -2 \end{pmatrix} \begin{pmatrix} x \\ y \end{pmatrix}$$

$$\begin{pmatrix} x' \\ y' \end{pmatrix} = \begin{pmatrix} 1 & -2 \\ 5 & -1 \end{pmatrix} \begin{pmatrix} x \\ y \end{pmatrix}$$

$$\begin{pmatrix} x' \\ y' \end{pmatrix} = \begin{pmatrix} 1 & -2 \\ 1 & 4 \end{pmatrix} \begin{pmatrix} x \\ y \end{pmatrix} + \begin{pmatrix} -8 \\ 10 \end{pmatrix}$$

$$\begin{pmatrix} x' \\ y' \end{pmatrix} = \begin{pmatrix} 4 & -5 \\ 5 & -4 \end{pmatrix} \begin{pmatrix} x \\ y \end{pmatrix} + \begin{pmatrix} 3 \\ 6 \end{pmatrix}$$

$$\begin{cases} x' = x - y \\ y' = x^2 - y \end{cases}$$

$$\begin{cases} x' = xy - 2 \\ y' = x - 2y \end{cases}$$

14 Semi-Analytic Spin-Up (SASU) of Coupled Carbon-Nitrogen Cycle Models

Xingjie Lu
Sun Yat-sen University, Guangzhou, China
Jianyang Xia
East China Normal University, Shanghai, China

This chapter introduces the concept of spin-up in land biogeochemical models. Spin-up is a standard initialization procedure, which is commonly included in protocols of model intercomparison projects. Conventional spin-up, repeatedly running the model with recursive environmental forcing to achieve a steady state, is very time consuming. Semi-analytic spin-up (SASU) enabled by the matrix representation significantly improves the computational efficiency of spin-up for land biogeochemical models. The implementation of SASU in the Community Atmosphere and Biosphere Land Exchange (CABLE) model is demonstrated. The SASU method has been proven to save about 92.4% and 86.6%, respectively, of the computational time for spin-up of the global carbon-only and coupled carbon-nitrogen models.

WHAT IS SPIN-UP?

Spin-up is a standard initialization procedure used by land surface models, atmosphere models, ocean models, or earth system models. Generally speaking, spin-up provides initial values of state variables, which are very critical to the definite solution. For instance, you might have heard of the so-called butterfly effect in weather/climate forecasting: slightly different initial values might cause different or even completely opposite results.

The initialization procedure is an essential step in simulation of land biogeochemical models. Model intercomparison projects are often carried out in order to elaborate understanding of carbon cycle dynamics and characterize intermodel uncertainty. These exercises require a standardized protocol in order to ensure that results from different participating models can be compared on an equal basis. Model intercomparison projects have for instance been undertaken to assess the anthropogenic impact on land biogeochemical cycles. The industrial revolution triggered much of the current anthropogenic impact, such as increasing fossil fuel emissions, land use change, and increasing nitrogen deposition. It represents the starting point of a period when humans have played a more important role than ever before in impacting the Earth's geology and ecosystems. Land biogeochemical cycles are experiencing a transition period unique in the history of the Earth. In the protocol of model intercomparison projects, the simulation of the transition period in biogeochemical cycles from industrial revolution to present-day is often called the historical simulation. One of the primary goals for the historical simulation is to hindcast the ecosystem changes in response to anthropogenic activity since the onset of industrialization around the 19th century. Initial values matter to the trend of ecosystem carbon storage. However, determining these initial values can be a tricky issue when we have little information on the initial state. Direct observations for variables such as soil temperature, moisture, carbon, and nitrogen storage are not generally available until very recent times. Thus, in these protocols, *steady state* is commonly used to determine initial values for the historical simulation.

Steady state is achievable by most land biogeochemical models as long as we run the model long enough. Chapter 1 provided the theoretical foundation; that is, for given forcing, ecosystem carbon storage tends to grow towards the same steady state regardless of the initial carbon storage once a model structure is defined and parameter values are given. The unique steady state does not rely on the initial value, as the land carbon cycle converges to a steady state that is determined by carbon input and residence time. Thus, the procedure of generating steady state can be standardized across different models.

Conventionally, the steady state of carbon storage can be achieved by repeatedly running the model with recursive environmental forcing. Most model intercomparison projects include a protocol to achieve the steady state. For examples, Multiscale Synthesis and Terrestrial Model Intercomparison Project (MsTMIP) protocol requires that the spin-up uses a random climate driver data package, constant 1801 land cover, constant pre-industrial atmospheric CO_2 concentration, and

DOI: 10.1201/9781032711126-18

constant nitrogen deposition to force the model. Under such forcing conditions, ecosystem carbon storage will ultimately reach a quasi-steady state. To detect whether steady state has been achieved, MsTMIP specifies that the 100-yr mean interannual change in total ecosystem carbon stocks for consecutive years must be below 1 gC m^{-2} yr^{-1} for 95% of grid cells. The carbon stocks accumulated at steady state are used as the initial values in the transient historical simulation from 1801 to 2010. This approach, which takes advantage of a model's native dynamics to drive the system towards the steady state, is called native dynamics spin-up (ND), and is the most widely used by current models.

DEVELOPMENT OF SPIN-UP APPROACHES

Several other spin-up approaches have also been developed. The motivation to develop alternative spin-up approaches to ND in land biogeochemical modeling is to improve the computational efficiency of the spin-up. For most models, the ND approach usually takes hundreds to thousands of simulation years to reach the steady state, which is computationally very expensive. Even this poor level of efficiency is becoming much lower as ongoing development makes models more and more complex. For example, vertically-resolved soil carbon modules are developed to represent carbon vertical mixing and the soil freeze-thaw impact on soil decomposition in permafrost regions. While accounting for these additional processes better represents the nature of the heterogeneity, the extremely slow turnover rates for deep soil carbon in permafrost regions substantially slow down the spin-up. A more efficient spin-up approach is urgently needed.

Spin-up approaches that modify ND but allow the state variables to naturally accumulate to the steady state without any arbitrary adjustments to the state variables themselves are called *ad hoc* approaches. A punctuated nitrogen addition approach (PN) manipulates the rate of new mineral nitrogen addition (Thornton et al., 2002; Thornton and Rosenbloom, 2005). PN is proposed as a solution to reduce spin-up times based on the observation that the time taken by a coupled carbon-nitrogen biogeochemical model to reach a steady state under ND is mainly controlled by the new nitrogen addition. PN is faster than ND but the higher nitrogen inputs also result in larger carbon pools at steady state. Therefore, an additional ND needs to be carried out after PN, so that PN results conform to ND results. The accelerated decomposition approach (AD) was first implemented into the Biome-BGC model (Thornton and Rosenbloom, 2005). AD arbitrarily increases the decomposition rate during spin-up to allow the system to approach a steady state faster. AD is informed by the experience that scaling of litter and soil carbon decomposition produce a new steady state, which is linearly related to the scaling constant (Thornton and Rosenbloom, 2005). AD is much faster than ND, but it also generates lower steady state carbon and nitrogen pools, because interactions with the soil passive pool become much stronger. To achieve the final steady state, additional ND after AD is required, which is called the post-AD process. The post-AD process is sometimes much longer than AD itself. AD has been implemented into CLM version 4 (Koven et al., 2013) and the following versions (CLM4.5 and CLM5) with vertically-resolved soil carbon.

In contrast to *ad hoc* approaches, spin-up approaches allowing arbitrary change in state variables are called generalized optimization approaches. The semi-analytic spin-up approach (SASU) updates the state variables with steady state estimates from an analytic solution of the model. The SASU approach was firstly implemented in the Community Atmosphere-Biosphere-Land Exchange (CABLE) model for carbon and nitrogen cycle spin-up (Xia et al., 2012). For carbon-cycle-only simulations, SASU saves well over 90% of the computational cost for CABLE site-level and global simulations. For carbon-nitrogen coupled simulation, SASU saves over 80% of the computational cost. The SASU approach has the additional advantage that it enables parameter sensitivity analysis of the biogeochemical parameters (Huang et al., 2018b), as well as data assimilation for the estimation of these parameters (Tao et al., 2020), which was prohibitive due to large computational cost before SASU was proposed.

Matrix representations have been derived for many land biogeochemical models, in part to enable them to use the SASU approach for spin-up. In the last decade matrix representations have been published for TECO (Jiang et al., 2017; Luo et al., 2017), CABLE (Xia et al., 2013), LPJ-GUESS (Ahlström et al., 2015), ORCHIDEE (Huang et al., 2018b), CLM3.5 (Hararuk et al., 2015; Hararuk et al., 2014; Rafique et al., 2016), CLM4 (Rafique et al., 2017), CLM4.5 (Huang et al., 2018a), and CLM5 (Lu et al., 2020).

The accelerated spin-up (ASU) approach is implemented in the Terrestrial Ecosystem Model (TEM). ASU is similar to SASU and also belongs to the generalized optimization approach, but is applied to the traditional as opposed to the matrix form of the model. ASU numerically solves the model's steady state considering the seasonal cycle of carbon and nitrogen storage. Qu et al., (2018) found that ASU saved 90% and 85% of computational costs for TEM site-level and North American regional simulations, respectively.

SEMI-ANALYTIC SPIN-UP

The SASU approach is built upon the mathematical foundation of biogeochemical cycles in terrestrial ecosystems, introduced in Chapter 1. The biogeochemical cycling of carbon in an ecosystem is usually initiated with plant photosynthesis, which transfers CO_2 from the atmosphere into an ecosystem. The carbon assimilated through photosynthesis is partitioned into compartments of plant biomass, such as leaf, root, and woody biomass. Plant biomass lost through phenological turnover, mortality, or damage by herbivores etc. becomes litter, entering metabolic, structural, and coarse woody debris (CWD) litter pools. The litter carbon is partially released to the atmosphere as CO_2 respired by decomposing microbes, and partially converted to soil organic matter (SOM) in fast, slow, and passive pools (Figure 14.1).

The mean carbon residence time varies greatly among different pools, from several months in leaves and roots to hundreds or thousands of years in some woody tissues and SOM fractions (Torn et al., 1997). These carbon processes

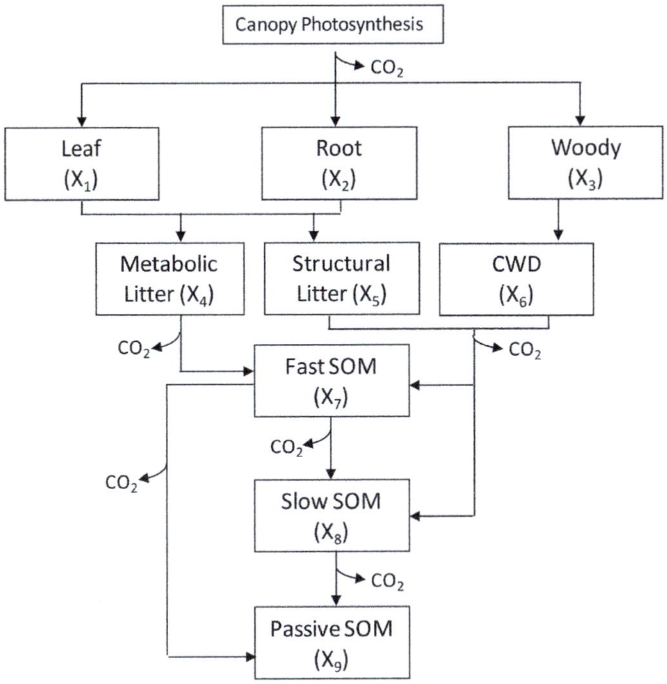

FIGURE 14.1 Diagram of the carbon processes of CABLE model on which model Equations 14.1–14.3 are based. SOM stands for soil organic matter.

in terrestrial ecosystems can be mathematically expressed by the following carbon balance equation in a matrix form (Luo et al., 2017; Luo et al., 2003):

$$\frac{dX(t)}{dt} = B\mu(t) + A\xi(t)KX(t) \quad 14.1$$

Taking the CABLE model (Wang et al., 2011) as one example, $X(t) = (X_1(t), X_2(t), \ldots, X_9(t))^T$ is a 9 × 1 vector describing pool sizes for the nine carbon pools leaf, wood, root, metabolic litter, structural litter, CWD, fast SOM, slow SOM, and passive SOM, respectively. A and K are then 9 × 9 matrices given by:

$$A = \begin{bmatrix} -1 & 0 & 0 & 0 & 0 & 0 & 0 & 0 & 0 \\ 0 & -1 & 0 & 0 & 0 & 0 & 0 & 0 & 0 \\ 0 & 0 & -1 & 0 & 0 & 0 & 0 & 0 & 0 \\ a_{41} & a_{42} & 0 & -1 & 0 & 0 & 0 & 0 & 0 \\ a_{51} & a_{52} & 0 & 0 & -1 & 0 & 0 & 0 & 0 \\ 0 & 0 & 1 & 0 & 0 & -1 & 0 & 0 & 0 \\ 0 & 0 & 0 & a_{74} & a_{75} & a_{76} & -7 & 0 & 0 \\ 0 & 0 & 0 & 0 & a_{85} & a_{86} & a_{87} & -1 & 0 \\ 0 & 0 & 0 & 0 & 0 & 0 & a_{97} & a_{98} & 0 \end{bmatrix} \quad 14.2$$

$$K = \text{diag}(k) \quad 14.3$$

where A denotes the carbon transfer matrix, in which a_{ij} represents the fraction of carbon transfer from pool j to i. The $diag(k)$ is a 9 × 9 diagonal matrix with diagonal entries given by vector $k = (k_1, k_2, \ldots, k_9)^T$, components k_j (j = 1, 2, …, 9) quantify the fraction of carbon left from pool X_j (j = 1, 2, …,9) after each time step. $\xi(t)$ is an environmental scalar accounting for effects of soil type, temperature, and moisture on carbon decomposition. $B = (b_1, b_2, b_3, 0, \ldots, 0)^T$ represents the partitioning coefficients of the photosynthetically fixed carbon into different pools. $\mu(t)$ is the input of carbon via plant photosynthesis. Other models may have a slightly different set of biomass, litter, and soil carbon pools, but in general, Equation 14.1 can adequately summarize C cycle processes in most land models (Luo et al., 2015; Luo et al., 2017; see also Chapter 1).

Equation 14.1 cannot be directly solved to obtain the steady-state values of carbon pools because the environmental scalar $\xi(t)$, ecosystem carbon influx $U(t)$, possibly matrix A, and vector B, vary with time and driving variables. Since carbon influx involves fast processes, its steady-state value U_{SS} can be obtained from a short ND spin-up, which we call initial spin-up. The initial spin-up uses recursive meteorological forcing to drive modeled carbon and nitrogen dynamics. Thus, it is possible to calculate averaged values of the environmental scalar ($\underline{\xi}$), the carbon transfer ($\underline{A}$), and partitioning ($\underline{B}$) coefficients within one cycling period of the meteorological variables. With carbon input at steady state μ_{ss} and the mean values of the time-varying variables ($\underline{\xi}$, $\underline{A}$, and $\underline{B}$), we can analytically calculate the steady-state carbon pool sizes $X_{SS,C}$ by letting the right-hand side of Equation 14.1 equal zero as:

$$X_{SS,C} = -\left(\underline{\xi}\underline{A}K\right)^{-1}\underline{B}\mu_{ss} \quad 14.4$$

We can divide $X_{SS,C}$ by C/N ratio in each pool to obtain steady-state nitrogen pool sizes $X_{SS,N}$. Equation 14.4 stands for the steady state in the absence of soil nitrogen feedback to NPP.

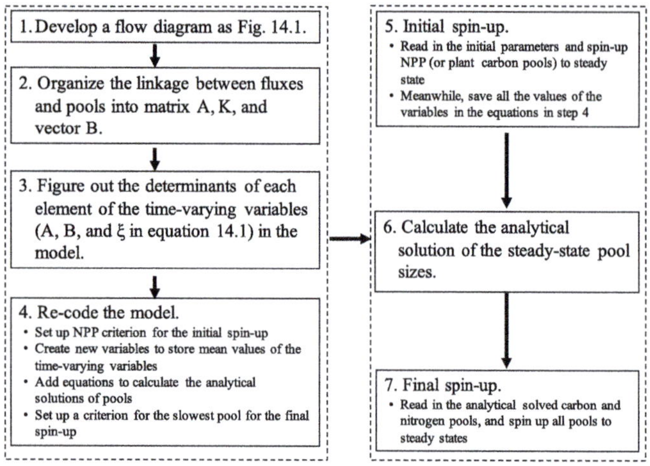

FIGURE 14.2 Procedure of the semi-analytical spin-up (SASU).

Reproduced from Xia et al., (2012).

Final steady state will be achieved by the full procedure, which will be described in the next section.

THE PROCEDURE OF SEMI-ANALYTIC SPIN-UP IN CABLE

The procedure of SASU implementation in a land surface model is demonstrated by the case of CABLE. In general, the procedure consists of four preparation and three execution steps: (1) development of flow diagram; (2) organization of matrix equation; (3) identification of the time-varying elements; (4) coding the analytic solution; and three modeling steps: (5) initial spin-up; (6) analytic solution of steady state; and (7) final ND spin-up. In more detail, these seven steps entail (Figure 14.2):

1 Developing a flow diagram reflecting the linkages between ecosystem carbon pools in the target model structure, as shown in Figure 14.1 (also see Chapters 1-4 of Unit 1).

2 Organizing CABLE into a matrix equation by encoding the linkage between carbon pools and fluxes in the carbon transfer matrices *A* and *K*, and plant carbon partitioning coefficients in vector *B*. Values for non-zero elements in vector *B*, and matrices *A* and *K* were assigned based on parameter values in the original CABLE model.

3 Figuring out how the time-varying elements of *A*, *B*, and ξ in Equation 14.1 are determined in the model.

4 Coding the analytic solution of the biogeochemical steady state in CABLE, which includes two steps: (a) creating new variables to store the mean values of the time-varying parameters; and (b) creating equations to calculate the analytical solutions of each pool according to the structures of matrix *A*, *K*, and vector *B*.

5 Performing the initial spin-up by running the model using repeated meteorological forcing until NPP (or all plant pools) reach steady state (U_{SS}). Run the model until the mean change in NPP over each loop is smaller than 0.01% per year. Meanwhile, the values of all the time-varying parameters in Equation 14.3 are updated by the mean values from initial spin-up.

6 Calculating the analytical solution of the steady states of carbon and nitrogen pools. The steady-state carbon pools are solved by setting carbon influx equal to efflux for each pool (Equation 14.4). Nitrogen pools are obtained by dividing the steady-state carbon pools by their C/N ratios at the end of the initial spin-up.

7 Performing the final spin-up by using the analytically solved carbon and nitrogen pools as initial values until the steady-state criterion for the soil carbon pools is met. Here our steady-state criterion will be that the change in the passive soil carbon pool (ΔC_{pass}) within each simulation year is smaller than 0.5 gC m^{-2} yr^{-1}. This criterion was chosen following a recommendation of Thornton and Rosenbloom (2005). According to the difference in turnover rate, a slower pool needs a longer time to reach steady state during the spin-up. When the criterion for recognizing that steady state has been achieved is small enough the final spin-up is determined by the dynamic of the slowest carbon pool.

COMPUTATIONAL EFFICIENCY

In our example using CABLE, the traditional spin-up method spends 2780 and 5099 simulation years for carbon-only and coupled carbon-nitrogen simulation, respectively, before the change in the slowest carbon pool meets the steady-state criterion (ΔC_{pass} < 0.5 gC m^{-2} yr^{-1}; Figure 14.3). After implementing SASU, the initial spin-up takes 200 simulation years to achieve steady states of plant carbon pools for both the carbon-only model and the coupled carbon-nitrogen

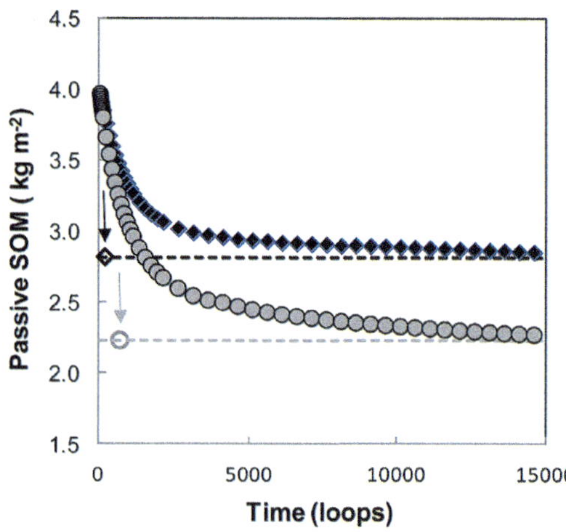

FIGURE 14.3 Dynamics of global mean passive soil carbon pool in carbon-only (filled black diamond) and coupled carbon-nitrogen (filled gray circle) simulations with traditional method and the SASU framework (dotted lines) using the CABLE model. The arrows and open symbols show the times and values estimated with the SASU method for carbon-only (black) and coupled carbon-nitrogen coupled (gray) simulations.

Adapted from Xia et al., (2012).

model. This allows the sizes of all SOM pools to be calculated analytically after this initial spin-up (step 6 above). With the SASU method, all carbon pools in the carbon-only model reach steady states after analytical calculation without a need for any final spin-up (step 7 above; black arrow in Figure 14.3). In the coupled carbon-nitrogen model, SASU needs another 483 simulation years to obtain the steady states of all pools (gray arrow in Figure 14.3) after the analytical calculation. This means that the SASU method saved about 92.4% and 86.6% of the computational time for spin-up of the global carbon-only and coupled carbon-nitrogen models to steady states, respectively, in the case of the CABLE model (Figure 14.3).

With the traditional spin-up method, the passive SOM pool continued to decrease after it reached the steady-state criterion (Figure 14.3). Because of the slow turnover rate of this pool, thousands of additional simulation years are needed for the traditional method to reach a steady state of the passive SOM pool. In contrast, the analytical solution obtained by the SASU method allowed this steady state to be accurately predicted after only 200 simulation years (Figure 14.3).

The results of this exercise using CABLE show that SASU can reduce the bottleneck of biogeochemical model spin-up by around 90%. That means spin-up with the SASU method can be around ten times as fast as the traditional method. The computational efficiency with the SASU method is higher than the accelerated decomposition method identified by Thornton and Rosenbloom (2005) for site-level spin-up in an evergreen needle-leaf forest. A similar analytic spin-up method developed by Lardy et al., (2011) accelerates spin-up with a pasture model (PaSim) by up to 20 times as well.

Liao et al. (2023) applied SASU to Community Land Model version 5 (CLM5) and examined its computational efficiency and accuracy (Figure 14.4). At the Brazil site, SASU is computationally seven times more efficient than (or saved up to 86% computational cost in comparison with) the traditional native dynamics (ND) spin-up to reach the same steady state. Globally, SASU is computationally eight times more efficient than the accelerated decomposition spin-up and 50 times more efficient than ND.

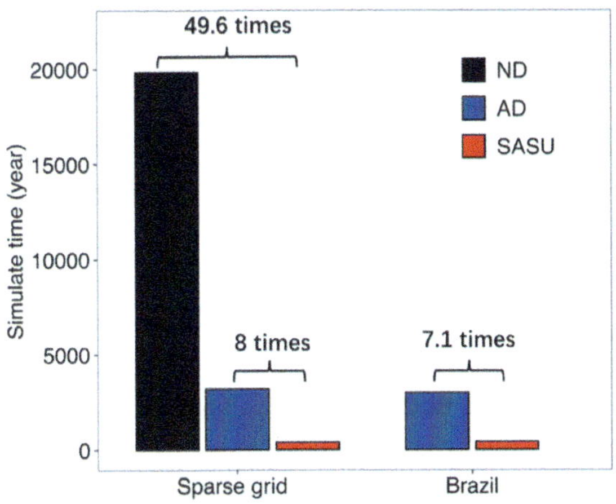

FIGURE 14.4 Model years needed for reaching equilibrium for different spin-up methods at 400 global grid cells and the Brazil site.

The SASU method can be easily implemented for biogeochemical models at site, regional, and global scales once a biogeochemical model is converted to a matrix model to enable analytical calculation of steady states of carbon and nutrient pools together with initial and final spin-ups.

SUGGESTED READINGS

Cuijuan Liao, Xingjie Lu, Yuanyuan Huang, Feng Tao, David M Lawrence, Charles D Koven, Keith W Oleson, William R Wieder, Erik Kluzek, Xiaomeng Huang, Yiqi Luo. (2023). Matrix Approach to Accelerate Spin-Up of CLM5. *Journal of Advances in Modeling Earth Systems*, *15*: MS003625.

Xia J, Luo Y, Wang Y-P, Weng E, Hararuk O. (2012). A Semi-Analytical Solution to Accelerate Spin-Up of a Coupled Carbon and Nitrogen Land Model to Steady State. *Geoscientific Model Development*, *5*: 1259–1271.

QUIZ

Select one option from the given answers

1 Which statement for the spin-up is **NOT** true?
   a Spin-up provides an initial state for model simulation.
   b Spin-up uses atmosphere CO_2 concentration at pre-industrial level.
   c Spin-up can speed up the historical simulation.
   d Spin-up in land biogeochemical models aims for the steady state.

2 The motivation to develop different spin-up approaches in land biogeochemical model is to
   a improve the simulation results.
   b improve the computational efficiency.
   c improve model structure.
   d improve model diagnostic capacity.

3 Semi-analytic spin-up (SASU) in CABLE calculates the steady state of carbon storage by setting
   a carbon storage to zero.
   b carbon influx to zero.
   c carbon influx equal to efflux.
   d carbon efflux to zero.

4 Which statement about CABLE spin-up efficiency is **NOT** correct.
   a The CABLE spin-up efficiency for the carbon-only model is higher than for the carbon-nitrogen coupled model.
   b SASU can be up to ten times as fast as traditional spin-up.
   c In the coupled carbon-nitrogen model, SASU does not need the final spin-up after the analytical calculation.
   d With the traditional spin-up method, the passive SOM pool continues to decrease after it reaches the steady-state criterion.

15 Time Characteristics of Compartmental Systems

Carlos A. Sierra
Max Planck Institute for Biogeochemistry, Jena, Germany

To understand carbon flows through terrestrial ecosystems, it is very important to use metrics to quantify the time carbon spends in the entire system and in particular compartments. In this chapter, we introduce the concepts of age and transit time as two fundamental metrics that characterize the speed at which carbon flows through ecosystems. Age is defined as the time carbon atoms spend in an ecosystem, from when they enter through photosynthesis until they are observed in a particular compartment. Transit time is defined as the time carbon atoms require to pass through the entire ecosystem, from the time they enter through photosynthesis until they are lost in gas, liquid, or solid form. We review here mathematical formulas for computing age and transit time in compartmental systems, distinguishing between formulas for autonomous systems in equilibrium and nonautonomous systems moving along a known trajectory.

INTRODUCTION

One of the advantages of representing models in the compact form of compartmental systems is that we can derive system-level diagnostics that help to better understand system dynamics. Differences in process representations, parameterizations, or sizes of compartments required to represent a system can be compared using simple aggregated metrics at the level of the entire system.

Two important system-level diagnostics for describing compartmental systems are the concepts of system age and transit (residence) time (Bolin and Rodhe 1973; Sierra et al. 2017). We define *system age* as the age of all atoms or particles inside the system, from the time they entered t_e until the time of observation t. *Transit time* is defined as the average time required for atoms or particles to traverse the system from their arrival time until they leave in the output flux. In other words, system age characterizes the age structure of all the atoms or particles in the system, while transit time characterizes the age structure of all atoms or particles in the output flux (Figure 15.1).

It is also possible to characterize the age structure of the atoms or particles inside each pool or compartment. We define *pool age* as the time elapsed since the atoms or particles entered the system until the time of observation t inside a pool i (Figure 15.1). Therefore, the system age is the aggregated result of the pool ages for all pools.

Since the age and transit time concepts are defined for all individual atoms or particles inside a system, we can also think about them in terms of distributions that quantify the proportional distribution of the mass in age classes. Therefore, these distributions can be characterized by statistics such as the mean, standard deviation, and quantiles such as the median.

In the following sections, we will introduce mathematical formulas to quantify age and transit time distributions for two separate cases, (1) autonomous systems in equilibrium, and (2) nonautonomous systems. Note that we will not be making a distinction between linear and nonlinear, because for case (1), linear and nonlinear systems in equilibrium can be treated similarly since the vector of states does not change once the equilibrium is reached and the system behaves similarly as a linear system. For case (2), the formulas rely on a linearization of the specific trajectory of a nonlinear system. Therefore, we will first introduce the linearization strategy and then provide the formulas for the linear nonautonomous case.

AGE AND TRANSIT TIME DISTRIBUTIONS FOR AUTONOMOUS SYSTEMS IN EQUILIBRIUM

The derivation of the formulas for age and transit time distribution of linear autonomous systems in equilibrium was originally introduced in Metzler and Sierra (2018). For their derivation, we were able to show that linear compartmental systems are analogous to absorbing continuous-time Markov chains. This means that linear compartmental systems can also be interpreted in a stochastic sense, with the deterministic system of differential equations representing the macroscopic behavior of entire masses, and the absorbing Markov chains representing the stochastic behavior of individual atoms of particles with respect to their age. For details about the stochastic process and derivation of formulas, interested readers can refer to Metzler and Sierra (2018) for additional details.

Let's consider linear autonomous systems introduced in Chapter 7, of the form of Equation 7.2, with an equilibrium point given by Equation 7.4. Let's also consider the 1-norm of a vector, defined as $\|v\|_1 = \|v\| := \sum_{i=1}^{n} |v_i|$, which is simply the sum of all the elements in the vector. We say that the random

DOI: 10.1201/9781032711126-19

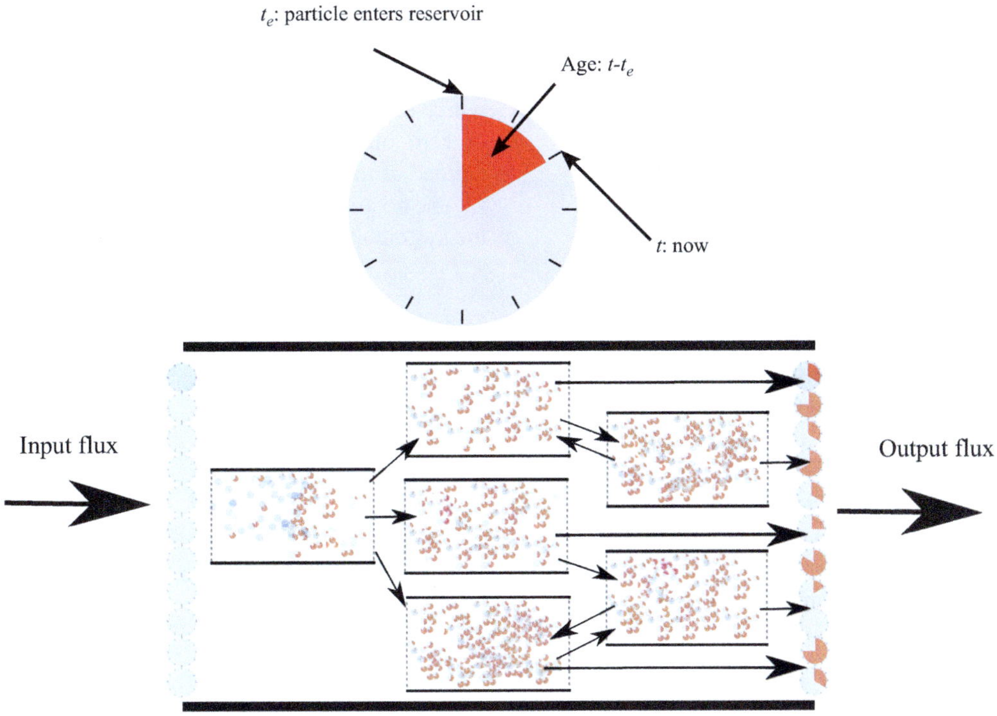

FIGURE 15.1 Graphical representation of the concepts of system age, transit time, and pool age. Mass entering a compartmental system can be conceptualized as being composed of small particles or atoms, each of them with a 'clock' that measures the time they have been in the system since they entered. All particles in the input fluxes have an age of zero. If we collect all particles inside the system at any given time and organize this information as a distribution of ages, we obtain the system age distribution of particles inside the system. If we collect the particles inside a specific pool and organize particles according to their age, we obtain the pool age distribution. Collecting particles in the output flux and organizing this information as a distribution of ages provides the transit time distribution.

Figure extracted from Sierra et al. (2017).

variable age a that measures age of atoms or particles in the system is distributed according to a Phase-Type (PH) distribution of the form:

$$f(a) = z^{\mathsf{T}} e^{a\mathbf{B}} \eta = z^{\mathsf{T}} e^{a\mathbf{B}} \frac{\boldsymbol{x}^*}{\|\boldsymbol{x}^*\|}, a \geq 0$$

Note that this density distribution is composed of three terms: the vector of fractional release coefficients, the matrix exponential of the compartmental matrix evaluated at each value of age, and the proportional distribution of mass at steady state. Since the fractional release coefficients can be computed directly from **B**, we can say that the system age distribution follows a Phase Type distribution with two parameters: the probability vector of mass at steady state, and the transition rate matrix generated by the compartmental matrix. This can be abbreviated as $a \sim \mathrm{PH}(\boldsymbol{\eta}, \mathbf{B})$.

The mean or expected value $\mathbb{E}$ of the system age distribution can be obtained as:

$$\mathbb{E}[a] = -\mathbf{1}^{\mathsf{T}} \mathbf{B}^{-1} \eta = \frac{\|\mathbf{B}^{-1} \boldsymbol{x}^*\|}{\|\boldsymbol{x}^*\|},$$

where **1** is a vector containing ones, and T is the transpose operator.

To obtain the pool-age density distribution, we define first a diagonal matrix with the steady-state values for each compartment as $\mathbf{X}^* := diag(x_1^*, \ldots, x_n^*)$. The vector-valued function that returns the age distribution for each pool is then given by:

$$\boldsymbol{f}(a) = (\mathbf{X}^*)^{-1} e^{a\mathbf{B}} \boldsymbol{u}, \ a \geq 0,$$

and the mean age for each pool:

$$\mathbb{E}[\boldsymbol{a}] = -(\mathbf{X}^*)^{-1} \mathbf{B}^{-1} \boldsymbol{x}^*.$$

The density distribution of the random variable transit time τ is also Phase-Type distributed, with the probability vector given by the proportional distribution of the input flux $\boldsymbol{\beta}$, and the compartmental matrix as the transition rate matrix; i.e., $\tau \sim \mathrm{PH}(\boldsymbol{\beta}, \mathbf{B})$. It can be obtained as:

$$f(\tau) = z^{\mathsf{T}} e^{t\mathbf{B}} \boldsymbol{\beta} = z^{\mathsf{T}} e^{t\mathbf{B}} \frac{\boldsymbol{u}}{\|\boldsymbol{u}\|}, \tau \geq 0,$$

with mean transit time given by:

$$\mathbb{E}[\tau] = \mathbf{B}^{-1}\beta = \frac{\|x^*\|}{\|u\|}.$$

Notice that the mean transit time is given by the ratio between the total mass at steady state and the total input flux.

AGE AND TRANSIT TIME DISTRIBUTIONS FOR NONAUTONOMOUS SYSTEMS

We consider now the nonlinear nonautonomous compartmental system introduced in Chapter 7 of the form of Equation 7.9, for which we can always find a unique numerical solution of the form $\boldsymbol{x}(t, t_0, \boldsymbol{x}_0)$. To obtain time-dependent age and transit time distributions for this system, we will use the known solution to construct an equivalent linear nonautonomous system with the exact same solution. Details of the approach are presented in Metzler, Müller, and Sierra (2018).

Plugging in the known solution $\boldsymbol{x}(t) = \boldsymbol{x}(t, t_0, \boldsymbol{x}_0)$ into a new linear version of the system, we can define a new vector of inputs as $\tilde{\boldsymbol{u}}(t) := \boldsymbol{u}(\boldsymbol{x}(t), t)$, and a new compartmental matrix as $\tilde{\mathbf{B}}(t) := \mathbf{B}(\boldsymbol{x}(t), t)$. Then, we obtain a linear nonautonomous compartmental system of the form:

$$\dot{\boldsymbol{y}}(t) = \tilde{\boldsymbol{u}}(t) + \tilde{\mathbf{B}}(t)\cdot \boldsymbol{y}(t),\ t > t_0,\ \boldsymbol{y}(t_0) = \boldsymbol{x}_0,$$

which has a unique solution $\boldsymbol{y}(t, t_0, \boldsymbol{y}_0)$. Since we assume that the original nonlinear system of Equation 7.9 also has a unique solution, both solutions must be identical; i.e., $\boldsymbol{y}(t, t_0, \boldsymbol{y}_0) = \boldsymbol{x}(t, t_0, \boldsymbol{x}_0)$. We can then use the solution of a nonlinear nonautonomous system to construct an equivalent linear nonautonomous compartmental system of the general form of Equation 7.7, which has a general solution given by Equation 7.8.

Age Distributions

We assume now that the initial content $\boldsymbol{x}_0$ has an initial age distribution $\boldsymbol{f}_0(a)$ such that $\boldsymbol{x}_0 = \int_0^\infty \boldsymbol{f}_0(a)\,da$. This initial age distribution is then perturbed by the time-dependent mass inputs and process rates of the system, generating a time-dependent age distribution of the form

$$\boldsymbol{f}(a,t) = \boldsymbol{g}(a,t) + \boldsymbol{h}(a,t),$$

where the term $\boldsymbol{g}(a, t)$ is the time evolution of the age distribution of the initial mass in the system, and $\boldsymbol{h}(a, t)$ is the time evolution of the age distribution of mass that enters the system after t_0.

The nonautonomous age distribution of the initial mass is given by:

$$\boldsymbol{g}(a,t) = 1_{[t-t_0,\infty)}(a)\cdot \Phi(t,t_0)\cdot \boldsymbol{f}_0\left(a-(t-t_0)\right)$$

where the indicator function $1_S(a)$ of a set S equals 1 if $a \in S$, or zero otherwise. The state transition operator $\Phi(t, t_0)$ is defined as in Chapter 7.

The nonautonomous age distribution of the mass that enters the system after t_0 is given by:

$$\boldsymbol{h}(a,t) = 1_{[0,t-t_0)}(a)\cdot \Phi(t,t-a)\cdot \boldsymbol{u}(t-a)$$

To obtain the age distribution of the entire system, we simply sum the densities over all pools as:

$$f(a,t) = \|\boldsymbol{f}(a,t)\|.$$

Transit Time Distributions

To obtain transit time distributions in the nonautonomous case, it is necessary to distinguish between the concepts of backward versus forward transit times. The backward transit time is defined as the age of particles in the output flux at the time of release from the system t_r. Using the fractional release coefficients, it is possible to obtain the vector of outflow rates at time t_r as:

$$z_j(t_r) = -\sum_{i=1}^{n} B_{ij}(t_r),\ j = 1,2,\ldots,n.$$

The backward transit time distribution can be obtained as:

$$f_{\mathrm{BTT}}(a,t_r) = z^T(t_r)\cdot \boldsymbol{f}(a,t_r)\ t_r \geq t_0.$$

Now, the forward transit time is defined as the age of an atom or particle that enters the system at an entering time $t_e > t_0$ and exits at time $t_r = t_e + a$. The forward transit time distribution can be obtained as:

$$f_{\mathrm{FTT}}(a,t_e) = z^T(t_e + a)\cdot \boldsymbol{f}(a,t_e + a).$$

Both distributions are tightly connected, with the forward transit time distribution of particles entering at time t_e being equal to the backward transit time distribution of the particles being released from the system at time t_r, i.e.,:

$$f_{\mathrm{FTT}}(a,t_e) = f_{\mathrm{BTT}}(a,t_r).$$

FINAL REMARKS

The compartmental system representation also unveils analogies between deterministic systems that conserve mass with stochastic systems that conserve probabilities. This stochastic representation can be used to obtain formulas for the age of particles or atoms in the compartmental systems.

With this approach, we derived formulas for the age of mass inside a compartmental system (system age), and the age of mass in the output flux (transit time). The concept of age can be very valuable to assess how old carbon and biogeochemical elements can be in an ecosystem. The concept of transit time can be very useful to understand how fast biogeochemical elements are processed inside an ecosystem, integrating all transfers and transformations that may take place.

There are other opportunities to further explore carbon cycle models in a stochastic setting. This could be particularly useful for studying, for example, the macroscopic properties at larger scales where patterns emerge by the action of microorganisms acting at microscopic scales. Also, the compartmental system representation may help to integrate concepts from other disciplines such as graph theory or control theory to address a number of questions not being explored yet in carbon cycle science.

SUGGESTED READING

A general introduction to the concepts of ages and transit times can be found in Bolin and Rodhe (1973). More specific results for the derivation of formulas and the computation of ages and transit times can be found in Rasmussen et al. (2016) for the mean of their distributions in nonautonomous systems, for complete distributions in autonomous systems in Metzler and Sierra (2018), and for complete distributions in nonautonomous systems in Metzler, Müller, and Sierra (2018).

QUIZ

1. Give examples of systems where the mean transit time is higher than the mean system age.
2. Give examples where the mean system age is higher than the mean transit time.
3. In what type of systems are the mean system age, the mean transit time, and the turnover time equal?

16 Practice 4

Efficiency and Convergence of Semi-Analytic Spin-Up (SASU) in TECO

Xingjie Lu
Sun Yat-sen University, Guangzhou, China

This practice aims to help the reader understand the convergence and efficiency of semi-analytic spin-up (SASU) for different carbon cycle representations in a model. By conducting the spin-up of the TECO model using native dynamics (ND) and the SASU method, we explore the convergence of different spin-up approaches, and the spin-up efficiency under different carbon input and soil turnover rates. We will also look at a weak nonlinear parameterization that assumes carbon input and soil turnover rates are functions of carbon pool sizes, verifying whether the two approaches converge and comparing their spin-up efficiencies.

SASU TO IMPROVE COMPUTATIONAL EFFICIENCY OF SPIN-UP OF BIOGEOCHEMICAL MODELS

Spin-up in biogeochemical models is an essential initialization procedure, which sets up the initial carbon and nitrogen pool sizes at a steady state for a model (see Chapter 14). When running a complex biogeochemical model, the spin-up is usually the most time-consuming procedure. So, the efficiency of spin-up becomes an important topic in biogeochemical model development.

Semi-analytic spin-up (SASU) is a recently developed technique to improve spin-up efficiency. The SASU approach builds on the fact that a matrix model of the carbon cycle can be semi-analytically solved to obtain steady-state values of pool sizes (Xia et al. 2012). Once biogeochemical models are presented in a matrix form, SASU can be easily implemented.

So far, SASU has been incorporated into the TECO, CABLE, ORCHIDEE, and CLM5 models. By implementing SASU, the spin-up computational efficiency is improved by 70% to 93% for different models under various configurations, including the most complicated, CLM5. CABLE was the first model to implement SASU for its carbon (C)-only and carbon-nitrogen (C-N) coupled versions. Results showed that SASU saves 92.4% of computational cost for C-only CABLE and 86.6% for the C-N version, when run at the global scale. Site simulations showed even greater improvement in the spin-up efficiency.

SPIN-UP IN THE SIMPLIFIED TECO MODEL

We will apply native dynamics spin-up (ND) and semi-analytic spin-up (SASU) on a simplified version of the TECO model in *Exercises 1–3* in CarboTrain. The simplified TECO model was used in the Unit 3 practice, Chapter 12. It includes seven pools, constant C input fluxes and turnover rate parameters. In *Exercise 1*, we are going to run ND and SASU for TECO.

Exercise 1
Run ND (default) spin-up for TECO matrix model (Exercise 1.1), then change the spin-up approach to SASU (Exercise 1.2). Learn whether carbon storage driven by the two alternative spin-up approaches converges to the same steady state. How fast is the SASU spin-up at reaching steady state compared with ND?

Exercise 1.1. Follow the steps below to run TECO spin-up using ND approach in CarboTrain:

- a Select **Unit 4**
- b Select **Exercise 1**
- c Select **Default ND**
- d Select **Output Folder**
- e Press **Run Exercise**
- f Check results in Output Folder. Time-dependent variation in carbon input, each pool size, residence time, total ecosystem carbon storage, total ecosystem carbon storage capacity, and carbon storage potential in `output.xls`. Figures are in `results.png`. Take note of **years for steady state**, and **Total_C_10000** (the total carbon at year 10,000) and complete the first row of Table 16.1, below.

Exercise 1.2. Follow the steps below to run TECO spin-up using the SASU approach in CarboTrain:

- a Select **Unit 4**
- b Select **Exercise 1**
- c Select **Default SASU**
- d Select **Output Folder**
- e Press **Run Exercise**

DOI: 10.1201/9781032711126-20

TABLE 16.1
Comparison of results from Exercise 1

Name	Description	C_{input}	$k_{passsoil}$	C_{init}	C_{10000}	T_{eq}
Ex 1.1	Default ND	0.00002245	1.5478×10^{-6}	0		
Ex 1.2	Default SASU	0.00002245	1.5478×10^{-6}	0		
Ex 1.3	High C_{input} ND	0.00004490	1.5478×10^{-6}	0		
Ex 1.4	High C_{input} SASU	0.00004490	1.5478×10^{-6}	0		
Ex 1.5	High $k_{passsoil}$ ND	0.00002245	3.0956×10^{-6}	0		
Ex 1.6	High $k_{passsoil}$ SASU	0.00002245	3.0956×10^{-6}	0		

Note: C_{input} = carbon input fluxes (gC m^{-2} s^{-1}); $k_{passsoil}$ = turnover rate of soil carbon (day^{-1}); C_{init} = initial total carbon pool size (gC m^{-2}); C_{10000} = total carbon storage at year 10000 (gC m^{-2}); T_{eq} = spin-up convergence time (year).

f Check results in Output Folder. Time-dependent variation in carbon input, each pool size, residence time, total ecosystem carbon storage, total ecosystem carbon storage capacity, and carbon storage potential in `output.xls`. Figures are in `results.png`. Take note of **years for steady state**, and **Total_C_10000** and complete the second row of Table 16.1 below.

Questions:
Does carbon storage converge to the same equilibrium in ND versus SASU? Which spin-up approach is faster at achieving equilibrium?

SPIN-UP WITH DIFFERENT MODEL PARAMETERS

Previous studies on CABLE spin-up have shown that the C-only and C-N coupled model versions differed in efficiency and equilibrium for both ND and SASU spin-up. We may speculate that the differences in parameters and model structures may be the causes. In *Exercises 1.3–1.6*, we consider idealized cases, and use the simplified TECO model to study whether or how the parameters influence the spin-up efficiency and equilibrium.

Run ND (default) spin-up for the TECO matrix model with increasing C input and passive pool turnover rate (Exercises 1.3 and 1.5), then change the spin-up approach to SASU (Exercises 1.4 and 1.6). Learn whether carbon storage driven by the two spin-up approaches converges to the same steady state. How is spin-up efficiency affected by different parameters or carbon input? Complete Exercises 1.3–1.6 and fill in the remaining rows of Table 16.1.

Exercise 1.3. Tune the carbon input and run TECO spin-up using the ND approach in CarboTrain:

a Select **Unit 4**
b Select **Exercise 1**
c Select **High carbon input ND**
d Select **Output Folder**
e Press **Open source code**
f Change the input_fluxes to 0.0000449 at line 46 and save the code
g Press **Run Exercise**
h Check results in Output Folder. Time-dependent variation in carbon input, each pool size, residence time, total ecosystem carbon storage, total ecosystem carbon storage capacity, and carbon storage potential in `output.xls`. Figures are in `results.png`. Take note of **years for steady state**, and **Total_C_10000** and enter results in Table 16.1, below.

Exercise 1.4. Tune the carbon input and run TECO spin-up using the SASU approach in CarboTrain:

a Select **Unit 4**
b Select **Exercise 1**
c Select **High carbon input SASU**
d Select **Output Folder**
e Press **Open source code**
f Change the input_fluxes to 0.0000449 at line 46 and save the code
g Press **Run Exercise**
h Check results in Output Folder. Time-dependent variation in carbon input, each pool size, residence time, total ecosystem carbon storage, total ecosystem carbon storage capacity, and carbon storage potential in `output.xls`. Figures are in `results.png`. Take note of **years for steady state**, and **Total_C_10000** and enter results in **Table 16.1**, below.

Questions:
Compared to the default cases (Exercises 1.1 and 1.2), do carbon input increases change the spin-up time for steady state? Do carbon pool sizes using the two spin-up approaches still converge when carbon input increases? Are the steady states the same when carbon inputs are higher? Why/why not?

Exercise 1.5. Tune the passive soil turnover rate and run TECO spin-up using the ND approach in CarboTrain:

a Select **Unit 4**
b Select **Exercise 1**
c Select **High K passive ND**
d Select **Output Folder**

e Press **Open source code**
f Change the last column in variable temp to 0.00000309564 at line 34 and save the code
g Press **Run Exercise**
h Check results in Output Folder. Time-dependent variation in carbon input, each pool size, residence time, total ecosystem carbon storage, total ecosystem carbon storage capacity, and carbon storage potential in `output.xls`. Figures are in `results.png`. Take note of **years for steady state**, and **Total_C_10000** and enter results in **Table 16.1**, below.

Exercise 1.6. Tune passive soil turnover rate and run TECO spin-up using the SASU approach in CarboTrain:

a Select **Unit 4**
b Select **Exercise 1**
c Select **High K passive SASU**
d Select **Output Folder**
e Press **Open source code**
f Change the last column in variable temp to 0.00000309564 at line 34 and save the code
g Press **Run Exercise**
h Check results in Output Folder. Time-dependent variation in carbon input, each pool size, residence time, total ecosystem carbon storage, total ecosystem carbon storage capacity, and carbon storage potential in `output.xls`. Figures are in `results.png`. Take note of **years for steady state**, and **Total_C_10000** and enter results in **Table 16.1**, below.

Questions:
Compared to the default cases (Exercises 1.1 and 1.2), do increases in passive soil turnover rate change the spin-up time for steady state? By comparing the **delta_C** in result.png, are results from SASU more stable? Do carbon pool sizes using the two spin-up approaches still converge when passive soil turnover rate increases? Are the steady states the same when soil carbon turnover rate is higher? Why/why not?

SPIN-UP IN A WEAK NONLINEAR SYSTEM

Nonlinearity is characteristic for many terrestrial biogeochemical models. For example, the canopy photosynthesis (i.e., carbon input) is usually dependent on leaf area, which is a function of leaf carbon pool size. In some models, the soil carbon turnover rate is pool size dependent. The following exercise examines how nonlinearity impacts the spin-up.

Exercise 2
Run ND (default) spin-up for the TECO matrix model with foliage nonlinearity and soil turnover rate nonlinearity (Exercises 2.1 and 2.3), then change the spin-up approach to SASU (Exercises 2.2 and 2.4). Learn whether carbon storage using the two alternative spin-up approaches converges to the same steady state. How long is the spin-up time for a model with nonlinearity in foliage or soil? Complete Exercises 2.1–2.4 and enter results in Table 16.2.

Exercise 2.1. Enable foliage nonlinearity in TECO and then run TECO using the ND spin-up in CarboTrain:

a Select **Unit 4**
b Select **Exercise 2**
c Select **Nonlinear foliage C with ND**
d Select **Output Folder**
e Press **Open source code**
f Add following lines at line 53 (remember to copy the indentation as well):

```
def input_fluxes(t, y):
    tmp = 0.00000449 + 0.00002245 / 2 * (1 - np.exp(-0.5 * y[0] * 0.008)) / 0.5
    return tmp
```

The lines added here describe the nonlinear relation between carbon input fluxes and foliage carbon (see Figure 16.1), which is formulated by:

$$C_{input} = f\left(C_{foliage}\right) = C_{input,min} + C_{input0} \cdot \frac{1 - e^{-k_d \cdot C_{foliage} \cdot SLA}}{k_d \cdot LAI_0} \qquad 16.1$$

where $C_{input,\,min} = 0.00000449$ gC m^{-2}s^{-1}, $C_{input0} = 0.00002245$ gC m^{-2}s^{-1}, $k_d = 0.5$, $SLA = 0.008$ m^2(gC)$^{-1}$, $LAI_0 = 2$ m^{2}m^{-2}, $C_{foliage}$ is the foliage carbon.

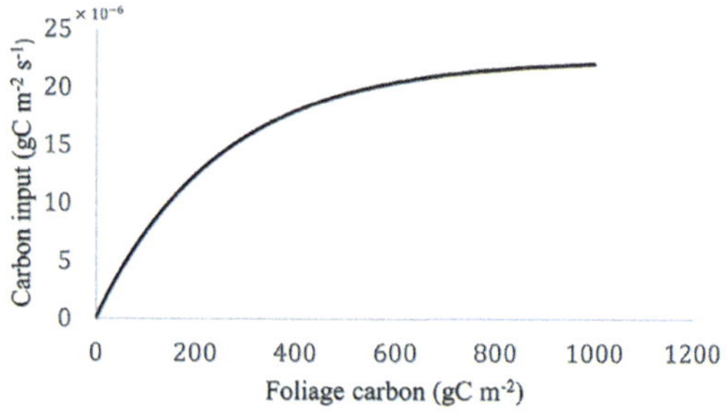

FIGURE 16.1 The nonlinear relation between carbon input and foliage carbon.

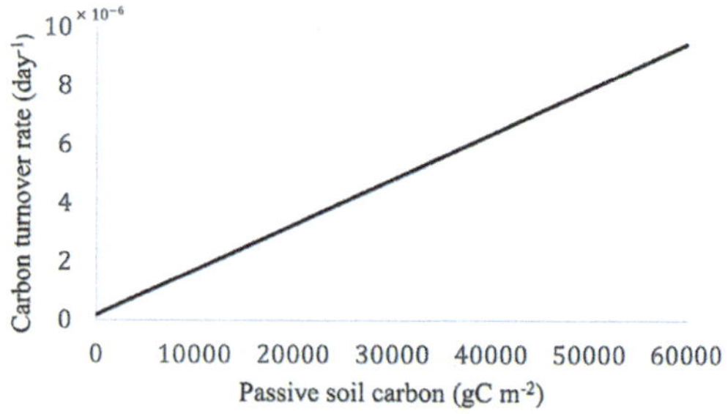

FIGURE 16.2 The relation between passive soil turnover rate and passive soil carbon.

a Press **Run Exercise**
b Check results in Output Folder. Time-dependent variation in carbon input, each pool size, residence time, total ecosystem carbon storage, total ecosystem carbon storage capacity, and carbon storage potential in `output.xls`. Figures are in `results.png`. Take note of **years for steady state**, and **Total_C_10000** and complete the third row of Table 16.2, below.

Exercise 2.2. Enable foliage nonlinearity in TECO and then run TECO spin-up using SASU approach in CarboTrain:

a Select **Unit 4**
b Select **Exercise 2**
c Select **Nonlinear foliage C with SASU**
d Select **Output Folder**
e Press **Open source code**

Add following lines at line 53 (remember to copy the indentation as well):

```
def input_fluxes(t, y):
    tmp = 0.00000449 + 0.00002245 / 2 * (1 - np.exp(-0.5 * y[0] * 0.008)) / 0.5
    return tmp
```

The lines added here were explained in Step f of Exercise 2.1.

a Press **Run Exercise**
b Check results in Output Folder. Time-dependent variation in carbon input, each pool size, residence time, total ecosystem carbon storage, total ecosystem carbon storage capacity, and carbon storage potential in `output.xls`. Figures are in `results.png`. Take note of **years for steady state**, and **Total_C_10000** and enter results in Table 16.2, below.

Questions:
Compared to the default cases (Exercises 1.1 and 1.2), does foliage nonlinearity increase the spin-up time for steady state? Does carbon storage using the two spin-up approaches still converge when foliage nonlinearity is enabled? Are the steady states the same between the two spin-up approaches?

Exercise 2.3. Enable nonlinearity in passive soil turnover rate and run TECO using the ND spin-up approach in CarboTrain:

a Select **Unit 4**
b Select **Exercise 2**
c Select **Nonlinear soil C with ND**
d Select **Output Folder**
e Press **Open source code**
f Replace line 53 with the following lines (remember to copy the indentation as well):
def fun_K(t, y): K[6][6] = temp[6] /86400 * (y[6] / 10000.0 + 0.1) return K mod = GeneralModel(times, B, A, fun_K, iv_list, input_fluxes)

The lines revised here describe the relation between passive soil turnover rate and passive soil carbon (see Figure 16.2), which is formulated by:

$$k_{passsoil} = f\left(C_{passsoil}\right) = k_0 \cdot \left(\frac{C_{passsoil}}{10000.0} + 0.1\right) \qquad 16.2$$

where k_0 = 0.00000154782 day^{-1}, $C_{passsoil}$ is the passive soil carbon storage.

a Press **Run Exercise**
b Check results in Output Folder. Time-dependent variation in carbon input, each pool size, residence time, total ecosystem carbon storage, total ecosystem carbon storage capacity, and carbon storage potential in `output.xls`. Figures are in `results.png`. Take note of **years for steady state**, and **Total_C_10000** and enter results in Table 16.2, below.

Exercise 2.4. Enable nonlinearity in passive soil turnover rate and run TECO using the SASU spin-up in CarboTrain:

a Select **Unit 4**
b Select **Exercise 2**
c Select **Nonlinear soil C with SASU**
d Select **Output Folder**
e Press **Open source code**

TABLE 16.2
Comparison of results from Exercises 1 and 2

Name	Description	C_{input}	$k_{passsoil}$	C_{init}	C_{10000}	T_{eq}
Ex 1.1	Default ND	0.00002245	1.5478×10^{-6}	0		
Ex 1.2	Default SASU	0.00002245	1.5478×10^{-6}	0		
Ex 2.1	Nonlinear foliage C with ND	$f(C_{leaf})$	1.5478×10^{-6}	0		
Ex 2.2	Nonlinear foliage C with SASU	$f(C_{leaf})$	1.5478×10^{-6}	0		
Ex 2.3	Nonlinear soil C with ND	0.00002245	$f(C_{passsoil})$	0		
Ex 2.4	Nonlinear soil C with SASU	0.00002245	$f(C_{passsoil})$	0		

Note: C_{input} = carbon input fluxes (gC m^{-2} s^{-1}); $k_{passsoil}$ = turnover rate of soil carbon (day^{-1}); C_{init} = initial total carbon pool size (gC m^{-2}); C_{10000} = total carbon storage at year 10000 (gC m^{-2}); T_{eq} = spin-up convergence time (year). $f(C_{foliage})$ = nonlinear relation between carbon input flux and foliage carbon (unitless) (see Exercise 2.1 or Exercise 2.2 for details). $f(C_{passsoil})$ = nonlinear relation between passive soil turnover rate and passive soil carbon storage (unitless) (see Exercise 2.3 or Exercise 2.4 for details).

f Replace line 53 with the following lines (remember to copy the indentation as well):
def fun_K(t, y): K[6][6] = temp[6] /86400 * (y[6] / 10000.0 + 0.1) return K mod = GeneralModel(times, B, A, fun_K, iv_list, input_fluxes)
The lines revised here were explained in Step f of Exercise 2.3.

g Press **Run Exercise**

h Check results in Output Folder. Time-dependent variation in carbon input, each pool size, residence time, total ecosystem carbon storage, total ecosystem carbon storage capacity, and carbon storage potential in `output.xls`. Figures are in `results.png`. Take note of **years for steady state**, and **Total_C_10000** and enter results in Table 16.2, above.

Questions:

Compared to the default cases (Exercises 1.1 and 1.2), does nonlinearity in soil turnover rate increase the spin-up time? Does carbon storage using the two spin-up approaches still converge when nonlinearity in soil turnover rate is enabled? Are the steady states the same between the two spin-up approaches?

Unit Five

Traceability and Benchmark Analysis

17 Overview of Traceability Analysis

Jianyang Xia
East China Normal University, Shanghai, China

This chapter provides an overview of the traceability framework as a method for identifying the key sources of uncertainty in land carbon cycle modeling. The analytical framework is built upon derivations of the matrix equation introduced in Chapter 1. This framework provides a systematic way to decompose the uncertainty of land carbon cycle projections into traceable components. Thus, traceability analysis can facilitate the understanding of carbon cycling, its drivers and controlling processes within a model and across models with different carbon-cycle structures, external climate forcings, and scientific assumptions. This chapter also introduces several examples to demonstrate the application of traceability analysis to model evaluation.

A KEY CHALLENGE FOR EARTH SYSTEM MODELS: IDENTIFICATION OF UNCERTAINTY SOURCES

Carbon cycle schemes incorporated in earth system models (ESMs) and offline land surface models are widely used to simulate terrestrial biogeochemistry and its feedbacks to climate change. The processes underpinning the terrestrial carbon (C) cycle constitute one of the most uncertain components, and this uncertainty has become a bottleneck in earth system modeling. The model-to-model variation in future projections of the global land C sink remained large from the third assessment report of the Intergovernmental Panel on Climate Change (IPCC), published in 2001, to the sixth report, published in 2021, despite an intervening period of two decades of research and model development. The new ESM projections and analysis in Phase 6 of the Coupled Model Intercomparison Project (CMIP6) show that uncertainty on the recent past and potential future evolution of the land carbon cycle, relevant to the assessment and understanding of global climate change, is still a prominent feature in the sixth IPCC report. Thus, one key challenge for earth system science is how to reduce the large disagreement in carbon cycle predictions among models of the terrestrial biosphere incorporated in ESMs.

To tackle this challenge, we need to answer one question: why are terrestrial carbon cycle predictions different among models? In the past three decades, numerous model intercomparison projects (MIPs) have been established, in part to shed light on this question. Although those MIPs have made important contributions to model evaluation and synthesis, they have been limited by design in their ability to explore the sources of model uncertainty. Thus, our question can be refined into how we can trace model uncertainty back to its sources, such as model structures, parameters, and external forcings.

TRACEABILITY FRAMEWORK: DESIGN AND KEY COMPONENTS

In Chapter 1 we noted that the terrestrial carbon cycle has four fundamental properties: (1) photosynthesis as the primary carbon influx pathway; (2) the transfer of assimilated carbon among different C pools (e.g., plant litter and soil); (3) the rate of transfer of this carbon controlled by the donor pool; and (4) the process of carbon transfer from the donor pool to the recipient pool, which can be described by first-order kinetics. These four fundamental properties, with minor variations, have emerged in most models using a pool-and-flux structure. Thus, the modeled terrestrial carbon cycle can be tracked by the equation:

$$\frac{dX(t)}{dt} = Bu(t) + A\xi(t)KX(t) \qquad 17.1$$

where the $\frac{dX(t)}{dt}$ characterizes the dynamic of terrestrial carbon storage, u is the carbon inputs from photosynthesis, and B is the partitioning coefficients to live carbon pools (e.g., leaf, wood, and root). Thus, the term $Bu(t)$ represents the partitioning of the assimilated carbon from photosynthesis among different plant carbon pools. The second term, $A\xi(t)$ $KX(t)$, describes the movement and exit rates of carbon atoms along their transferring paths (Xia et al. 2013; Luo et al. 2017). A is a transfer coefficients matrix to represent the movements of carbon atoms among multiple carbon pools. K represents the exit rates of different carbon pools. $\xi(t)$ modifies the decay rate of different pools by adding influences from environmental factors (temperature, moisture, nutrients, soil texture, and so on). Different models may have different combinations of plant and soil pools and different values in vector B or matrices A and C, but they can be unified by this matrix equation.

At steady state, terrestrial carbon reservoirs reach their maximum storage capacity and there are no further net carbon exchanges. Therefore, the steady-state land carbon storage (X_{ss}) can be solved by letting Equation 17.1 equal 0 (i.e., $dX(t)/dt = 0$):

DOI: 10.1201/9781032711126-22

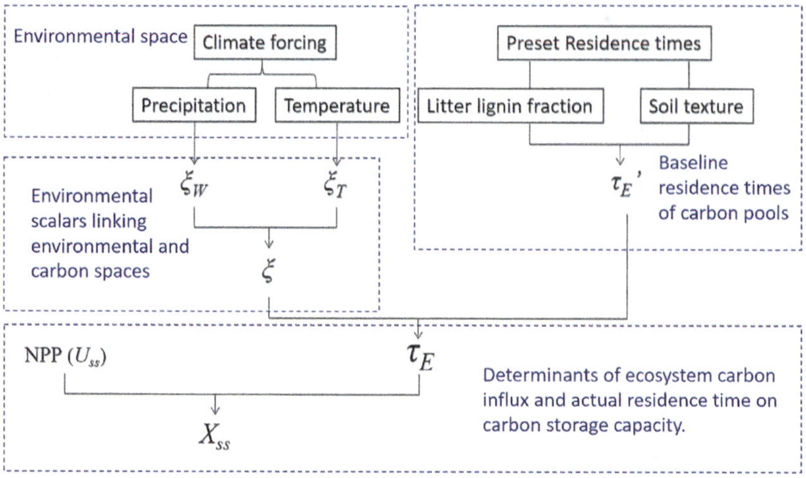

FIGURE 17.1 The traceability framework for decomposing steady-state terrestrial carbon storage into its traceable components.

$$X_{ss} = (-A\xi K)^{-1} BU_{ss} \qquad 17.2$$

where X_{ss} is a vector containing the steady-state pool sizes of all carbon pools, and U_{ss} is the carbon influx at steady state. The term $(-A\xi K)^{-1}B$ measures the ecosystem carbon residence time (τ_E), which is an important ecosystem property involving multiple processes, including carbon allocation (the B matrix), carbon transfer network (the A matrix), decomposition processes (the K matrix) and modifications from environmental factors (the ξ matrix). Here, net primary productivity (NPP) is treated as carbon inputs, and the X_{ss} can then be decomposed into NPP and τ_E:

$$X_{ss} = NPP \times \tau_E \qquad 17.3$$

As mentioned above, τ_E is determined by four items: $(-A\xi K)^{-1}B$. Here, A, K, and B are the intrinsic model properties, while ξ represents external influences from environmental factors. We thus rearrange these items and express τ_E as:

$$\tau_E = (-A^{-1}K^{-1}B)\xi^{-1} \qquad 17.4$$

We further merge $(-A^{-1}K^{-1}B)$ and define it as the baseline carbon residence time (τ_E'). Then, τ_E can be decomposed into τ_E' and ξ^{-1}:

$$\tau_E = \tau'_E \xi^{-1} \qquad 17.5$$

where τ_E' is determined by model intrinsic properties, associated with plant trait, soil attributes, number of C pools and how C is transferred among these pools at different cycling rates. As defined above, ξ represents the modifying effects of environmental factors as a fractional value applied multiplicatively to the baseline C residence time. If, as an example, we consider temperature and water availability as the main environmental factors scaling process rates in our system, we may divide ξ into a temperature scalar (ξ_T) and a water scalar (ξ):

$$\xi = \xi_T \xi_W \qquad 17.6$$

Through the above mathematical rearrangements of Equation 17.2, the modeled land carbon storage at steady state is decomposed into its determinative components as illustrated in Figure 17.1. X_{ss} is first decomposed into NPP and τ_E. Then, τ_E can be further traced into τ_E' and ξ, while ξ is divided into ξ_T and ξ_W. This framework offers a systematic way to decompose the steady-state land carbon cycle in a form that fits the structure of many common land carbon models. Most recently, advances in terrestrial carbon cycle theory (Luo et al. 2017) have extended the capability of this framework to trace transient-state land carbon storage. This is explored further in Chapter 18.

BENEFITS OF TRACEABILITY ANALYSIS FOR IDENTIFYING MODEL UNCERTAINTY SOURCES

Land carbon cycle schemes in most earth system models can be mathematically represented in the form of Equation 17.1. Thus, the traceability framework can facilitate the evaluation of land carbon cycle models and the ESM frameworks to which they are coupled by identifying the sources of simulation difference within and between models. It should be noted that the uncertainty of land carbon cycling in ESMs not only stems from model structure and parameters, but also is affected by climate outputs from the atmospheric components of ESMs (Ahlström et al. 2017). Below are some cases which have shown the application of traceability analyses for model evaluation. More application cases are presented in Chapter 18 and one recent study (Hou et al. 2023).

Because the CABLE model can be applied with or without carbon-nitrogen coupling enabled, the traceability analysis

CASE 1 UNDERSTANDING TERRESTRIAL CARBON CYCLE VARIATIONS WITHIN A MODEL

Modern global land-surface models account for a vast array of different processes, making it difficult to trace features of the results of simulations to the models' constituent processes and interactions to aid understanding and evaluation. For example, all global land carbon models simulate the spatial variability of ecosystem C storage, but it is often unclear how the geographic distribution of ecosystem C storage across a global map is determined. Xia et al. (2013) introduced the framework of traceability analysis and applied it to the Australian Community Atmosphere Biosphere Land Exchange (CABLE) model to help understand differences in modeled carbon processes among biomes and as influenced by nitrogen processes.

They first estimated ecosystem carbon capacity among different biomes in CABLE and traced it to the ecosystem residence time (τ_E) and carbon influx (i.e., NPP or U_{ss} in Equation 17.1). Results indicated that different biomes showed different patterns. For example, evergreen broadleaf forests have a very high NPP but a short carbon residence time. The tundra biome has a very low NPP but a long carbon residence time. Some barren biomes, such as deserts, have low carbon storage capacity because of low NPP and short carbon residence time (Figure 17.2). Using Equation 17.5 they then further decomposed the ecosystem carbon residence time into baseline carbon residence time and environmental scalars. They found that tundra and evergreen broadleaf forests have similar baseline carbon residence times, but the environmental scalars are much lower in tundra than evergreen broadleaf forests. The lower value of the environmental scalar signifies stronger environmental limitations on decomposition rates of organic carbon, which implies that tundra has much longer actual carbon residence time than evergreen broadleaf forests, even though the baseline residence time is similar. The environmental space of temperature and water scalars among biomes in the CABLE model was then considered, which led to the insight that the spatial difference in environmental control of carbon residence time is mainly driven by the temperature scalar rather than the water scalar in CABLE.

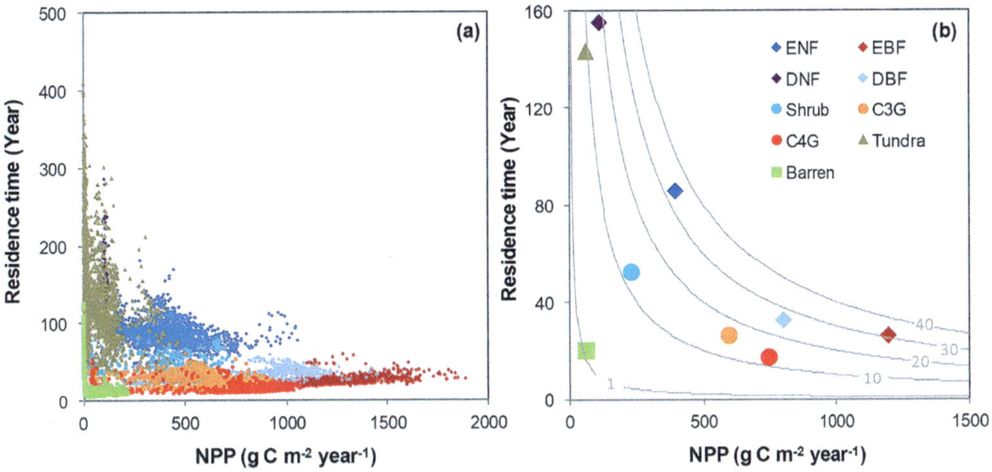

FIGURE 17.2 Determinants of ecosystem C storage capacity by NPP and residence time. Values of all grid cells are plotted in panel (a). In panel (b), the hyperbolic curves represent constant values of ecosystem carbon storage capacity. ENF – Evergreen needle leaf forest; EBF – Evergreen broadleaf forest; DNF – Deciduous needle leaf forest; DBF – Deciduous broadleaf forest; Shrub – Shrub land; C3G – C_3 grassland; C4G – C_4 grassland; and Barren – barren/sparse vegetation.

can be used to evaluate how the incorporation of nitrogen cycle processes can affect carbon cycling within the model. The analysis showed that incorporating the nitrogen cycle into CABLE reduces ecosystem carbon storage capacity in all biomes in comparison with the carbon-only model. Specifically, the CABLE model simulates lower NPP in woody biomes but shorter carbon residence time for non-woody biomes when the nitrogen cycle is switched on (Xia et al., 2013).

The traceability framework could be applied in a similar way to analyze the carbon cycle impacts of other types of external forcings, such as different CO_2 scenarios, disturbance regimes, and land use/cover changes. As shown in Ahlström et al. (2015), this approach can help to diagnose which processes are most important in determining the model uncertainty under given external forcings. This in turn can identify processes for priority attention in evaluating, revising, or improving the carbon cycle model.

CASE 2 INTERMODEL COMPARISONS OF TERRESTRIAL CARBON CYCLE SIMULATIONS

Although model intercomparison projects have shown that the ensemble means of multiple models fit data well, the intermodel difference in carbon cycle pools and fluxes is usually large. To better understand the sources of variations in modeled carbon storage capacity among models, Rafique et al. (2016) used the traceability framework to compare two land carbon cycle models, CLM-CASA' and CABLE. As shown in Figure 9.2, CABLE and CLM-CASA' both showed a distinctive structure of the ecosystem carbon cycle, like the number of carbon pools, NPP partitioning coefficients, decomposition rates, and carbon transfer coefficients among different pools, which resulted in different baseline carbon residence times. Due to more NPP partitioning into roots and wood, the baseline carbon residence time in CABLE was longer than in CLM-CASA'. Despite difference in model structure, the traceability analysis showed that CABLE and CLM-CASA' simulate similar global mean soil carbon storage capacity. This is because CABLE has lower NPP but longer carbon residence time compared to CLM-CASA'. The longer carbon residence time in CABLE mainly results from the baseline carbon residence time rather than the environmental scalars.

Overall, this case indicated that the major factors contributing to the differences between the two models were primarily due to parameter settings related to photosynthesis, carbon input, baseline residence times, and environmental conditions. The application of traceability analysis to intermodel comparisons is useful for developing a model with different versions or interpreting the different predictions of land carbon cycling by different models under the same climate forcings.

CASE 3 EVALUATING IMPACTS OF EXTERNAL FORCINGS ON TERRESTRIAL CARBON CYCLE SIMULATIONS

One important uncertainty source in land carbon cycle studies stems from the differences in baseline and projected climate fields generated by different climate and earth system models. When a land carbon model is forced by output from different climate models, the resultant predictions of the land carbon cycle also differ – a source of uncertainty propagating *through* the carbon cycle model *from* the forcing climate model. By applying traceability analysis on the LPJ-GUESS model when forced with 13 different climate datasets from CMIP5 general circulation models, Ahlström et al. (2015) studied the impact of climate model uncertainty on the land carbon cycle. LPJ-GUESS is a global dynamic vegetation-ecosystem model that is based on detailed representation of vegetation structure, demography, and resource competition. To understand how ecosystems around the world respond to future projections of atmospheric CO_2 concentrations and climate, transient and steady-state simulations were performed forced by output fields from different climate models under the RCP8.5 radiative forcing scenario. The authors then quantified the relative contributions of the three groups of processes – NPP, vegetation dynamics and turnover, and soil decomposition – to future carbon uptake uncertainties. They achieved this by fitting the traceability framework as an emulator to each of the 13 LPJ-GUESS simulations. Because the emulator has a common structure, it was possible to 'exchange' the key carbon cycle processes among the 13 simulations, allowing the importance of each in explaining the variability among simulations to be derived (Figure 17.3). Further details about the model and simulations can be found in Ahlström et al. (2015). Since there is only one carbon model involved in this study, we can be sure that all the differences in the carbon cycle projections stem from the external climate forcing. The results showed that NPP, vegetation turnover, and soil decomposition rate respectively explain 49%, 17%, and 33% of the uncertainty in carbon uptake by terrestrial ecosystems globally under the RCP8.5 future scenario (Figure 17.3c).

SUMMARY

The traceability analysis provides a relatively new approach to the evaluation of model uncertainties impacting simulations of the terrestrial carbon cycle. The framework builds on the fundamental properties of the terrestrial carbon cycle, which are reflected broadly by different models despite differences in structure and process detail. Equation 17.1 provides the theoretical basis for the traceability analysis. The application cases described illustrate how traceability analysis can benefit the understanding of variations in terrestrial carbon cycling within a model, the intercomparison of terrestrial carbon cycling among models, evaluation of the contributions of external forcings to carbon-cycle uncertainty, the assessment of newly incorporated processes into carbon cycle models, and the development of online tools for quick and consistent model evaluation. The application cases presented in this chapter mainly focus on the steady-state ecosystem C storage. In Chapter 18 we will explore how the traceability analysis can be adapted to apply to the transient dynamics of the land carbon cycle in models.

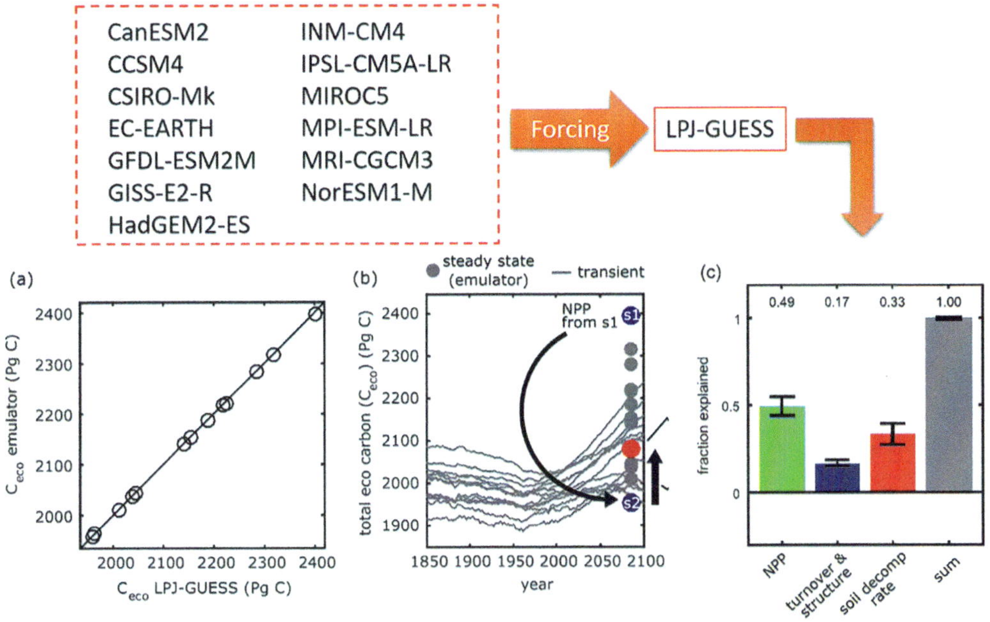

FIGURE 17.3 Emulator performance and example of experimental design. (a) Comparison between global ecosystem carbon stock (C_{eco}) at steady state as simulated by LPJ-GUESS and emulator solution of global C_{eco} at steady state for the 13 simulations and emulator solutions. The small deviations from the 1:1 relationship (black line) are mainly due to the stochastic internal variability in LPJ-GUESS where a true steady state is never reached, in contrast to the emulator, which solves the steady-state conditions analytically. (b) Illustration of experimental design. Gray curves are time trajectories of C_{eco} from transient simulations with LPJ-GUESS; circles denote corresponding emulator-computed steady-state values of C_{eco}. (c) Global partitioning of steady-state C_{eco} uncertainties. From this panel, we can see that NPP is the largest contributor to the difference in modeled land carbon uptake, accounting for almost 50% of variability among simulations with different forcings.

Figure reproduced from Ahlström et al. (2015).

CASE 4 ASSESSMENT OF NEW PROCESSES IN LAND CARBON MODELS

Although the integration of more process detail into a model can increase its utility, it also tends to increase the number of interacting processes and feedbacks influencing the model output, and the relationship of the model output to the model forcing data. As a result, it can become more difficult to understand or evaluate how the newly incorporated processes influence the behavior and performance of the model. For example, recognizing the important role that nitrogen (N) availability plays for the dynamics of the world's ecosystems, many current models that originally included only a carbon cycle have been enhanced to incorporate a nitrogen cycle that interacts with the model's carbon processes and state. The availability of nitrogen can strongly affect both ecosystem carbon input and mean carbon residence time. Nevertheless, the detailed structure of the C-N coupling scheme varies greatly among different models. How these diverse representations of C-N interactions affect carbon cycle modeling remains unclear. Thus, Du et al. (2018) incorporated three different C-N coupling schemes, derived from the TECO-CN, CLM4.5, and O-CN models, into the carbon-only version of the Terrestrial ECOsystem (TECO) model and then used the traceability framework to evaluate their impacts on the carbon cycle. The three C-N coupled frameworks are different in many aspects, including C:N stoichiometry in plants and the soil, plant N uptake strategies, down-regulation of photosynthesis under N deficit, and the pathways of N acquisition.

As shown in Figure 17.4a, each N process is simulated with different assumptions by different models. The results showed that each of the integrated C-N coupling schemes reduced the carbon storage capacity compared with the carbon-only version of TECO. However, the magnitude of the reduction varies among the three schemes, i.e., −23%, −30%, and −54% for TECO-CN, CLM4.5, and O-CN, respectively. The reduced carbon storage capacity was driven by reduced NPP (−29%, −15%, and −45%) and mean C residence time (9%, −17%, and −17%) (Figure 17.4b). The differences in these results for different N cycle implementations in TECO indicate that adding interactive nitrogen dynamics to a carbon cycle model could generate new uncertainty sources for carbon cycle-climate feedbacks. This example illustrates that the traceability analysis can improve our understanding of the impacts of newly incorporated processes on an existing carbon cycle model.

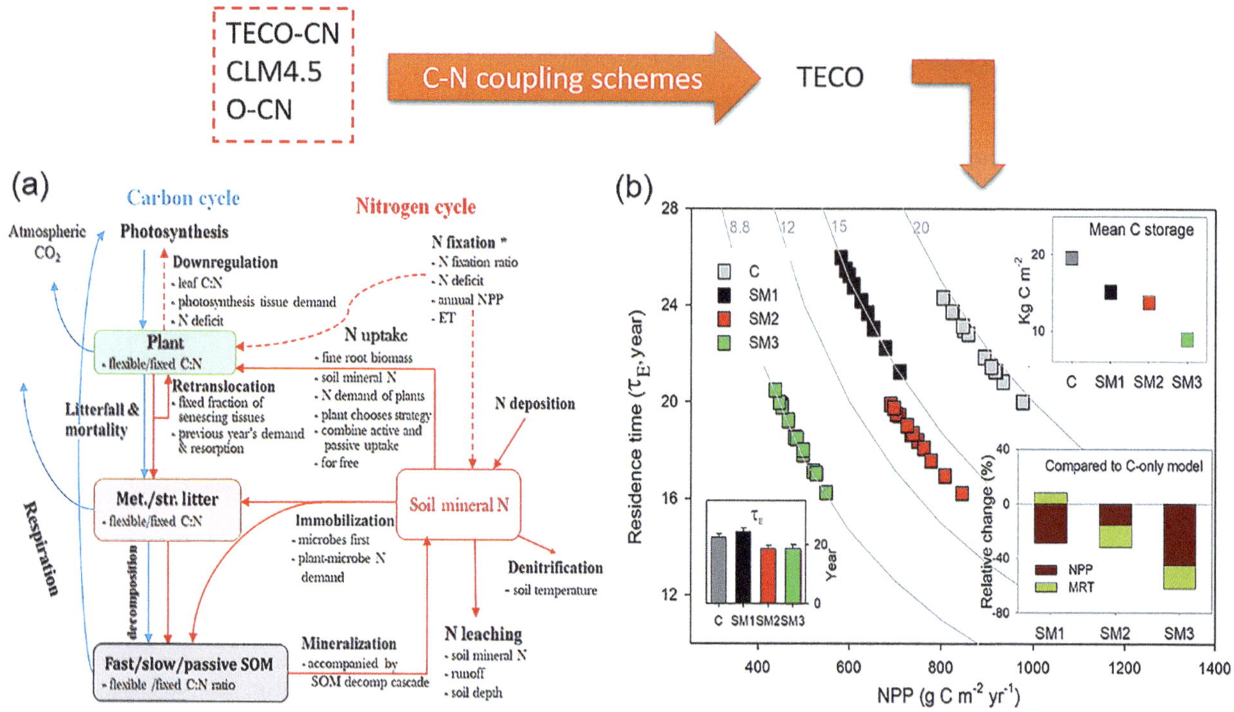

FIGURE 17.4 Schematic diagram of the terrestrial ecosystem carbon and nitrogen coupling model and its traceable components under different C-N coupling schemes. (a) The major carbon and nitrogen pool-and-flux structure in a terrestrial ecosystem, with alternative assumptions of the N processes represent in SM1 (TECO-CN), SM2 (CLM4.5), and SM3 (O-CN) C-N coupling schemes. Light blue arrows indicate C-cycle processes and red arrows show N-cycle processes. Met./Str. litter – metabolic and/or structural litter; SOM – soil organic matter. *Set N fixation as an option when the plant N uptake is not enough for growth in terms of C investment in SM1, but go directly to soil mineral N pool in SM2 and SM3. (b) Simulation of annual ecosystem carbon storage capacity for 1996 to 2006 at Duke Forest by carbon in flux (NPP, *x*-axis) and ecosystem residence time (τ_E, *y*-axis) in the TECO model framework with three C-N coupling schemes (SM1, SM2, and SM3) and in the TECO C-only model (C). The inserted panel in the left bottom corner shows the τ_E in SM1, SM2, SM3, and the C-only model; the top right panel shows the mean ecosystem carbon storage simulated among SM1, SM2, SM3, and the C-only model; the right bottom panel shows the relative change in the simulated NPP and τ_E among the three schemes compared with the C-only model.

CASE 5 ACCELERATING THE PACE OF MODEL EVALUATION VIA ONLINE TOOLS

Increasing model complexity not only leads to a large divergence in simulations of the land carbon cycle by different models, but also tends to increase the computational consumption of model runs, which can become a bottleneck for model evaluations. As illustrated above, the traceability framework allows a carbon cycle model to be simplified and generalized into several traceable components, which can be further decomposed to quantify the structural sources of the uncertainty built into the full version of the model. Thus, an online traceability analysis system for model evaluation (TraceME) was built to accelerate the pace of model evaluation of the land carbon cycle, with earth system models in mind.

The TraceME system is a cloud-based platform (Zhou et al. 2021). It provides user-friendly interfaces and scientific workflow to automatically perform the model evaluation. On the website (http://www.traceme.org.cn/, last access: October 2020), the user can select the data of interest, and submit the task through the browser to complete the traceability analysis. The collaborative framework of TraceME provides a convenient data-sharing platform for all users to filter data of interest from the entire system for traceability analysis. Once a task is requested through a web browser, the scientific workflow of TraceME is triggered and it will execute the corresponding processes, such as data preprocessing, traceability analysis, and evaluation. The submitted data will be systematically decomposed into traceable components for quantifying the variance contributions of these components to land carbon dynamics. After the submitted task is completed, TraceME provides a visual interface to show and download the results in the forms of figures and Network Common Data Form-format (NetCDF) files. These files can be used to perform further analysis. TraceME is a convenient tool to evaluate models based on the traceability analysis framework. It has, for example, been applied to evaluate land carbon dynamics in CMIP6 earth system models (Zhou et al. 2021).

The traceability framework is developed for the terrestrial carbon cycle, but it can be extended to include nutrient and water processes. Further applications include the integration of benchmark analysis (see Chapter 19) with traceability analysis, the connection between structurally traceable components and model parameters, and the application of traceability analysis to understand climate-carbon cycle feedbacks in coupled land-atmosphere simulations with earth system models.

SUGGESTED READING

Xia, J., Luo, Y., Wang, Y.-P., & Hararuk, O. (2013). Traceable components of terrestrial carbon storage capacity in biogeochemical models. *Global Change Biology*, *19* (*7*), 2104–2116

QUIZ

1 In the steady-state traceability analysis, which variable is first decomposed into NPP and carbon residence time?
   a. Soil C stock
   b. Vegetation biomass
   c. Ecosystem total C stock
   d. Ecosystem C storage capacity
2 When the nitrogen cycle is incorporated into a carbon cycle model, which components in the terrestrial carbon cycle will be changed? What happened in the CABLE model?
3 Which model has the longer ecosystem carbon residence time, CABLE or CLM-CASA'? Why?
4 Describe the external forcings which can contribute to the large model uncertainty of the terrestrial carbon cycle.
5 How do TECO-CN, CLM4.5, and O-CN models differ in their approach to simulating the coupling between the terrestrial carbon and nitrogen cycles? How do these differences impact carbon cycle dynamics in the TECO model?

18 Applications of the Transient Traceability Framework

Lifen Jiang
Cornell University, Ithaca, USA

This chapter introduces the traceability analysis of transient carbon storage, which is a modification of the original traceability framework that relies on an assumption of steady state. Transient traceability analysis is particularly useful to address the origin and drivers of carbon flow of a system in a state of transition towards a new steady state – a transient system. We will illustrate how the transient traceability can be applied to address scientific questions with two examples. One tracks the differences in modeled carbon storage between two forest ecosystems, the second compares and contrasts outcomes of a set of Model Intercomparison Projects (MIPs) encompassing multiple land carbon models.

INTRODUCTION

Simulations by Earth system models in the Coupled Model Intercomparison Phase 5 project (CMIP5) showed that differences among the models entail large uncertainty in land carbon (C) storage (Jones et al. 2013). The spread of simulated future land C change across the models is even greater than the spread across the four radiative forcing scenarios, when the ensemble averages for each scenario are compared. Arora et al. (2020) compared model results from two CMIP phases (CMIP6 vs. CMIP5) and they found that the model mean values of the carbon-concentration and carbon-climate feedback parameters and their multi-model spread under a 1% per year CO_2 increase experiment did not change significantly across the two CMIP phases for both land and ocean. Moving forward, a big challenge is how to understand and reduce the uncertainty across models to achieve more reliable predictions.

Although land C models have become increasingly complex in recent decades with more and more processes incorporated, most current models have the same theoretical foundation and therefore share some of the same general properties, as described in Chapters 1 and 2. These shared theoretical foundations and properties enable many land C models to be represented or approximated in matrix form, as demonstrated in Chapter 5. With the matrix representations of the C cycle models, we are able to decompose the modeled land C storage into different traceable components, which is the unified diagnostic system for uncertainty analysis, an overview of which was provided in Chapter 9. Based on the common properties and matrix representations of land C models, Xia et al. (2013) developed a traceability framework to decompose steady-state C storage into traceable components. In Chapter 17 we saw how the framework could be applied to investigate differences in carbon storage across biomes and how differences in model structure influence the simulated effects of nitrogen cycling and land use change on the carbon cycle.

The traceability framework in its original form depends on a steady-state assumption for ecosystem carbon stocks. However, due to climate change and past disturbances, most ecosystems are not at steady state. The non-steady state is called transient state and the challenge is how to trace transient C storage dynamics. In this chapter, we first introduce a general framework for transient traceability analysis. Then, we illustrate how the transient traceability framework can be applied to address scientific questions using two examples. Our first example uses transient traceability analysis to track differences in C storage between two forest ecosystems. The second applies transient traceability analysis to three model intercomparison projects (MIPs) to identify the sources for the uncertainty in modeled carbon storage dynamics within each MIP and across the three MIPs.

A TRACEABILITY FRAMEWORK FOR TRANSIENT LAND CARBON STORAGE DYNAMICS

In order to realize the traceability of transient C storage, Luo et al. (2017) conducted a theoretical analysis to extend the steady-state traceability framework to work for systems in a transient state. The key was to add another term called C storage potential. This new term was introduced in Chapter 9 and is further discussed below. The core equation for traceability of transient C storage is from Luo et al. (2017) and as follows:

$$X(t) = \left(-A\xi(t)K\right)^{-1} B(t)u(t) - \left(-A\xi(t)K\right)^{-1} X'(t) \quad 18.1$$

where $X(t)$ is individual pool size at time t, which is a vector in a multi-pool model; A is a matrix of transfer coefficients between C pools; $\xi(t)$ is a diagonal matrix of environmental scalars to reflect the control of physical and chemical properties, e.g., temperature, moisture, nutrients, litter quality, and soil texture, on C cycle processes; K is a diagonal matrix of exit rates from donor pools, which encapsulates mortality rates for plant pools and decomposition coefficients for litter and soil pools; B is a vector of allocation coefficients of C input to each pool; $u(t)$ is C input, i.e., NPP or GPP; and $X'(t)$

DOI: 10.1201/9781032711126-23

is net change of any individual C pool at time t, which is a vector for a multi-pool model. The sum of X' of all individual C pools corresponds to net ecosystem production (NEP), or the sign opposite of net ecosystem exchange (NEE).

In this equation, the inverse of the product of $-A$, $\xi(t)$ and K, i.e., $(-A\xi(t)K)^{-1}$, is named chasing time, which is a matrix representing the timescale for the net C pool change to be redistributed in the network consisting of all C pools. The product of chasing time and the allocation coefficient B, i.e., $(-A\xi(t)K)^{-1}B(t)$, is the residence time of individual pools. The product of the residence time and C input is the maximum C that individual pools or the whole ecosystem can store at a time, which is defined as C storage capacity, X_c, that is, $X_c = (-A\xi(t)K)^{-1}B(t)u(t)$. The second term in Equation 18.1 – the product of chasing time and net C pool change $((-A\xi(t)K)^{-1}X'(\xi t))$ – represents redistribution of net change of individual C pools in the network. This redistribution of net C pool change indicates the potential of an individual pool or the whole ecosystem to gain or lose C. Therefore, it is named the C storage potential, X_p. So, we can derive another equation from the above descriptions as follows:

$$X(t) = X_c(t) - X_p(t) \qquad 18.2$$

This is the overall equation of the transient traceability framework of land carbon storage. To demonstrate how this transient traceability framework is a useful tool, we will walk through two example applications. The first application uses the framework exactly as described above to investigate the differences in carbon storage between two well-studied forest ecosystems, Duke Forest and Harvard Forest, and analyze the underlying mechanisms that explain these differences. The second application uses this general framework in combination with some other methods to decompose modeled land carbon storage in three MIPs: CMIP5, Trends in Net Land-Atmosphere Carbon Exchange (TRENDY), and Multiscale Synthesis and Terrestrial Model Intercomparison Project (MsTMIP). Decomposing differences in land carbon storage across models within each MIP and between the three MIPs helps us to understand why the models simulate different outcomes, and what features of the models' structure or parameters may be responsible for these differences.

TRANSIENT TRACEABILITY ANALYSIS OF CARBON STORAGE AT DUKE FOREST AND HARVARD FOREST

In this case study, we will go over an application of the matrix approach to trace differences in carbon storage dynamics between Duke Forest and Harvard Forest. Duke Forest, located in North Carolina, USA (35°58′41″N, 79°5′39″W), is evergreen needleleaf forest and the dominant tree species at this site is *Pinus taeda* (loblolly pine). Harvard Forest, located in Massachusetts, USA (42°32′16″N, 72°10′17″W), is deciduous broadleaf forest dominated by *Quercus rubra* (red oak) and *Acer rubrum* (red maple). These two study sites have contrasting ecosystem types and there are plenty of measured data at both sites available for calibrating our model.

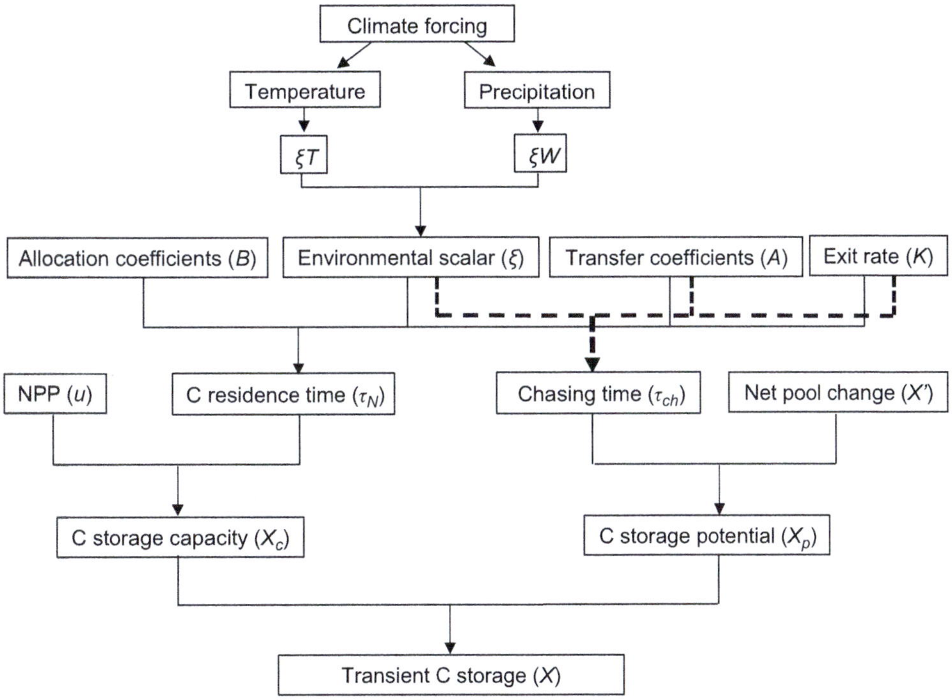

FIGURE 18.1 Schematic diagram of the traceability framework to analyze transient carbon storage dynamics of terrestrial ecosystems. ξ_W and ξ_T are water and temperature scalars, respectively. Dashed lines show the components that determine chasing time τ_{ch}

Adapted from Jiang et al. 2017.

The framework for transient traceability of land C storage dynamics is shown in Figure 18.1. Using this framework, we can decompose transient C storage into C storage capacity and C storage potential. C storage capacity is the product of NPP and C residence time. C storage potential is the product of chasing time and net C pool change. Further, chasing time is jointly determined by environmental scalars for temperature and precipitation, transfer coefficients, and exit rate. C residence time is jointly determined by the environmental scalars, transfer coefficients, exit rate, and allocation coefficients. Environmental scalars can be derived from climate forcing.

The model used is the TECO model, which has been described in Chapters 2 and 5. The procedure for this transient traceability analysis is as follows. We first calibrate the TECO model with GPP data for the two sites downloaded from the AmeriFlux website (available at http://ameriflux.lbl.gov/). We then run TECO to steady state by recycling ten years of forcing data from 1850 to 1859. Climate forcing data, including air and soil temperature, precipitation, photosynthetically active radiation, vapor-pressure deficit, and relative humidity are derived from an offline run of the Community Land Model version 4.5 (CLM4.5) for both historical (1850–2005) and RCP8.5 future (2006–2100) simulations. After that, we run the model in forward simulation mode from 1850 to 2100. We output each component (X, A, ξ, K, B) in Figure 18.1 and calculate transient C storage using Equations 18.1 and 18.2. We then verify the calculated transient C storage. That is, we compare direct model output of C storage with C storage calculated by the transient traceability framework. They are almost identical in both forests (Figure 18.2), confirming that the transient traceability framework works very well to reproduce the full model simulations.

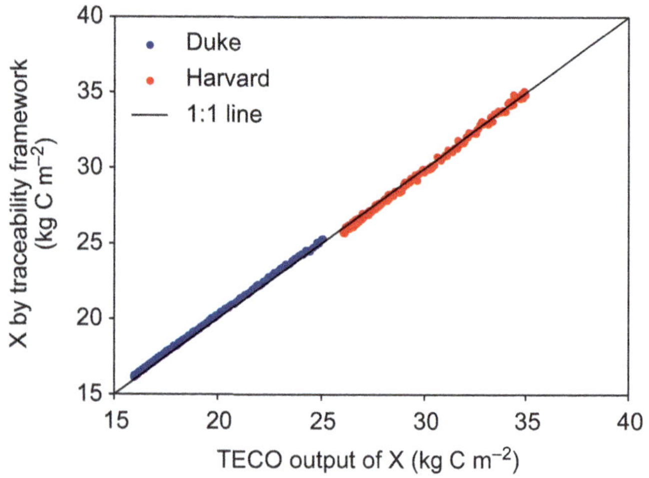

FIGURE 18.2 Correlation between direct model output of carbon storage (X, the sum of all carbon pools) by the Terrestrial ECOsystem (TECO) model and as calculated by the traceability framework for Duke and Harvard Forests

Adapted from Jiang et al. 2017.

The results for transient C storage, C storage capacity, and C storage potential are shown in Figure 18.3. The trajectories of transient C storage (X, or rather, the sum of the elements [pools] of the vector X), C storage capacity, X_c, and C storage potential, X_p, over time are similar between the two ecosystems, all increasing with time. Moreover, X closely tracks X_c in both ecosystems and X_p only accounts for a very small proportion of X, which indicates that transient C storage in these two ecosystems is predominated by the maximum C storage, i.e., carbon storage capacity (X_c), while carbon storage in response to climate change is relatively small. The most important difference between the sites is that Harvard Forest has higher X and X_c than Duke Forest (Figure 18.3a). Panel b shows the change of the three variables in the two ecosystems by the end of 2100, which are averages of the last ten years' C storage minus averages of the first ten years' values.

The components of transient C storage are shown in Figure 18.4. The two components of C storage capacity, that is, NPP and C residence time, are shown in panels a and b,

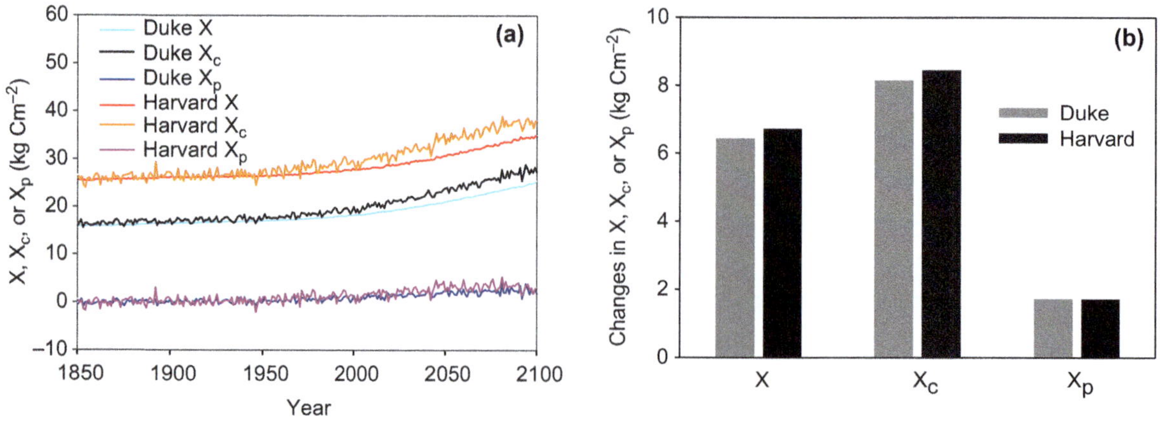

FIGURE 18.3 Transient total carbon storage (X, the sum of all carbon pools), carbon storage capacity (X_c), carbon storage potential (X_p) in Duke Forest and Harvard Forest and their changes by the end of the 21st century under the RCP8.5 scenario.

Adapted from Jiang et al. 2017.

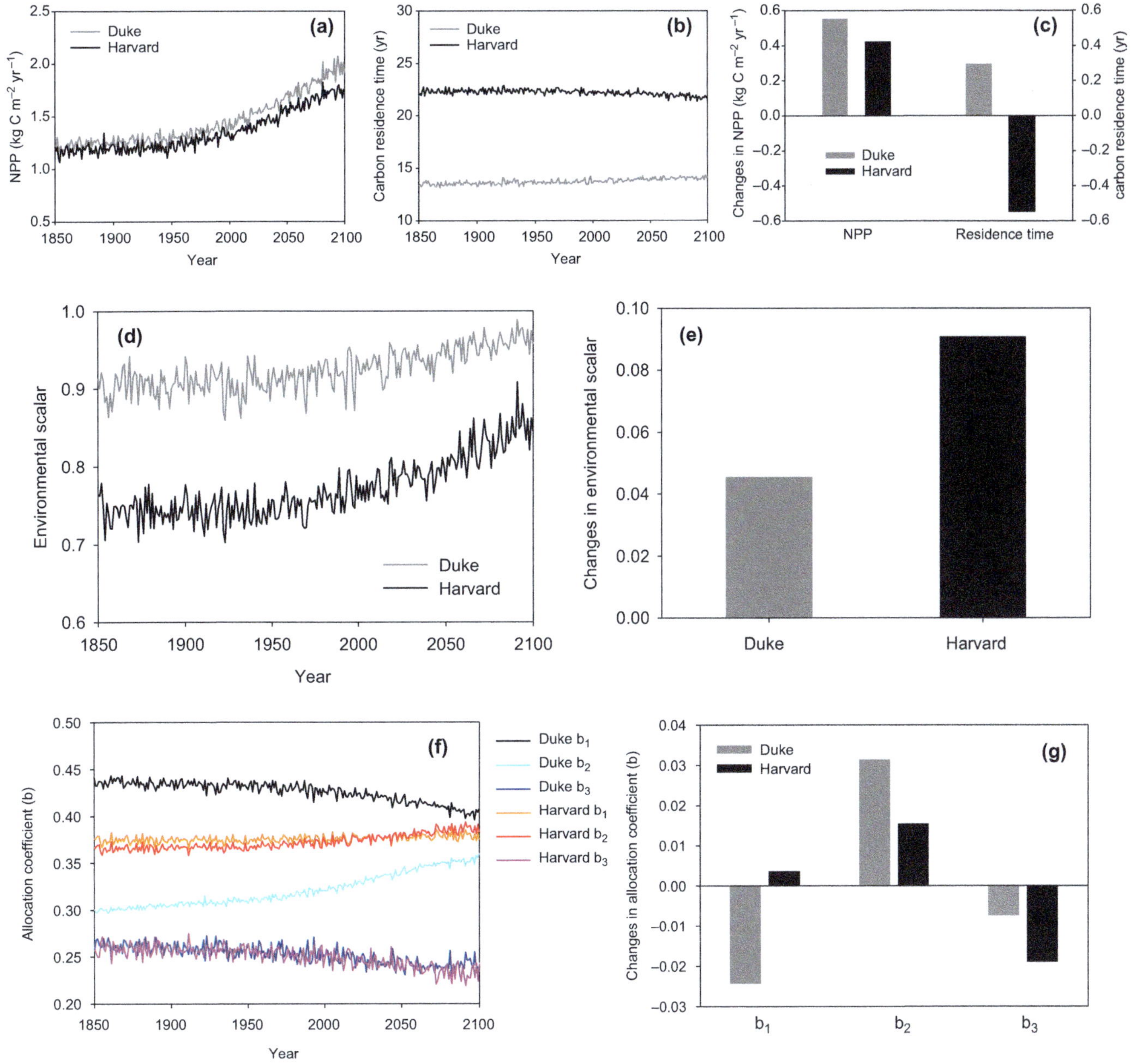

FIGURE 18.4 Net primary production (NPP), ecosystem carbon residence times, environmental scalars, and allocation coefficients of NPP to leaf (b_1), wood (b_2) and root (b_3) in Duke Forest and Harvard Forest, and their changes by the end of the 21st century.

Adapted from Jiang et al. 2017.

respectively. NPP is similar between the two ecosystems, but residence time shows different trends. In Duke Forest, C residence time increases over time, but in Harvard Forest, C residence time decreases. Panel c shows the changes of NPP and C residence time in these two ecosystems at the end of the 21st century compared to 1850.

Environmental scalars increase in both forests (Figure 18.4d, e), which signifies steadily decreasing environmental limitations on C processes. Allocation coefficients show different trends between the two ecosystems (Figure 18.4f). In Duke Forest, b_1 (allocation to leaf) and b_3 (allocation to root) both decline with time, but b_2 (allocation to wood) increases greatly. In Harvard Forest, allocation to leaves and wood both slightly increase, but allocation to roots decreases. The substantial increase in wood allocation at Duke Forest may explain why C residence time in this ecosystem increases; wood usually has longer residence time than leaves and roots. Similarly, panel g shows the changes of allocation coefficients by the end of the 21st century in these two ecosystems.

As shown in Equation 18.1 and Figure 18.1, C storage potential, X_p, is codetermined by the chasing time and net C pool change, X', which equates to NEP or NEE at ecosystem scale. Figure 18.5 shows the correlation between NEP and C storage potential in these two ecosystems. The coefficients of determination, R^2, are high in both Duke Forest (0.80) and Harvard Forest (0.79). This indicates that X_p is mostly determined by NEP rather than chasing time. Chasing time, represented by the slopes of the linear regressions between X_p and NEP, is an indicator of approximate time needed for transient C storage to reach C storage capacity. Chasing time in Duke Forest is shorter than that in Harvard Forest. In Duke Forest, it takes approximately 19 years for the changed carbon pool to be redistributed in the network. In Harvard Forest, this time is around 28 years. Having shorter chasing time, X_p in Duke Forest is lower than that in Harvard Forest.

To summarize this case study, the transient traceability framework can decompose modeled transient C storage into a few traceable components. This helps us to understand the mechanisms for modeled C dynamics in response to climate change in Duke Forest and Harvard Forest. For example, the difference in carbon storage capacity (X_c) between the two ecosystems is mostly caused by the difference in inherent C residence time. In addition, the contrasting responses of C residence time to climate change between the two ecosystems can be attributed to the different responses of allocation of NPP to plant parts (leaves, wood, and roots). This application demonstrates that the traceability framework can be used to understand how and why different ecosystems respond to climate change differently. Similarly, it can also be used to address how other global change drivers (such as land use change and elevated CO_2) affect land C storage dynamics across ecosystems in simulations with ecosystem models. When applied to global land models, it can also help investigate the differences across biomes under different environmental scenarios.

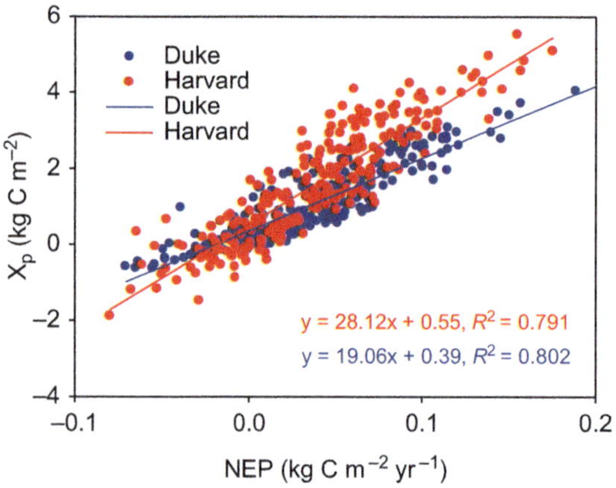

FIGURE 18.5 Correlation between net ecosystem production (NEP) and C storage potential (X_p) in Duke Forest and Harvard Forest.

Adapted from Jiang et al. 2017.

TRANSIENT TRACEABILITY ANALYSIS OF LAND CARBON STORAGE IN MODEL INTERCOMPARISON PROJECTS

Another important application of the transient traceability framework is to be used in MIPs to identify sources of uncertainty and thereby help improve model development. For example, Hou et al. (2023) conducted a MIP with eight land models (i.e., TEM, CENTURY4, DALEC2, TECO, FBDC, CASA, CLM4.5, and ORCHIDEE) in the matrix form to identify sources of uncertainty using both the traceability analysis and parameter manipulation. Baseline residence time and environmental scalars were identified as the major sources of model uncertainty (see detailed description in Chapter 9). Zhou et al. (2018) applied the transient traceability framework but with a modified method to diagnose the causes of uncertainty in modeled global annual land carbon storage within and across three MIPs: CMIP5, TRENDY, and MsTMIP.

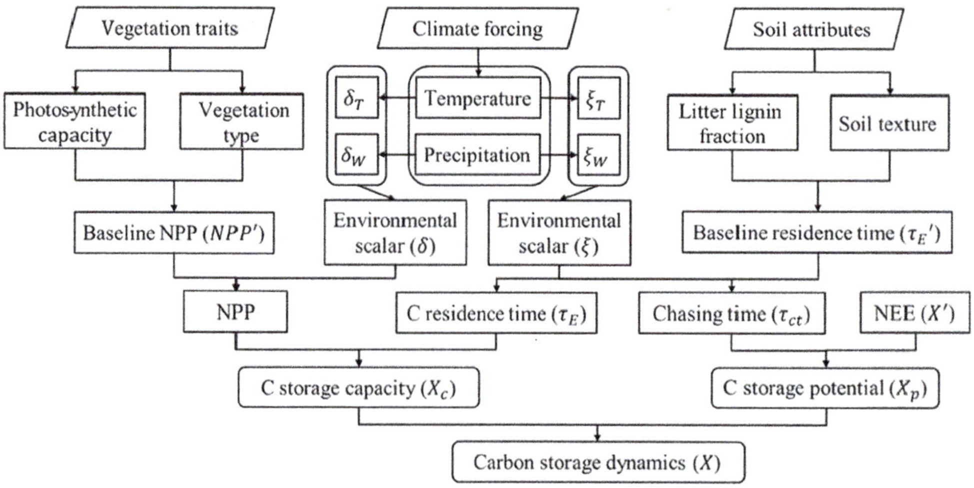

FIGURE 18.6 Schematic diagram of the transient traceability framework by Zhou et al. (2018) © American Meteorological Society. Used with permission. This framework traces the modeled transient carbon storage dynamics to carbon residence time, NPP, carbon storage potential, and their source factors.

The transient traceability analysis of carbon storage at Duke Forest and Harvard Forest introduced above is called authentic transient traceability analysis, that is, the modeled differences of C storage among ecosystems or among models can be traced back to differences in respective components in Equation 18.1, and finally to individual processes or parameters in the models. However, the application of authentic transient traceability analysis to MIPs requires much time to figure out the structures and parameterizations of all involved models in order to recode each in matrix form for a thorough model intercomparison. Due to the challenge in acquiring all the details of the involved models, in their analysis, Zhou et al. (2018) applied the transient traceability framework in combination with another technique, variance decomposition, to identify the underlying causes for uncertainty in simulated land carbon storage within and across three MIPs. We called this kind of analysis post-MIP transient traceability analysis to be distinguished from the authentic transient traceability analysis. Figure 18.6 shows the schematic diagram of this transient traceability analysis within and among the three MIPs.

In this post-MIP traceability analysis, they found that models differ a lot in the global annual carbon residence time, NPP, and carbon storage potential (Figure 18.7a–c). Usually, within each model, relative year to year variation in carbon residence time is much smaller than that of NPP. In addition, NPP in those models with a coupled nitrogen cycle (e.g., BNU-ESM, CESM1(BGC), and NorESM1-Me in CMIP5, and CLM4, CLM4VIC, ISAM, and DLEM in MsTMIP) is lower compared to other ESMs without a coupled nitrogen cycle. Carbon residence time and NPP show smaller variations across the nine dynamic global vegetation models in TRENDY in comparison to models in CMIP5 and MsTMIP. As a result, the variations in carbon storage capacity in TRENDY are not as large as in CMIP5 and MsTMIP.

The global annual carbon storage and carbon storage capacity also vary considerably across the models and the temporal dynamics of carbon storage and carbon storage capacity of the models are highly diverse (Figure 18.7d–f). The large range of carbon storage across those models is closely related to that of carbon storage capacity. The interannual trends of carbon storage in the models of the three MIPs are mainly affected by the carbon storage potential, because the sign and magnitude of carbon storage potential determine the direction and rate of carbon storage change, respectively. The interannual variability of carbon storage in TRENDY is much smaller than that in CMIP5 and MsTMIP.

Carbon residence time and NPP are further traced to their baseline values and environmental scalars. The differences in carbon residence time (or NPP) across the models are codetermined by the differences in baseline carbon residence time (or baseline NPP) and the environmental scalars (Figure 18.8). The baseline carbon residence time and baseline NPP among the models in the three MIPs can differ as much as threefold. In contrast, the environmental scalars are more convergent among the models for both carbon residence time and NPP. That indicates that the large differences in carbon residence time and NPP across models are due mainly to the differences in their baseline carbon residence time and baseline NPP, not much being caused by environmental scalars.

Finally, by adopting a variance decomposition method, Zhou et al. (2018) quantified the relative contribution of each component to the simulated global annual carbon storage for each MIP and for all the three MIPs together. The results revealed that variations in transient carbon storage are dominated by carbon residence time and NPP, and carbon storage potential only contributes less than 1% (Figure 18.9). Moreover, the baseline carbon residence time and baseline NPP contribute more than 90% to the variations in carbon residence time and NPP, respectively. In contrast, the contributions by temperature and water scalars to the variations in the carbon residence time and NPP are both less than 5%. As a consequence, the variations in simulated transient carbon storage across the models can be primarily attributed to the differences in models' baseline carbon residence time and baseline NPP.

The post-MIP approach adopted by Zhou et al. (2018) is a novel approach that provides an alternative way to understand the causes of the uncertainty of multiple models when authentic traceability analysis is not able to be realized due to the effort required to accommodate detailed information on the original models. Such post-MIP traceability analysis can be automatically run with a public platform, TraceME (v1.0), which is an online traceability analysis system for model evaluation on land carbon dynamics (Zhou et al. 2021). While post-MIP traceability analysis offers useful insight, it would be helpful to apply the authentic traceability analysis, as in the first case of this chapter, to MIPs to help identify the specific model components and assumptions that dominate model uncertainty and focus attention on those issues in need of closer scrutiny to improve model behavior.

After identification of the causes by which the models differ in their behavior, modelers can then use observational data to determine which models are more accurate than others in representing the actual processes. This is the realm of benchmark analysis, which will be introduced in detail in Chapter 19. In this way, model performance can be greatly improved towards more realistic projections.

SUMMARY

This chapter has demonstrated how the transient traceability framework can be applied to address different scientific questions with two case studies. This recently developed framework has the potential to be used more broadly in ways similar to the overview of steady-state traceability analysis of land carbon storage in Chapter 17. The ultimate goal of the transient traceability framework is to enhance our understanding of how terrestrial ecosystems respond to various environmental changes and to better incorporate such understanding in models to predict the future status of land carbon storage.

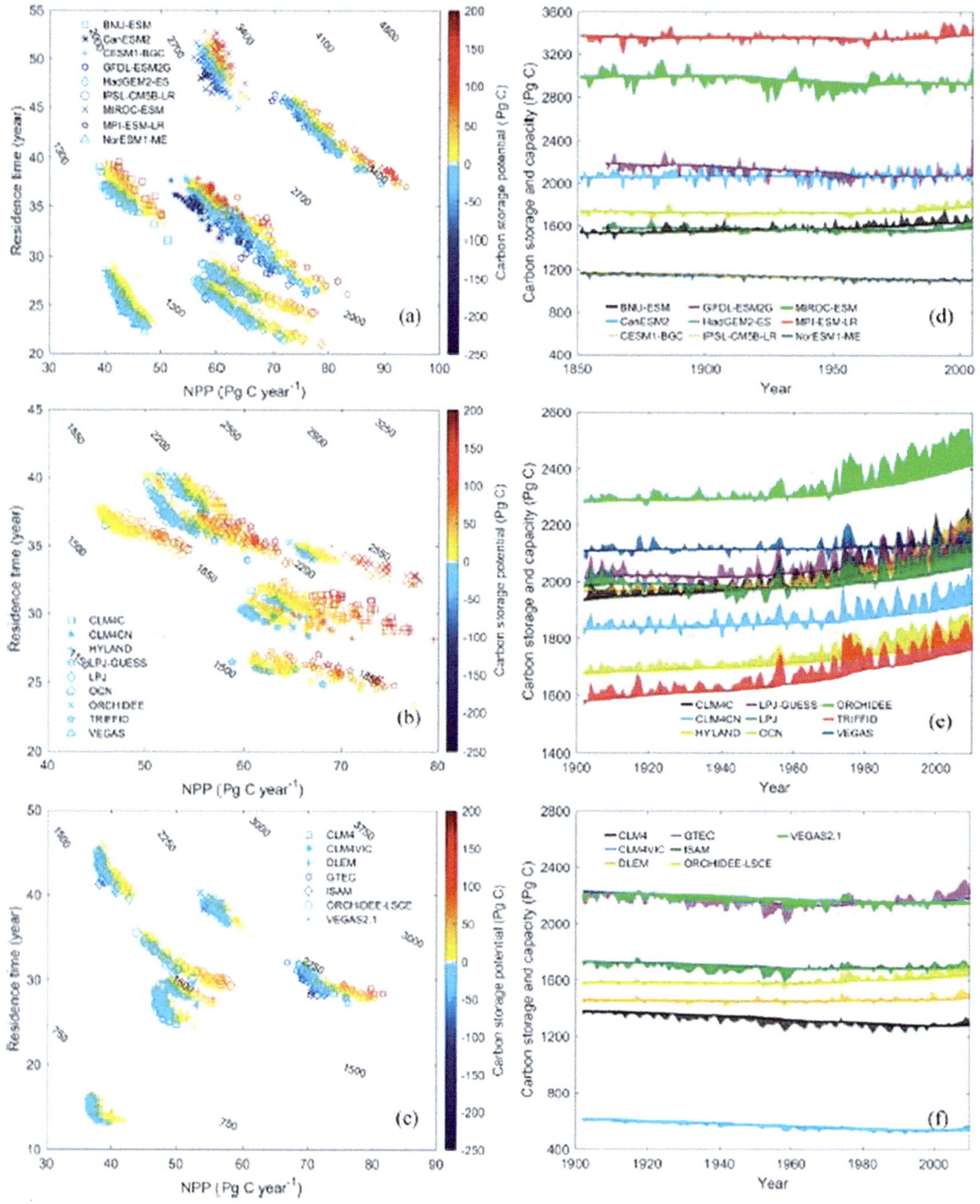

FIGURE 18.7 The 3D model output space (carbon residence time, NPP, and carbon storage potential), and time series of annual carbon storage (solid lines) with the shaded outlines indicating the year-to-year fluctuations due to changes in carbon storage capacity for the models in CMIP5 (a and d), TRENDY (b and e), and MsTMIP (c and f). The points in (a)–(c) represent the global annual values for the three variables. The contour lines in (a)–(c) represent the carbon storage capacity. Shading in (d)–(f) shows the values of the carbon storage potential for the models (positive above the solid lines, and negative below the solid lines).

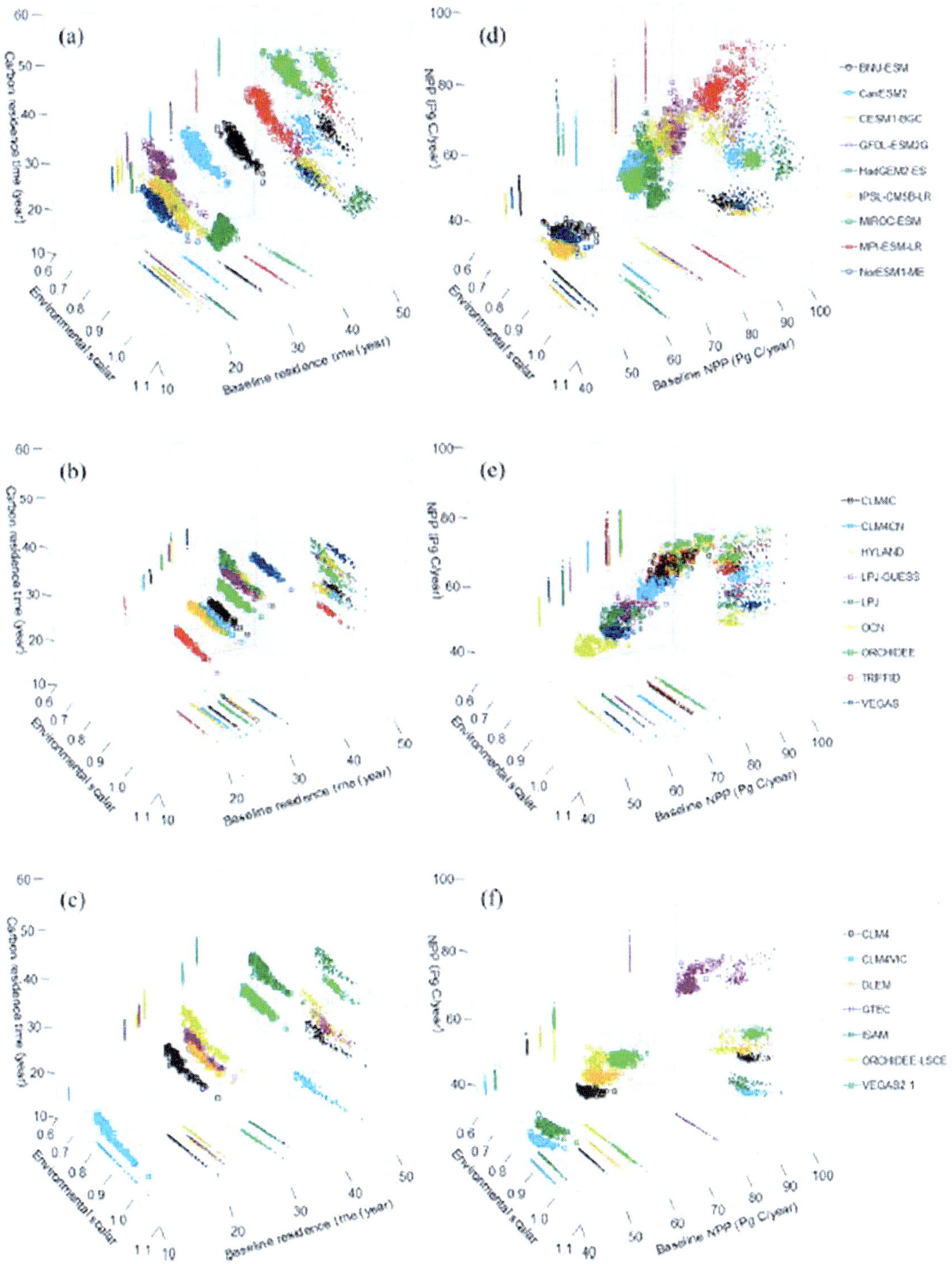

FIGURE 18.8 Decomposition of the carbon residence time into the baseline carbon residence time and the environmental scalar and decomposition of annual NPP into the baseline NPP and the environmental scalar for CMIP5 (a and d), TRENDY (b and e), and MsTMIP (c and f). The environmental scalar is a product of the temperature and water scalars, which convert the baseline carbon residence time and baseline NPP into their actual values.

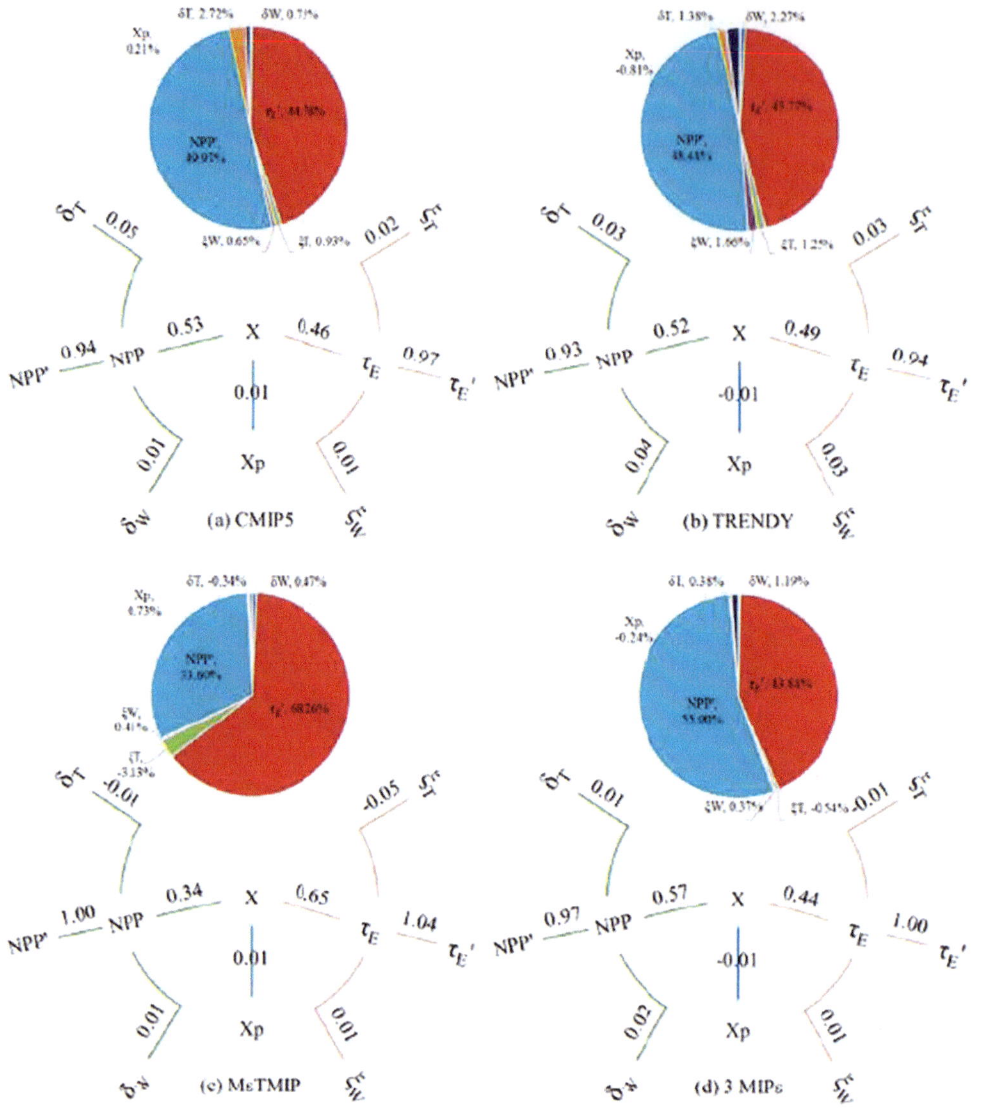

FIGURE 18.9 Variance decomposition of the carbon storage based on global annual data from models in the three MIPs. First, the variation of the carbon storage X is decomposed into that of the carbon residence time τ_E, NPP, and the carbon storage potential X_p. Second, variations of the carbon residence time and NPP are decomposed into their baseline values (τ'_E and NPP′) and the temperature (ξ_T and δ_T) and water (ξ_W and δ_W) scalars. Positive/negative values mean positive/negative contributions of the variables to the variation of carbon storage.

SUGGESTED READINGS

Jiang LF, Shi Z, Xia JY, Liang JY, Lu XJ, Wang Y, Luo YQ (2017) Transient traceability analysis of land carbon storage dynamics: procedures and its application to two forest ecosystems. *J Adv Model Earth Syst 9*:2822–2835.

Zhou S, Liang JY, Lu XJ et al. (2018) Sources of uncertainty in modeled land carbon storage within and across three MIPs: Diagnosis with three new techniques. *J Clim 31*:2833–2851.

QUIZ

1 Is transient C storage determined by C storage capacity and C storage potential? Why/why not?

2 Is carbon storage potential always positive? Why/why not?

3 Carbon storage potential is co-determined by:
   a. Carbon residence time
   b. Chasing time
   c. Net C pool change
   d. NPP

4 What scientific questions do you think the transient traceability framework can address? How can it be applied in each case?

19 Benchmark Analysis

Yiqi Luo
Cornell University, Ithaca, USA

Forrest M. Hoffman
Oak Ridge National Laboratory, Oak Ridge, USA

Tremendous progress has been achieved in the development of land models and their inclusion to account for land-atmosphere feedbacks in earth system models (ESMs). However, we still have insufficient knowledge on the performance skills of these land models, individually and in comparison to one another. This chapter introduces benchmark analysis, which is a procedure to measure performance of models against a set of defined standards. The benchmark analysis includes: (1) defining targeted aspects of model performance to be evaluated; (2) testing model performance in comparison with a set of benchmarks; (3) measuring model performance skill through quantitative metrics; and (4) evaluating model performance and offering suggestions for future model improvement.

INTRODUCTION

Over the past decades, tremendous progress has been achieved in the development of land models and their inclusion in earth system models (ESMs). State-of-the-art land models now account for biophysical processes (exchanges of water and energy) and biogeochemical cycles of carbon, nitrogen, and trace gases. They also simulate vegetation dynamics and disturbances. When coupled as components in ESMs, land models now allow simulation of land-atmosphere biophysical interactions and carbon-climate feedbacks. These models are now used for policy-relevant assessment of climate change and its impact on ecosystems or terrestrial resources, for instance through the IPCC process. However, there are still many gaps in our knowledge of the performance skills of these land models, especially when embedded in ESMs. Verifying the performance skills of land models would promote confidence in their predictions of future states of ecosystems and climate, and identify those models whose predictions are more likely to be accurate, where ensemble members diverge.

Model performance has traditionally been evaluated via comparison with observed data sets. 'Validation' by plotting model data side-by-side with observed data, or computing mismatch metrics such as root-mean-square-error, is traditionally the most common approach to model evaluation (Oreskes, 2003; Rykiel, 1996; see also Chapter 2). However, a land model typically simulates hundreds of biophysical, biogeochemical, and ecological processes at regional and global scales over hundreds of years. It would be unrealistic to undertake validation of so many processes at all spatial and temporal scales, even if observations were available. The complex behavior of these interacting processes can be realistically understood only if we holistically assess land models and their major components. Benchmark analysis is an approach that has been recently developed to evaluate the performance of land models.

Benchmark analysis is a standardized evaluation of one system's performance against defined reference data (i.e., benchmarks) that can be used to diagnose the system's strengths and deficiencies for future improvement (Luo et al., 2012). Benchmark analysis has been recently applied to evaluate land models against observations (Collier et al., 2018). A benchmark analysis has four elements: (1) targeted aspects of model performance to be evaluated; (2) benchmarks as defined reference data to evaluate model performance; (3) a scoring system of metrics to measure relative performance among models; and (4) evaluated performance of models and future improvement (Figure 19.1).

ASPECTS OF LAND MODELS TO BE EVALUATED

Land models typically simulate many processes. Although individual studies may assess only a few aspects of model performance, a comprehensive benchmark analysis is required to evaluate all these major components when land models are integrated with ESMs. The performance of a model should be evaluated for its baseline simulations over broad spatial and temporal scales, and modeled responses of land processes to global change.

The baseline state for biogeochemical cycles includes simulated global totals, spatial distributions, and temporal dynamics of gross primary production (GPP), net primary production (NPP), vegetation and soil carbon stocks, ecosystem respiration, litter production, litter mass, and net ecosystem production. For example, the International Land Model Benchmarking (ILAMB) project evaluates biomass, burned area, GPP, leaf area index (LAI), global net ecosystem carbon balance, net ecosystem exchange (NEE), ecosystem respiration, and soil carbon (Collier et al., 2018).

DOI: 10.1201/9781032711126-24

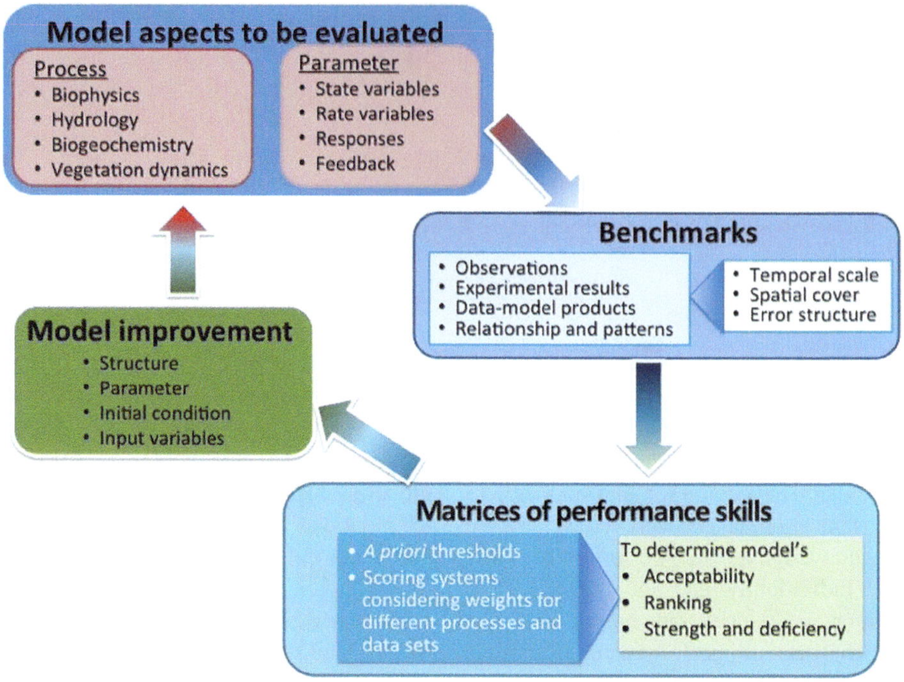

FIGURE 19.1 Schematic diagram of the benchmarking framework for evaluating land models. The framework includes four major components: (1) defining model aspects to be evaluated, (2) selecting benchmarks as standardized references to evaluate models, (3) developing a scoring system to measure model performance skills, and (4) stimulating model improvement.

Adopted from Luo et al., 2012.

To reliably predict future states of ecosystems under a changing environment, land models have to realistically simulate responses of land processes to disturbances and global change. Major global change factors include rising atmospheric CO_2 concentration, increasing land use and surface air temperature, altered precipitation amounts and patterns, and changing nitrogen (N) deposition. The direct effects of these global change factors are relatively easily benchmarked since we have general knowledge of how ecosystems respond to rising atmospheric CO_2 concentration, increasing temperature, altered precipitation, and changing nitrogen deposition. However, indirect effects of these factors on ecosystem carbon processes are not well understood, although many field experiments have been conducted. Thus, it is more difficult to benchmark model performance in predicting future states of ecosystems.

REFERENCE DATA SETS AS BENCHMARKS

A comprehensive benchmarking analysis usually uses a set of benchmarks, against which land model performance can be evaluated (Table 19.1). Benchmarks could consist of direct observations, results from manipulative experiments, data-model products, or data-derived functional relationships. Direct observations and experimental results are generally accepted to be the most reliable benchmarks for model performance and are typically referred to as reference data. Reference data that are often used for benchmarking biogeochemical models include global data products of GPP, NPP, soil respiration, ecosystem respiration, plant biomass, and soil carbon. When they are used in a benchmarking analysis, reference data sets are usually assessed and weighted for their degree of certainty, scale appropriateness, and overall importance of the constraint or process to model predictions (Collier et al., 2018). The ILAMB project evaluates eight variables using a variety of reference data as listed in Table 19.1.

Land models can also be evaluated on their simulated variable-to-variable relationships in comparison with relationships in observations. For example, model representations of the relationships that GPP exhibits with precipitation, evapotranspiration, and temperature are often assessed. Such variable-to-variable relationships are quantified over a time period from reference data sets and used as benchmarks for the relationships diagnosed in models. This approach is particularly effective to understand the consistency between the observed and simulated sensitivity of ecosystem responses to climate change.

BENCHMARKING METRICS

A comprehensive benchmarking study usually uses a suite of metrics across several variables to holistically assess model performance at the relevant spatial and temporal scales. Many statistical measures are available to quantify mismatches between multiple modeled and observed variables. Five metrics were developed for ILAMB to evaluate model

TABLE 19.1
Reference data sets used to measure ecosystem and carbon cycle performance

Variables	Reference data sets	Description
Biomass	Tropical (Saatchi et al., 2011)	forest carbon stocks in tropical regions across three continents
	NBCD2000 (Kellndorfer et al., 2013)	aboveground biomass and carbon baseline data in North America
	USForest (Blackard et al., 2008)	U.S. forest biomass
Burned area	GFED4S (Giglio et al., 2010)	variability and long-term trends in burned area
GPP	Fluxnet (Lasslop et al., 2010)	net ecosystem exchange, photosynthesis, and respiration
Leaf area index	AVHRR (Myneni et al., 1997)	global land cover, leaf area index and FPAR
	MODIS (De Kauwe et al., 2011)	leaf area index product for a region of mixed coniferous forest
Global NECB	GCP (Le Quéré et al., 2016)	global carbon budget 2016
Net ecosystem exchange	Fluxnet (Lasslop et al., 2010)	net ecosystem exchange, photosynthesis, and respiration
Ecosystem respiration	Fluxnet (Lasslop et al., 2010)	net ecosystem exchange, photosynthesis, and respiration
Soil carbon	HWSD (Todd-Brown et al., 2013)	Harmonized World Soil Data
	NCSCDV22 (Hugelius et al., 2013)	organic carbon storage to 3m depth in soils of the northern circumpolar permafrost region.

NECB = net ecosystem carbon balance; FPAR = fraction of photosynthetically-active radiation.

performance. The five metrics are to measure bias, root-mean-square-error (RMSE), phase shift, interannual variability, and spatial distributions (Collier et al., 2018).

The bias measures differences between the mean value of the reference data and that of the model over the same time period and the same spatial area. For example, the bias of GPP between the reference data and the model (e.g., Community Land Model version 4.5, CLM4.5) is calculated between their respective means in each grid cell where both reference data and modeled values are available. To account for the bias due to the variability at any given spatial location, the bias is nondimensionalized as a relative error to measure the bias score.

RMSE is computed as the square root of the mean square error between modeled values and the reference data over a time period. The RMSE is normalized by the centralized RMSE of the reference data set to get a relative error as a score. By scoring the centralized RMSE, the bias is removed from the RMSE, allowing the RMSE score to be focused on an orthogonal aspect of model performance.

The phase shift is evaluated for the annual cycle of many data sets that have monthly variability by comparing the timing of the maximum of the annual cycle of the variable at each spatial cell across the time period of the reference data set. The phase shift is calculated as the difference between the reference and model data sets by subtracting their respective maximum values in days.

The interannual variability in model simulations is evaluated by removing the annual cycle from both the reference data and the model. A score is then computed as a function of their differences over space.

The spatial distribution of any time-averaged variable is evaluated by computing the standard deviation of modeled values over space normalized by the standard deviation of the reference data. The spatial correlation is also calculated for the period mean values of reference data and modeled values. A score is assigned applying a penalty for large deviation of the normalized standard deviations and the spatial correlation from a value of 1.

The overall score for a given variable and data product is a weighted sum of the five metrics, producing a single scalar score for each variable for every model or model version. Readers who are interested in details of these metrics may study the paper by Collier et al. (2018).

PERFORMANCE OF THREE CLM VERSIONS AND FUTURE IMPROVEMENTS

The metrics for bias, RMSE, seasonal cycle phase, spatial distribution, interannual variability, and variable-to-variable assessments were applied to evaluate three CLM versions (CLM4 vs. CLM4.5 vs. CLM5) under two forcing data sets (GSWP3v1 vs. CRUNCEPv7) (Lawrence et al., 2019). The quality of the simulations across model generations was found to be generally improving. CLM5 outperforms CLM4 for the majority of assessed variables (Figure 19.2). The improvements from CLM4.5 to CLM5 were relatively subtle in that several variables show improvement (e.g., biomass, burned area, LAI, net ecosystem carbon balance, NEE, and ecosystem respiration) but others show degradation (e.g., soil carbon).

The functional relationships were also assessed between two variables (e.g., precipitation vs. GPP or LAI) (Figure 19.3). CLM5 performed better than CLM4 or CLM4.5 for the relationships between GPP and climate variables. However, the relationship between GPP and surface air temperature slightly degraded from CLM4.5 to CLM5.

The ILAMB benchmark analysis provides some insights into model development. An improvement or degradation trend between two CLM versions can result from a mix of scores for individual metrics. The degradation in the simulations of soil carbon stocks from CLM4.5 to CLM5 may be partially

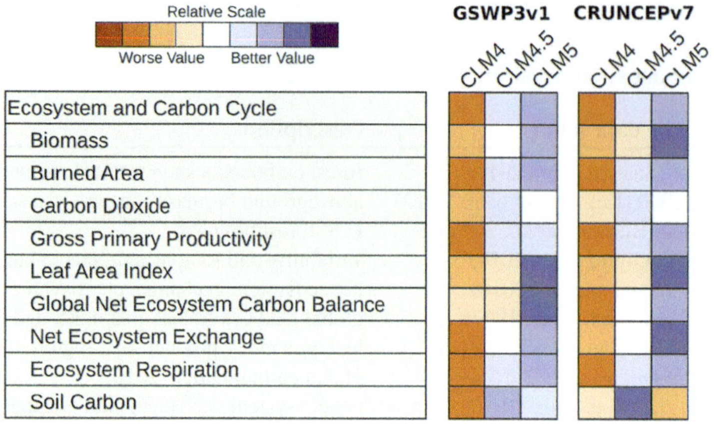

FIGURE 19.2 Evaluation of performance of CLM4, CLM4.5, and CLM5 under two sets of forcing, GSWP3v1 and CRUNCEPv7. A stoplight color scheme is used to indicate aggregate performance for each model by variable.

Adopted from Lawrence et al., 2019.

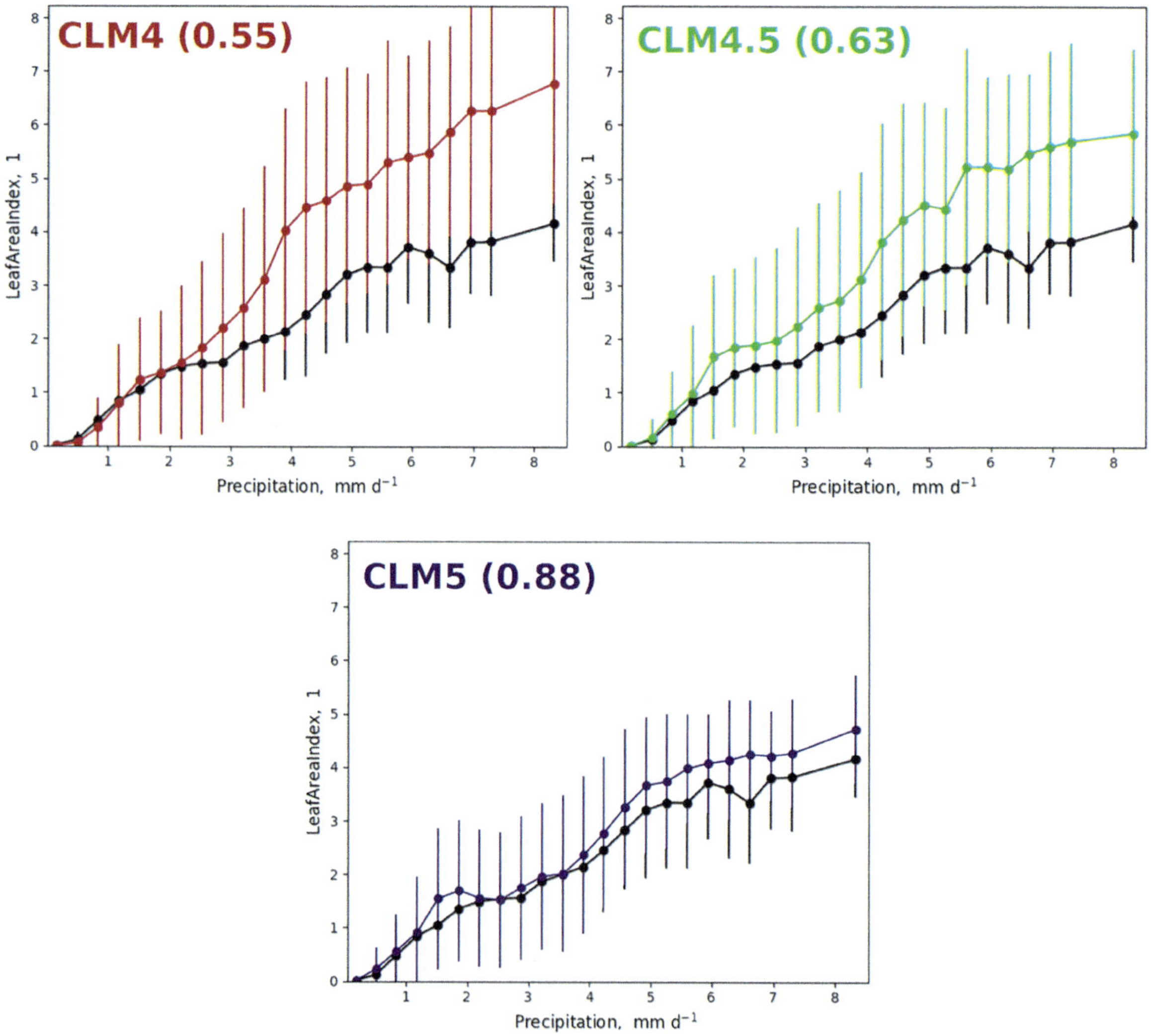

FIGURE 19.3 Variable-to-variable comparison between annual precipitation and LAI for CLM4, CLM4.5, and CLM5 under the GSWP3v1 forcing. The black line is the observationally derived relationship. Error bars indicate the ±1 standard deviation of LAI for all grid cells that lie within that precipitation bin. Values in parentheses are the scores for that comparison.

Adopted from Lawrence et al., 2019.

linked to high uncertainty in the observational estimates. Another metric evaluates the models against apparent soil carbon turnover time, showing an improvement from CLM4.5 to CLM5. The disagreement between two metrics of soil carbon may suggest the need for future improvement of observationally constrained estimates.

Model performance depends on three elements: model structure, parameterization, and forcing (see Chapters 21 and 29). The model structure that simulates soil carbon dynamics in CLM is primarily based on first-order kinetics. Although this model structure has been questioned, it has been demonstrated that almost all data sets from studies of litter decomposition and soil incubation are consistent with it (see Chapter 1). Model parameterization is likely the main cause of the model-data mismatch. Chapter 38 discusses methods to improve model parameterizations of CLM5 to improve model performance.

CONCLUSIONS

A four-component benchmark analysis was outlined: (1) identification of aspects of models to be evaluated; (2) selection of benchmarks as standardized references to evaluate models; (3) a scoring system to measure model performance skills; and (4) evaluation of model performance to inform model improvement. The International Land Model Benchmarking (ILAMB) project has developed an open-source model benchmarking software package to score model performance. ILAMB has developed a suite of reference data sets as benchmarks, five metrics plus variable-to-variable relationships as the scoring system to evaluate models or model versions. The ILAMB package has been applied to perform comprehensive model assessment across a wide range of land variables. Such benchmark analysis offers insights into strengths and weaknesses of different models or model versions for identifying future improvements.

SUGGESTED READINGS

Collier, N., F. M. Hoffman, D. M. Lawrence, G. Keppel-Aleks, C. D. Koven, W. J. Riley, M. Mu, and J. T. Randerson (2018), The International Land Model Benchmarking (ILAMB) System: Design, Theory, and Implementation, *J. Adv. Model. Earth Syst.*, *10*(*11*):2731–2754, doi:10.1029/2018MS001354.

Luo, Y. Q., J. T. Randerson, G. Abramowitz, C. Bacour, E. Blyth, N. Carvalhais, P. Ciais, D. Dalmonech, J. B. Fisher, R. Fisher, P. Friedlingstein, K. Hibbard, F. Hoffman, D. Huntzinger, C. D. Jones, C. Koven, D. Lawrence, D. J. Li, M. Mahecha, S. L. Niu, R. Norby, S. L. Piao, X. Qi, P. Peylin, I. C. Prentice, W. Riley, M. Reichstein, C. Schwalm, Y. P. Wang, J. Y. Xia, S. Zaehle, and X. H. Zhou (2012), A Framework for Benchmarking Land Models, *Biogeosci.*, *9*(*10*):3857–3874, doi:10.5194/bg-9-3857-2012.

QUIZ

1. What are the similarities and differences between model validation and benchmark analysis?
2. How does benchmark analysis evaluate model performance?
3. What variables in carbon cycle models would you choose to be evaluated by a benchmark analysis?
4. What data sets do you think would be important to be used as benchmarks to evaluate models?
5. What five metrics does the ILAMB package use to score model performance?

20 Practice 5

Traceability Analysis for Evaluating Terrestrial Carbon Cycle Models

Jianyang Xia and Jian Zhou
East China Normal University, Shanghai, China

The practice is designed to help you learn traceability analysis to identify sources of model uncertainty in predicting terrestrial carbon (C) storage. All practices are performed in the training software CarboTrain. With this tool, you will apply traceability analysis to simulation results from a matrix form model (called authentic traceability analysis) and to model intercomparison projects (MIP) without matrix models (i.e., post-MIP traceability analysis). The authentic traceability analysis will show you how simulation results from a matrix model are explained by traceable components over space and among biomes. The post-MIP traceability analysis can help you understand the sources of uncertainty among different models.

INTRODUCTION

Traceability analysis provides an approach to divide the simulated land carbon dynamic into several traceable components, such as carbon (C) storage capacity, gross primary productivity (GPP), C residence time, and environmental scalars. This practice offers two exercises. The first uses authentic traceability analysis in which simulation results by the CABLE matrix model are explained by traceable components hierarchically over space and among biomes. The authentic traceability analysis for the first exercise is described in detail in Chapter 17, based on the study by Xia et al. (2013). The second exercise uses post-MIP traceability analysis to attribute variations in modeled land C storage among three CMIP6 models (i.e., CESM2, CNRM-ESM2-1, and IPSL-CM6A-LR) over 1980–2000 to different sources. The post-MIP traceability analysis is described in detail in Chapter 18. More information is available in papers by Zhou et al. (2018, 2021). The authentic traceability analysis can pinpoint model uncertainty to individual processes and/or specific parameter values but requires matrix models. In comparison, the post-MIP traceability analysis can be applied to any modeling results to understand sources of uncertainty but may not be able to trace uncertainty to specific processes and/or parameters.

The exercises are performed in the training software CarboTrain by selecting Unit 5 and Exercises 1–2. Click *Run Exercise* to generate the figures.

Exercise 1 Authentic traceability analysis

This exercise helps you learn to do traceability analysis with a matrix model derived from the Community Atmosphere-Biosphere-Land Exchange (CABLE) model. CABLE is one of the global land models used for simulating terrestrial biogeochemical and biophysical processes. The C cycle diagram of the CABLE model is shown in Chapter 14, Figure 14.1. CABLE has nine carbon pools, including the plant pools (leaf, root, and wood), litter pools (metabolic and structural litter as well as coarse woody debris) and three soil pools (microbial biomass, slow and passive soil organic matter). The matrix form of the CABLE model we will use here was derived by Xia et al. (2013). In this exercise, the spatial resolution is 1 × 1°. The land grid cells are categorized into nine biomes including Evergreen Needleleaf Forest (ENF), Evergreen Broadleaf Forest (EBF), Deciduous Needleleaf Forest (DNF), Deciduous Broadleaf Forest (DBF), C_3 Grassland (C3G), C_4 Grassland (C4G), Tundra, and Barren/sparse vegetation (Barren).

In the main window of CarboTrain (Figure 20.1), choose Unit 5 and Exercise 1, and then set the output path of the running results by clicking the "**Set Output Folder**" button. After you have done all the above steps, click '**Run Exercise**' to start the exercise.

A pop-up window will show up with the message "Task submitted!" after you click the 'Run Exercise' button as shown in Figure 20.2. After clicking 'OK', the task will run automatically on your computer.

The command window shows the running processes of the task and you can see which step of the program is complete as shown in Figure 20.3. When the task is finished, another pop-up window will show up with the message "Finished!" as shown in Figure 20.2. The software will generate the running results to the output path you set before.

Figure 20.4 shows an example of the output results after the task is completed. Enter the output path you set before, you will see a directory named unit5_exercise1. Enter this directory and find the files as shown in Figure 20.4.

There are four data files in the folder unit5_exercise1/dataSource. The simulated global distribution of total ecosystem C storage capacity by the CABLE model is recorded in data_global_ctot.csv. The 'nan' values represent ocean grids or non-vegetated lands. The global map of the total ecosystem C storage capacity is shown in Figure 20.5a.

The file `data_npp_resTime.xlsx` provides the simulated net primary productivity (NPP) and ecosystem C residence time in each land grid cell. Figure 20.5b shows

DOI: 10.1201/9781032711126-25

FIGURE 20.1 Steps to run Exercise 1.

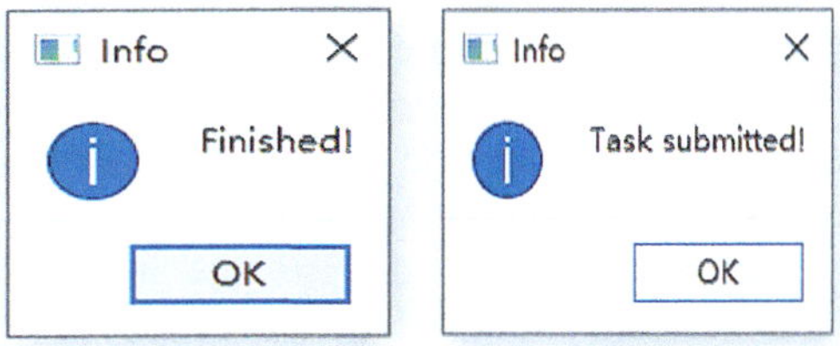

FIGURE 20.2 Tips at the beginning and end of the task.

how NPP and ecosystem C residence time together determine the spatial difference in ecosystem C storage capacity in the CABLE model. For example, ENF has an intermediate NPP (0.39 kg C m^{-2} yr^{-1}) and a relatively long C residence time (86.4 years), leading to the largest total ecosystem carbon storage capacity (34.1 kg C m^{-2}) among the nine biomes. By contrast, Tundra has a small ecosystem C storage capacity (8.7 kg C m^{-2}) due to the low NPP (0.1 kg C m^{-2} yr^{-1}), though its C residence time is long (141.2 years).

The file `residence_components.xlsx` contains simulated results of total ecosystem C residence time, baseline residence time, and environmental scalar in each land grid cell. Figure 20.5c shows how baseline C residence time and environmental scalar jointly determine the global distribution of ecosystem C residence time in the CABLE model. It is clear that the order of ecosystem C residence time among biomes is different from that of baseline C residence time. In CABLE, ecosystem C residence time changes with biome type as DNF (163.3 years) > Tundra (141.2 years) > ENF (86.4 years) > Shrubland (52.6 years) > DBF (33.3 years) > C3G (26.6 years) > EBF (26.3 years) > Barren (20.4 years) > C4G (17.5 years).

The `environmentalScalars.xlsx` file provides the mean annual precipitation, mean annual temperature, water scalar, and temperature scalar in each land grid cell. As shown in Figure 20.5d–e, the environmental scalars link the climate forcings directly to terrestrial C cycle processes in the CABLE model. Figure 20.5e shows that the temperature scalar varies systematically among biomes in the CABLE model, whereas the mean water scalar is distributed in a narrow range from 0.65 in EBF to 0.87 in DNF.

The figures of this exercise can be found in unit5_exercise1/output_figs. More details about the CABLE model and simulations in this exercise are provided in Chapter 17.

Questions:

1 Which environmental factor, water or temperature, contributes more to the difference in ecosystem C residence time among biomes in the CABLE model? Why?
2 Which biome has the longest baseline C residence time? Why?
3 Would you expect the incorporation of nitrogen cycling to influence the simulated ecosystem C storage compared to the C-only model? How can this question be explored by applying the authentic traceability analysis with the CABLE model?

Exercise 2 Post-MIP traceability analysis

In this exercise we will practice performing uncertainty analysis of carbon cycle modeling after simulations were done without models being converted to matrix equations (see Chapter 18). Because the outputs of transient land C storage rather than long-term steady state carbon storage (i.e., carbon storage capacity) are available from CMIP6 models, we compare the simulation results of the three CMIP6 models over 1980–2000.

Launch CarboTrain and select Unit 5 and Exercise 2. Open the *Config 5* tab, where you can customize the spatial and

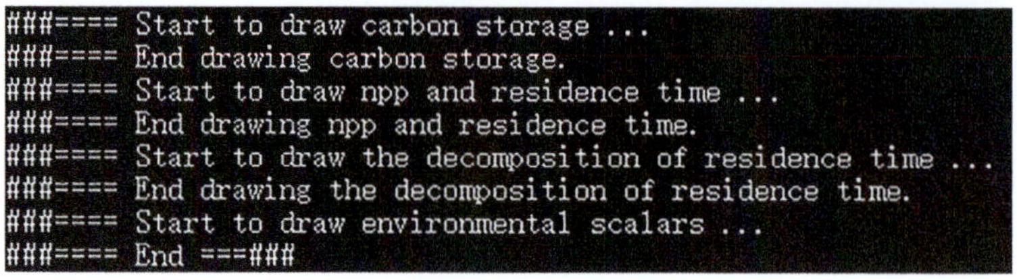

FIGURE 20.3 Prompt in CMD interface when the program is running.

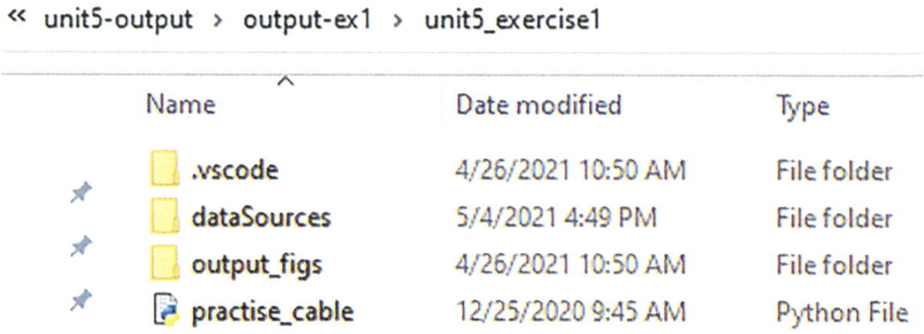

FIGURE 20.4 Output results after the task is complete.

temporal ranges of the running results by entering latitude and longitude and temporal ranges. Please note that the data entered here should be within the allowable range (−90 to 90 for latitude, 0 to 360 for longitude, 1980 to 2000 for the temporal range). Set the output path by clicking *Set Output Folder*. Finally, click *Run Exercise* to start the traceability analysis of the three CMIP6 models. (Note that the task running time is related to the selected ranges and the configuration of the computer.)

The command window shows the running processes of the task as shown in Figure 20.6. When the task is finished, results will appear in the output path you set before (Figure 20.7).

Here are descriptions of the files and the workflow of the model evaluation system in CarboTrain. All files can be found in or under the unit5_exercise2 folder in the output directory you specified earlier. The package consists of a file named Preset.txt, three Python scripts (Main_traceme.py, AnnualTAT.py, RegionTAT.py), and three subfolders (data, results, R_docs). Preset.txt contains a specification of the CMIP6 model outputs stored in the data folder. Traceability analysis requires the outputs of each model to include total C storage (vegetation C, soil C and/or litter C), GPP, NPP, temperature, and precipitation. After the task is submitted, the Main_traceme.py script reads the information in Preset.txt to preprocess the data. The preprocessed data is transferred to the AnnualTAT.py and RegionTAT.py scripts to perform temporal and spatial traceability analysis, respectively. The R_doc folder contains the R language script used to calculate the variance contribution of different components to C storage in the temporal traceability analysis. The results of the traceability analysis are output in the results folder.

We will now step through using the traceability analysis to analyze the differences in land C storage among three CMIP6 models (i.e., CESM2, CNRM-ESM2-1, and IPSL-CM6A-LR). In the results folder, you can find the results of the temporal traceability analysis on the three CMIP6 models over 1980–2000. First, you can find differences in the time series of simulated C storage among the models (temporal-1-Carbon-Dynamic.png; Figure 20.8a). It is clear that the IPSL-CM6A-LR model has the lowest land C storage among the three models for the simulation period. The traceability analysis decomposes the C storage into C storage capacity and potential (Figure 20.8a). The result showed that the lowest land C storage in IPSL-CM6A-LR was due to the lowest C storage capacity rather than the C storage potential. Then, ecosystem C storage capacity is further decomposed into NPP and C residence time (temporal-2-NPP-ResidenceTime.png; Figure 20.8b). We can see that the lowest ecosystem C storage capacity in IPSL-CM6A-LR was driven by the shortest ecosystem C residence time among the three models. Furthermore, NPP is decomposed into GPP and C use efficiency (CUE) (temporal-3-GPP-CUE.png; Figure 20.8c). Ecosystem C residence time can be decomposed into baseline C residence time and environmental scalars (temporal-4-Envs-baselineResidenceTime.png; Figure 20.8d). The simulated environmental scalars have not been provided for each model, so we cannot here further evaluate how climate forcings influence the baseline C residence time. However, among the three earth system models in this exercise, it is clear that the model parameterization of C allocation, C age, and C transfers among pools are the most important contributors to the large difference in global land C dynamic in Figure 20.8a.

In the results/nc-files folder, you can access the NetCDF-format data of the temporal and spatial traceability analysis for each model. In the results/figures folder, you can find the outputs that show the results of the spatial traceability analysis on the three models over 1980–2000. Each figure includes the global distribution of a variable simulated by

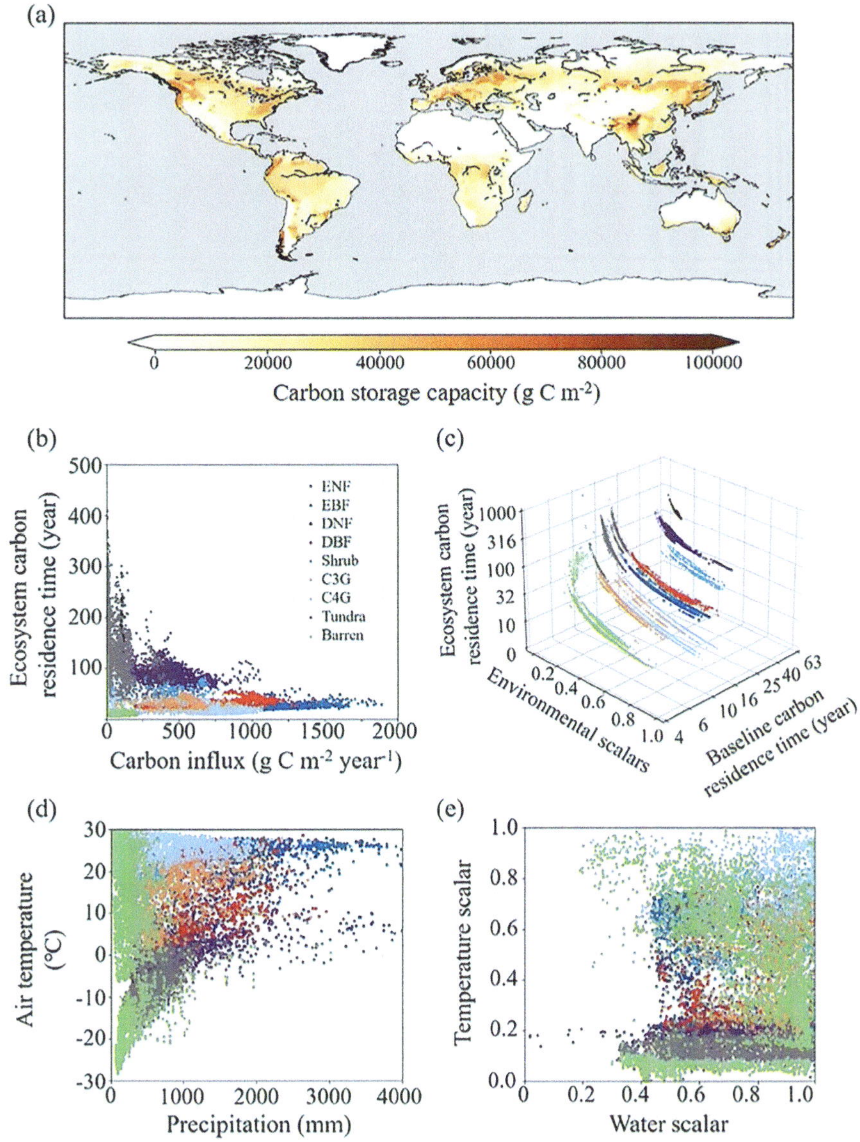

FIGURE 20.5 Simulated spatial distribution of terrestrial carbon storage capacity and its traceable components by the CABLE model. (a) Spatial distribution of total ecosystem C storage capacity; (b) determination of the ecosystem carbon storage capacity by NPP and ecosystem residence time in various biomes; (c) dependence of ecosystem C residence time on baseline C residence time and the environmental scalar in various biomes; (d) global distribution of major biomes in relation to annual temperature and precipitation; (e) global distribution of major biomes in relation to water and temperature scalars.

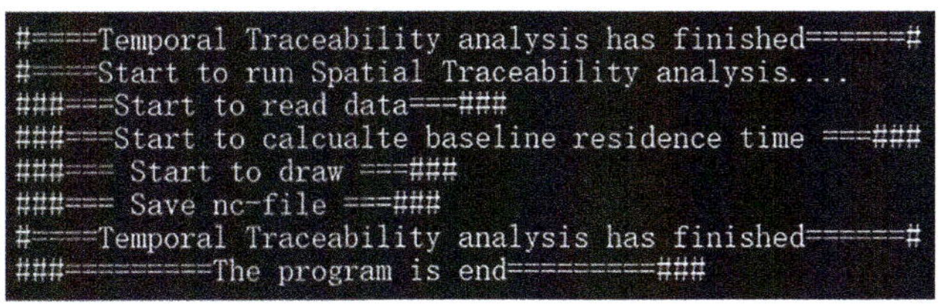

FIGURE 20.6 Prompt in CMD interface when the program is running.

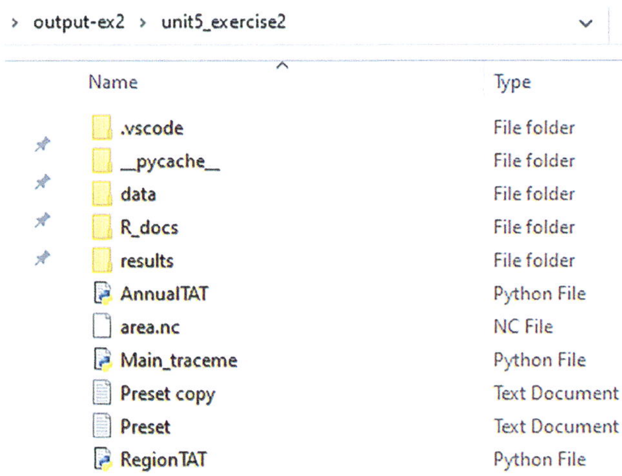

FIGURE 20.7 Output results after the task is done.

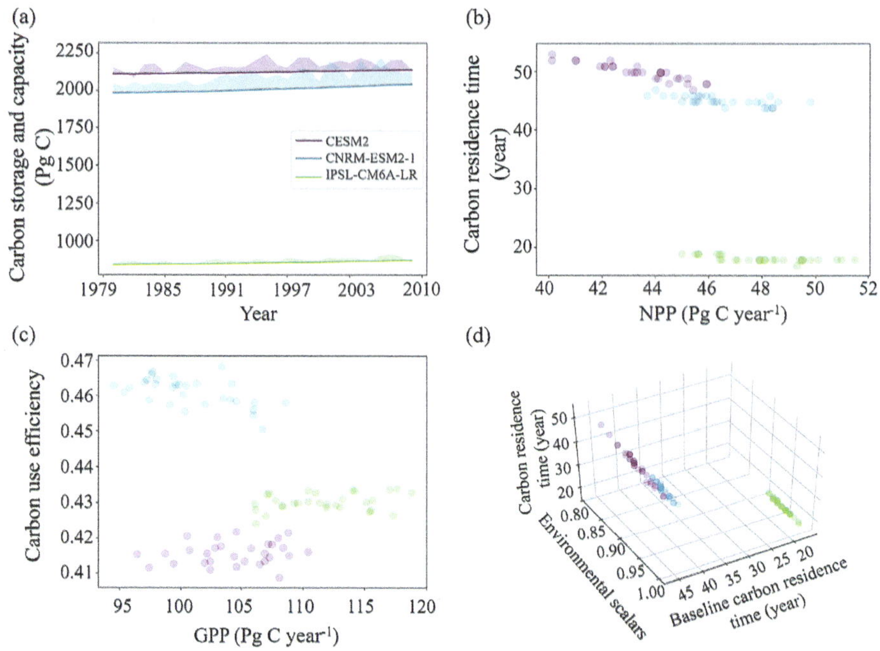

FIGURE 20.8 Results of the temporal traceability analysis on three CMIP6 models from 1980 to 2000. (a) Land C storage decomposed into C storage capacity and potential. (solid lines: C storage; shaded outlines: C storage capacity; shades: C storage potential (positive above and negative below the solid lines)); (b) C storage capacity decomposed into NPP and residence time; (c) NPP determined by GPP and CUE; (d) residence time decomposed into baseline residence time and environmental scalar.

each model and the standard deviation of that variable. First, you can get the global distribution of the simulated C storage by each model and the global distribution of standard deviation of simulated ecosystem C storage among the models. The global distributions of all traceable components in Figure 20.8 are mapped. Figure 20.9 shows that the lowest baseline C residence time in IPSL-CM6A-LR is widely distributed in the Northern Hemisphere, except for the Tibetan Plateau. This analysis is helpful for informing developers of the IPSL-CM6A-LR model on how they may further improve their parameterization of ecosystem baseline C residence time.

QUIZ:

1. Among the three earth system models examined in this exercise, why does the IPSL-CM6A-LR model simulate the lowest land C storage? Which traceable component contributes most to the low values for C storage?
2. Based on Figure 20.9, can you figure out which region has the largest variation in baseline C residence time among the three models? Based on the output of the post-MIP traceability analysis, can you generate a global map of the standard deviation of baseline C residence time among the three models?
3. If we want to apply the authentic traceability analysis to a model-intercomparison project (e.g., CMIP6), what additional information or modeling outputs are needed from each model?

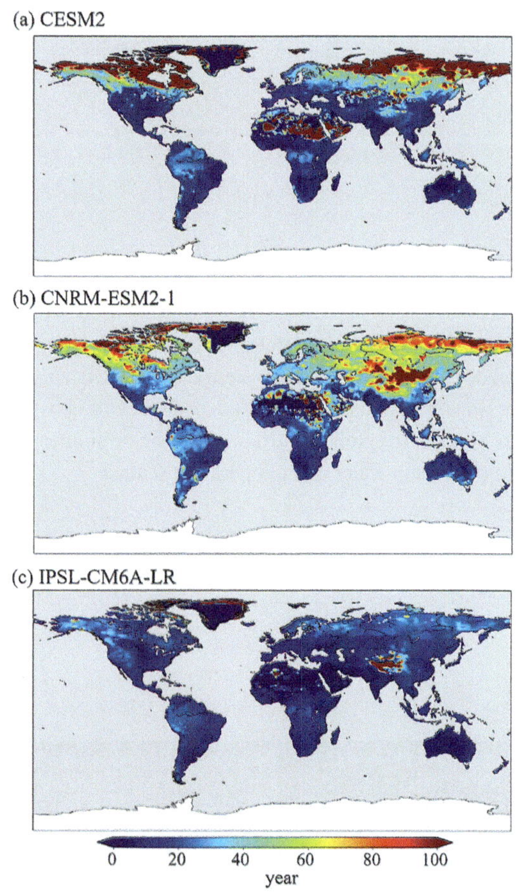

FIGURE 20.9 Global distribution of baseline carbon residence time in three CMIP6 earth system models over 1980–2000.

Unit Six

Introduction to Data Assimilation

21 Data Assimilation

Introduction, Procedure, and Applications

Yiqi Luo
Cornell University, Ithaca, USA

Realistic prediction of ecosystem responses to climate change requires not only a perfect model structure to represent the real-world processes but also parameterization to constrain model specifications and external forcing variables to reflect the environment that an ecosystem experiences to perform its functioning. Data assimilation is a statistical approach to model parameterization. This chapter introduces data assimilation, mainly focusing on concepts, procedure, and applications. We relate data assimilation to regression analysis to show a seven-step procedure: defining a research objective, having data, using one model, measuring data-model mismatches, minimizing the mismatches via global optimization, estimating parameters, and predicting ecosystem changes.

INTRODUCTION OF DATA ASSIMILATION

Data assimilation is a statistically rigorous approach to model parameterization. The latter is one of the three essential elements of realistic model prediction. To realistically predict ecosystem responses to climate change, we need the model structure to represent the real-world processes that control system functions. We also need model specification through parameterization to constrain model predictions (Figure 21.1). Moreover, the external forcing variables have to reflect the physical, chemical, and biological environment that an ecosystem experiences to perform its functioning. Ideally, all three elements have to be perfectly aligned before a model can well predict ecosystem responses.

In reality, we spend much more time developing process-based models than thinking about parameterization or forcing. When prediction of a model does not match well with observations, we usually look at model structure and often ignore parameterization and forcing. In fact, parameterization cannot be totally ignored. To make a model work, we have to tune parameters. But we have not learned much from parameter tuning in the past several decades. Data assimilation is a relatively new approach that can be used to rigorously estimate parameter values and offers a new way to learn about model parameterization. We will learn about data assimilation in the training Units 6 and 7.

External forcing will be explored in Unit 8. In short, we need to have a data-model consistent system, which can be realized by an interactive Ecological Platform for Assimilating Data (EcoPAD) or similar workflow systems, to have realistic external forcing variables to drive models for ecological forecasting.

THE NEED FOR DATA ASSIMILATION

To fully understand the need of data assimilation, we should first understand process-based modeling. Traditional modeling is often called simulation modeling or forward modeling. Simulation modeling usually needs to develop a process-based model first. The model usually means a model structure with a series of equations to represent processes in a system and assigned parameter values. Then, we use forcing variables to drive the model. The forcing variables for an ecological model usually include radiation, temperature, and precipitation. When the forcing variables are used to drive the model, the model generates outputs, such as carbon storage or net ecosystem exchange. The model outputs are sometimes compared with data for validation. This simulation approach has been extensively used by ecologists for about 60 years. It is very powerful for exploring ideas and hypotheses in ecology.

Here is an example of a simulation modeling study done at the Free Air CO_2 Enrichment (FACE) Duke Forest project (Luo and Reynolds 1999). The project had three FACE rings and three control rings plus one prototype ring. In the FACE rings, the CO_2 concentration was 200 parts per million higher than that in the three ambient rings. One of the ideas for the FACE study was to show how a forest would be when CO_2 concentration increases to the elevated level in the middle of this century. The hypothesis was that forest carbon sequestration in the elevated CO_2 rings would be an estimate when the atmospheric CO_2 concentration was gradually increasing to that level.

Dr. James Reynolds and I worked together to test this hypothesis using a simulation modeling approach (Figure 21.2). We used an ecosystem model to test that hypothesis. We simulated two scenarios. One was to have CO_2 concentration gradually increasing as shown in panel A or abruptly increased as in panel D of Figure 21.2. When the atmospheric CO_2 concentration is gradually increasing, both photosynthesis and ecosystem respiration are gradually increasing as shown in panel B. But respiration lags behind due to the carbon that enters the ecosystem through photosynthesis and has to stay in the ecosystem for a while before it is released via respiration. The difference between photosynthesis and respiration is the

DOI: 10.1201/9781032711126-27

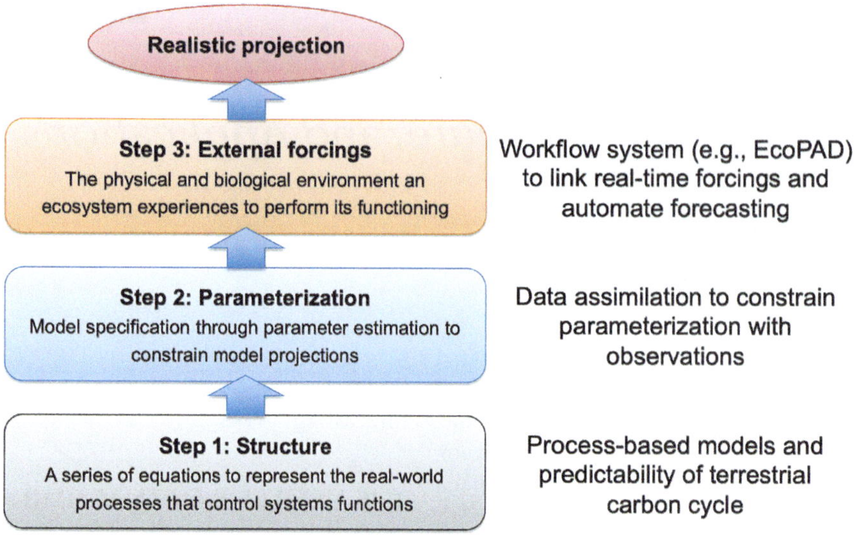

FIGURE 21.1 Three elements required for realistic model prediction of ecosystem responses to global change. Parameterization is equally important as model structure and forcing in predicting states of ecosystems under global change. Data assimilation offers a statistically rigorous approach not only to fit a model better with data, but also enable it to evaluate which, how, how much, and why parameters change (Luo and Schuur 2020). The workflow system Ecological Platform for Assimilating Data (EcoPAD) to link real-time forcing variables for automating forecasting and predictability of terrestrial carbon cycle is discussed in Chapter 29.

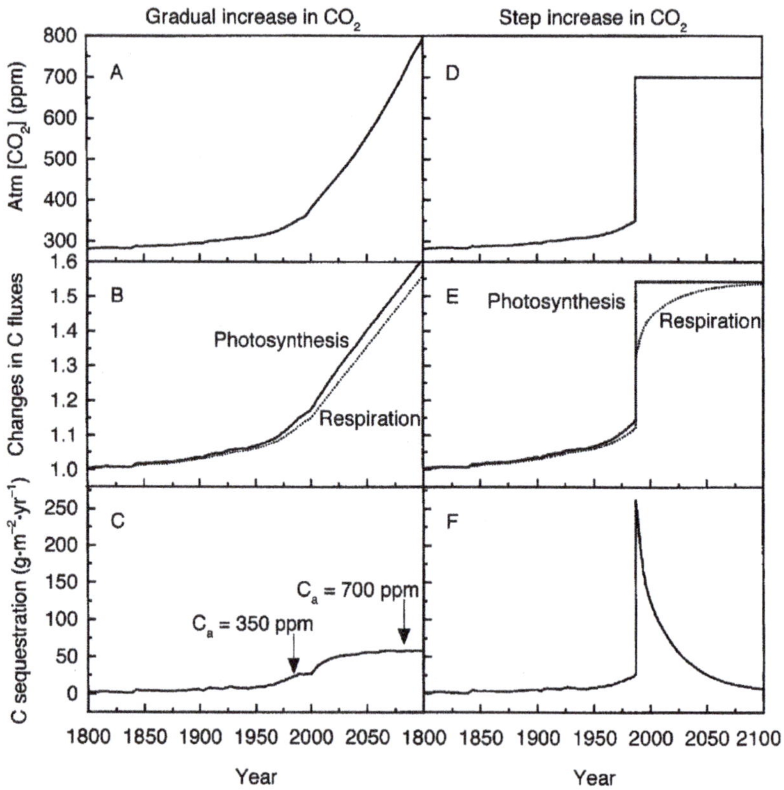

FIGURE 21.2 Simulation model to evaluate responses of ecosystem carbon (C) processes to a gradual vs. step increase in atmospheric [CO_2]. In response to the gradual increase in [CO_2] (A), both photosynthesis and respiration increased gradually (B), leading to a gradual increase in carbon sequestration (C). In response to the step CO_2 increase (D), photosynthesis immediately increases, but respiration slowly increases (E), leading to a high rate of carbon sequestration right after the step increase in CO_2, followed by decline.

Adopted from Luo and Reynolds 1999.

ecosystem carbon sequestration as shown in panel C. The modeled carbon sequestration gradually increases under the gradual CO_2 increase scenario in the real world. In contrast, photosynthesis abruptly increases in response to a step increase in CO_2 concentration as in the FACE experiment, but ecosystem respiration gradually increases, leading to a pulse increase in carbon sequestration followed by a gradual decline as shown in panel F. This is a simulation modeling study. It is quite powerful to contradict the original hypothesis that carbon sequestration observed in FACE plots can be extrapolated to the real world when CO_2 concentration is gradually increasing.

However, when simulation models are used for prediction or forecasting, the model and data usually do not match well. Predictions of soil carbon density by eleven models used in the Coupled Model Intercomparison Project Phase 5, CMIP5, do not match with the observation-based carbon density from the homogenized world soil carbon database (HWSD) (Luo et al. 2015). None of the model predictions match with observations well. The mismatches between the models and observations are partly due to the lack of data constraints of either model structures or parameterization.

Data assimilation is also needed to integrate information contained in both model and data. Modeling is one approach to scientific inquiry mainly through process thinking. Data are obtained from field or laboratory research, which is also an approach to scientific inquiry through snapshot records of ecosystem states at the time when the observation was made. The two approaches acquire information on different aspects of an ecosystem. The information acquired from recording the state of the ecosystem is highly complementary to that of process understanding. Integration of model and data will help gain the best knowledge from imperfect data and imperfect models (Luo et al. 2011).

When data assimilation is used for integration of data and models, it starts with data and associated uncertainty (or noise). Data is combined with modeled values to inversely infer model structures and/or parameter values. Therefore, data assimilation is sometimes called inverse analysis, data-model fusion, inverse modeling, multiple constraints, or inference analysis, depending on situations when this technique is applied.

SEVEN-STEP PROCEDURE OF DATA ASSIMILATION

A data assimilation study is usually conducted by following a seven-step procedure. I will explain the seven steps mainly using the example in the study by Xu et al. (2006). You may go back and forth between this chapter and the paper by Xu et al. (2006). This seven-step procedure is further explained in the practice of Unit 6 (i.e., Chapter 24) and other chapters.

Data assimilation is fundamentally a statistical analysis. It is very similar to regression analysis in terms of the procedure.

When we conduct a regression analysis to understand a relationship, for example, between arm length vs. height, we usually have to go through seven steps (the left side of Figure 21.3). First, we need to decide what the objective is. In this case, we try to predict height from arm length. Second, we need to collect data of the arm length and height of all individuals in a sample of a population we are going

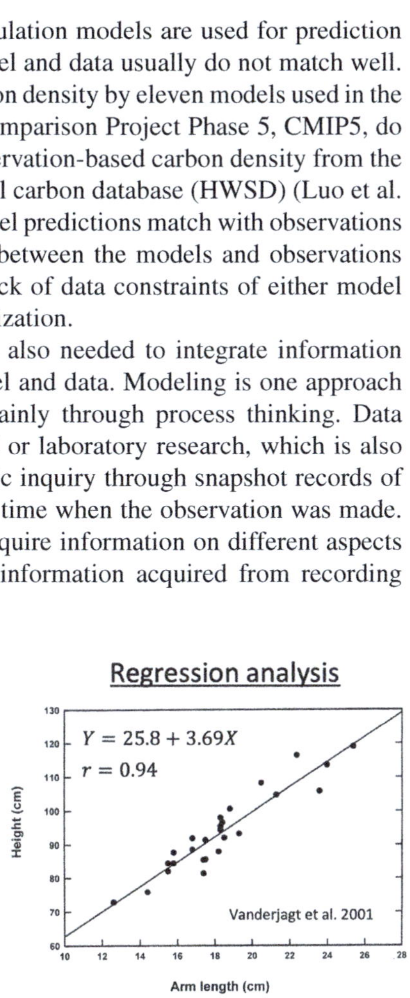

1. Using arm length (X) to predict height (Y) (**objective**)
2. Collect **data** of X and Y of all people in a group
3. Select an **equation** $Y = a + bX$
4. Measure **mismatch** $(y - \hat{y})$, $\hat{y}$ is an estimate of y
5. Minimize $\sum(y - \hat{y})^2$ to **optimize estimation** of a and b
6. Obtain the optimized **parameters** a and b values
7. Use the optimized equation to do **prediction**

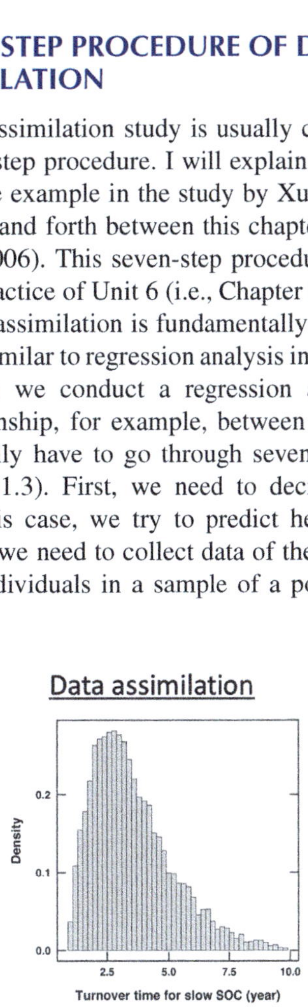

1. **Objective**: To predict ecosystem responses to global change
2. **Data** sets
3. **Model**
4. Cost function as a measure of **mismatches**
5. Global **optimization**
6. Estimated **parameters**
7. **Prediction**

FIGURE 21.3 A seven-step procedure for both regression analysis and data assimilation. Data assimilation is a statistical approach and has a similar procedure with regression analysis. A key measure is fitness between data and model values in regression and posterior probability distributions of estimated parameters in data assimilation, although the data-model fitness is also important for data assimilation.

to investigate. Third, we need to select a regression equation. In this case, a linear model is adequate. Fourth, we need to measure mismatches between observed heights, y, and estimated heights $\hat{y}$ from the regression line. Fifth, we will minimize the mismatch using the method of least squares. Sixth, we will obtain the optimized estimates of regression coefficients a and b of the linear model. The seventh step is to predict a height from the arm length of one individual. For example, the regression line between heights and arm lengths from a sample in Nigeria reported from a paper by Vanderjagt et al. (2001) shows $y = 25.8 + 3.69x$, where x is the length of arm and y is the height (Figure 21.3). In short, the seven steps of conducting regression analysis are: (1) setting up an objective; (2) collecting data; (3) selecting an equation; (4) measuring mismatches; (5) optimization by minimizing the sum of squares of the residuals; (6) estimation of parameters; and (7) prediction (Figure 21.3).

The procedure to conduct a data assimilation study is similar to a regression analysis. To conduct a data assimilation study, we also need to: first, set up an objective, such as predicting ecosystem response to global change; second, have data sets; third, have a model; fourth, develop a cost function to measure mismatches between observations and model values; fifth, have global optimization to minimize the cost function; sixth, estimate parameter values; and seventh, do prediction.

We use a study done by Xu et al. (2006) to show the seven steps. First, the objective of the study was to predict terrestrial carbon sequestration in response to elevated CO_2 concentration. Second, the study used six data sets at the ambient CO_2 treatment and six data sets at the elevated CO_2 treatment. Third, the study used the Terrestrial ECOsystem (TECO) model. Fourth, mismatches between observed and modeled values were measured by a cost function. Fifth, the optimization method used in the study was the Markov Chain Monte Carlo with Metropolis-Hasting algorithm. Sixth, the estimated parameters were decomposition coefficients of litter and soil organic carbon. Seventh, prediction was on carbon sequestration in nine pools of the TECO model at both the ambient and elevated CO_2 treatments.

Although the procedure is very similar to the regression analysis, there are some new concepts, such as mapping functions, a cost function, Markov Chain Monte Carlo (MCMC) sampling series, and the Metropolis-Hasting criterion. Those new concepts are explained in Chapter 22 and illustrated with practice examples in Chapter 24.

Let us go over each of the seven steps (Figure 21.3). Note that cited symbols, page numbers, figures, and tables in this section all refer to those in the paper by Xu et al. (2006).

1. Step 1 is to define an objective. The general objective of the study was to predict ecosystem responses to elevated CO_2 concentration with uncertainty quantified. You can find this objective in paragraph 2 on page 1. A more specific objective of the study was to estimate parameter values, $c_1 \ldots c_7$, in Table 1 on page 3 of the paper by Xu et al. (2006) and then predict carbon pool changes at ambient and elevated CO_2 treatments in Duke Forest as described in paragraph 4 on page 2. Parameters, $c_1 \ldots c_7$, represent decomposition of organic carbon from litter and soil pools.
2. Step 2 is to have data sets. The data sets used in the study are soil respiration, woody biomass, foliage biomass, litterfall, soil carbon, and mineral carbon at the ambient and elevated CO_2 treatments. Those data sets are described in paragraph 6 on page 3 and Table 2 on page 4 of the paper by Xu et al. (2006). We also need to use standard deviations of the six data sets in data assimilation.
3. Step 3 is to have a model. This study uses the Terrestrial ECOsystem (TECO) model. The TECO model is depicted in Figure 1 and described in equation 1 on page 2 of the paper by Xu et al. (2006). Please note that that figure has a typo. Pool X_1 should be a non-woody pool, which includes foliage and fine root biomass. One more point is that the TECO model in this paper was described by the matrix equation, which is the same equation as we studied in Units 1–5 with slightly different notations.
4. Step 4 is to define a cost function. The cost function is to measure mismatches between observed and modeled values. The mismatches are described by equation 5 for individual data sets and equation 6 for all the six data sets on page 4 of the paper by Xu et al. (2006). The modeled values have to be mapped to observations via a mapping function Φ, which has six elements, respectively, for six data sets, as described in equation 3 on page 4. The mapping function is further illustrated in the practice session in Unit 6 of this book (i.e., Chapter 24). The cost function is equivalent to the likelihood function in equation 8 on page 4 of the paper by Xu et al. (2006).
5. Step 5 is to minimize the cost function via global optimization. The optimization was done with Markov Chain Monte Carlo (MCMC) sampling series. The MCMC has two phases, a proposing phase and a moving phase. The proposing phase is to generate new parameter sets. The moving phase is to examine if the newly proposed parameters should be accepted or not using a Metropolis-Hastings criterion. The two phases of MCMC are described in paragraphs 12 and 13 on pages 4 and 5 of the paper by Xu et al. (2006). The study uses five parallel runs to test convergence of sampling series with the Gelman-Rubin method as described in paragraphs 14 and 15 on page 5 of the paper by Xu et al. (2006).
6. Step 6 is to estimate parameter values after the cost function is minimized in step 5. The estimated parameter values are described in paragraph 17 on page 5, depicted in Figures 3 and 4, and Table 3 on pages 6–8 of the paper by Xu et al. (2006). Estimated parameters have three types of distributions, or histograms, as shown in the mid-columns in Figures 3 and 4. The three types of histograms are bell-shaped for parameters c_1, c_2, and c_4, edge-hitting for c_6, and flat for c_3, c_5, and c_7 in Figure 4 on page 7 of the paper by

Xu et al. (2006). The bell-shaped histograms indicate that those parameters are well constrained by data. The flat-shaped histograms mean that those parameters are not constrained by data. In other words, the six data sets do not have much information that is relevant to constrain those parameters. The edge-hitting histograms mean that those parameters may be strongly correlated with other parameters or caused by other reasons.

7 Step 7 is to use the estimated parameters from step 6 in the TECO model to predict future states of ecosystems. The predicted carbon pool changes in the year 2010 at the ambient and elevated CO_2 treatments are described in paragraph 17 on page 5, Figure 9 on page 11, and Table 4 on page 12 of the paper by Xu et al. (2006). Uncertainty in model prediction is quantified with cumulative density functions in Figure 9 and confidence intervals in Table 4. CO_2 effects are represented by the predicted changes in pool sizes as indicated by the differences between the solid lines at elevated CO_2 and dashed lines at ambient CO_2 in Figure 9. As you can see from the paper by Xu et al. (2006), elevated CO_2 had the largest effects on woody biomass but no effects on the passive soil carbon pool.

Again, the seven steps to conduct a data assimilation study are setting up an objective, collecting data sets, having a model, and defining a cost function, using a global optimization method to minimize the cost function, estimating parameter values, and predicting.

The seven-step data assimilation is essential to estimate parameter values and constrain model predictions through parameterization. The parameterization, in turn, is essential toward realistic model prediction.

SCIENTIFIC VALUES OF DATA ASSIMILATION

Data assimilation is a statistical tool. Its scientific values are realized when it is used to estimate parameter values, select alternative model structures, quantify uncertainty, and/or evaluate values of different data sets to constrain model prediction. The study by Xu et al. (2006) has illustrated how data assimilation was used to estimate parameters and quantify uncertainty of estimated parameters and model predictions. Data assimilation has been used to quantify uncertainty associated with different model structures. For example, the study conducted by Shi et al. (2018) evaluated three soil carbon models: a classic or conventional model, a microbial model, MIMICS, and a vertically resolved model, CLM4.5, with three data sets: topsoil organic carbon in 0–30 cm, subsoil organic carbon in 30–100 cm, and microbial biomass carbon. The three data sets can constrain subsets of parameters for all the three models while complex models generate larger uncertainties in predictions even with the same data sets to constrain parameters.

Data assimilation also can be used to evaluate values of different data sets to constrain model predictions. The values of data sets are measured by information content according to the Shannon information index and quantified from probability density functions of predicted changes (Weng and Luo 2011). In principle, data sets contain information mainly on the parameters that are related to relevant processes. For example, flux data, such as net ecosystem exchange and gross primary production, can usually constrain parameters related to flux processes, such as leaf area index and maximal carboxylation rate. In comparison, pool data, such as plant biomass and soil carbon content, usually have information to constrain pool-related parameters, such as carbon transfer coefficients between pools.

Data assimilation is often used to test alternative model structures. For example, four alternative models were evaluated for their representation of the priming effect on soil organic carbon decomposition with 84 data sets from 26 studies (Liang et al. 2018). Scientifically, the study evaluates whether the soil priming effect leads to net loss or net gain of soil organic carbon by adding new carbon input. Priming is a term to describe stimulation of old soil organic carbon (SOC) decomposition by new carbon addition. This priming effect usually results from stimulated microbial growth by new carbon addition.

The study requires data to be collected from isotope-labeled carbon addition experiments. Those studies have to provide SOC content, the added amount of new carbon, multiple measurements of CO_2 emission rates from total SOC samples and from labeled new carbon substrate, from which the CO_2 emission rates from old SOC can be estimated.

The study uses four models, including conventional soil carbon dynamic model, new-old carbon interactive model, Michaelis-Menten model, and reverse Michaelis-Menten model. Data assimilation is used to integrate the 84 data sets with the four models to evaluate data-model matches by comparing observed with modeled cumulative CO_2 emission rates from the total SOC, old SOC, and new carbon substrate (Figure 21.4). The conventional model can fit observed carbon releases from new, old, and total SOC extremely well. The interactive model can fit observations quite well, too. However, the fitting is not good for Michaelis-Menten and reverse Michaelis-Menten models, especially with observed carbon releases from old and total SOC.

The model-data fitting also shows that the conventional model fits the best and the Michaelis-Menten model fits the worst. A deviance information criterion (DIC) was used to select alternative model structures. DIC penalizes the model severely by the number of parameters. Among the four models, DIC is the smallest for the interactive models but largest for the conventional model as the latter has 12 parameters although the fitting of the model with data is the tightest. Based on the DIC, the most parsimonious model is the interactive model.

Then, the interactive model was used to estimate priming effect, replenishment, and net change in carbon after adding new carbon to the soil incubation experiments. Nearly 54% of the newly added carbon stays in the soil, stimulated old carbon decomposition via the priming effect is nearly 10% of the newly added carbon. As a consequence, SOC has a net

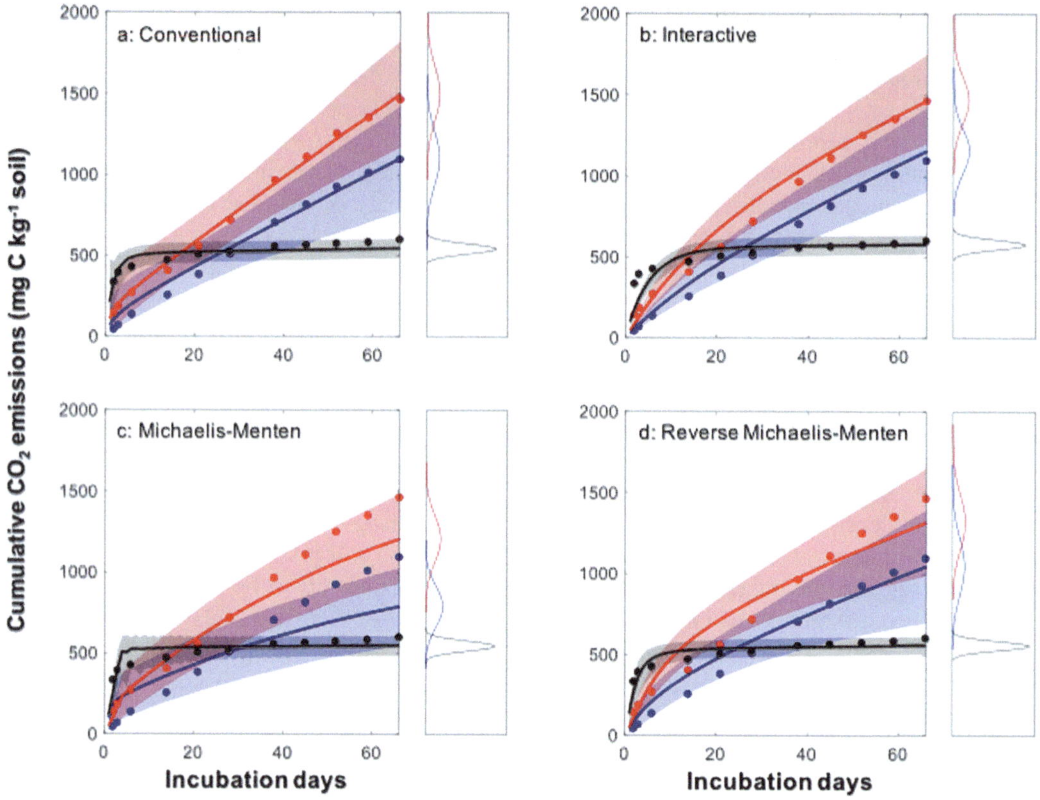

FIGURE 21.4 An example showing the performances of different models in simulating cumulative CO_2 emissions from old and new C substrates. Dots and lines are observations and model simulations, respectively. Shaded areas are the simulated ranges from 2.5th to 97.5th percentiles (i.e., 95% range). Blue and red are CO_2 emissions from old C at the control and new C addition treatments, respectively; Black is CO_2 emissions from added new C. The distributions of simulated cumulative CO_2 emissions at the end of experiment are also shown in each panel.

Adopted from Liang et al., 2018.

gain of 32% of the added new carbon. This analysis indicates that priming does happen but is unlikely to lead to a net loss of SOC.

This study used data assimilation to address a highly controversial issue on priming effect. The study shows that adding new carbon to soil does result in a priming effect but eventually results in net carbon gain. The increase in SOC is related to nitrogen content in the added substrates. The study shows how data assimilation was used to select alternative model structures and suggests that a two-pool interactive model is the most parsimonious model to represent a SOC priming effect.

Overall, data assimilation is a statistically rigorous approach to model parameterization, which is essential to generate realistic model prediction. Data assimilation is usually conducted by following a seven-step procedure. The seven steps are: (1) setting up an objective; (2) collecting data sets; (3) having a model; (4) defining a cost function; (5) using a global optimization method to minimize the cost function; (6) estimating parameter values; and (7) predicting. The scientific values of data assimilation can be realized when it is used to estimate parameters, select alternative model structures, evaluate values of data sets in constraining model predictions, and quantify uncertainty in model prediction.

SUGGESTED READING

Xu, T., L. White, D. Hui, and Y. Luo. 2006. Probabilistic inversion of a terrestrial ecosystem model: Analysis of uncertainty in parameter estimation and model prediction. *Global Biogeochemical Cycles*, *20*, GB2007. doi:10.1029/2005GB002468

QUIZ

1 What are the three elements, which all have to be perfectly lined up to realistically forecast future states of an ecosystem? Why?
2 Data assimilation is a method
   a to integrate data with model.
   b to calibrate a model with data.
   c to use statistical principles for analyzing data.
   d often used for parameter estimation.
   e all the above
3 Why is simulation modeling good for exploring ideas but not for prediction or forecasting?

4 Why are field research and modeling scientifically complementary? It is because (can choose more than one answer):
   a one makes measurement and the other uses computer.
   b they are two approaches to scientific inquiry of a research subject in different ways.
   c the field research records the state of an ecosystem whereas the modeling explores relationships among processes.
   d data from field research can be better interpreted from process understanding whereas model forecast can be better constrained with data.
5 What are the seven steps of conducting a data assimilation study?
6 A cost function is (choose one answer)
   a a function to calculate cost of training.
   b to measure mismatches between modeled and observed values for all the data sets.
   c a data set to be used in data assimilation.
   d a model to be optimized through data assimilation.
7 Posterior probability density function is (choose one answer)
   a to indicate carbon density after field measurement.
   b to indicate probability of a parameter value before data assimilation.
   c a function of time from prior to posterior.
   d to indicate a relative likelihood of an estimated parameter value after data assimilation.
8 Why do we need a global instead of local optimization method for data assimilation?

22 Bayesian Statistics and Markov Chain Monte Carlo Method in Data Assimilation

Feng Tao
Cornell University, Ithaca, USA

Data assimilation is an effective way to integrate observations into models. We will demonstrate how parameters in a model may be estimated by data assimilation in such a way that model simulations best fit observations. Data assimilation based on Bayesian inversion is used to retrieve posterior distributions of model parameters from observations. The Markov Chain Monte Carlo (MCMC) method is applied as a numerical method to home in on the parameter set that maximizes goodness of fit between model outputs and measurements.

INTRODUCTION

In this chapter, we will first give a brief introduction to Bayesian statistics, which is the theoretical foundation of data assimilation. By learning the Bayes' theorem, you will understand why we can get knowledge from the data. After familiarizing ourselves with the theory, we will then learn how to apply the theory to an algorithm, namely the Markov Chain Monte Carlo (MCMC) method, which is useful for optimization problems, like fitting a model with multiple parameters to observational datasets. At the end of the chapter, we will discuss how to achieve stable optimization results with the MCMC algorithm, the so-called simulation convergence problem. This chapter is a brief introduction to these topics to help readers build an intuitive understanding of data assimilation. If you want to know more about rigorous theoretical proofs of theorems and equations mentioned in this chapter, you are encouraged to refer to the suggested readings at the end.

BAYES' THEOREM

The Bayes' theorem, in plain words, is a method for calculating the validity of thinking (Horgan 2016). The "thinking" can be one's hypotheses, claims, or propositions. The results (i.e., the updated validity of thinking) given by Bayes' theorem are based on the best available evidence, such as the data, observation, or any information one can obtain. Bayesian thinking happens every day in our lives. Before we collect any related information and update our thinking over a particular question, we all do some initial thinking on the validity of the question, no matter if our attitude is initially skeptical or credulous. When we get new related data or information from the real world, we will naturally update our thinking over the question. This evidence-based updated thinking is the primary character of Bayes' theorem.

We can mathematically express Bayes' theorem as:

$$P(A|B) = \frac{P(B|A)P(A)}{P(B)} \propto P(B|A)P(A) \qquad 22.1$$

If we take events happening in our lives alphabetically as A and B, the probability of event A happening can be expressed as $P(A)$. The probability of event B happening can be expressed as $P(B)$. We use $P(A|B)$ to indicate the probability of event A happening given the fact that event B has happened. Analogously, $P(B|A)$ represents the probability of event B happening given the fact that event A has happened.

The Bayes' theorem intuitively expresses that the probability of event A to happen given event B has already happened is proportional to the possibilities of event B happening where event A has already happened, and, at the same time, event A happening. When we calculate the probability of event A happening given event B has already happened (i.e., $P(A|B)$), we can assume event B has been well observed. The probability of event B happening (i.e., $P(B)$) then becomes a constant. $P(A|B)$ is now proportional to a product of $P(B|A)$ and $P(A)$. We may know some information about A and can transfer such information into an initial thinking (i.e., $P(A)$). Based on the initial thinking, we can also get a sense of how likely it is for event B to happen given A has happened (i.e., $P(B|A)$). Mathematically we call $P(A)$ our prior knowledge, while $P(B|A)$ is termed the likelihood. Combining the prior knowledge and likelihood, we update our thinking as posterior knowledge (i.e., $P(A|B)$).

We will use an example to illustrate how Bayes' theorem works. Suppose you are in an international meeting and receive a flyer advertising the training course "New Advances in Land Carbon Cycle Modeling", which will train people in carbon cycle modeling and data assimilation. You are interested and prepared to apply. Before the application, however, you feel concerned that your lack of experience in simulation modeling may hinder your success in the training course. You may wonder what the probability of your success in the training course will be, given you have no previous modeling

DOI: 10.1201/9781032711126-28

experience. By emailing the course coordinator, you learn that according to previous records, 80% of past trainees attending the training course succeeded in understanding the training material and finished the practices. In terms of the background of all trainees, half of them had no experience in modeling before the course. Among the successful trainees, 43.75% had no experience in modeling before the course.

We can conceptualize this issue into Bayes' theorem. We can define the event A as one's success in the training course and the event B as the trainee having no modeling experience before the training course. We now know $P(B) = 0.5$ (half of the trainees have no modeling experience before the training course); $P(A) = 0.8$ (80% of the trainees succeeded in the training course); and $P(B|A) = 0.4375$ (among the successful trainees, 43.75% had no experience in modeling before the training). By substituting all the factors into Equation 22.1, we get $P(A|B) = 0.7$. The result implies that even if a trainee did not have any experience in modeling before the training course, the chance of success is 70%, which is still pretty large.

The question could go the other way. Suppose you have prior experience in modeling, you may ask what the possibility of your being successful in the training course is. To answer this question, we need to calculate $P(A|\neg B)$, where $\neg B$ indicates the event of B not happening, and $P(\neg B) = 1 - P(B)$. According to Equation 22.1,

$$P(A \mid \neg B) = \frac{P(\neg B \mid A)P(A)}{P(\neg B)} = \frac{\left[1 - P(B \mid A)\right]P(A)}{1 - P(B)} \quad 22.2$$

when we substitute all the probability numbers we obtained from the coordinator into Equation 22.2, we get $P(A|\neg B) = 0.9$. So, if you have previous experience in modeling, the chance of success in the training course increases from 70% to 90%.

We used discrete events and their possibilities in the above example to illustrate Bayes' theorem. In simulation modeling, we use parameters that are continuous values to represent ecological processes. We assume that if we choose the correct parameter values, observations in the real world (e.g., soil organic carbon content) can then be predicted by the model. In this context, we use probability distributions of parameters (as expressed by the probability density function) to show all possible values of one parameter and the likelihood of occurrence for different values within that overall distribution. The Bayes' theorem will help us, in a reverse way, find the most likely distribution of parameter θ to best describe the observed data x in a model:

$$p(\theta \mid x) = \frac{p(x \mid \theta)p(\theta)}{p(x)} \propto p(x \mid \theta)p(\theta) \quad 22.3$$

where the $p(\theta)$ is the prior probability density function of parameter θ; $p(x|\theta)$ represents the likelihood that we use the model with θ set to different values from its distribution, and correctly predict the observation x. Combining the prior distribution $p(\theta)$ with the likelihood $p(x|\theta)$ will give the posterior probability density function of θ (i.e., $p(\theta|x)$).

Bayes' theorem explicitly offers the possibility of using observational data to refresh our prior knowledge. The term $p(x|\theta)$ in Equation 22.3 shows how likely we are to observe data x under different values of parameter θ from its distribution. This indicates if we propose different θ values and record them with their likelihoods (i.e., $p(x|\theta)$), we will eventually obtain the posterior distribution describing the optimized parameter values that best fit the observations (Figure 22.1).

MARKOV CHAIN MONTE CARLO METHOD

In data assimilation, the Markov Chain Monte Carlo (MCMC) method offers an algorithm that screens a range of possible parameter values under specific prior knowledge and retrieves the posterior distribution of the parameter. The MCMC method is composed of two parts, namely a Markov Chain and the Monte Carlo method. A Markov Chain is a stochastic model that describes a sequence of possible events. The probability of each event in a Markov Chain depends only on the state attained in the previous event. The Monte Carlo method, on the other hand, represents a class of algorithms that sample events from a Markov Chain to fit an optimization target. Various algorithms exist for sampling. Here, we will focus on the Metropolis-Hastings algorithm.

We begin with an example to show the workflow of the MCMC method. Suppose that we are organizing the training

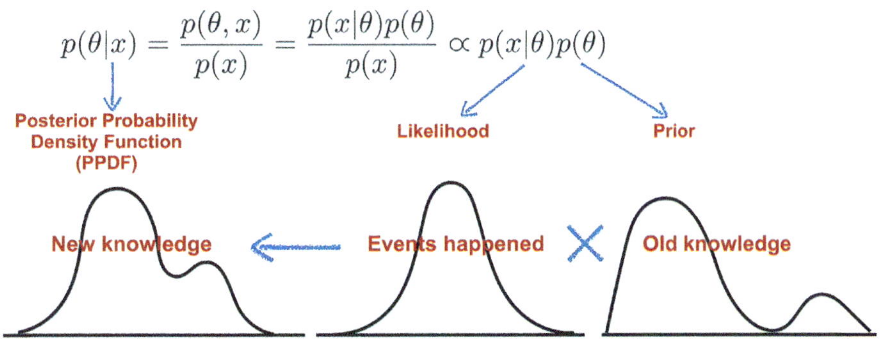

FIGURE 22.1 Schematic diagram of the Bayesian inference process.

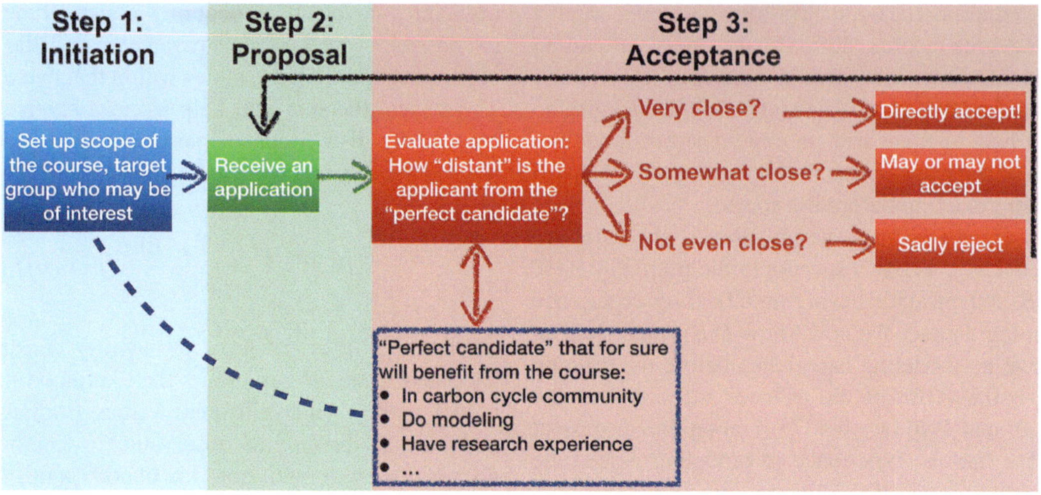

FIGURE 22.2 An example of using the MCMC method to admit candidates based on their application details.

course "New Advances in Land Carbon Cycle Modeling". The admission process is critical to the final outcome of the course. By scanning the profiles of applicants, we need to select those applicants who will most likely benefit from the training. The question at this stage is then formulated as "through which algorithm shall we select the participants from among all the candidates?"

First, we may need to specify the topical scope of the training course so that we can define the target group who may be of interest (Figure 22.2). A clear and well-defined scope will help define the "perfect candidate" who will definitely benefit from the training. For example, we may accept people who are doing research in the carbon cycle field, familiar with basic simulation modeling, and who would like to integrate models and data to enhance their understanding of ecological processes.

After defining the scope of the training course, we may send out the flyers to advertise the course. We will then receive applications from people who are interested in and believe that they can benefit from the course. All the applicants now become the potential candidates as trainees. We now need to evaluate the match between the backgrounds of candidates and our training scope. The characteristics of the "perfect candidate" in our mind will function as a reference in such evaluation.

We make decisions based on the assessments of candidates. Three possibilities will be on the table. We may find that the background of the applicant is very close to that of our "perfect candidate", which means the applicant will most likely benefit from the training course. In this case, we will admit the applicant without hesitation. Conversely, we may also receive applications where the background of the applicant does not fit the scope of the training at all. We will directly reject those applications. Another possibility is that the background of the applicant does not perfectly fit our "perfect candidate", but is still close enough to get likely benefits from our course. Instead of direct rejections, we may consider an algorithm to decide whether to accept or reject the applicant. For example, we may set new criteria to evaluate the fitness of these applicants to the training course and eventually admit those who meet the criteria.

Four steps together make up the admission process. The first step initiates the process. We need to set the scope of the training course and determine the target group who will be of interest. In the second step, applications provide all the possibilities we can choose from to fit our target of maximizing the overall benefit of the training. We then make decisions in the third step to accept or reject the applications according to some standards. Finally, in step four, we repeat the procedure from steps two and three for each candidate.

The MCMC method follows the same four steps to optimize parameters to fit model simulations with observations (Figure 22.3). In step one, we initiate the algorithm by first determining the optimization target (i.e., the observations). We set a prior range of the possible parameter values that remains consistent with the ecological meaning of the investigated process, and simultaneously, allows sufficient flexibility in model simulations.

The second step generates the Markov Chain. We propose a sequence of candidates of parameter values so that we can select the candidates by a certain algorithm (in our example, the Metropolis-Hastings algorithm). A proposal distribution will be selected and serves to generate new candidate parameter values. We then apply the proposed parameter values in the model and get the simulation results.

We use two proposal distributions in step two, which are the uniform distribution in the test run and the normal distribution in the formal run (Haario et al. 2001). When we have no prior knowledge about the parameter, a uniform distribution will be a reasonable start. The uniform distribution does not contain any shape information of the parameter distribution, but only sets the upper and lower limits of parameter values, which we determined based on our general understanding of the parameter in step one. The kth proposed parameter value is expressed as:

$$\theta^k = \theta^{k-1} + r\frac{\theta^{max} - \theta^{min}}{D} \tag{22.4}$$

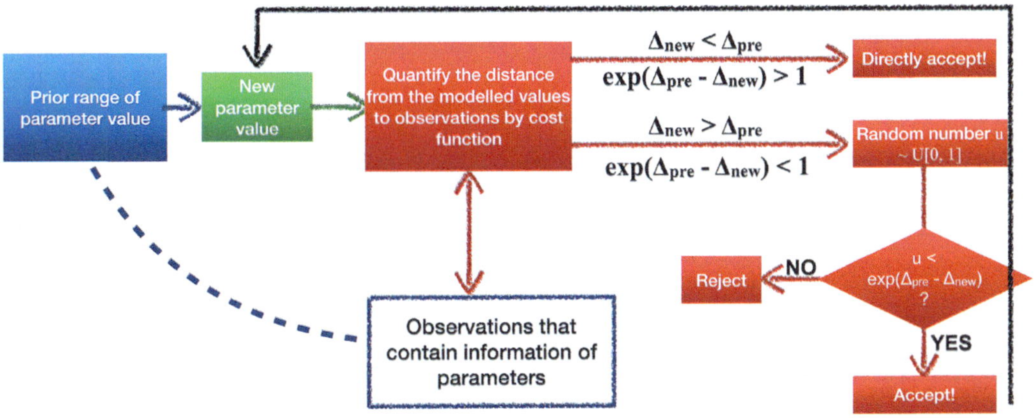

FIGURE 22.3 Workflow of the MCMC method. The Metropolis-Hastings algorithm is used to accept or reject proposed parameter values in a Markov Chain.

where r is a random value drawn from the uniform distribution in the interval between 0 and 1; D indicates the maximum step size in the proposal. For example, $D = 5$ indicates that the maximum step of the kth proposed parameter value (θ^k) from the k-1th proposed parameter value (θ^{k-1}) is 20% of the prior range (i.e., $\theta^{max} - \theta^{min}$).

After the test run of MCMC with a uniform proposal distribution, we obtain some preliminary information on the investigated parameter. The following formal run will then use the results of the test run as its prior knowledge so that we can more efficiently get the stable posterior distribution of the parameter. A normal distribution will be assumed in the formal run. When we investigate multiple parameters, we assume that the parameters follow a multivariate normal distribution. The parameter values will be proposed as:

$$\theta^k = \begin{cases} \theta^{k-1} + N\left(0, \operatorname{cov}\left(\theta^{testrun}\right)\right) & k \le k_0 \\ \theta^{k-1} + N\left(0, \operatorname{cov}\left(\theta^{formalrun}\right)\right) & k > k_0 \end{cases} \quad 22.5$$

where we calculate the covariance of parameters by the results of the test run at the starting stage of the formal run (i.e., before the k_0^{th} iteration). After some iterations of the formal run, we continuously update the covariance information from the formal run results. The point k_0 is an empirical value we may set based on experience, depending on how fast we think the formal run can fully utilize the prior information in the test run.

In step three, we apply the Metropolis-Hastings algorithm to accept or reject proposed parameter values. We evaluate how the simulated results based on the proposed parameter values fit the observations and decide whether to accept the proposed parameter values or not. A cost function serves to quantify the degree of discrepancy of the predicted values of the target variable x in the simulations relative to the observations of x. We express the cost function of the kth proposed parameter (θ^k) as:

$$\Delta_k = \frac{1}{2\sigma^2}\left[\sum_{i=1}^{n}\left(x_{i,mod} - x_{i,obs}\right)^2\right]_{\theta^k} \quad 22.6$$

where $\left[\sum_{i=1}^{n}\left(x_{i,mod} - x_{i,obs}\right)^2\right]_{\theta^k}$ describes the distance of the simulation results from observations with the sample size of n (e.g., ten-year record of net primary productivity of a grassland site); σ^2 is the variance of observations, which we can either calculate from measurements or from empirical knowledge. Equation 22.6 expresses the cost function for a single data set. When multiple data sets are used (e.g., ten-year record of net primary productivity, soil organic carbon content, and soil respiration of a grassland sites), the mismatches of individual data sets are added together, optionally with weightings to emphasize the importance of certain variables relative to the others, to obtain the cost function.

Now we make use of Bayes' theorem. From Equation 22.3, we can define the likelihood of the parameter value (θ^k) in fitting model simulations ($x_{i,\ mod}$) to observations ($x_{i,\ obs}$) as:

$$L\left(\theta^k\right) = p\left(\theta^k \mid x\right) \propto p\left(x \mid \theta^k\right) p\left(\theta^k\right) \quad 22.7$$

We use the cost function in a monotonically decreasing exponential form to describe $p(x \mid \theta^k)$:

$$L\left(\theta^k\right) \propto \exp\left(-\Delta_k\right) \quad 22.8$$

Using Equation 22.8 to quantify the likelihood requires several assumptions in Bayesian statistics (Craiu and Rosenthal 2014). From the data optimization perspective, Δ_k expressed in Equation 22.6 offers a measurement of the deviation of model simulations from observations. We connect such a metric with the likelihood of the proposed parameter values: we assign a high likelihood to proposed parameter values when the cost function result is small and *vice versa*.

We use the likelihood ratio of consecutively proposed parameter values in the Markov Chain to make the acceptance decision. The probability of accepting the kth proposed parameter value (θ^k) based on the results of k–1th proposed parameter value (θ^{k-1}) can be formulated as:

$$P\left(\theta^{k-1}, \theta^k\right) = \min\left\{1, \frac{L\left(\theta^k\right)}{L\left(\theta^{k-1}\right)}\right\} \quad 22.9$$

Substituting Equation 22.8 into Equation 22.9 gives:

$$P\left(\theta^{k-1}, \theta^k\right) = \min\left\{1, \exp\left(\Delta_{k-1} - \Delta_k\right)\right\} \quad 22.10$$

When the cost function with θ^k is smaller than that with θ^{k-1}, this tells us that the results of the present simulation are closer to the observations than the previous one. By Equation 22.10, we get $P(\theta^{k-1}, \theta^k) = 1$, which indicates we will definitely accept the proposed parameter value θ^k. Conversely, if the cost function calculated from proposed parameters θ^k is higher than that of θ^{k-1}, this tells us that the simulation results by θ^k, in comparison to observations, is worse than that by θ^{k-1}. We then get a $P(\theta^{k-1}, \theta^k)$ result in the interval between 0 and 1. In this case, we may or may not accept the proposed θ^k, depending on a random chance. In practice, we compare $P(\theta^{k-1}, \theta^k)$ with a random value u that follows the uniform distribution $u \sim U(0, 1)$. We accept the proposed θ^k when $P(\theta^{k-1}, \theta^k) > u$. Otherwise, we will reject the proposed θ^k.

It may seem strange that we sometimes accept the proposed θ^k, instead of a direct rejection when a proposed set of parameters yields simulation worse than the previous set (i.e., the new cost larger than the previous one, $\Delta_k - \Delta_{k-1} > 0$). In optimization, the "bad" results can be helpful in finding the global minimum of the cost function. The response surface of the cost function can be extremely nonlinear for a multivariate, high-dimension model. If we do not give a chance to accept a "bad" set of parameters, we may easily get trapped in a local optimum instead of the global optimum. Therefore, we assign a possibility of acceptance, subject to chance, when $P(\theta^{k-1}, \theta^k)$ is in the interval between 0 and 1.

CONVERGENCE OF MCMC RESULTS

When we do MCMC, we need to confirm that the results are the same even if we start from different points along the Markov Chain. This is called testing for convergence (Gelman et al. 2014). In theory, the MCMC should converge as long as the Markov Chains are long enough. That means that the parameters from different independent MCMC simulations will share the same or very similar posterior distributions.

The estimated parameters in the MCMC simulation, however, do not necessarily resemble each other at the very beginning among different Markov Chains. In the test run and starting stage of the formal run, accepted parameter values may experience a different pathway to find the global optimum. However, after sufficient iterations of MCMC simulation (e.g., 10,000 iterations), the observations will eventually constrain the parameter to the same space, where the model simulates most closely to the measurements entering the cost function. By plotting the sampling series, we can visually determine when the accepted parameter values are stably constrained and set the early stage of the sampling series as the burn-in period. The exact length of burn-in period can be empirical. The first half of the accepted parameter chain is a safe setting for burn-in period given sufficient iterations in MCMC simulation (e.g., 100,000 iterations). In the final analysis, we exclude the burn-in results and only use the results after convergence to generate the posterior distribution.

Both visual and statistical assessment can be used to determine the convergence of MCMC results. The degree of overlap among different MCMC sampling series is a rough indicator of the convergence. We can also quantify the convergence statistically. The Gelman-Rubin (G-R) statistics are one option. The G-R statistics quantify the differences among different runs (B_i) and the fluctuation of accepted parameter values within the same run (W_i):

$$\begin{cases} B_i = \dfrac{N}{K-1}\sum_{k=1}^{K}\left(\bar{c}_i^{\cdot,k} - \bar{c}_i^{\cdot\cdot}\right)^2 \\ W_i = \dfrac{1}{K(N-1)}\sum_{k=1}^{K}\sum_{n=1}^{N}\left(c_i^{n,k} - \bar{c}_i^{\cdot,k}\right)^2 \end{cases} \quad 22.11$$

where i denotes parameters investigated in the study; K is the number of parallel runs; N is the length of each run; $c_i^{n,k}$ represents the nth accepted value of parameter i in the kth parallel run after the burn-in period. The G-R statistic is then defined as:

$$GR_i = \sqrt{\frac{W_i(N-1)/N + B_i/N}{W_i}} \quad 22.12$$

Once convergence is reached, GR_i should approximately approach 1.

In summary, this lecture introduces two fundamental underpinnings of data assimilation. The first, the Bayesian inference, sets the theoretical foundation of improving model performance by observations. The second, the Markov Chain Monte Carlo method, provides a numerical method to assimilate observational data into a model based on an optimization of the goodness of fit of the model output to the observational data. In the next chapter, using examples from different ecological studies we will show you how data assimilation can be deployed to advance understanding of the land carbon cycle.

SUGGESTED READINGS

Haario, H., E. Saksman, and J. Tamminen. 2001. An adaptive Metropolis algorithm. *Bernoulli*, *7*, 223–242.

Xu, T., L. White, D. Hui, and Y. Luo. 2006. Probabilistic inversion of a terrestrial ecosystem model: Analysis of uncertainty in parameter estimation and model prediction. *Global Biogeochemical Cycles*, *20* (2).

QUIZ

1 Briefly describe the Bayes' theorem.
2 Why do we need a test run before the formal run when we know little about the prior?
3 Explain why we need to sometimes accept parameters that have a larger cost function value (Δ) than the previously accepted ones.
4 Why do we discard the results from the burn-in period?
5 We need to generate a random value (u) to compare with the results exp(Δpre – Δnew) when the cost function value is larger than the previous one (Δnew > Δpre). Why do we accept the proposed parameter values only when u < exp(Δpre – Δnew), instead of u > exp(Δpre – Δnew)?

23 Application of Data Assimilation to Soil Incubation Data

Junyi Liang
China Agricultural University, Beijing, China

Jiang Jiang
Nanjing Forestry University, Nanjing, China

Soil incubation is a widely used technique in studying soil organic carbon cycling. Integrating soil incubation data with soil carbon models can potentially reveal mechanisms of soil carbon dynamics underlying observations. This chapter aims to illustrate how data assimilation is applied to analyze data from soil incubation experiments using soil carbon models. After a brief introduction to soil incubation experiments and soil carbon models, a three-pool model is used to illustrate the seven-step procedure of data assimilation for the analysis of soil incubation data. Two critical aspects, different cases in the optimization step, and the dependence of parameter acceptance rate on cost function are described with detailed examples.

SOIL INCUBATION EXPERIMENTS

Soil incubation experiments are commonly carried out for studying soil organic carbon and nitrogen cycling processes. Typically, fresh soil samples are collected from the field, crushed, sieved, and mixed before a certain amount of soil is placed in a container (e.g., a Mason jar). The container is then exposed to different treatments of, for example, temperature and moisture. For carbon decomposition studies, carbon dioxide (CO_2) emission rate (or respiration rate) is usually measured repeatedly during the incubation period. Data from soil incubation studies are usually plotted by either CO_2 emission rates or cumulative CO_2 emission over time. Compared with *in situ* observations in the field, soil incubation experiments allow ready control of environmental factors and reduce heterogeneity and confounding effects from many processes and factors. Thus, results from such experiments can facilitate mechanistic understanding of soil carbon processes, such as the turnover rates and temperature sensitivity of soil organic carbon decomposition. However, incubation experiments usually isolate the soil from plants and other components in ecosystems, and thus may show different behavior from soils in a full ecosystem setting.

To illustrate how soil incubation data may be used to inform modeling via data assimilation, this chapter takes as an example the study by Liang et al. (2015), which compares different methods for estimating temperature sensitivity of soil organic carbon decomposition. The data are from a study by Haddix et al. (2011). Fresh soils were sampled from the field and transported to the laboratory. In the laboratory, the soil samples were sieved, and visible roots and rocks were picked out. From each soil sample, subsamples were taken and incubated in jars at different temperatures. The first seven days were treated as pre-incubation to minimize the possible artificial influences of field sampling, sieving, and transportation. After the pre-incubation, CO_2 emission rates were measured as the rate of CO_2 concentration increase in the head space of the jars over time. CO_2 emission rates were measured daily during the first two weeks of incubation, weekly for the next two weeks, and every four weeks thereafter. Overall, there were 36 sampling occasions over the 588-day incubation period. Cumulative CO_2 emissions can be calculated by adding together the daily amounts of emitted CO_2 from day 0 (Figure 23.1). These example data will be used for the following illustration in this chapter.

Conventionally, total CO_2 emissions and/or CO_2 emission rates during the incubation period are directly compared to reveal the effect of different treatments, such as temperature or moisture levels. For example, temperature sensitivity impacts the decomposition of soil organic carbon under climate warming. In the past, the temperature sensitivity of soil organic carbon decomposition was directly calculated as the CO_2 emission rate at a higher temperature, R_{high}, divided by that at a lower temperature, R_{low} (Rey and Jarvis 2006). This estimate usually underestimates the temperature sensitivity after the initial incubation stage because greater decomposition results in less substrate at high than low temperatures at the same point of incubation time. In addition, the direct comparisons of total CO_2 emissions and/or CO_2 emission rates may not reveal processes underlying soil carbon dynamics, such as turnover rates of different carbon components and dependences of temperature sensitivity on substrate quality. When soil carbon models, which explicitly represent such processes, are integrated with data from soil incubation experiments via data assimilation, we can potentially learn more about the process responses underlying the observed soil carbon dynamics.

DOI: 10.1201/9781032711126-29

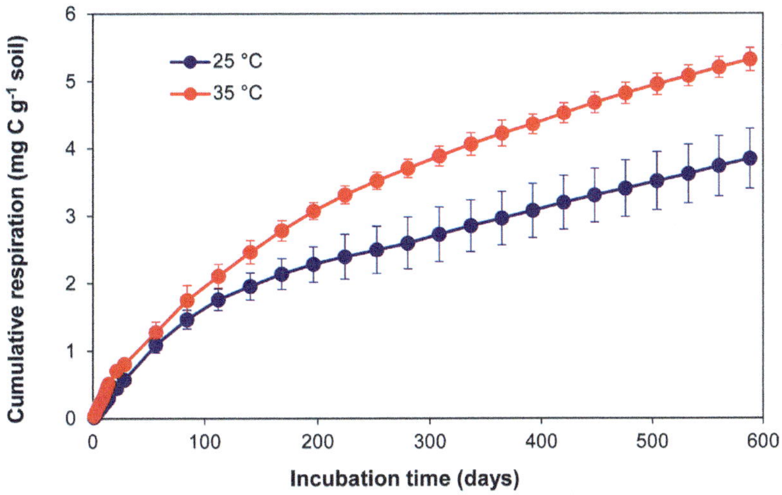

FIGURE 23.1 Cumulative CO_2 emission (R) at 25°C and 35°C. Data originally from Haddix et al. (2011) and used in Liang et al. (2015).

Modified with permission from *Soil Biology and Biochemistry*: Liang et al. (2015).

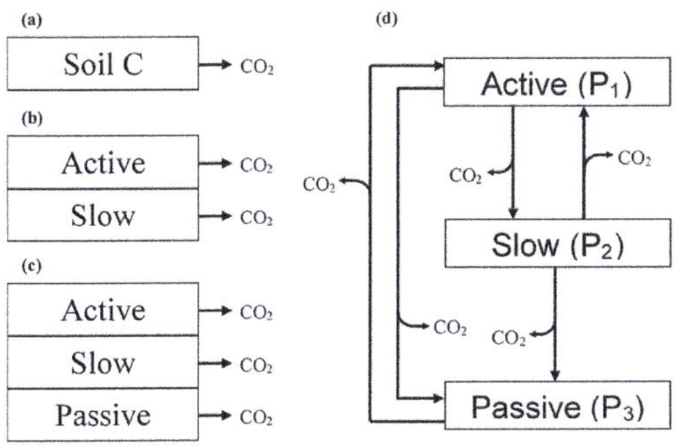

FIGURE 23.2 Schemes of soil carbon models using first-order kinetics.

Modified with permission from *Soil Biology and Biochemistry*: Liang et al. (2015).

SOIL CARBON MODELS

The general framework and underlying principles of many current soil carbon models were presented in Chapters 1 and 2. Generally, first-order kinetics are used to simulate soil carbon cycling (Stanford and Smith 1972, Andren and Paustian 1987). The simplest soil carbon decomposition model simulates soil carbon dynamics as a single pool (Figure 23.2a). However, one-pool models usually perform poorly, since soil organic matter is compartmentalized into pools of different lability, due to various structures of substrates and protection mechanisms. As a result, multiple-pool models are most commonly used to simulate soil carbon dynamics. For example, a two-pool model divides soil carbon into active and slow pools (Figure 23.2b), while a three-pool model divides soil carbon into active, slow, and passive pools (Figure 23.2c). The two- and three-pool models simulate soil carbon cycling without transfers among pools. Another type of model simulates transfers among pools (Figure 23.2d). Similar to ecosystem models discussed in previous chapters, soil carbon models can be described in a matrix form (see Chapter 5). The only difference when using soil carbon models to represent incubation experiments is that they have initial soil carbon pools size(s) at the very beginning of the experiment, but do not have carbon inputs.

Here we will take the three-pool model without transfer (Figure 23.2c) as our example to illustrate the procedure of data assimilation for analysis of soil incubation data. The three-pool model can be described as:

$$\frac{dX(t)}{dt} = -KX(t) \tag{23.1}$$

where

$$K = diag(k_i) = \begin{pmatrix} k_1 & 0 & 0 \\ 0 & k_2 & 0 \\ 0 & 0 & k_3 \end{pmatrix}, \tag{23.2}$$

and

$$X(t) = \begin{pmatrix} X_1(t) \\ X_2(t) \\ X_3(t) \end{pmatrix} \tag{23.3}$$

k_1, k_2, k_3 are the turnover rates of the active, slow and passive pools, and X_1, X_2, X_3 are their pool sizes, respectively.

APPLICATION OF DATA ASSIMILATION TO SOIL INCUBATION DATA

Our data assimilation example follows the seven steps introduced in Chapter 21. Before we start, let us revisit the seven-step procedure again: (1) defining an objective; (2) preparing data; (3) choosing a model; (4) using a cost function; (5) applying an optimization method; (6) estimating parameters; and (7) generating predictions.

1 *Defining an objective*. The objective of the study by Liang et al. (2015) is to compare different methods for estimating temperature sensitivity of soil organic carbon decomposition. They first reviewed published methods for temperature sensitivity. The temperature sensitivity is usually expressed as Q_{10}, measuring the proportional change in soil carbon decomposition rate for a 10 K warming. They found that many studies directly compare CO_2 emissions at different incubation temperatures, estimating an apparent temperature sensitivity of soil organic carbon decomposition. However, this method does not provide the intrinsic temperature sensitivity of different soil carbon components. As introduced earlier, soil carbon models consider soil carbon dynamics in different pools depending on their turnover rates (k), which can be used to represent substrates with different lability. How the parameter k changes with temperature can inform the intrinsic temperature sensitivity of different soil carbon components. Therefore, a specific objective of the study is to estimate the intrinsic temperature sensitivities of different carbon pools in soil models.

2 *Preparing data*. As mentioned above, the data come from an incubation experiment by Haddix et al. (2011). For the illustration of this chapter, data from the 25°C and 35°C treatments (Figure 23.1) are used. Means and standard deviations of measurements from the incubation experiment are needed for data assimilation. The data are organized as shown in Table 23.1.

TABLE 23.1
Cumulative CO_2 emission during the incubation organized for data assimilation

Incubation time (days)	25°C		35°C	
	Mean	SD	Mean	SD
1	0.025	0.002	0.039	0.003
2	0.048	0.004	0.076	0.001
3	0.072	0.003	0.112	0.003
4	0.099	0.007	0.147	0.004
5	0.121	0.007	0.182	0.006
6	0.142	0.009	0.214	0.008
7	0.165	0.007	0.250	0.009
8	0.180	0.006	0.267	0.023
9	0.212	0.005	0.306	0.019
10	0.226	0.004	0.347	0.014
11	0.244	0.003	0.389	0.021
12	0.271	0.010	0.434	0.010
13	0.288	0.011	0.487	0.015
14	0.312	0.016	0.528	0.015
21	0.459	0.026	0.718	0.014
28	0.585	0.055	0.819	0.053
56	1.099	0.113	1.290	0.155
84	1.478	0.139	1.764	0.219
112	1.770	0.161	2.121	0.176
140	1.968	0.200	2.477	0.175
168	2.149	0.229	2.797	0.151
196	2.294	0.265	3.087	0.126
224	2.406	0.334	3.326	0.132
252	2.508	0.353	3.535	0.129
280	2.609	0.386	3.719	0.135
308	2.738	0.407	3.900	0.146
336	2.865	0.384	4.076	0.168
364	2.977	0.395	4.235	0.192
392	3.088	0.400	4.373	0.148
420	3.207	0.398	4.535	0.150
448	3.314	0.399	4.691	0.155
476	3.414	0.418	4.829	0.159
504	3.525	0.429	4.962	0.159
532	3.630	0.435	5.087	0.154
560	3.745	0.445	5.210	0.156
588	3.853	0.444	5.326	0.172

3 *Choosing a model*. Models are chosen dependent on the study objective. Liang et al. (2015) uses four models to compare different estimates of Q_{10}. In practice, the length of the incubation experiment is an important aspect to consider when choosing which model to use. If the length of experiment is relatively short, for example, days to months, the one-pool or two-pool models may be appropriate. If the incubation lasts longer, for example, years, the three-pool model may be better to fit data. Here, our purpose is to illustrate how to use data assimilation to analyze soil incubation data, and the example experiment lasts for 588 days. Therefore, the three-pool model is chosen. The three-pool model has a total of eight to-be-determined parameters, including the initial fractions of active and slow pools, f_1 and f_2, decomposition rates of organic carbon in three pools at 25°C, k_1, k_2, k_3, and corresponding temperature sensitivity parameters, q_1, q_2, and q_3. Details of these parameters are shown in Table 23.2. The initial fraction of the passive pool, f_3, does not need to be estimated as it can be directly calculated as $1 - f_1 - f_2$. The turnover rates of carbon pools at 35°C can be calculated as $k_i \times q_i$. If your experiment does not have treatments at different temperatures, just ignore those temperature sensitivity parameters. In that case, the model has five parameters to be estimated. To conduct data assimilation, the prior probability density functions (PDFs) of the parameters are needed, which represents the prior knowledge ahead of data assimilation. When there is not much prior knowledge about the parameter distributions, the prior PDFs can be specified as uniform distributions over parameter ranges, for example based on available literature (Table 23.2).

4 *Cost function*. Before the cost function is estimated, a mapping function is needed to relate the simulation results to their corresponding measurements. Looking at Table 23.1, the 15th measurement is conducted in day 21. However, the 15th model simulation is CO_2 emission in day 15, and the model simulation for day 21 is the 21st result. Our mapping function will gather a subset of model results that correspond to

TABLE 23.2
A description of model parameters and the ranges of their prior uniform distributions of the three-pool models used in the chapter

Parameter	Unit	Description	Prior uniform distribution	
			Minimum	Maximum
f_1	Unitless	Initial fraction of the active pool	1.00×10^{-2}	1.00×10^{-1}
f_2	Unitless	Initial fraction of the slow pool	1.00×10^{-1}	6.00×10^{-1}
k_1	d^{-1}	Turnover rate of the active pool	1.00×10^{-3}	2.00×10^{-2}
k_2	d^{-1}	Turnover rate of the slow pool	1.00×10^{-5}	5.0×10^{-4}
k_3	d^{-1}	Turnover rate of the passive pool	1.00×10^{-6}	5.0×10^{-5}
q_1	Unitless	Temperature sensitivity of the active pool	1.00	3.00
q_2	Unitless	Temperature sensitivity of the slow pool	1.00	4.00
q_3	Unitless	Temperature sensitivity of the passive pool	1.00	5.00

each measurement occasion. We can build a function between the incubation time, t, and the rank of measurements, i: $t = f(i)$, using the first column of Table 23.1. The incubation time corresponding to a specific measurement can be derived via the mapping function. For example, the incubation time of the 36th measurement is day 588 ($588 = f(36)$). With this mapping function (i.e., $t = f(i)$), you can easily locate the model results corresponding to the 36th measurement is in day $f(36)$ (i.e., day 588).

After deriving the subset of model results that correspond with observations, the cost function, J, which measures the magnitude of mismatch of the simulation results from observations, can be specified. We will adopt the formula for J introduced in Chapter 22. Figure 23.3 shows a snapshot of the J values from iteration 101 to 200 when running data assimilation using the three-pool model and data in Table 23.1. The cost function is used in the optimization step to determine whether to accept or reject proposed parameter sets, which is further discussed in the next step.

5 *Optimization.* Let us follow the introduction to the Markov Chain Monte Carlo (MCMC) method in Chapter 22 to conduct the optimization. The value of the cost function in iteration m is compared with that in iteration m-1. There are basically two cases when comparing the values of cost function, $J_m \leq J_{m-1}$ and $J_m > J_{m-1}$. Examples below are used to show the two cases.

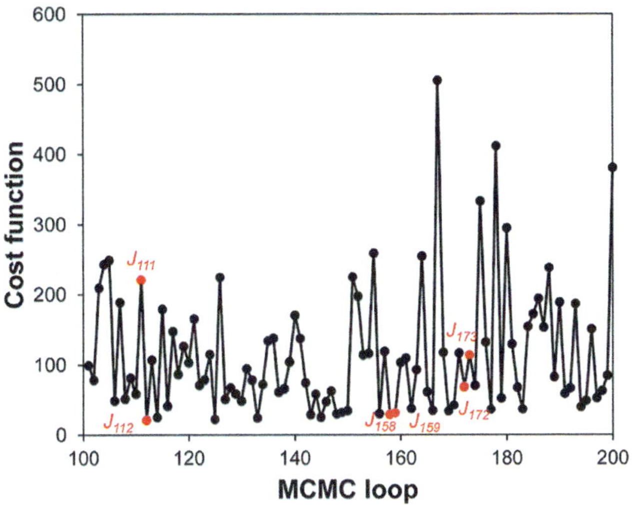

FIGURE 23.3 A snapshot of the cost function values from iteration 101 to 200.

Case 1: In Figure 23.3, $J_{112} = 21.3$ and $J_{111} = 221.2$, respectively. 21.3 is smaller than 221.2, meaning the mismatch between model results and measurements is reduced, i.e., the model simulation is improved. In this case, the proposed parameters are accepted.

Case 2: $J_{173} = 114.1$ and $J_{172} = 69.2$, respectively (Figure 23.3). 114.1 is bigger than 69.2, meaning the mismatch between model results and measurements is increased, i.e., the model simulation is worsened. In this case, the exponential of (J_{172}-J_{173}) is calculated. Here the value is 3×10^{-20}. The value is then compared with a randomly selected number between 0 and 1. If the randomly selected number is smaller than 3×10^{-20}, the proposed parameters are accepted. Otherwise, the proposed parameters are rejected. Given the number to be compared is a random value between 0 and 1, the chance for accepting proposed parameters is (3×10^{-18})%. Let us take a look at another example, $J_{159} = 31.9$ and $J_{158} = 29.6$, respectively (Figure 23.3). 31.9 is slightly bigger than 29.6. Therefore, we calculate the exponential of (J_{158}–J_{159}), which is 1×10^{-2}, suggesting there is a chance of 1% to accept the proposed parameters.

Recall that the purpose of data assimilation is to derive global optimizations for parameters. Allowing a chance to accept proposed parameters even with the increased mismatch between simulations and observations can avoid the parameter estimations getting trapped in local optimizations. But the chance

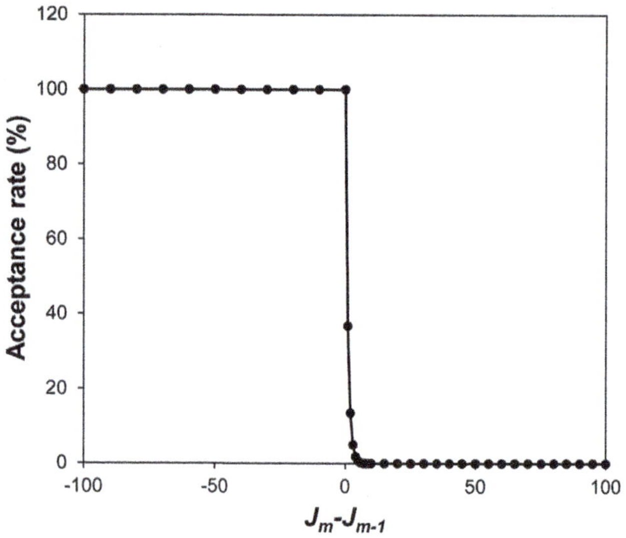

FIGURE 23.4 Dependence of parameter acceptance rate on the change of cost function ($J_m - J_{m-1}$). If the mismatch between simulations and observations decreased ($J_m - J_{m-1} \leq 0$), the acceptance rate is 100%. If $J_m - J_{m-1} > 0$, there is a chance to accept proposed parameters, and the acceptance rate decreases steeply with the mismatch increase.

to accept proposed parameters decreases steeply with the increase in simulation-observation mismatch (Figure 23.4).

6 *Estimating parameters*. Now we have all the accepted parameters after finishing 100,000 MCMC iterations. Excluding results in the burn-in period (see Chapter 22), we have posterior distributions for the selected parameters (Figure 23.5). In the figure, the distributions of parameters related to the active and slow pools have significant single peaks, which means these parameters are well constrained. The distributions of parameters related to the passive pool, by contrast, are not as good, suggesting the data may not have enough information to constrain them.

From the posterior distributions, we can derive the maximum likelihood estimates (MLEs) of parameters. MLEs represent the parameter value at which the distribution peaks. In the example, the Q_{10} distributions peak at 1.22 and 1.76 for the active and slow pools. We can use these values to represent the intrinsic temperature sensitivities at 25°C of the corresponding pools. The Q_{10} distribution of the passive pool is not well constrained, and we can use the mean value, 2.67, as a reasonable guess of its intrinsic temperature

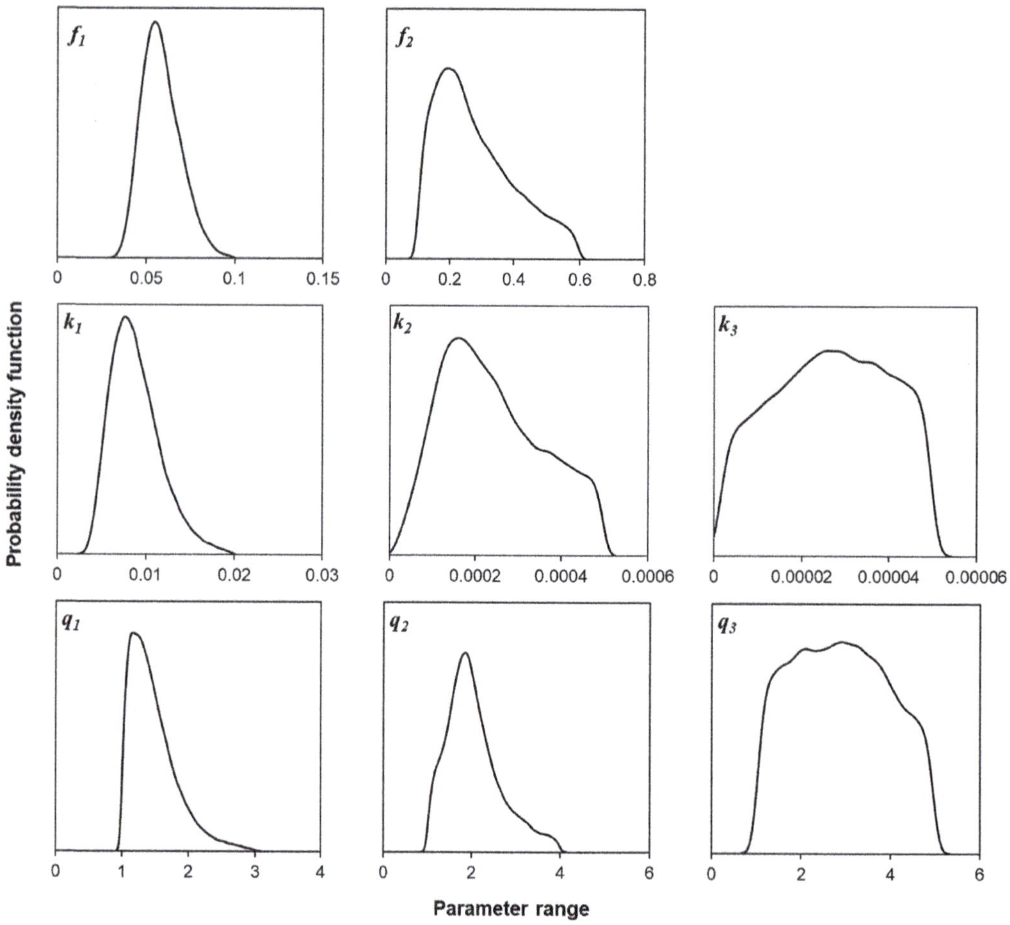

FIGURE 23.5 Posterior probability density functions of the parameters in the three-pool model.

Modified with permission from *Soil Biology and Biochemistry*: Liang et al. (2015).

sensitivity at 25°C. The temperature sensitivity increases with the decrease of substrate lability, suggesting that carbon pools with longer turnover times are more vulnerable to warming.

7 *Generating predictions*. Predictions of soil carbon decomposition can be generated with the derived parameters and the model. With the generated predictions, we can evaluate the model performance by comparing observations and simulations. In Figure 23.6, if all dots are right on the 1:1 line (i.e., y = x), it means the model simulations are exactly equal to observations, which of course is very unlikely. In practice, we can use a regression line to describe the comparison of model simulations and observations. The regression lines at 25°C and 35°C are y = 0.9919x and y = 0.9998x, or very close to y = x which would represent perfect agreement. This indicates that the constrained model fits the data very well.

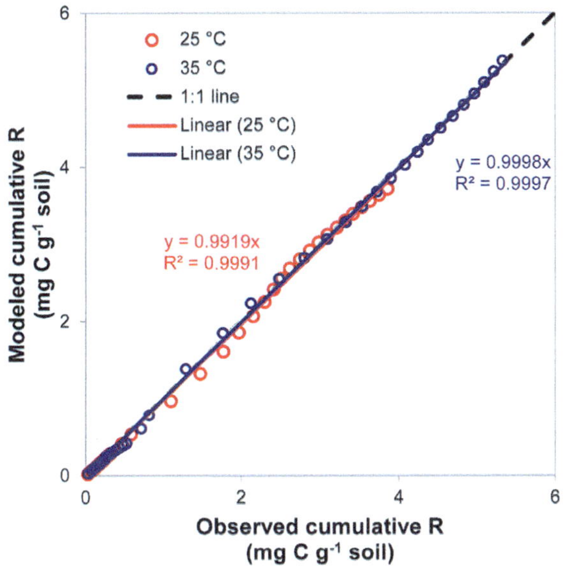

FIGURE 23.6 Comparison of observed and modeled cumulative CO_2 emissions (R) in the incubation experiment. All the dots distribute around the 1:1 line, suggesting the model simulations are well matched with observations.

With the constrained model, we can also simulate the dynamics of different carbon pools during the incubation. Figure 23.7 shows that the active pool dominates the CO_2 emission at the early stage and is depleted very fast. At the end of incubation, the cumulative CO_2 emission from the active pool is similar between the 25°C and 35°C incubations. The contributions of the slow and passive pools to CO_2 emissions gradually increase after the active pool is depleted.

SUMMARY

This chapter introduced the application of data assimilation to soil incubation experiments. A variety of models can be chosen to simulate soil carbon cycling depending on the scientific question and the length of incubation. Assimilating soil incubation data with models can help understand processes underlying the observed soil carbon dynamics, such as turnover rates and temperature sensitivities of substrate with different qualities.

SUGGESTED READING

Liang, J., D. Li, Z. Shi, J. M. Tiedje, J. Zhou, E. A. G. Schuur, K. T. Konstantinidis, and Y. Luo. 2015. Methods for estimating temperature sensitivity of soil organic matter based on incubation data: A comparative evaluation. *Soil Biology & Biochemistry 80*:127–135.

QUIZ

1 How can soil carbon models and data assimilation help with the analysis of soil incubation data?
2 Why is the length of incubation experiments an important consideration when choosing a model?
3 Suggest why some parameters can be well constrained and other cannot, and how this information guides future experimental design.

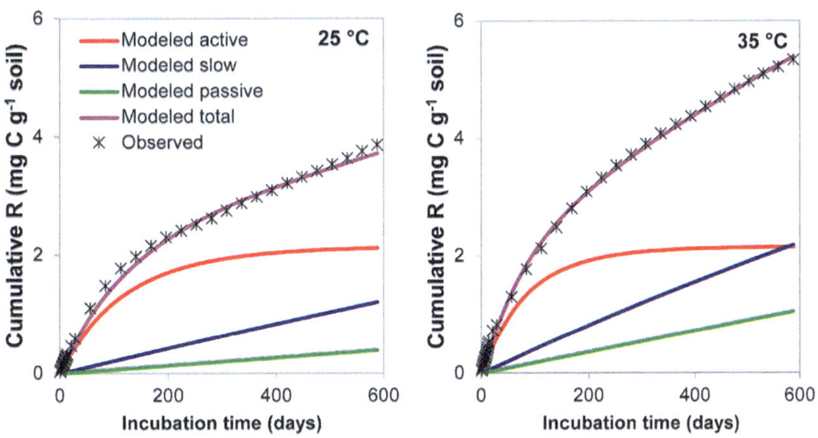

FIGURE 23.7 Observed and modeled cumulative CO_2 emissions (R) from individual and total pools at two incubation temperatures.

Modified with permission from *Soil Biology and Biochemistry*: Liang et al. (2015).

24 Practice 6

The Seven-Step Procedure for Data Assimilation

Xin Huang
National Center for Atmospheric Research, Boulder, USA

This chapter will guide you to learn the seven-step procedure of data assimilation by replicating a published study on the assimilation of observational data into the TECO model to calibrate decomposition rate parameters for multiple soil organic matter pools. We will first review the seven steps in Chapter 21 and learn how to program these steps using code examples in Python. Then we will perform three exercises to reproduce figures from an earlier paper on the study with the CarboTrain toolkit. It is recommended that you go over the source code of the three exercises in CarboTrain. You are expected to program a data assimilation algorithm with your own model or data sets after this practice.

INTRODUCTION

The seven steps in data assimilation (DA) are described in detail in Chapter 21. They are: (1) defining an objective; (2) preparing data; (3) choosing a model; (4) using a cost function; (5) applying an optimization method; (6) estimating parameters; and (7) generating predictions (Figure 24.1). This practice will use an example from a study by Xu et al. (2006) to introduce how to program these seven steps. The example of the DA study by Xu et al. (2006) is programmed in a Python file 'Probabilistic_inversion.py'. We will use this Python program to illustrate each of the seven steps in a DA study.

STEP 1: DEFINING AN OBJECTIVE

Defining an objective usually means to decide target parameters to be estimated by DA. The target parameters chosen in the study by Xu et al. (2006) are decomposition rates to represent fractions of carbon (C) leaving seven soil organic matter (SOM) pools of the Terrestrial ECOsystem (TECO) model (Table 24.1). The TECO model will be described in Step 3.

STEP 2: PREPARING DATA

The DA study in Xu et al. (2006) integrated six observations (i.e., soil respiration, woody biomass, foliage biomass, litterfall, C in forest floor, and C in forest mineral soil) under ambient and elevated CO_2 treatments. All the 12 data sets are saved in six data files under the '/input' folder. There are three rows in each data file. The first row is the time series, the second is observation under ambient CO_2 treatment and the final row is observation under elevated CO_2 treatment. Figure 24.2 shows the Python code for reading the six data files and calculating the six observation variances. The variable *ninput* with a value of 1 or 2 is to indicate which CO_2 treatment is applied in the Python program. For example, if *ninput* has a value of 1, the variables *soilResp, woody, foliage, litterfall, forestFloor*, and *forestMineral* save the six observations from the ambient CO_2 treatment. All these six observations are collected in the *obsList* variable and the corresponding six observation variances are saved in the *varList* variable. These variables will be used in Step 4.

STEP 3: MODEL

A terrestrial ecosystem (TECO) model is used (Xu et al., 2006). The TECO model has a seven-pool structure and the fractions of C exiting each pool each day are the parameters to be estimated (Table 24.1, the variable *c* in Figure 24.3a). This step involves two program fractions in Probabilistic_inversion.py: defining model simulation and running model simulation (Figure 24.3). A function *run_model* (Figure 24.3a) defines the TECO model simulation with a set of parameter values (variable *c*). The core codes of model simulation (lines 65–68) are to program the differential equation:

$$\frac{dX(t)}{dt} = \xi(t)ACX(t) + BU(t) \tag{24.1}$$

where $X(t)$ is a 7 × 1 vector to represent the seven C pool sizes at time t; $\xi(t)$ is a scaling function describing environment effects at time t; A is a 7 × 7 matrix representing the C transfer coefficients among the seven pools; C is a 7 × 7 diagonal matrix with diagonal elements describing the fraction of C leaving each pool or the parameters to be estimated; $B = (0.25, 0.3, 0, 0, 0, 0, 0)^T$ is a vector accounting for the partitioning coefficients of C input to non-woody and woody biomasses; $U(t)$ is the C input from photosynthesis at time t.

The variable *Nt* (Figure 24.3a) is five-year simulation time on a daily time step, with a value of 1825. The variable *xsimu* will accrue the seven C pool sizes over the *Nt* simulation days. Rather than the seven C pool sizes, the return values of the *run_model* function are six simulated data sets (i.e., *woody_simu, foliage_simu, litterYr, forestFloor_simu, forestMineral_simu*, and *soilResp_simu*) to match with the six observational datasets. Mapping operators are used for this purpose and we will learn about these operators in Step 4.

To run the model simulation, a new set of parameter values (variable *c_new*) needs to be passed to the *run_model*

DOI: 10.1201/9781032711126-30

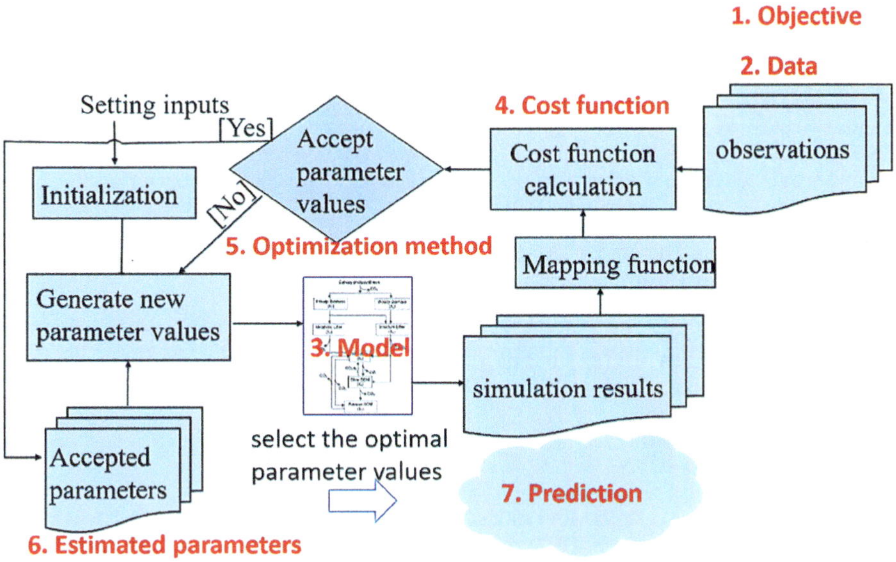

FIGURE 24.1 Seven-step procedure of the data assimilation (DA) as illustrated in A model-independent data assimilation (MIDA) module. **Adopted from Huang et al., 2021.**

TABLE 24.1
Seven parameters to be estimated

Parameter	Description	Unit	Min (×10^{-4})	Max (×10^{-3})
c_1	Fraction of C leaving pool 1	$gCg^{-1}\,d^{-1}$	1.764	2.95
c_2	Fraction of C leaving pool 2	$gCg^{-1}\,d^{-1}$	0.548	0.274
c_3	Fraction of C leaving pool 3	$gCg^{-1}\,d^{-1}$	54.79	27.34
c_4	Fraction of C leaving pool 4	$gCg^{-1}\,d^{-1}$	5.48	2.74
c_5	Fraction of C leaving pool 5	$gCg^{-1}\,d^{-1}$	27.4	6.85
c_6	Fraction of C leaving pool 6	$gCg^{-1}\,d^{-1}$	0.548	0.274
c_7	Fraction of C leaving pool 7	$gCg^{-1}\,d^{-1}$	0.0137	0.00913

Based on the study by Xu et al., (2006)

```
## Step 2: 6 Data Sets and their variances
## initialize observation-related varibles
soilResp=np.loadtxt('./Source_code/unit_6/input/SoilRespiration.txt')[[0,ninput],]
woody=np.loadtxt('./Source_code/unit_6/input/Woody.txt')[[0,ninput],]
foliage=np.loadtxt('./Source_code/unit_6/input/Foliage.txt')[[0,ninput],]
litterfall=np.loadtxt('./Source_code/unit_6/input/Litterfall.txt')[[0,ninput],]
forestFloor=np.loadtxt('./Source_code/unit_6/input/ForestFloor.txt')[[0,ninput],]
forestMineral=np.loadtxt('./Source_code/unit_6/input/ForestMineral.txt')[[0,ninput],]
## collect the 6 data sets into an observation list
obsList=[woody[1,],foliage[1,],litterfall[1,],forestFloor[1,],forestMineral[1,],soilResp[1,]]
## variance of 6 observations
# ddof=1 non-bias estimator for sample variance
varList=[np.var(obs, ddof=1) for obs in obsList]
```

FIGURE 24.2 The Python code for reading six observations from text files and calculating their variances.

function (Figure 24.3b). During this process, the parameter values in variable *c_new* will be assigned to variable *c* in Figure 24.3a. Variable *simuList* saves the return values of the model simulation (i.e., the six simulated data sets). In Step 4, the cost function will calculate the mismatch between the simulated data sets (*simuList)* and the observed ones (*obsList* in Step 1).

STEP 4: COST FUNCTION

The cost function quantifies the discrepancy between simulated data sets (i.e., *simuList*) and observed ones (i.e., *obsList*). In the TECO model simulation, function *run_model* calculates the seven C pool sizes over the course of five years, with results saved in variable *xsimu*. To be comparable with six observed data sets (*obsList*), *xsimu* needs to be

(a)

```
## Step 3: TECO model
def run_model(c):
    """Forward run model.
    Global args: tau, b, u, x0, Nt, cbnScale"""
    ## x stores 7 simulated carbon pools simulated in 5 years (daily scale)
    x=np.zeros([7,Nt+1], dtype=float)
    ## multiple matrix A and diagnol matrix  C
    AC = np.matrix([[-c[0], 0, 0, 0, 0, 0, 0],
          [0, -c[1], 0, 0, 0, 0, 0],
          [0.7123 * c[0], 0, -c[2], 0, 0, 0, 0],
          [0.2877 * c[0], c[1], 0, -c[3], 0, 0, 0],
          [0, 0, 0.45 * c[2], 0.275 * c[3], -c[4], 0.42 * c[5], 0.45 * c[6]],
          [0, 0, 0, 0.275 * c[3], 0.296 * c[4], -c[5], 0],
          [0, 0, 0, 0, 0.004 * c[4], 0.03 * c[5], -c[6]]])
    x[:,0] = x0
    ## simulate for 5 years
    for i in range(1,Nt+1):
       ## Eq. 1 in Xu et al., (2006)
       x[:,i]=np.dot((np.eye(7)+AC*tau[i-1]),x[:,i-1])+np.asarray(b)*u[i-1]*cbnScale
    ## simulated carbon pools with 7 rows and Nt columns
    xsimu=x[:,range(1,Nt+1)]
    ::::::
    ::::::
    return [woody_simu, foliage_simu, litterYr, forestFloor_simu, forestMineral_simu,
        soilResp_simu]
```

(b)

```
            ## Step 3: TECO model
            ## running model, return [woody_simu, foliage_simu, litterYr, forestFloor_simu,
                forestMineral_simu, soilResp_simu]
            simuList = run_model(c_new)
```

FIGURE 24.3 The Python code for (a) defining a function of TECO model simulation; (b) calling the function to run the model.

```
## Step 3: TECO model
def run_model(c):
   ::::::
    ## simulated carbon pools with 7 rows and Nt columns
    xsimu=x[:,range(1,Nt+1)]
    ## use mapping function to covert 7 carbon pools to 6 simulated data sets
    ## get yearly simulated litterfall
    litterDaily=tau[:-1] * np.dot(phi_litterfall, xsimu)
    litterYr = [sum(litterDaily[range((i-1)*365, i*365)]) for i in range(1,6)]
    ## get simulated soilResp
    # convert float to int. In python, index starts from 0.
    soilTime=soilResp[0,:].astype(int)-1
    soilResp_simu=np.asarray(tau)[soilTime]*np.dot(phi_slResp,xsimu
        [:,(soilTime)])+0.25*(1-b[0]-b[1])*u[soilTime]
    ## get simulated woody biomass
    woodyTime=woody[0,:].astype(int)-1
    woody_simu = np.dot(phi_woodBiom, xsimu[:, woodyTime])
    ## get simulated foliage biomass
    foliageTime=foliage[0,:].astype(int)-1
    foliage_simu=np.dot(phi_foilageBiom, xsimu[:, foliageTime])
    ## get simulated Forestfloor biomass
    forestFloorTime = forestFloor[0, :].astype(int) - 1
    forestFloor_simu = np.dot(phi_cForestFloor, xsimu[:, forestFloorTime])
    ## get simulated ForestMinearal biomass
    cMineralTime = forestMineral[0, :].astype(int) - 1
    forestMineral_simu = np.dot(phi_cMineral, xsimu[:, cMineralTime])
    return [woody_simu, foliage_simu, litterYr, forestFloor_simu, forestMineral_simu,
        soilResp_simu]
```

FIGURE 24.4 Python code for mapping seven pool sizes (*xsimu*) to six simulated data sets (*woody_simu, foliage_simu, litterYr, forestFloor_simu, forestMineral_simu*, and *soilResp_simu*) from the TECO model simulation.

(a)

```
## The mappings
phi_slResp=[0.25*c[0],0.25*c[1],0.55*c[2],0.45*c[3],0.7*c[4],0.55*c[5],0.55*c[6]]
phi_woodBiom=[0,1,0,0,0,0,0]
phi_foilageBiom=[0.75,0,0,0,0,0,0]
phi_litterfall=[0.75*c[0],0.75*c[1],0,0,0,0,0]
phi_cMineral=[0,0,0,0,1,1,1]
phi_cForestFloor=[0,0,0.75,0.75,0,0,0]
```

(b)

```
## only update 2 mapping when new parameter values generate
phi_slResp=[0.25*c_new[0],0.25*c_new[1],0.55*c_new[2],0.45*c_new[3],0.7*c_new
    [4],0.55*c_new[5],0.55*c_new[6]]
phi_litterfall=[0.75*c_new[0],0.75*c_new[1],0,0,0,0,0]
```

FIGURE 24.5 The Python code for (a) defining the mapping operators; (b) updating mapping operators according to parameter values (*c_new*).

```
        ## Step 3: TECO model
        ## running model, return [woody_simu, foliage_simu, litterYr, forestFloor_simu,
            forestMineral_simu, soilResp_simu]
        simuList = run_model(c_new)

        ## Step 4: Cost Function
        J_new=sum([sum((simuList[i]-obsList[i])**2)/(2*varList[i]) for i in range(6)])
```

FIGURE 24.6 Python code for calculating the mismatch between simulated (*simuList*) and observed datasets (*obsList*) based on cost function.

converted to six simulated data sets (*simuList*) with mapping operators (i.e., *phi_litterfall, phi_slResp, phi_woodBiom, phi_foilageBiom, phi_cForestFloor, phi_cMineral* in Figures 24.4 and 24.5). Before mapping, function *'run_model'* needs to update the values of *Phi_slResp* and *Phi_litterfall* mapping operators with parameter values *c* (Figure 24.5b). After mapping, the mismatch between *simuList* and *obsList* is calculated according to:

$$P(Z|c) \propto exp\left\{-\sum_{i=1}^{6}\frac{1}{2\sigma_i^2}\sum_{t\in obs(Z_i)}\left[Z_i(t)-\varphi_i X(t)\right]^2\right\} \quad 24.2$$

where $P(Z|c)$ is conditional probability density of observations Z on parameters c, i.e., the likelihood function of parameter c; σ_i is the ith observation variance; $obs(Z_i)$ is the time series of ith observation; $Z_i(t)$ is the ith observation at time t; φ_i is the mapping operator; $X(t)$ is the simulated seven pool sizes at time t; $\varphi_i X(t)$ is the ith simulated data set after mapping.

The value of mismatch is saved in the variable *J_new* (Figure 24.6). Step 5 (Optimization) will use the value of *J_new* to decide whether the set of parameter values should be accepted or not.

STEP 5: OPTIMIZATION METHOD

Our study uses the Metropolis-Hasting method to draw parameter samples from their prior distribution. This method iteratively executes two phrases (i.e., proposing phase and moving phase) until a preset iteration number (e.g., 20,000) is reached. The proposing phase is implemented by function *GenerateParamValues*, which generates new parameter values (i.e., variable *cNew*) based on the current accepted values (i.e., variable *c_opt*) (Figure 24.7a). During this process, function *GenerateParamValues* will call another function *isQualified* to assess whether the newly proposed parameter values (*cNew*) are in the reasonable parameter range or not. If the *isQualified* function returns *TRUE*, *cNew* is a reasonable new set of parameter values and the generation of new parameter values will stop. Otherwise, function *GenerateParamValues* will discard this set of parameter values, generate new parameter values, assign these values to *cNew*, and call function *isQualified* again. Following the proposing phase, *cNew* will be used to update mapping operators (i.e., *Phi_slResp* and *Phi_litterfall*) and run model simulations through calling function *run_model* (Figure 24.7b). The mathematical mechanism behind the proposing phase is available in Xu et al. (2006).

The moving phase decides whether the new parameter values (i.e., *cNew*) are accepted or not (Figure 24.8).

The value of *J_new* (i.e., mismatch between the simulated (*cNew*) and observed datasets) is compared with *J_record[record]* (i.e., the mismatch using the previously accepted parameter value *c_record[record]*). The *cNew* will be accepted if *J_new* is smaller than *J_record[record]*, or the value *exp* (*J_record*[*record*] – *J_new*) is larger than a random number (*randNum*) from the uniform distribution between 0 and 1. If accepted, the new parameter values are saved into an array, *c_record*. The corresponding mismatch will be saved in another array, *J_record*. The count of accepted

(a)

```
def GenerateParamValues(c_op):
    """Generate new parameter values based on eigvalue and eigvectors
    Global args: eigD, eigV, cmin, cmax, paramNum"""
    flag = True
    while (flag):
        # Normally distributed pseudorandom numbers
        randVector = np.random.randn(paramNum)
        cT = randVector * np.sqrt(eigD)
        cNew = np.dot(eigV, (np.dot(eigV.T, c_op) + cT))
        if (isQualified(cNew)):
            flag = False
    return cNew

def isQualified(c):
    """Decide whether the new parameter values exist in [cmin, cmax] value interval
    Global args: paramNum, cmin, cmax"""
    flag = True
    for i in range(paramNum):
        if(c[i] > cmax[i] or c[i] < cmin[i]):
            flag = False
            break
    return flag
```

(b)

```
## Proposing step: generate a new set of parameter values based on current accepted
    parameter values
c_new = GenerateParamValues(c_record[:, record])
## only update 2 mapping when new parameter values generate
phi_slResp=[0.25*c_new[0],0.25*c_new[1],0.55*c_new[2],0.45*c_new[3],0.7*c_new
    [4],0.55*c_new[5],0.55*c_new[6]]
phi_litterfall=[0.75*c_new[0],0.75*c_new[1],0,0,0,0,0]
## Step 3: TECO model
## running model, return [woody_simu, foliage_simu, litterYr, forestFloor_simu,
    forestMineral_simu, soilResp_simu]
simuList = run_model(c_new)
```

FIGURE 24.7 Python codes for (a) defining the function of proposing phase in Metropolis-Hasting algorithm; (b) calling the function to generate parameter samples.

```
## Step 4: Cost Function
J_new=sum([sum((simuList[i] - obsList[i])**2) / (2 * varList[i]) for i in range(6)])
delta_J = J_record[record] - J_new
## Moving step: to decide whether the new set of parameter values will be accepted or not
randNum = np.random.uniform(0, 1, 1)
if (min(1.0, np.exp(delta_J)) > randNum):
    ## accept the new set of parameter values and update relevant variables
    record += 1
    c_record[:, record] = c_new
    J_record[record] = J_new
    ## print out the acceptance rate
    print('simu=' + str(simu) + ' accepted=' + str(record))
```

FIGURE 24.8 The Python code of the moving phase in Metropolis-Hasting algorithm.

parameter values, *record*, is then increased by 1. Therefore, *c_record[record]* and *J_record[record]* indicate the current last element in the two arrays, which also represent the currently accepted parameter values and corresponding mismatch. If the new parameter values from this iteration are not accepted, they are discarded. Whether *cNew* is accepted or not, the next iteration always uses the currently accepted parameter values (*c_record[record]*) for the proposing phase to generate a new set of parameter values.

STEP 6: ESTIMATE PARAMETERS

The outputs of DA are accepted parameter values (*c_record*), corresponding mismatches (*J_record*), the number of accepted parameter values (*record*), the optimal parameter values through maximum likelihood estimation (*bestC*), and the simulated data sets given the optimal parameters (*bestSimu*). All these outputs are saved to text files through calling function *write_io_file* (Figure 24.9). All parameter values accepted are

```
def write_io_file(outDir):
    """write outputs to files.
    Global args: c_record, J_record, record, bestC, bestSimu"""
    np.savetxt(outDir+'/mismatch_accepted.txt', J_record[1:record])
    np.savetxt(outDir+'/param_accepted.txt', c_record[:, 1:record])
    np.savetxt(outDir+'/accepted_num.txt', [record])
    np.savetxt(outDir+'/bestParam.txt', bestC)
    np.savetxt(outDir+'/Woody_bestSimu.txt',bestSimu[0])
    np.savetxt(outDir+'/Foliage_bestSimu.txt',bestSimu[1])
    np.savetxt(outDir+'/Litterfall_bestSimu.txt',bestSimu[2])
    np.savetxt(outDir+'/Forestfloor_bestSimu.txt',bestSimu[3])
    np.savetxt(outDir+'/ForestMineral_bestSimu.txt',bestSimu[4])
    np.savetxt(outDir+'/SoilResp_bestSimu.txt',bestSimu[5])
```

FIGURE 24.9 The Python code for saving DA outputs to text files.

saved in 'param_accepted.txt', all mismatches with parameter values accepted in 'mismatch_accepted.txt', the number of accepted parameter values in 'accepted_num.txt', the optimal parameter values in 'bestParam.txt', and simulation outputs given the optimal parameter values in *'xxx_* bestSimu.txt' (e.g., Woody_CestSimu.txt) in the output directory folder (*outDir*).

Posterior distributions of parameters are the constrained parameter range after DA, which are often used for estimating the parameter uncertainty. An R script is provided to plot the posterior distributions with all accepted parameter values in 'param_accepted.txt'. The peak of the distribution represents the optimal parameter value. The simulated data sets with these optimal parameter values are also compared with observations in the R script. After DA, 'Probabilistic_inversion.py' runs the R script automatically and saves plots into the '/figures' folder.

STEP 7: PREDICTION

Parameter uncertainty as expressed by the posterior distribution of parameters will translate into prediction uncertainty characterized by the cumulative probability distribution of simulated C pool sizes. Prediction in the study of Xu et al. (2006) uses two functions: *prediction* and *forward_run* (Figure 24.10a). The function *prediction* is the start point of the prediction step. First this function generates 12,000 samples (*c_all*) from the accepted parameter values saved under *outDir* folder. Then function *prediction* uses each sample (*c*) of these 12,000 sets of parameter values for model forward simulation through calling function *forward_run*. The variable *x_record* saves all simulated pool sizes (Line 128 in Figure 24.10a). Finally, the results of prediction (i.e., *x_record*) will be written to the file 'prediction.txt' in the *outDir* folder. To call function *prediction*, we only need to provide the directory of DA results *(outDir)* as shown in Figure 24.10b.

With each set of parameter values (*c*) sampled in function *prediction*, function *forward_run* executes the model simulation over the 10 years following the DA period (i.e., 2001 to 2010) using Equation 24.1. This function is similar to function *run_model* in Step 3 except that the simulation times are different. The environmental scalar (*tau_forescast*) and C input from photosynthesis (*u_forest*) from 1996 to 2000 were replicated twice to provide environmental 'forcing' for 2001 to 2010. Variable *x* stores the seven simulated C pool sizes in each daily simulation step. The return value of *forward_run* is the simulated pool sizes at the end of year 2010 (*xsimu*). Line 128 in Figure 24.10a shows an example of calling *forward_run* in function *prediction*.

EXERCISES WITH CARBOTRAIN TOOLBOX

The following three exercises using the CarboTrain toolbox will help readers become familiar with the DA methodology described above. Exercise 1 is to conduct DA with the TECO model separately for ambient and elevated CO_2 treatments. Based on the results, we will generate figures similar to Figures 3, 4, and 8 in Xu et al. (2006). Exercise 2 uses the optimized parameter values from Exercise 1 to predict soil C pool sizes. The expected figure is similar to Figure 9 in Xu et al. (2006). The Posterior distributions of parameters c_1, c_2, and c_4 are bell-shaped in Figures 3 and 4 of Xu et al. (2006). In Exercise 3 we will re-conduct DA under the ambient CO_2 treatment using enlarged prior parameter ranges for parameters c_3, c_5, and c_7 . Results similar to Figure 5 in Xu et al. (2006) are expected.

Exercise 1

1. Open CarboTrain ➔ A dialog appears (Figure 24.11) ➔ Select Unit 6 and Exercise 1 ➔ Choose ambient CO_2 ➔ Choose an output directory on your computer ➔ Click 'Run Exercise' ➔ A dialog appears (Figure 24.12a) ➔ Click 'OK'
2. A series of outputs will be printed in another window (Figure 24.13). The variable *simu* is the number of total executed simulations while the *accepted* variable means the number of simulations with accepted parameter values. These numbers will keep increasing until *simu* reaches a value approximating 20,000. A dialog will pop up to notify that DA is finished. Execution time is about 20 minutes depending on the specifications of your computer.
3. After DA, a dialog appears to notify that the task is complete (Figure 24.12b). Open the output directory you specified in Step 1. There will be four folders: '/practice_1_ambient', ' /practice_1_elevated', '/practice_2', '/practice_3_ambient'. Open

(a)

```
def prediction(outDir):
    nsample = 12000
    # check whether two files exist
    if(os.path.isfile(outDir+"/param_accepted.txt") and
        os.path.isfile(outDir+"/accepted_num.txt")):
        c_record = np.loadtxt(outDir+"/param_accepted.txt")
        record = int(np.loadtxt(outDir+"/accepted_num.txt"))-1
        # generate 12,000 random integer ranging from 1 to record
        sampleId = np.random.randint(1,record,nsample)
        c_all = c_record[:,sampleId] # 12,000 samples of p(c|Z)
        x_record = np.zeros([7,nsample], dtype=float)
        for i in range(nsample):
            c = c_all[:,i]
            x_record[:,i] = forward_run(c)
            print('the '+str(i+1)+'th sampling, total '+ str(nsample))
        # save the predicted pool sizes
        np.savetxt(outDir+'/prediction.txt', x_record)
    else:
        # Can't find param_accepted.txt and accepted_num.txt in the output folder
        print('Warning: Please finish exercise 1 first!')

def forward_run(c):
    """prediction with the parameter values c
    Global args: tau, b, u, x0, Nt, cbnScale
    """
    ## environmental scalar and C input duplicate to represent 2000 to 2010
    Nt_forecast = 365*10
    tau_forecast = np.tile(tau,2) # repeat twice
    u_forecast = np.tile(u,2)
    ## x stores 7 simulated carbon pools from 2000 to 2010 (daily scale)
    x=np.zeros([7,Nt_forecast+1], dtype=float)
    ## multiple matrix A and diagnol matrix  C
    AC = np.matrix([[-c[0], 0, 0, 0, 0, 0, 0],
          [0, -c[1], 0, 0, 0, 0, 0],
          [0.7123 * c[0], 0, -c[2], 0, 0, 0, 0],
          [0.2877 * c[0], c[1], 0, -c[3], 0, 0, 0],
          [0, 0, 0.45 * c[2], 0.275 * c[3], -c[4], 0.42 * c[5], 0.45 * c[6]],
          [0, 0, 0, 0.275 * c[3], 0.296 * c[4], -c[5], 0],
          [0, 0, 0, 0, 0.004 * c[4], 0.03 * c[5], -c[6]]])
    x[:,0] = x0
    ## simulate carbon pools from 2000 to 2010
    for i in range(1,Nt_forecast+1):
        ## Eq. 1 in Xu et al., (2006)
        x[:,i]=np.dot((np.eye(7)+AC*tau_forecast[i-1]),x[:,i-1])+np.asarray(b)*
            u_forecast[i-1]*cbnScale
    ## simulated carbon pools at 2010
    xsimu=x[:,Nt_forecast]
    return xsimu
```

(b)

```
if ninput == 1 and sys.argv[2] == "ParamRange.txt" and enable_prediction == 1:
    prediction(outDir)
elif ninput == 2 and sys.argv[2] == "ParamRange.txt" and enable_prediction == 1:
    prediction(outDir)
```

FIGURE 24.10 Python code for (a) defining functions for prediction; (b) calling functions for prediction.

the '/practice_1_ambient' folder. You will find text files and figures generated in the subfolder '/figures'. These files and figures are results of DA under the ambient CO_2 treatment.

4 Repeat Steps 1–3 but choose elevated CO_2 in Step 1. Do not change the output directory. After DA, open the '/practice_1_elevated' folder in the output directory. The files and figures in this folder are results of DA under the elevated CO_2 treatment. Compare the results with Figures 3, 4, and 8 in Xu et al. (2006).

Questions:

What is the acceptance rate in each exercise? Which parameters are well constrained? Which observations contribute to the constrained parameter values? How many shapes are there in the posterior distributions (e.g., bell shape)? What are the meanings behind these different shapes?

Exercise 2

1 In the main window of CarboTrain, select Unit 6 and Exercise 2 ➔ Use the default output folder as for Exercise 1➔ Click 'Run Exercise'.

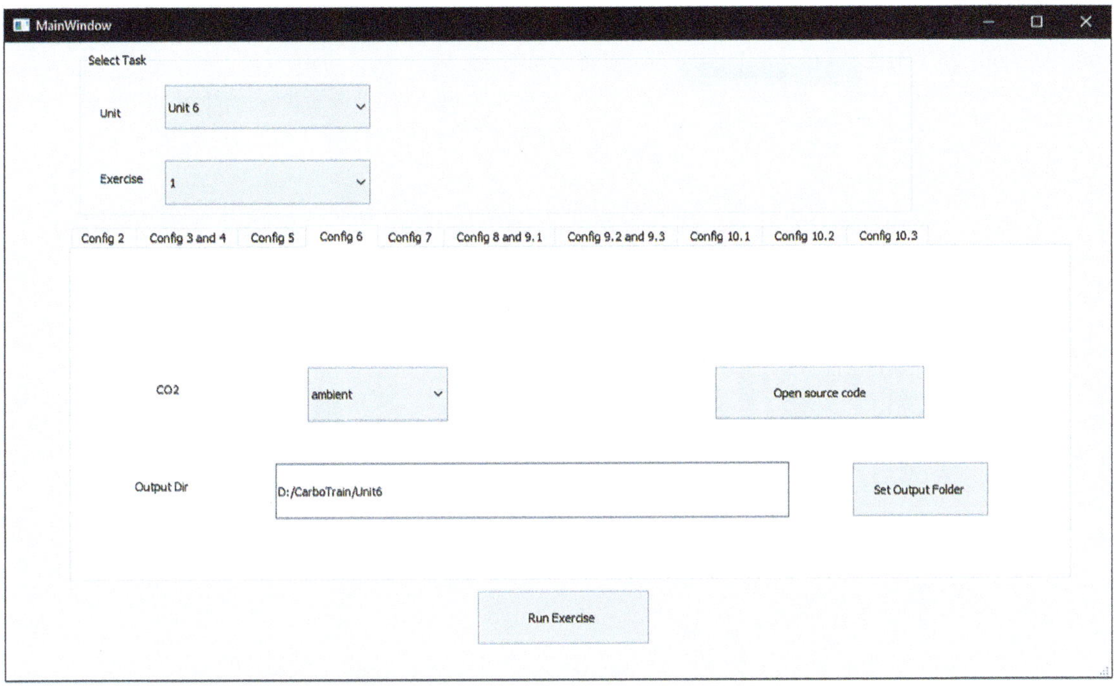

FIGURE 24.11 The main window in CarboTrain toolbox for this practice.

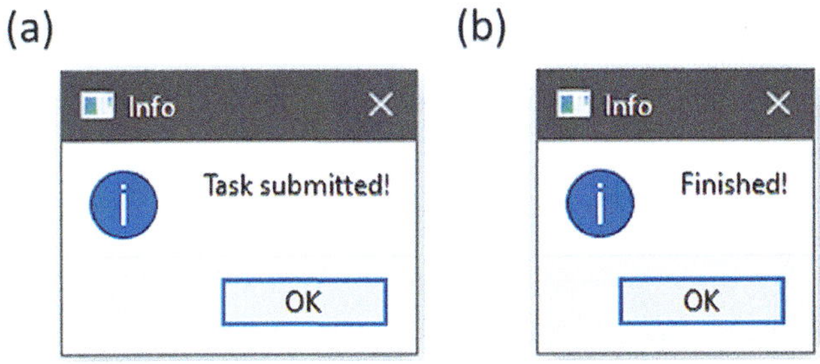

FIGURE 24.12 The dialog notifies (a) submitting a new task; (b) finishing a task.

2 Open the output folder and find a figure generated in the '/practice_2' folder. Compare this figure with Figure 9 in Xu et al. (2006).

Questions:

Does elevated CO_2 positively influence C accumulation (i.e., larger pool sizes)? Do poorly constrained parameters (i.e., c_3, c_5, and c_7) influence the predicted sizes of associated pools (i.e., X_3, X_5, and X_7)?

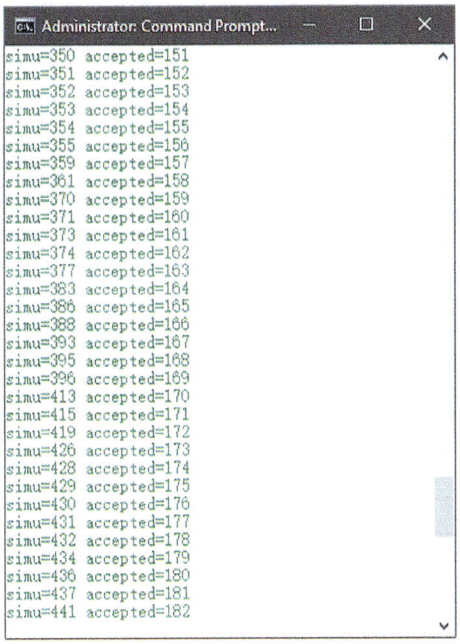

FIGURE 24.13 Window in CarboTrain toolbox to display the progress of the DA experiment.

Exercise 3

1 Repeat Steps 1–3 of Exercise 1 but choose ambient CO_2 and Exercise 3 in the CarboTrain main window (Figure 24.11).

2 After DA, go to the output directory and open the '/practice_3_ambient' folder. The files and figures generated in this folder are results of DA using enlarged prior parameter ranges for c_3, c_5, and c_7. Compare the results with Figure 5 in Xu et al. (2006).

Questions:

Did the posterior distributions of c_3, c_5, and c_7 change after their prior parameter ranges were extended? What else can we do to further constrain the parameter uncertainty?

SUGGESTED READINGS

Xu, T., L. White, D. Hui, and Y. Luo. 2006. Probabilistic inversion of a terrestrial ecosystem model: Analysis of uncertainty in parameter estimation and model prediction. *Global Biogeochemical Cycles*, *20*(2).

Huang, X., D. Lu, DM. Ricciuto, PJ. Hanson, AD. Richardson, XH. Lu, ES. Weng, S. Nie, LF. Jiang, EQ. Hou, IF. Steinmacher, and YQ. Luo. 2021. A model-independent data assimilation (MIDA) module and its applications in ecology. *Geoscientific Model Development*, *14*: 5217–5238.

Unit Seven

Data Assimilation with Field Measurements and Satellite Data

25 Model-Data Integration at the SPRUCE Experiment

Daniel Ricciuto
Oak Ridge National laboratory, Oak Ridge, USA

This chapter is intended as a brief introduction to carbon-cycle modeling and field measurements at the Spruce and Peatland Response Under Changing Environments (SPRUCE) experiment in a forested wetland in northern Minnesota. The goal is to familiarize the reader with the study in preparation for subsequent training and example applications of data assimilation into models using SPRUCE data.

INTRODUCTION

SPRUCE is a large-scale, decade-long experiment designed to assess the response of a northern peatland bog ecosystem, which contains a large amount of carbon belowground, to changes in atmospheric temperature and carbon dioxide (CO_2) concentrations that approximate possible conditions in the latter half of the 21st century. Peatlands have been identified as vulnerable ecosystems that potentially have large feedbacks to the global carbon cycle, and as a result may affect climate change. Although peatlands comprise a relatively small fraction, about 3%, of global land area, they are estimated to contain at least one-third of all carbon on the land surface. Therefore, understanding the response of these systems to warming, changing moisture conditions, and rising atmospheric CO_2 is critical to our ability to predict future climate using coupled Earth system models.

One key uncertainty is how carbon may leave the system to the atmosphere, as either CO_2 or as methane gas (CH_4). This is critically important because CH_4 is a much more potent greenhouse gas than CO_2, with more than 25 times greater warming potential over a 100-year timeframe. The saturated peat, biogeochemical environment, and the types of vegetation in peatlands are conducive for anaerobic decomposition (breakdown of organic matter in the absence of oxygen) and therefore high levels of CH_4 emissions compared to other ecosystems. While CO_2 fluxes from aerobic decomposition (when oxygen is available) are sensitive to temperature, CH_4 production and emissions may be even more sensitive to warming than CO_2. On the other hand, drying associated with warming conditions may reduce methane production and increase methanotrophy (consumption of CH_4 by microbes in the peat). Observations of CO_2 and CH_4 fluxes from SPRUCE are informing model parameter values and structural representations of key processes in mechanistic models to improve predictive understanding of the system.

SPRUCE researchers are also very interested in how the experimental treatments impact the different types of peatland vegetation in terms of physiology, phenology (timing and length of growing season), reproduction, and mortality. The most extreme level of warming in the experiment is consistent with climate model projections at the end of the century under the strongest greenhouse gas emissions scenarios. It effectively shifts the climate of SPRUCE to that currently experienced in central Missouri, which is hundreds of kilometers to the south.

SITE DESCRIPTION

The SPRUCE experiment is located in a forested wetland called the S1 bog, which is part of the United States Forest Service (USFS) Experimental Forest in North Central Minnesota. The S1 system is a raised-dome ombrotrophic bog, meaning that it is rain-fed, nutrient-poor, and has a perched water table that is disconnected from the influence of regional groundwater. The bog has a mean annual air temperature of 3.3°C and mean annual precipitation of 768 mm. The bog experiences cold winters: snow cover generally persists from late autumn until early spring, and ice layers form in the peat that may persist until May or June. Within the bog, there is microtopography consisting of raised areas (hummocks) and sunken areas (hollows). Hummock areas are generally 15–20 cm higher than hollows and are almost never inundated. In typical years, the water table generally ranges from as low as 20–30 cm below the hollow surface in dry conditions to 10–15 cm above the hollow surface in wet conditions. At the beginning of the study, both hummock and hollow surfaces of the bog were nearly completely covered with *Sphagnum* mosses. Above that, there is a woody shrub layer dominated by two species: *Rhododendron groenlandicum* (Labrador tea) and *Chamaedaphne calyculata* (leatherleaf). There are two main types of trees on the S1 bog, *Picea mariana* (black spruce) and *Larix laricena* (larch). Existing trees were cleared in strip cuts in 1974 for a different experiment, and new trees have been regrowing since then. This area with relatively young, short trees within the strip cut areas is ideal for the SPRUCE experiment because the enclosures do not have to be as large or as costly to expose these trees to whole-ecosystem warming.

SPRUCE uses an open-top enclosure design (Figure 25.1) in which the whole-ecosystem warming and elevated CO_2 treatments are conducted (Hanson et al., 2017a and 2017b). Air warming is accomplished using propane-fired furnaces in combination with blowers distributed around the enclosure.

DOI: 10.1201/9781032711126-32

Peat warming is accomplished using resistance heaters that heat depths between 2 and 3 meters below the surface. A corral system isolates the hydrology within the enclosures, so that the water table within may be lower or higher than the surrounding bog.

SPRUCE has a total of ten enclosures with five different levels of warming ranging from no added heat (+0°C) to +9°C in 2.25°C increments. Half of the enclosures have ambient CO_2 concentrations (about 400 parts per million) at the five warming levels, while the other half have an elevated CO_2 concentration target of +500 parts per million (ppm) over ambient, typically ranging between 800 and 900 parts ppm at the same warming levels. Elevated CO_2 is supplied only during daytime and in the growing season when photosynthesis is occurring. No water vapor is added, causing reduced relative humidity and increased vapor pressure deficit in the warmed enclosures. Extensive measurements within the enclosures include meteorological conditions, peat temperature and water table depth, CO_2 and CH_4 fluxes using a large-collar chamber measurement (Hanson et al., 2016), species-level vegetation biomass and productivity, phenology cameras, and porewater chemistry. In addition to the active warming treatments, there are passive enclosure effects including increased longwave radiation input, decreased shortwave radiation, and changes in air flow, resulting in additional temperature increases between 1°C and 2°C at all warming levels compared to outside the enclosures. Two additional plots without enclosures are also measured to record the evolution of the ecosystem under ambient conditions. The study's regression-based design allows for the determination of response curves over a range of conditions and facilitates comparison with model outputs.

Lengthy preparation was required for the large-scale experiment. Pretreatment characterization at the S1 bog began in 2009, involving extensive peat and vegetation measurements. Construction for the treatment experiments began in 2012, beginning with roadwork and other infrastructure development. The corral systems were then built, followed by the construction of the enclosures. Whole-ecosystem warming began in August 2015, and elevated CO_2 treatments began in June 2016. The experiment is anticipated to run through 2025. A number of key science questions were posed at the beginning of the experiment:

- Are peatland ecosystems and organisms vulnerable to atmospheric and climatic change?
- At what rate will ancient carbon be released from accumulated peat in response to deep belowground warming, and what is the relative release of CO_2 compared with CH_4?
- What are the interactions between ecosystem responses to warming and the availability of nutrients and water?
- How does elevated CO_2 modify ecosystem responses to warming and the availability of nutrients and water?

MODELING FOR THE SPRUCE EXPERIMENT

Although the scale of SPRUCE is unprecedented for a peatland ecosystem manipulation experiment, it is difficult to know if the answers to these questions at the study site will be consistent across other peatland systems. The project investigators therefore envision mechanistic modeling as a key method to extrapolate the results of SPRUCE to other

FIGURE 25.1 Top-down view of a SPRUCE enclosure.

systems. This requires a multi-scale modeling framework that can be applied from site to global scales.

The United States Department of Energy (DOE), which provides the primary funding for SPRUCE, has heavily invested in such a modeling framework which we are using. In 2014, the DOE initiated development of a new Earth system model, the Energy Exascale Earth System Model (E3SM). This model branched from the well-known Community Earth System Model (CESM). The land component of E3SM, known as ELM, is a land-surface model that includes cycling of water, carbon, nitrogen, and phosphorus. ELM has been used extensively in coupled E3SM (Burrows et al., 2020), uncoupled (land-only simulations driven by observed atmospheric forcings), and at the site-level including eddy covariance and ecosystem manipulation sites like SPRUCE. However, the default version of ELM lacks key processes necessary to simulate peatland carbon, water, and nutrient dynamics accurately. A SPRUCE-specific version, ELM-SPRUCE, was recently developed (Shi et al., 2015, 2021). It incorporates wetland hydrology, microtopography, and plant functional types that are not currently represented in ELM including *Sphagnum* mosses. ELM-SPRUCE also includes a more mechanistic CH_4 model that explicitly represents microbial populations involved in CH_4 production and consumption. ELM-SPRUCE can help to answer the above research questions, test other hypotheses about the impact of environmental change at SPRUCE, and eventually simulate broader peatland regions on a global scale.

In addition to ELM-SPRUCE, other modeling groups are also simulating the experimental treatments at the site. The Terrestrial ECOsystem (TECO) model was introduced in Chapter 2. It has well-developed methods for model-data integration. A version of the TECO model has been developed for application at SPRUCE (Ma et al., 2017). This model, TECO-SPRUCE, is being used to make projections of CO_2 and CH_4 fluxes under the experimental treatments and to estimate the magnitudes of prediction uncertainties that result from uncertainties in forcings and model parameters (see Chapter 26).

In general, having projections from multiple models is an important way to understand the impact of structural uncertainty, which reflects the different ways in which different model frameworks may represent processes. Some models may be more complex than others by including more processes, or they may use different equations or algorithms to represent a specific process. A model intercomparison study focused on SPRUCE is currently under development, and SPRUCE data are being made available to interested modeling groups.

MODEL VALIDATION AND UNCERTAINTY QUANTIFICATION

While having projections from multiple models is useful, quantitatively estimating within-model uncertainty is key to knowing the confidence in our predictions, and for understanding what measurements may be most useful in constraining these predictions further. Model uncertainty quantification (UQ) is a key goal of the SPRUCE modeling work, and it has been used in both the ELM-SPRUCE and TECO-SPRUCE models. An important part of UQ is estimating how uncertain model parameters contribute to uncertainty in predictions such as CO_2 fluxes or stocks. A model such as ELM-SPRUCE is very complex and contains well over 100 uncertain model parameters. An example of an uncertain parameter would be the sensitivity of heterotrophic respiration to temperature (Q_{10}), for which published values in the literature may range between 50% lower or higher than the assumed model default value. This parameter uncertainty is likely to cause large uncertainties in the predicted response of net CO_2 fluxes to experimental warming. Ultimately, we would like to calibrate such parameters and reduce their uncertainty using measurements; for example, the within-enclosure CO_2 flux information could be used to constrain the Q_{10} value at SPRUCE. However, as the number of uncertain parameters grows, the computational expense of model calibration rises exponentially as it requires a larger and larger number of model simulations (also referred to as ensemble members) to sample the potential parameter space. Therefore, we usually need to identify the most important parameters first using sensitivity analysis.

Sensitivity analyses are usually conducted for a set of model outputs, or quantities of interest (QoIs). For example, a QoI might be the site-averaged net ecosystem exchange over a 10-year period, or the average date of leaf-out in spring. The contribution of each parameter to the variance of a QoI may be estimated in a sensitivity analysis using a model ensemble in which multiple parameters are varied randomly. Fortunately, a smaller number of ensemble members is necessary for sensitivity analysis than for calibration. The objective of the sensitivity analysis is to reduce the number of calibration parameters to a reasonable number, usually around 20 or less. In the ELM-SPRUCE model, we conduct the sensitivity analysis first by identifying minimum and maximum possible values for each parameter. This can be done by surveying the literature, trait databases such as TRY and the Fine Root Ecology Database (FRED), or by making educated guesses about parameter uncertainty (e.g., +/- 25% from default values). An ensemble of model parameter values is then created by randomly sampling parameter values in these ranges. This model ensemble is used to create a surrogate model, or model response surface for each QoI. Many different approaches can be used to create surrogate models, including machine learning; here, we use polynomial functions. Using this approach, we found that about 2500 ensemble members can yield trustworthy sensitivities for about 65 uncertain parameters in ELM (Ricciuto et al., 2018). While running this number of simulations is computationally feasible in ELM on a mid-size computing cluster, other models may be more or less computationally expensive, allowing for different ensemble sizes.

Model calibration involves finding a set or sets of optimal parameters that best fit observations by minimizing a cost function (see Chapter 21). A cost function typically yields a single value that can integrate information from multiple types of observations (e.g., both CO_2 fluxes and biomass measurements) and from observations at multiple times. Observations may be weighted by their uncertainties or using other methods. Model calibration may also involve the estimation of parameter and prediction uncertainties. It may be accomplished using a variety of techniques. A preferred technique is Markov Chain Monte Carlo (MCMC), introduced in Chapter 22. MCMC has the desirable quality that it can calculate the full parameter posterior probability density functions (PPDFs) without making any prior assumptions about the functional forms of the distributions. These PPDFs may also be used to estimate prediction uncertainty for QoIs. However, it is a relatively expensive method that requires a large number of model evaluations (usually at least 10,000) and is not easily parallelized. In models that are fast to evaluate such as TECO-SPRUCE, MCMC may be used directly. However, in expensive models such as ELM-SPRUCE, it is first necessary to construct a surrogate model. An example of surrogate model calibration in ELM is given by Lu et al. (2018a). Similar methods are used as is done for the surrogates used in sensitivity analysis. However, the surrogate models introduce error into the calibration and therefore must be considerably more accurate to ensure a trustworthy calibration result.

Using ELM-SPRUCE, model calibration was performed with pre-treatment observations of vegetation biomass and productivity for four different plant functional types (Shi et al., 2021). The calibrated model parameters differed substantially from the default ELM parameters used in the global parameterization for boreal plant functional types. The calibrated model was then used to predict treatment responses at the site for the most extreme warming scenario of +9°C. The model predicted that black spruce trees would steadily decline in biomass and productivity with warming, while the shrubs and larch would increase slightly. The *Sphagnum* productivity was simulated to decline during dry periods and increase during wet periods. We can now begin to validate the model using treatment observations. There is some indication that the black spruce trees are actually responding negatively to the warming treatments, especially in comparison to the shrubs and larch trees. However, a recent study showed that *Sphagnum* productivity and biomass are rapidly declining, which was not predicted by ELM-SPRUCE. This may be in part due to a lack of certain processes in ELM-SPRUCE. For example, more productive shrubs may shade out the moss layer, which cannot be represented currently in ELM-SPRUCE because there is no competition for light in that model. It is also possible that the model predictions will improve when we begin to use the treatment data for model calibration. We did find that ELM-SPRUCE predicts the change in net carbon flux with temperature quite accurately, and that both model and observations indicate that warming causes a significant source of CO_2 and CH_4 to the atmosphere (Hanson et al., 2020). If one naively assumes that all peatland systems respond similarly over the entire globe, this would result in a large source of greenhouse gases and a positive feedback which would be large enough to further strengthen global warming. We will test this assumption in the model in the future by running global simulations. Interestingly, however, the observations do not yet indicate a strong effect of elevated CO_2 concentrations on vegetation biomass at SPRUCE. ELM-SPRUCE and TECO-SPRUCE before data assimilation both predict a strong fertilization effect. Reconciling this with observations will take additional empirical and model development work for ELM-SPRUCE and this will be informed by data assimilation using TECO-SPRUCE.

The experimental treatments at SPRUCE combined with the model-data integration framework in ELM-SPRUCE, TECO-SPRUCE, and other models provide a useful testbed for improving predictive understanding of peatland ecosystems. The interaction of the site hydrology, biogeochemistry, and multiple vegetation types under varying treatments makes obtaining accurate predictions particularly challenging compared to other study sites, for example, eddy covariance tower footprints which often have relatively homogeneous vegetation coverage. Recently it has been observed that responses of the different vegetation types to warming are not uniform; while shrubs are becoming more productive and growing more fine roots (Malhotra et al., 2020), the *Sphagnum* mosses are dying and sharply declining in areal coverage under strong warming (Norby et al., 2019). The growing season generally becomes longer with warming, but some vegetation types have a stronger phenology response than others (Richardson et al., 2018). Ideally, models of the SPRUCE system must be simple enough to ensure simulations are relatively inexpensive so that we can run parameter ensembles to explore prediction uncertainty. On the other hand, models must contain enough process realism to capture the divergent responses of different vegetation types and to predict the strong increases in CO_2 and CH_4 surface fluxes to the atmosphere. Much research remains to be done to predict how the SPRUCE S1 bog and other peatlands will respond to rapidly changing environmental conditions.

SUGGESTED READINGS

Hanson, Paul J., Natalie A. Griffiths, Colleen M. Iversen, Richard J. Norby, Stephen D. Sebestyen, Jana R. Phillips, Jeffrey P. Chanton, Randall K. Kolka, Avni Malhotra, Keith C. Oleheiser, Jeffrey M. Warren, Xiaoying Shi, Xiaojuan Yang, Jiafu Mao, and Daniel M. Ricciuto. 2020. Rapid net carbon loss from a whole-ecosystem warmed Peatland. *AGU Advances*. doi:10.1029/2020AV000163

Shi, X., Thornton, P. E., Ricciuto, D. M., Hanson, P. J., Mao, J., Sebestyen, S. D., Griffiths, N. A., and Bisht, G. 2015. Representing northern peatland microtopography and hydrology within the Community Land Model, *Biogeosciences*, *12*, 6463–6477. doi:10.5194/bg-12-6463-2015

QUIZ

1. Why are peatlands important for land-atmosphere feedbacks?
2. What is the gradient experimental design for this SPRUCE project?
3. What are the benefits of incorporating modeling approaches in such a large experimental study?
4. When modeling results are different from observations, what should we do?

26 Application of Data Assimilation to a Peatland Methane Study

Shuang Ma
University of California, Los Angeles, USA

Data assimilation is widely used in terrestrial ecosystem studies. This chapter illustrates the use of data assimilation with a site-level study to project peatland methane (CH_4) emission in response to warming. Wetland CH_4 emissions comprise one third of the global CH_4 source and remain the largest source of uncertainty in the global budget. Wetland CH_4 emission estimated by process-based models (bottom-up) are used as the prior information for atmospheric inversion estimates (top-down). It is thus important to constrain process-based models with *in situ* observations to improve both the bottom-up and top-down estimates. We give a brief background of methane modeling and then show the application of data assimilation in the methane model in seven steps.

UNCERTAINTY IN METHANE MODELING

Methane (CH_4) has 25 times the global warming potential of CO_2 over a 100-year scale (Myhre et al., 2013). It is directly responsible for approximately 20% of global warming since pre-industrial time (Forster et al., 2007). Wetlands are an important natural source of CH_4 emissions to the atmosphere, but constitute a principal source of uncertainty in the global atmospheric CH_4 budget (Saunois et al., 2020). Global wetland CH_4 emission estimated by process-based models have large disagreement compared with atmospheric inversion model ensembles (Saunois et al., 2020).

There are three major sources of uncertainty in the model estimated CH_4 emission. The first is the uncertainty in mechanisms that control biogeochemical processes due to the difficulty to acquire empirical data. For example, it is very difficult to measure the aerobic and anaerobic oxidation of methane. The redox potential effect on methane oxidation awaits more empirical data to be represented in models. The second source of uncertainty is the wide range of possible parameter values for methane-related processes. Flux-based measurements of Q_{10} (temperature sensitivity) of CH_4 release from different warming plots at one single site range from 2.12 to 32.16 (Gill et al., 2017). Manually tuning the parameter values to match the observed CH_4 fluxes could achieve the right answer with diverse combinations of parameter values – the problem of equifinality. The third uncertainty is a poorly mapped wetland extent and seasonal inundation. Current wetland maps are mainly based on inventory data and satellite observations. Inventory maps are limited due to the low spatial and temporal coverage, while satellite-based maps cannot capture the wetland area with dense vegetation cover.

It is critical to understand how wetland CH_4 emissions may respond to climate change, given the much larger warming potential of CH_4 compared to CO_2 over a 100-year scale (Myhre et al., 2013). Terrestrial biosphere models that include methane processes explicitly describe the CH_4 flux exchange through plant-mediated transport, diffusion, and bubbling (ebullition). These are the three major pathways of wetland CH_4 emission. The relative contributions of these three pathways to methane emissions under climate warming have not been unraveled either using experiments or modeling approaches. In most process-based methane models, these CH_4 emission pathways are calculated based on CH_4 concentration in each peat layer, which is primarily dominated by CH_4 production. If some of the parameters in CH_4 production, plant-mediated transportation, ebullition, and diffusion can be constrained by observational data, we may be able to improve model predictions both by improving accuracy and reducing uncertainty.

ASSIMILATION OF METHANE EMISSIONS DATA INTO THE TECO MODEL

The seven steps of data assimilation were introduced in Chapter 21. As an illustrative example, we will apply these steps to the assimilation of methane emissions data from *in situ* measurement into the TECO model.

1 Define the objective.
Our objective is to reduce the uncertainty of model estimations of how methane emissions vary in response to warming. Taking the example of a peatland methane study (Ma et al., 2017), data assimilation is used first to constrain parameters with observational data, thereby reducing uncertainty in methane prediction.
2 Prepare data.
Our data come from the Spruce and Peatland Responses Under Changing Environments (SPRUCE) experiment at a northern peatland site in Minnesota, USA. The SPRUCE project uses experimental warming and elevated CO_2 treatments to assess the responses of northern peatland ecosystems to future environmental conditions (Hanson et al., 2016). Here we will use *in situ* net CH_4 emission data from ambient plots to constrain parameters of a process-based methane model. CH_4 emission measurements were

DOI: 10.1201/9781032711126-33

acquired during the snow-free months using a portable open-path analyzer in a static chamber (1.13m^2 area), near monthly from 2011 to 2016 (Hanson et al., 2017a and 2017b). The 2010–2014 data are used for data assimilation and 2015–2016 are used for validation. In total, 45 daily CH_4 chamber measurement data points were integrated from ambient plots from 2011 to 2016. A mean and a standard deviation are calculated from all the measurements on the same day in each ambient plot.

3 Choose a model.

Our example uses a methane-enabled version of the Terrestrial ECOsystem (TECO) model, which incorporates a ten-layer vertical mixing CH_4 module (Ma et al., 2017). The TECO model has been calibrated to the SPRUCE site to study the carbon cycle and soil thermal dynamics (Jiang et al., 2018). A detailed description of TECO can be found in Weng and Luo (2008).

TECO-CH_4 explicitly considers the transient and vertical dynamics of CH_4 production, CH_4 oxidation, and CH_4 transport from belowground to the atmosphere. The structure and processes are adapted from a number of previous studies and models such as CLM4.5 (Riley et al., 2011), LPJ-WHyMe (Wania, Ross, and Prentice, 2010), Walter's model (Walter & Heimann, 2000), and TEM (Zhuang et al., 2004). All the above models assume that soil can be separated into aerobic and anaerobic layers divided by the water table. These models also assume that the majority of CH_4 oxidation occurs in the aerobic layers and rhizosphere, and that most methane production occurs in the anaerobic layers. Within each soil layer, CH_4 concentration is calculated as the mass balance of CH_4 production (gain), CH_4 oxidation (loss), CH_4 diffusion from adjacent layers, CH_4 ebullition (loss), and plant-mediated transport (loss). The flow diagram of TECO-CH_4 is shown in Figure 26.1 and further described below.

CH_4 production is determined by carbon availability represented by heterotrophic respiration, and by soil environmental conditions such as water table height and soil temperature. As in most methane models, CH_4 production only occurs when soil temperature is above 0°C and below 45°C. Given that CH_4 oxidation is largely controlled by CH_4 concentration, it is assumed to follow the Michaelis-Menten kinetics.

The CH_4 diffusion across soil layers follows Fick's law, which relates the diffusive flux to the gradient of the concentration, and Henry's Law, which resolves the diffusive flux at the liquid-atmosphere boundary. The net exchange between the surface soil layer and atmosphere is accounted as the diffusive part of CH_4 emission (or uptake). The methane flux at the bottom boundary is set to zero and the atmospheric CH_4 concentration at the soil surface (or water surface if the water table is at or above the soil surface) is set to standard atmospheric CH_4 concentration.

Air-filled aerenchyma tissues of plants act as a chimney to quickly emit CH_4 from the rhizosphere directly into the atmosphere. A portion of CH_4 is oxidized within the plant tissue during the transport. TECO-CH_4 uses a parameter (T_{veg}) to represent the ability to transport CH_4 through tissues at a plant community level. The growth of plants also affects the amount of gas transported through the influence of Leaf Area Index (LAI). Ebullition entails the formation of bubbles when the CH_4 concentration exceeds a certain threshold and directly emits into the atmosphere if the water table is above the soil surface, bypassing the aerobic zones that lead to CH_4 consumption. The bubbles are otherwise added to the

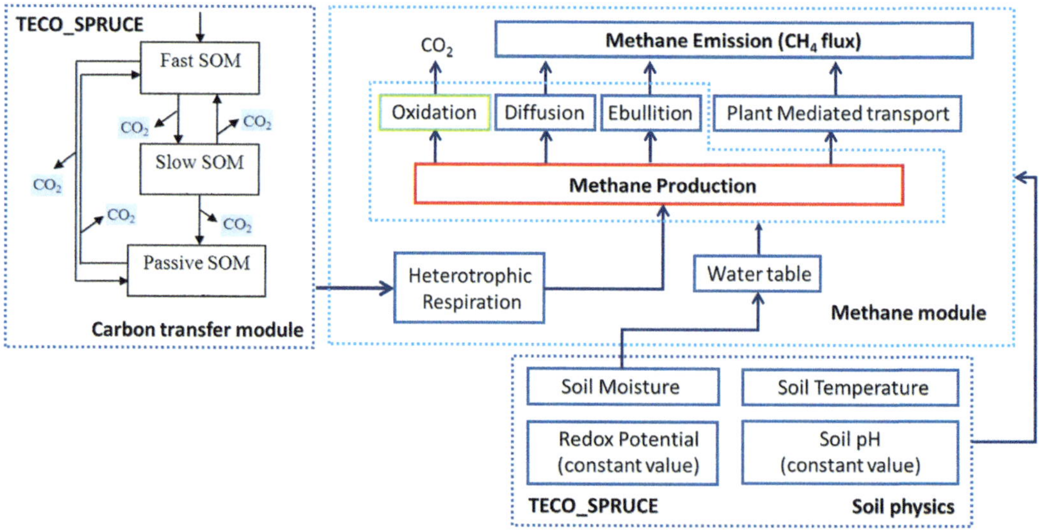

FIGURE 26.1 Flow diagram of CH_4 module and linkage to soil C model in TECO-CH_4.

Reproduced from Ma et al., (2017).

TABLE 26.1
Major parameters in CH_4 production, oxidation, diffusion, ebullition, and plant-mediated transportation in TECO-CH_4. Bold signifies parameters used for initial sensitivity test. Parameters with a range indicate the model is sensitive to their values and are used for data assimilation

Process	Parameters	Values	Range	Unit	Description	References
CH_4 production	**r_me**	0.65	(0.0,0.7)	—	Potential ratio of anaerobically mineralized C released as CH_4	Zhuang et al. (2004), Segers (1998), Zhu et al. (2014)
	$\boldsymbol{Q_{10\_pro}}$	7.2	(0.0,10)	—	Q_{10} for CH_4 production	Walter and Heimann (2000)
	$\boldsymbol{T_{opt\_pro}}$	20.0		°C	Optimum temperature for CH_4 production	Wilson et al. (2016)
CH_4 oxidation	$\boldsymbol{K_{CH4}}$	5.0	—	μmol L^{-1}	Michaelis-Menten coefficients	Walter and Heimann (2000), Zhang et al. (2002),
	$\boldsymbol{O_{max}}$	15.0	(3.0,45.0)	μmol L^{-1} h^{-1}	Maximum oxidation rate	Zhuang et al. (2004)
	$\boldsymbol{Q_{10-oxi}}$	2.0	—	—	Q_{10} for CH_4 oxidation	Walter and Heimann (2000), Meng et al. (2012)
	T_{opt_oxi}	10.0		°C	Optimum temperature for CH_4 production	Zhuang et al. (2004)
CH_4 diffusion	f_{tort}	0.66	—	—	Tortuosity coefficient	Walter and Heimann (2000)
	D_{air}	0.2	—	cm^2s^{-1}	Molecular diffusion coefficient of CH_4 in air	Walter and Heimann (2000)
	D_{water}	0.00002		cm^2s^{-1}	Molecular diffusion coefficient of CH_4 in water	Walter and Heimann (2000)
CH_4 ebullition	$\boldsymbol{(CH_4)_{thre}}$	750	—	μmol L^{-1}	CH_4 concentration threshold above which ebullition occurs	Walter and Heimann (2000), Zhu et al. (2014)
Plant-mediated transportation	$\boldsymbol{T_{veg}}$	0.7	(0.01,15.0)	—	Factor of transport ability at plant community level	Walter (1998), Zhuang et al. (2004)

Reproduced from Ma et al. (2017).

soil layer just above the water table and then diffuse through the upper layers if the water table is below the soil surface.

Once a model is chosen, the next step is to choose parameters for optimization. The performance of data assimilation is affected by the variety and amount of observational data as well as the parameters that are targeted for optimization. One common way to choose parameters is through a sensitivity test. In this study, we choose nine key parameters used in TECO-CH_4 for an initial sensitivity test (Table 26.1). Four of these parameters are revealed to be particularly important for the modeled variability in CH_4 emission, i.e., the emissions are sensitive to those parameters. We thus pick these parameters for data assimilation. The prior ranges of these parameters are estimated from published experimental measurements or empirical values used in CH_4 biogeochemistry models (Table 26.1).

4 Cost function.

As the data assimilation algorithm for this study, we will choose the adaptive Metropolis-Hastings Markov Chain Monte Carlo (MCMC). The approach was introduced in Chapter 22. Figure 26.2 shows the logic flow of data assimilation written into the source code of the TECO data assimilation framework. At each step in the chain, a new set of parameter values is chosen at random, the model generates results with these parameter values, and the disagreement to the observations is quantified using the cost function:

$$J_{new} = \sum_{i=1}^{n} \sum_{t=1}^{30} \frac{\left[Z_i(t) - X_i(t)\right]^2}{2\sigma_i^2(t)} \quad 26.1$$

The cost function aggregates a total model-data mismatch value (J_{new}) from all the 30 time points (since we have 30 near-monthly net CH_4 emission observation from 2011 to 2014). In this study n equals 1 as we have only one set of observation data (net CH_4 emission rate). We save modeled CH_4 emission as $X(t)$ when the corresponding observed CH_4 emission is available ($Z(t)$) at time t. The standard deviation ($\sigma(t)$) is considered as the confidence level of the observation. In this study, it reflects errors from instruments, measurement, and spatial-temporal heterogeneity. It is used in the cost function to adjust the weight of model-data mismatch for individual data points. In data assimilation, both mean and standard deviation from observations are very important and should always be carefully considered in practice.

5 Optimization.

If J_{new} (Step 4) passes the acceptance criteria, the proposed parameter values are saved. Acceptance depends on the value of the cost function relative to

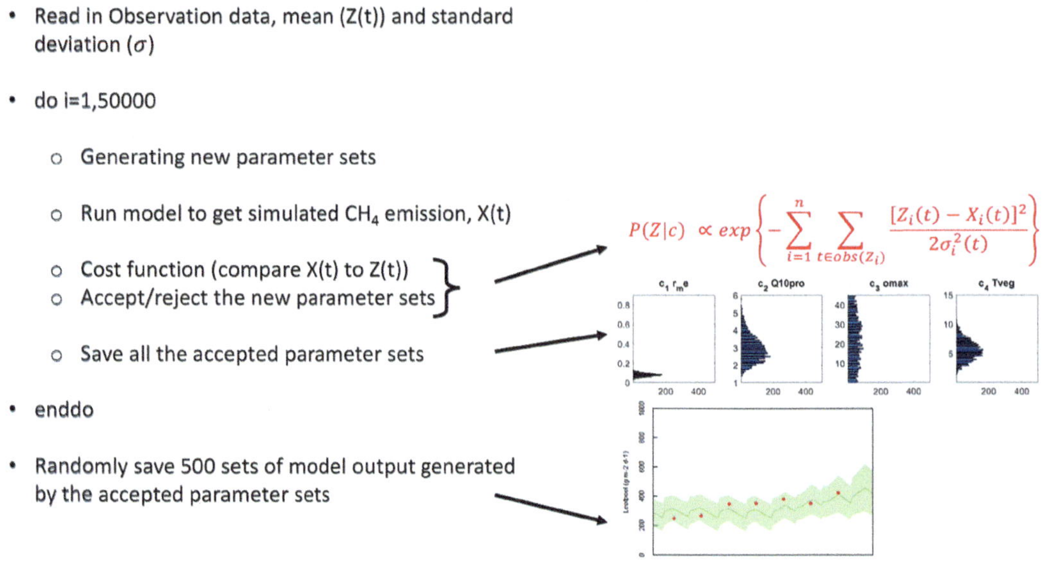

FIGURE 26.2 Logic flow of the TECO-CH_4 data assimilation framework.

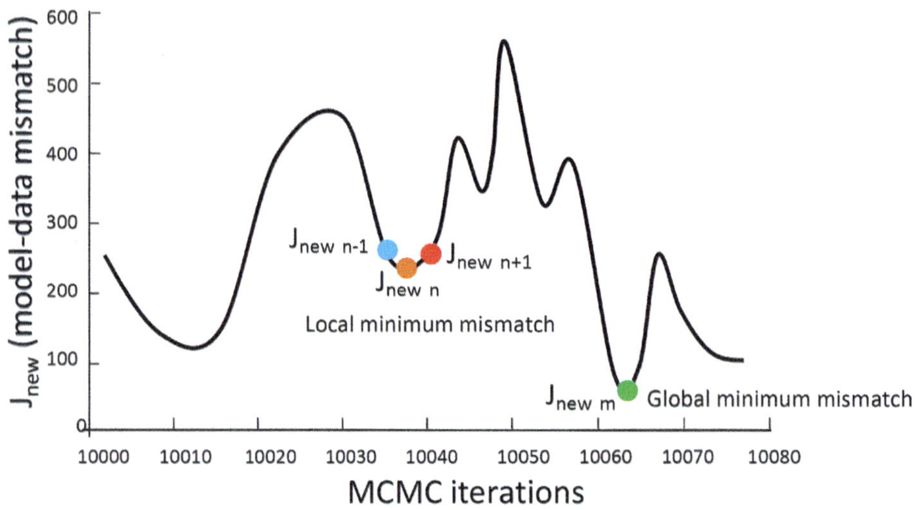

FIGURE 26.3 A diagram showing trajectory of updated model-data mismatch (J_{new}) on MCMC chain. Blue is J_{new} at $(n-1)$th iteration, orange dot is one of the local minimum mismatch point (J_{new} n), red is the next accepted J_{new} at $(n+1)$th iteration, green is global minimum mismatch (J_{new} m).

that (J_{last}) from the previous iteration of the MCMC. An initial value for J_{last} is set at 9,000,000. This is just a large initial value, to ensure that first proposed value is accepted and the iteration begins. At each step, if $J_{new} < J_{last}$, the proposed parameter values are accepted, J_{last} value is updated with J_{new} and used in the next round. After a warm-up period, parameter values start to converge (accepted parameter values during this period are discarded from the posterior distribution). Figure 26.3 shows the trajectory of updated model-data mismatch (J_{new}) on the MCMC. As the MCMC seeks the global minimum mismatch, the acceptance criteria also allow J_{new} to be accepted with a very small probability when $J_{new} > J_{last}$, so that the chain gets a chance to leave the local minimum mismatch point and reach the global minimum mismatch. See Chapter 22 for a further discussion of acceptance criteria in the MCMC.

6 Estimating parameters.
The posterior parameter distributions achieved by MCMC reveal that both of the CH_4 production-related parameters (the potential ratio of anaerobically mineralized C released as CH_4, and temperature sensitivity of CH_4 production) are well-constrained (Figure 26.4). By applying a linearized Q_{10} function to measured CH_4 emission fluxes, Gill et al. (2017) estimated the mean value of CH_4 flux Q_{10} to be 5.63 (2.92–10.52 with 95% confidence interval) at the same study site during the 2015 growing season. Our constrained Q_{10} range is 2.34–6.33 with 95% confidence interval, which overlaps with but has a narrower range than this estimate by Gill et al. (2017).

The other two parameters – maximum oxidation rate and factor of plant transport ability at community level – are not well constrained by the data. A longer/denser record of observation data and extra datasets such as peat CH_4 concentration are likely to be helpful for constraining these parameters.

7 Generating methane predictions.
To quantify model uncertainty due to parameter values, we can randomly draw sets of parameters from the posterior distribution and run the model with each, resulting in a distribution of outputs for CH_4 flux. Here, we will perform 500 simulations with different, randomly drawn, parameter sets, with forcing consisting of stochastically generated environmental variables (2017–2024) based on historical meteorology data. Results including a baseline historical simulation (2011–2016) are shown in Figure 26.5. It is apparent that the data constrained TECO-CH_4 simulations match both the magnitude and seasonal variations of CH_4 emission reasonably well for both the 2011–2014 and the 2015–2016 period.

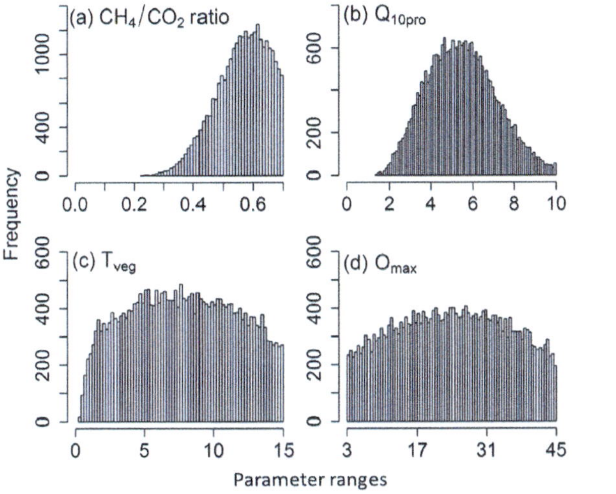

FIGURE 26.4 Posterior distributions of parameters of 50,000 samples from M-H simulation. (a) Potential ratio of anaerobically mineralized carbon released as CH_4; (b) Q_{10} for CH_4 production; (c) maximum oxidation rate; (d) factor of transport ability at plant community level.

Reproduced from Ma et al., (2017).

The historical part of the simulation reveals the prediction accuracy of the model, and its sensitivity to environmental forcing. The model tracks the interannual variability of the measurements, notably including the spike emission in 2016. By comparing the seasonal variation of CH_4 emission to environmental drivers, we find that wetland CH_4 emission is dominated by surface inundation and soil temperature. Soil temperature is the restricting factor below 10°C, but the water table level controls peak growing season CH_4 emission when soil temperature is well in the active range for methane processes. CH_4 emission is more sensitive to soil temperature during wet periods when the whole soil is inundated.

Data-constrained parameter probability distributions are then used to predict CH_4 response to warming. An increase of +2.25°C, +4.5°C, + 6.75°C and + 9°C in both air and soil temperature is added to drive the TECO-CH_4 model (a set of warming scenarios corresponding roughly to the experimental warming treatments at the SPRUCE site). We find exponential increase of CH_4 emission in response to warming, with four times increase under +9°C warming (Figure 26.6, panel a).

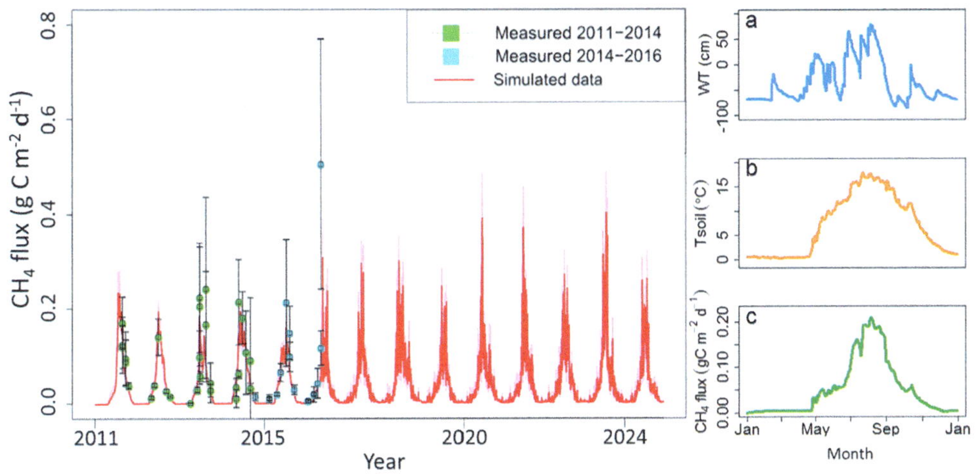

FIGURE 26.5 Simulation of CH_4 emission dynamics based on actual (2011–2016) and stochastically generated (2017–2024) weather forcing data. Green dots refer to observations from 2011 to 2014 which are used for data assimilation. Blue dots indicate observations from 2015 to 2016 which are used for model validation, and error bars indicate the standard deviation of each observation. Red line is simulated mean methane emission. The shading area corresponds to 1 standard deviation based on 500 model simulations with parameters randomly drawn from the posterior distribution. (a–c) The 2011 daily variation of water table, surface soil temperature, and methane emission.

Reproduced from Ma et al., (2017).

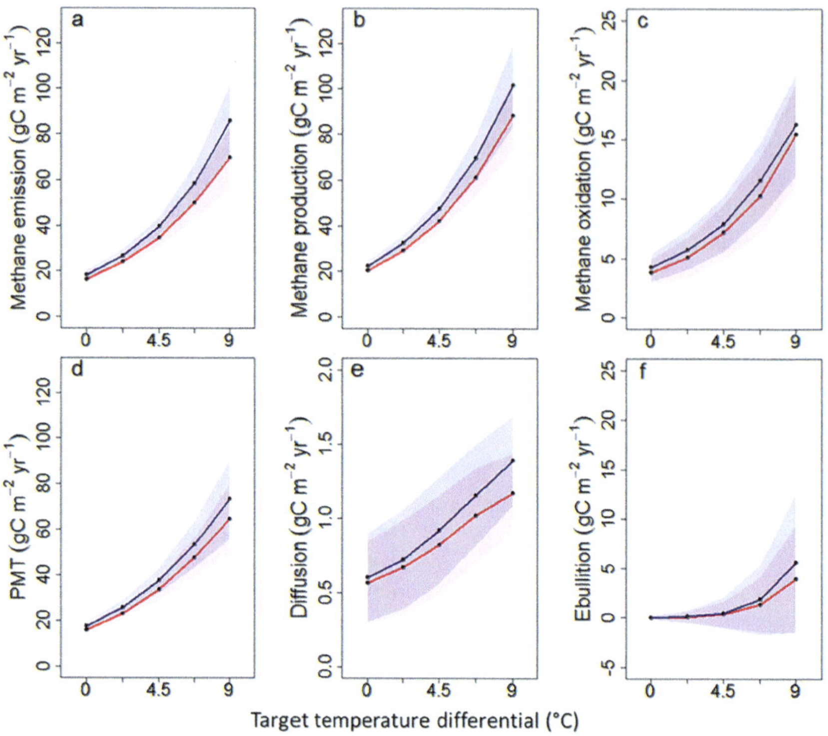

FIGURE 26.6 Responses of annual CH_4 emission to warming and elevated CO_2 (eCO_2). Red lines indicate CH_4 fluxes under warming treatments and 380 ppm CO_2, blue lines indicate CH_4 fluxes under warming treatments and 880 ppm CO_2. X-axes indicate the warming treatments of +0°C, +2.25°C, +4.5°C, +6.75°C and +9°C above ambient level. Shading area correspond to mean ± one standard deviation based on 500 randomly chosen model simulations with parameters drawn from the posterior distribution.

Reproduced from Ma et al., (2017).

The uncertainty in plant-mediated transport and ebullition increases most under warming and contributes to the overall change in CH_4 emission uncertainty (panels d–f).

In summary, this chapter shows how data assimilation is applied to reduce the uncertainty of modeled methane emission in a northern wetland ecosystem. The data assimilation approach used in this case study enabled parameter estimation and uncertainty quantification for forecasting methane fluxes in response to climate change.

SUGGESTED READING

Ma, S., Jiang, J., Huang, Y., Shi, Z., Wilson, R. M., Ricciuto, D., … Luo, Y. (2017). Data-constrained projections of methane fluxes in a northern Minnesota peatland in response to elevated CO_2 and warming. *Journal of Geophysical Research: Biogeosciences*, *122*, 2841–2861.

QUIZ

1. Give two examples of observation data streams that could be used to constrain a methane model.
2. What are the main sources of uncertainty in wetland methane emission in terrestrial ecosystem models?
   A. Model structure
   B. Parameter values
   C. Wetland extent/inundation map
   D. All the above
3. Could you still perform data assimilation if your observation data had gaps?
4. Is the CH_4 emission data able to constrain all the parameters in this study?

27 Global Carbon Cycle Data Assimilation Using Earth Observation

The CARDAMOM Approach

Mathew Williams
University of Edinburgh, Edinburgh, UK

The goal of this chapter is to explore the potential for data to support diagnostics and forecasting of the terrestrial carbon cycle via model-data fusion. This understanding will be built by explaining and exploring an existing framework, CARDAMOM, which is linked to an intermediate complexity model, DALEC. The key learnings will include the concept of ecological and dynamical constraints, the potential to generate emergent maps of key parameters, the role of observational error, and the key avenues for future research using earth observations.

INTRODUCTION

Challenges for Modeling

We begin with the premise that all models are wrong, but some are useful (see Chapter 2). Models are wrong because none fully describes the simulated system, being instead a simplified and incomplete representation. Models are constructed based on a series of hypotheses about the target system, and these hypotheses and their connections determine the model's structure. The structure represents the interacting components of a system (its state variables), and describes the processes that determine system evolution (changes to state variables). We recognize that the underlying hypotheses may be incorrect, over-simplified and incomplete, to varying degrees.

A core requirement for modeling is data, for calibration, validation, and forcing. Data may quantify exogenous forcing factors such as weather, which affect the rates of processes, such as photosynthesis, or may input stochastic adjustment such as disturbance, which can disrupt state variables. Data also support calibration of process parameters and the setting of initial conditions for state variables. For instance, measurements of leaf photosynthesis under varied conditions can be used to calibrate the rate parameters for electron transport and carboxylation. The data required for drivers and for calibration will be incomplete as not every process is measurable. This incompleteness leads to poorly determined parameters and missing forcing data. Available data will have errors, systematic or random. It is the interaction of hypothesis/structural error (e.g., missing processes) and data error/gaps that causes models to be wrong.

Despite these challenges, models are useful for testing theoretical understanding and providing practical support for management and decision-making. System models, the focus of this book, are particularly useful for understanding feedbacks between processes and state variables. These feedbacks occur over a variety of scales of time and space. For example, stomatal conductance responds to atmospheric conditions, which change on the scale of minutes to hours, and also responds to soil moisture, which changes on the scale of days to weeks. Soil moisture responds to external, stochastic factors such as precipitation, but also to the activity and penetration of plant roots, which might vary over months and years, and water demand from the total leaf area and its stomatal opening. Only process-based models can allow these complex hypothesized linkages to be made and explored. Models provide a means to diagnose and understand what controls changes in the system, identifying key timescales, feedbacks, and interactions. Models are capable of interpolation in space and time, and therefore of making forecasts for the components of the system that are represented.

Model Complexity

Forest carbon (C) models are structured on a variety of different hypotheses and levels of complexity. For instance, some simulate the dynamics of individual trees, others of cohorts of different ages, and some of pools of C in live and dead organic matter. These different representations of the forest system trade off realism versus simplicity. Modeling individual trees is more realistic, including the potential to simulate competition among them and to model adjustments to stand microclimate, that are known from observation and experiment to feed back to growth processes and therefore system dynamics. But this realism requires more hypotheses, for instance on competition, and therefore results in more model complexity. Complexity generates more potential connections to observations, with increased requirements for drivers and for calibration data. If these demands for data cannot be met then additional complexity may result in at best a poor characterization of model error and at worst a heavily biased model. Complex models tend to have more parameters, and this extended demand for parameterization becomes increasingly challenging to meet. A key challenge in model construction is to determine the appropriate and justifiable complexity.

DOI: 10.1201/9781032711126-34

Process rates in models are determined by influences (either *internal* from associated state variables or *external* from drivers such as weather), functional forms, and parameters. Internal influences generate interactive control, whereby the magnitude of a state variable affects processes that influence that or other state variables. Thus, the magnitude of leaf area influences the magnitude of photosynthesis and evapotranspiration in many models. In more complex models the leaf area of individual stems may affect the light conditions and therefore photosynthesis of other stems, influencing competition. External influences in carbon cycle models can include weather, management, or disturbance. For photosynthesis, downwelling solar radiation is a key control, provided as a driving variable for the model at the appropriate temporal resolution and specified units. Parameters determine how the magnitude of state variables and driving variables are connected to the magnitude of the process modeled. For instance, photosynthesis (gC m^{-2} d^{-1}) might be estimated in a simple linear model by multiplying the light energy absorbed by a canopy (MJ m^{-2} d^{-1}) by a calibrated light use efficiency parameter (LUE, gC MJ^{-1}). Functional forms determine how inputs and parameters are connected algebraically, for example, determining whether the response of the process is linear, saturating, or exponential.

Models strive to be realistic, general, and accurate. But there are complex trade-offs required in seeking these goals, particularly the need for models to be tractable, and so simple. There is ongoing debate about whether realistic (i.e., including all known processes) or general (i.e., globally applicable) models are necessarily more accurate. It is more practical for globally-applied models to be simpler, as a simple structure means leaner requirements for parameters and drivers. Mapping parameter variation globally is less demanding for simple models because they have fewer parameters, and these are usually easier and/or possible to obtain from the literature. For example, it is simpler to parameterize a LUE model for photosynthesis (as above), with its single parameter for calibration for each biome or land cover type, than to parameterize typical photosynthesis models in land surface schemes that might have >10 parameters for the same process. The potential advantage of the more complex model is that it is realistic, combining all the dominant controls on a process, rather than just a limited selection as in the simple model. Realism introduces more detailed processes and extra state variables, and makes more connections between them. Additional complexity adds parameters and often requires more complex functional forms. However, data sparsity can mean that parameters and functional forms are not well determined. In this case the complex model may be less accurate, particularly in making forecasts. With large numbers of parameters there is a risk of over-fitting, i.e., the complex model can be calibrated to match noise rather than signal in the data, given its many adjustable parameters. A complex model fitted to available data may make poorer forecasts than a simple model if the calibration data are biased or its errors are poorly quantified.

Model Error

While models are imperfect, they can be useful if their error can be quantified and understood. These errors can be determined through calibration and validation against observational data with robust error statistics. Validation allows testing for over-fitting. Observational error allows the modeler to weight the importance of data for calibration, and avoids over-fitting. Observational error can be incorporated into the model forecasts by propagating the error through the calibration process into parameter uncertainty. Once parameter uncertainty is quantified and included in model analyses and forecasts, models can become useful tools for generating understanding, constraining prediction, and supporting management and control. Alternate model structures, based on varied hypotheses, can be compared to understand the error associated with the model structure. Model structural error can be compared to error from the parameterization process.

Data provide objective and independent measures of the system of interest and the basis for evaluating and developing the hypotheses that create models. While models can provide useful theoretical tools for developing ecological thinking, for practical purposes related to diagnosing and forecasting global change effects it is vital that models are calibrated and validated using observations of key state variables and their controlling factors. The value of observational data is enhanced by clear description of their confidence intervals, which allows modelers to weight their importance robustly. Observational data, particularly time series, have exceptional value for evaluating understanding of dynamical systems. But the complexity and expense of collecting these data means that data replication is difficult and so uncertainty quantification is a challenge.

Data-Model Integration

Combining models with data has a long history in ecological science. Detailed studies of the photosynthetic process provided insights into the functional forms of the reactions and the critical parameters that led to robust models. Photosynthetic measurements at canopy scale were used to parameterize simple response functions by minimizing the sum of square differences between observation and model. Soil respiration models were derived empirically from large samples of respiration data under varying soil conditions. These data helped produce response models for key processes, but further work was required to produce system models with feedbacks. Combining ecophysiological process models (e.g., photosynthesis or respiration) with simulation of state variables that provide inputs to and respond to these processes has proved more challenging. Photosynthesis is a function of CO_2 concentration within the leaf, leaf area, leaf enzyme content, temperature, and irradiance. Leaf area is an ecological variable that is itself determined by allocation from C fixed by photosynthesis and phenological factors such as temperature, photoperiod, interaction with labile or nonstructural C stores, and so on. Photosynthesis varies on timescales of seconds to hours while leaf area varies on timescales of weeks, so the interaction between ecophysiology and phenology is complex. To forecast future photosynthesis requires coupling a model of photosynthesis and a model of phenology to determine interactions and codevelopment. Times series data are required for both these processes to support coupled model calibration.

Eddy covariance (EC) data have revolutionized the modeling of ecosystem C cycling by providing long time series of net CO_2 fluxes, the outcome of photosynthesis, phenology, and respiration. As time series have extended, EC data have allowed evaluation of model simulated diurnal and seasonal cycles in ecophysiology, and in some cases of succession and disturbance effects on C cycling. However, characterizing the uncertainty on eddy covariance measurements of net CO_2 exchange has proved very challenging and is an ongoing area of study. Converting high frequency measurements of wind velocity and CO_2 concentration into net CO_2 fluxes relies itself on a model that makes assumptions about atmospheric processes on a range of temporal and spatial scales that are highly dynamic. It is rare to have colocated eddy covariance systems to assess instrumental error. Understanding errors arising from the eddy covariance assumptions is still developing, particularly biases that may arise due to averaging timescales and complexity of terrain. To maximize the information content of these hard-won data for model improvement requires a renewed focus on error characterization, particularly in tropical systems where EC data are rare.

We have seen that for creating robust and useful ecosystem models the fusion of model and data must provide information on multiple processes operating over a range of timescales, from physiological to phenological. Data are particularly sparse for processes operating at longer timescales, such as the evolution of the large pools of C in wood and soil. Information for understanding these relatively slow changing state variables is limited. The signal for change in these pools is difficult to extract from eddy covariance data, which are dominated by the large gross fluxes of photosynthesis and respiration, rather than the slower rates of C changes to soil and wood pools. But it is the change in the large pools that ultimately determines whether ecosystems are C sources or sinks. So, there are open questions on how processes that govern these large pools of C can be calibrated effectively. Forest inventory data provide useful information. But, like EC, high-quality inventory data, particularly for soils, are rare and sparse, concentrated in a few locations globally. Soil and forest inventories rarely coincide. For global application, models need to be calibrated and validated with relevant data that describe the complex heterogeneity of our planet and its dynamics over annual to decadal timescales.

Earth observation (EO) data are now providing vast and expanding data sources for the terrestrial carbon cycle that offer exciting new opportunities for model calibration and validation. Leaf area index, a key auxiliary state variable related to photosynthesis, has had global and continuous products since ~2000 from optical satellites operated by NASA (MODIS) and ESA (European Space Agency Copernicus data sets), with new ESA Sentinel products promising even further refinement. However, while LAI products are now delivered at finer resolutions in space and time, the products have poorly quantified biases. Optical satellites also provide products related to burned area and to deforestation. Meanwhile radar and lidar satellites are providing intermittent regional maps of aboveground biomass. The errors on these products are also poorly developed and a subject of research. It is critical that EO products are calibrated and validated against detailed field data at relevant scales. But the strengths of these EO data are compelling, providing insights into ecological variables across the globe and their changes over time. If EO data can be used to calibrate and validate models there is the potential to test the generality of models at a new level of detail and across biomes and landscapes. Thus, a major research effort has been conducted to build approaches for model calibration and validation that use EO data. The rest of this chapter describes one of these efforts.

CARDAMOM AND DALEC – AN EXAMPLE FRAMEWORK FOR C CYCLE DIAGNOSTICS

CARbon DAta MOdel fraMework (CARDAMOM) is an approach constructed to quantify the interaction between C model structure and data, with a focus on earth observations and local to global application. CARDAMOM is designed to produce a probabilistic model calibration based on local observations, sensitive to their number, type, and error. This section aims to demonstrate the potential of model-data fusion to generate understanding of the terrestrial carbon cycle through evaluating the hypotheses contained in models and to diagnose C cycling and its uncertainty. In this example, CARDAMOM calibrates the DALEC ecosystem C model. The text first describes the DALEC model, then the CARDAMOM framework, and finally provides an example application, and some analysis of its results.

The Data Assimilation Linked Ecosystem Carbon (DALEC) Model

DALEC is a pool-based, mass balance model of the terrestrial C cycle, of intermediate complexity. The model runs on daily to monthly time steps, and can be applied across a broad range of spatial scales (from 1 ha to ~100 × 100 km). The version of DALEC described here has 17 parameters and six pools (Figure 27.1). DALEC's six state variables include live and dead pools of organic C. Live pools represent the vegetation as foliage, wood, and fine root C stocks, and a labile pool of C that flushes leaves at the start of the growing season. Dead pools represent litter and soil organic matter C stocks. These pools evolve according to inputs to and losses of C from each, which are determined by a range of processes. Inputs of C to the system are determined by photosynthesis, which is a function of weather variables (drivers), leaf area index and leaf traits. Leaf area index is determined from the foliar C pool and the leaf trait of leaf C mass per area (LCMA). The

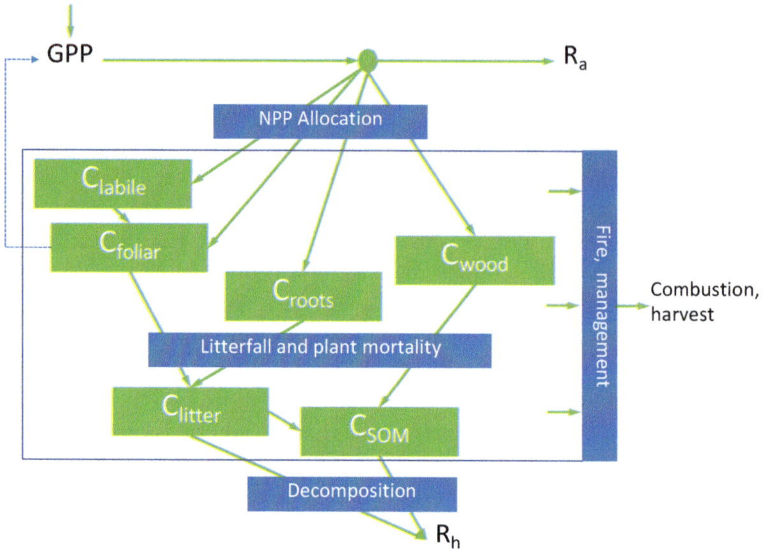

FIGURE 27.1 The DALEC model structure. Green boxes are pools of C in live and dead pools. SOM is soil organic matter. The labile C pool represents stored C which supports leaf flush at the start of the growing season. Green lines show C fluxes, identified by the text in the blue boxes. GPP is gross primary production. NPP is net primary production. R is respiration, either heterotrophic (h) or autotrophic (a). Fire and management can remove C from any of the pools by combustion or harvest, with prescribed loss fractions. Climate influences rates of GPP, R_h, and in some versions of DALEC influences the fluxes into and out of the labile pool. The mass of foliar C determines canopy leaf area index, which is a determinant of GPP – this influence is shown by the dashed blue arrow and generates feedback within the system.

foliar N parameter describes the potential of a unit leaf area to fix C, and is another important leaf trait. DALEC uses the aggregated canopy model (ACM) to simulate photosynthesis. ACM is a response surface model which relates climate, soil, and atmospheric drivers to internal variables to predict gross primary production (GPP). ACM is a simplified version of the process-based Soil-Plant-Atmosphere (SPA) model. The advantage of ACM over SPA is that it operates at daily time steps, and so is orders of magnitude faster than sub-daily SPA, while still capturing the sensitivity of photosynthesis to key ecological and physical variables.

Once GPP is determined, DALEC apportions a fraction of this to autotrophic respiration (R_a), leaving the remainder for net primary production, NPP. A simplified model of R_a, requiring a single parameter, is selected due to the lack of a robust process model – this is a key area for future research. NPP is then allocated to all plant live pools. A phenological model determines how allocation to the labile pool and foliage varies over annual cycles, and how labile C is used to flush leaves, requiring parameters to determine the start of leaf flush and of leaf senescence, and their duration. The remaining C is then allocated at fixed fractions to fine roots and wood. Wood and fine root C pools have a parameter that determines their life spans, the inverse of turnover rate. The generation of plant litter is a first order process determined by the mass of C in each foliage, root, and woody pool and its turnover rate.

Plant litter from foliage and fine roots enters the litter pools, and litter from wood enters the soil organic matter (SOM) pool. Both these pools have turnover rate parameters that determine the mineralization of these pools to CO_2. This turnover is also sensitive to air temperature, according to an exponential function with a calibrated parameter. Litter C also has a decomposition rate parameter that converts its C to SOM. Heterotrophic respiration is the sum of mineralization from both litter and SOM pools.

Disturbances such as fire can be imposed on the C stores in DALEC, so that if the drivers indicate such an event, a fraction of C in live and dead pools is removed from the ecosystem. Foliage, wood, and litter tend to have higher fractional losses from fire than roots and SOM. Harvest effects can also be imposed, again with pool-specific parameters that determine what fraction is removed and what is added to litter and SOM. In this application these disturbance parameters are derived from the literature and are not calibrated.

The Carbon Data Model Framework (CARDAMOM)

DALEC here has 23 unknowns. As each parameter and the initial condition of each pool must be known for the model to be run, CARDAMOM is used to calibrate these model parameters using available information at the location simulated (Figure 27.2). Calibrating a 23-dimensional hypervolume is highly challenging computationally. One approach to calibration involves differentiating the model code to determine the sensitivity of its outputs (predictions) to its parameters. This process allows the model to be inverted and parameters found that produce outputs consistent with independent measurements of those outputs. But in CARDAMOM we use a forward modeling approach. This avoids the challenging process of code differentiation. Instead, Markov Chain Monte Carlo (MCMC) methods are deployed.

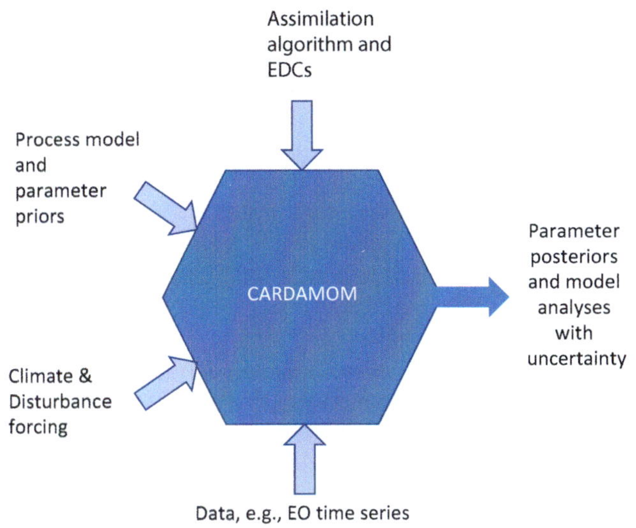

FIGURE 27.2 A schematic of the Carbon Data Model Framework, CARDAMOM, showing the inputs to the framework and the outputs. EDCs are ecological and dynamic constraints. The process model in this example is DALEC. The framework can be applied at a single site or over multiple pixels using earth observations (EO) as inputs.

MCMC was introduced in Chapter 22. The MCMC approach involves running a model many times, with varying parameters. The approach finds those parameter sets that generate model outputs (e.g., LAI time series) that match independent observations (e.g., from EO), weighted by their errors, and keep these sets. By running a model millions of times, the MCMC approach searches parameter space to find parameter sets consistent with observations and observational error. The approach to select or reject parameter sets is Bayesian. That is, the approach assesses the likelihood that model and data set are consistent. This likelihood is stored with each parameter set, and the search process ends when likelihoods have converged. By running separate chains of MCMC runs, starting from different initial parameter guesses, the approach can check that different chains arrive at consistent likelihoods. If chains do not converge, the approach is judged to have failed. Failure to converge can occur if there are problems with the data, their error specification, and due to model structural errors.

Note that the output of this Bayesian approach is a set of accepted parameter sets. Parameter values are correlated because of the model structure, which connects parameterized processes via pool interactions. We can use a sample of accepted parameter sets to generate statistics about each parameter's posterior distribution. For instance, it is typical to report the 90% confidence interval on each parameter posterior. It is also informative to inspect the covariance of parameters produced by MCMC approaches to understand process interactions within the model. An interesting result from the CARDAMOM approach when used at global scales was that the estimates of leaf life span and leaf mass per area were correlated. This result is consistent with expectations from the leaf economic spectrum but was emergent from the structure of DALEC and the climate and time series of LAI assimilated.

Prior parameter ranges are needed for Bayesian calibration, because the calibration process must be informed of the search range of each parameter. Setting the parameter priors for MCMC calibration approaches is a challenging task. If the prior is made too wide, then the search effort is enormous and the MCMC approach may not find a calibration that is accepted, as likelihoods are too low and do not converge. If the prior is made too narrow, calibration may exclude values that are actually realistic. This can lead to edge-hitting, where the posterior parameters are found to cluster at the upper or lower bound of the prior range. It is important to inspect the posterior distribution and compare to the prior to spot edge-hitting. In this case the prior range may be expanded and the calibration repeated. Another challenge in setting the prior is to inform the MCMC of the prior distribution across the range. The simplest approach is to set a uniform prior with equal likelihood of selecting any value in the range during the MCMC approach. But in some cases, it may make sense to set a Gaussian (i.e., normal, or bell-shaped) prior, which increases the likelihood of selecting some values within this range. A Gaussian prior may be chosen if one has independent knowledge about the value of the parameter.

Setting observational errors is another important task for Bayesian approaches, because these determine the likelihoods of parameter sets. Observational data, for instance earth observation products, should be provided with an estimation of uncertainty. However, in some cases errors are not available, and where these errors are provided they may be over-confident. Thus, caution suggests that relatively large uncertainties should be applied. But if observational uncertainties are too large then they do not provide constraint. In the absence of reliable observational uncertainties, we walk a fine line balancing between the need to maximize the information content of data without over-fitting to error-filled observations. It is also advisable to apply geometric errors rather than arithmetic errors to avoid zero crossing when data with values close to zero are assimilated. Thus, the uncertainty on LAI might be set at $\times/\div 1.5$, rather than ± 0.5.

To assist the process of searching dimensions of variability and to ensure that nonsensical solutions are avoided, CARDAMOM uses the concept of ecological and dynamics constraints (EDCs). EDCs ensure that model simulations are realistic and ecologically viable. EDCs reduce the uncertainty in the model parameters by rejecting estimations that do not satisfy various conditions applied to C allocation and turnover rates as well as trajectories of C stocks. For example, EDCs ensure that turnover rates of wood are always slower than for fine roots and foliage. They ensure that root:shoot biomass ratios are within broad but realistic bounds. EDCs can be used to favor parameters that result in pools close to steady state values. Ensuring a close-to-steady-state system means that pools and fluxes are broadly consistent with hypothesized processes and climate forcing. This assumption is broadly valid for low resolution studies with coarse grid cells (e.g., at $1 \times 1°$ grids). At finer resolutions (e.g., ha) increasing heterogeneity will include aggrading and recently disturbed systems with accumulating biomass, an ongoing challenge for CARDAMOM.

CARDAMOM is a codebase that integrates the DALEC model, the driving data sets, the observational data sets and the Bayesian MC algorithm (Figure 27.2). CARDAMOM can be applied at a single site, or at multiple sites, including pixel-by-pixel across a gridded land surface, even up to global scale. However, computational demands are high, so running CARDAMOM globally requires low resolution (1 × 1° pixels) at present. The outputs of CARDAMOM are accepted parameter sets for each pixel/location, consistent with local forcing and observations. From these parameter sets the full C cycle at each location can be reproduced probabilistically, by using these and the driving data to run multiple instances of DALEC. Typically, 500 randomly sampled parameter sets are stored from each of three chains in the MC process. From the resulting 1500 stored parameter sets a detailed distribution of uncertainty in processes, traits, carbon fluxes and stores, and their covariances, can be determined at pixel scale.

Innovations in the CARDAMOM Approach

The outputs of CARDAMOM have clear differences from typical C cycle model simulations. CARDAMOM produces large ensembles of simulations allowing uncertainty to be characterized, whereas typical models produce only one simulation, and so lack uncertainty estimates. CARDAMOM avoids a strict imposition of steady state conditions which typical models employ. The typical model is run under steady state drivers for hundreds of years or more, until all C pools reach a steady state. Then, from this steady state, adjustments to drivers are introduced, such as climate change, and the adjustment to stocks and fluxes are followed. In CARDAMOM there is no spin-up. For each location or pixel, EDCs ensure that the C cycle is in quasi-steady state if there is no clear information from assimilated data on whether the system is a source or sink. Thus, the ensemble of accepted parameter sets will result in some that are sinks, and some that are sources, spanning a balanced net ecosystem exchange (NEE) of CO_2. Thus, the ensemble will register uncertainty about source/sink characteristics if this information is not clear from assimilated data.

Typical ecosystem C models use the concept of plant functional types (PFT) to assign parameters. Thus, there are parameter sets for evergreen broadleaf forest, for C_3 grasslands, tundra, etc. that are the same in all such locations globally. Competition among PFTs is simulated by the model in each pixel to determine the relative coverage. In CARDAMOM parameter sets are independent for each pixel, determined only from information available in that pixel, its forcing, and EDCs. Thus, parameters differ spatially, adjusting along climate gradients and varying between continents depending on available information. Where information is limited, then parameter ensembles will be less constrained and will more closely match the prior information on parameter bounds. The information content of data in each CARDAMOM pixel can be determined by quantifying how much each parameter posterior is reduced in scale from the

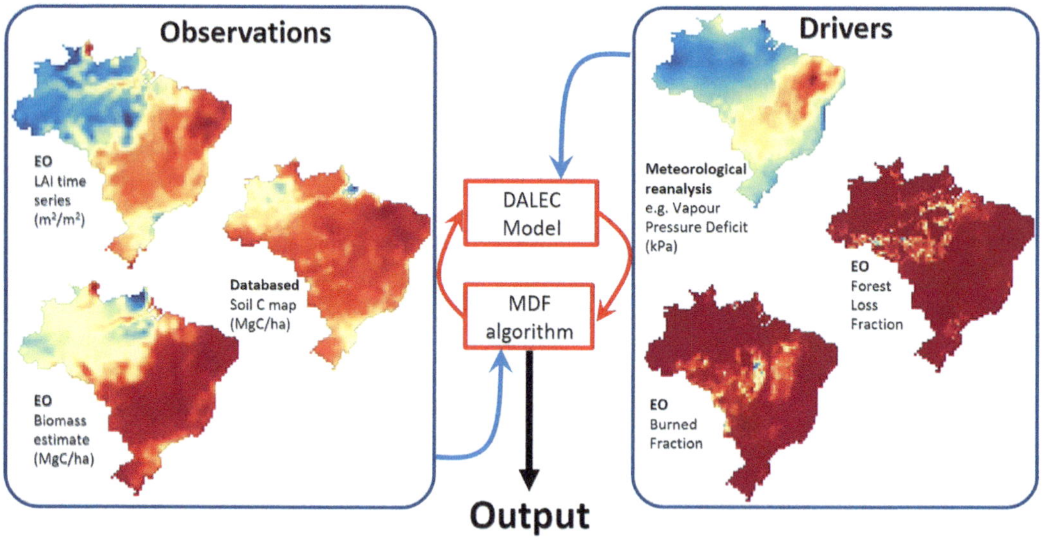

FIGURE 27.3 An example of the inputs used in generating a Brazilian model-data fusion (MDF) analysis of C cycling at 1° resolution monthly over the period 2001–17. The left-hand panel shows the observational data used to constrain the C cycle, including time series of leaf area index (LAI) from satellite observations, a map of soil C from interpolated field data, and a biomass map created from multiple satellite data. The right-hand panel shows observations used to drive process rates, such a vapor pressure deficit (an input to the GPP model in DALEC) and observations used to force land use change and fire impacts. Earth observations (EO) products are used to determine deforestation losses of C, and the burned fraction for combustion losses. The MDF algorithm runs very large ensembles in each grid cell using the drivers, and selects parameter sets that generate pools of C consistent with the observations and their errors. The output is a selection of these accepted parameter sets for each pixel in the analysis.

Analysis provided by T.L. Smallman.

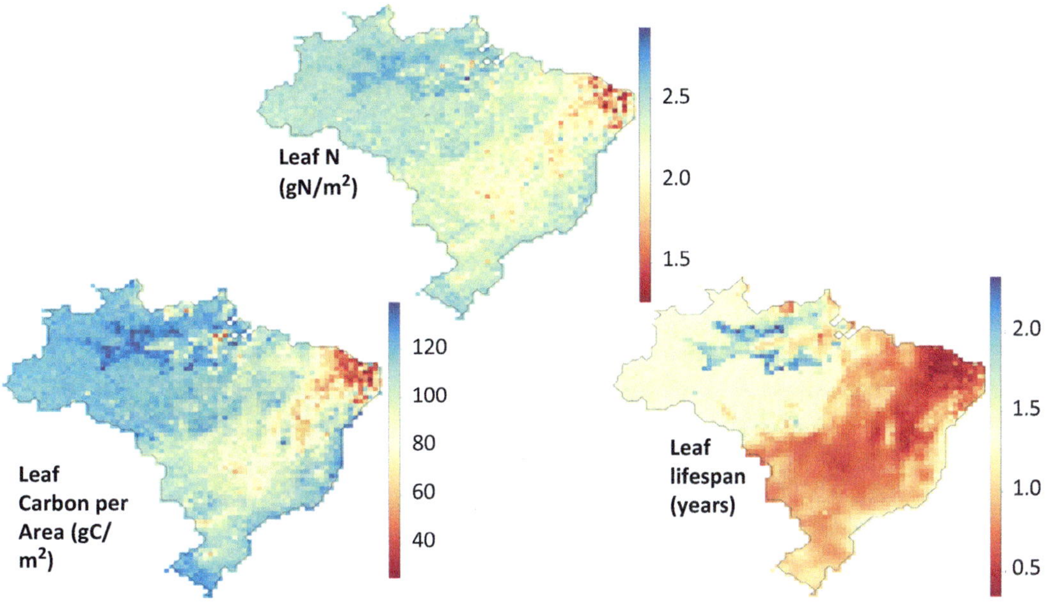

FIGURE 27.4 Retrieved median estimates of leaf traits for the Brazilian analysis using CARDAMOM. These traits are parameters within DALEC. Leaf N determines the maximum rate of photosynthesis. Leaf C per area determines the relationship between foliar C mass and leaf area index (LAI), and the investment requirement in C to construct leaf area. Leaf life span determines the turnover time of foliage, and therefore the investment of C required each year to maintain the canopy at the observed LAI. For each pixel a probability distribution for each parameter is determined.

Analysis provided by T.L. Smallman.

parameter prior. Likewise, comparing this prior-posterior metric among parameters provides insights into which processes have been constrained by available data, and which have not. For instance, it is typical for parameters related to foliage (e.g., leaf life span, canopy efficiency) to be more strongly constrained than those related to root turnover or litter mineralization. This difference is because CARDAMOM usually has available time series of satellite LAI data which provide useful information on foliar processes directly. Information on roots or litter is sparse from earth observation.

An Example of CARDAMOM

An example of CARDAMOM application at sub-continental scale is revealing. Here we have completed an analysis for Brazil during 2001–2017. The input data include Copernicus LAI time series, a biomass map, and a soil C map. Drivers include meteorological data, a forest loss product, and a burned area product (Figure 27.2). CARDAMOM produces parameter estimates, whose median values can be mapped to show the ecological spatial variation across Brazil (Figure 27.3). It is clear from inspection that there are emergent patterns resulting from the pixel-by-pixel analysis for leaf traits. One can see distinct differences in leaf N, leaf C mass per area, and leaf life span. The Amazon basin has patterns clearly different from Cerrado (bordering the Amazon to the SE), Caatinga (NE Brazil), and Atlantic Forest (E coast of Brazil), for example. Within the Amazon there is evidence of some parameter differences perhaps related to wetland areas. Boundaries between these key biomes are clearer or more blurred in some areas.

These biome patterns are also clear in parameters related to biomass stocks (Figure 27.4). Fractional allocation of NPP to leaves, wood, and roots shows clear emergent trade-offs across Brazil. In the Amazon, allocation to wood is the dominant fraction, reflecting the productive and competitive nature of tropical moist forests. Similar patterns can be observed in the narrow band of Atlantic Forest along the coast. In the dry forests of the Cerrado and even drier Caatinga the dominant allocation fraction is to roots, reflecting the water limited nature of these systems, and their disturbance by fire. Within the Amazon there is evidence of regional variability, with the northeastern area having relatively higher allocation to wood and lower allocation to leaves than other areas.

A key novelty of the CARDAMOM approach is that it generates emergent patterns in C processing, between and within biomes – for instance across the equatorial forest biome. Field studies are now providing better spatially resolved, bottom-up information on leaf and plant traits to guide model calibration. The top-down results from CARDAMOM can be evaluated against these independent data to check for consistency across biomes and continents (Figure 27.5).

The typical model evaluation approach is to test using data independent from the calibration data. For instance, CARDAMOM estimates of leaf mass per area across the globe, emergent from earth observations, climate drivers, and DALEC model structure, could be compared to interpolations of field observations. Areas of agreement and mismatches could be highlighted. The magnitude of mismatches could be compared using different versions of DALEC (e.g., with alternate phenology schemes), different climate forcing, and different earth observations of LAI to explore their validity.

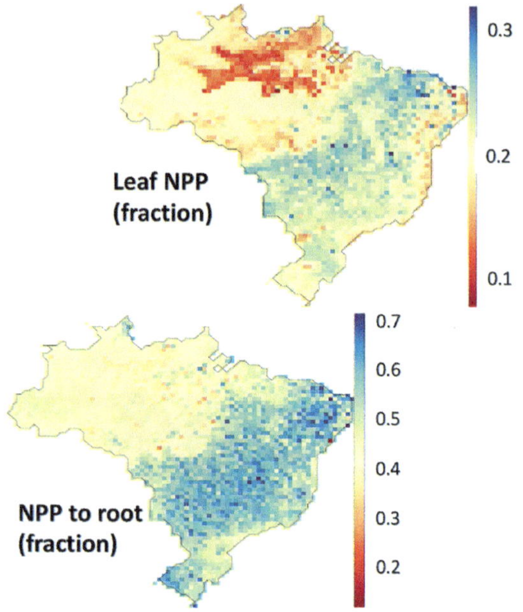

FIGURE 27.5 Retrieved median estimates of plant traits for the Brazilian analysis using CARDAMOM. Three of the panels show the fraction of net primary production (NPP) allocated to leaves, wood and fine roots (i.e., these fractions sum to 1). The final panel shows the wood residence time, indicating the longevity of C storage in live woody biomass. The patterns are emergent, as each pixel is analyzed independently to produce likely estimates of these parameters.

Analysis provided by T.L. Smallman.

But there is an alternate philosophy for model-data fusion approaches, which is to assimilate the field observations of leaf traits (i.e., not leave any for independent testing). We can evaluate the resulting analysis for consistency of model and data, and for its likelihood. The probabilistic approach used in CARDAMOM provides a clear quantification of model value. The information content of new data sets like a bottom-up traits map can be evaluated by comparing analyses with and without the new data. Thus, what is the impact on analytical likelihood of the new assimilate? Does this have a spatial or temporal impact?

KEY CHALLENGES AND OPPORTUNITIES FOR DATA ASSIMILATION

The earth's land surface is both dynamic and highly heterogeneous with variation in C stocks occurring at varied length scales across the globe according to biological processes and exogeneous factors linked to disturbance and management. Disturbance and management operate on a range of scales, but typically are more important at finer spatial resolutions. Thus, as analyses sharpen their resolution from 1 × 1° to 1 ha, for instance, disturbance and management become more important and ecosystem properties become more dynamic and potentially further from steady state. At 1 × 1° the annual rate of forest loss is relatively small compared to stocks, for typical forested landscapes. But in these same landscapes a 1 ha forest plot can lose most of its stocks in one year, due to fire or clearance, and rates of aggradation can also be fast.

There is a need to resolve forest C processes at fine resolutions (e.g., 1 ha) to support monitoring, reporting, and verification processes linked to global treaties for climate. For this reason, there are numerous satellite missions under construction or in orbit to provide data on forest structure at these resolutions. Frameworks such as CARDAMOM can assimilate these satellite observations to produce C flux estimates at similar resolutions. There are technical challenges relating to computing power requirements and data management. There are also algorithmic challenges to calibrate models for dynamic forest patches where disturbance and recovery from disturbance are common states. High temporal resolution data from satellites provide a means to monitor for rates of change and thereby to calibrate internal processing of C. Thus, repeated biomass stock estimates have been shown to provide insights into carbon allocation parameters, and even into key soil parameters, as these are downstream of wood dynamics. Further, the spatial distribution of biomass stocks also provides information on the local landscape dynamics. A steady state landscape will have a normal distribution of biomass stocks. A degrading landscape will be dominated by lower biomass with a long tail of higher biomass from remnant forest patches. Using this information will require new algorithms not currently used in CARDAMOM, to link information from neighboring pixels in a landscape analysis.

The atmospheric C community has a long history of using inversion approaches to identify source and sink regions from linking atmospheric transport models and atmospheric CO_2 concentration data in carbon cycle data assimilation system (CCDAS). There is a clear opportunity to link the ecological data assimilation in frameworks such as CARDAMOM to atmospheric approaches. In fact, links to atmospheric data would be valuable particularly because identifying whether a landscape is a source or sink is challenging. Future developments include the potential to link to atmospheric inversions for independent tests of net exchanges.

SUGGESTED READINGS

For more details on the CARDAMOM method read:

Bloom, A. A., and Williams, M. (2015). Constraining ecosystem carbon dynamics in a data-limited world: Integrating ecological "common sense" in a model-data-fusion framework. *Biogeosciences 12*: 1299–1315.

Bloom, A. B., Exbrayat, J.-F. van der Velde, I. R. Feng, L., and Williams, M. (2016). The decadal state of the terrestrial carbon cycle: Global retrievals of terrestrial carbon allocation, pools and residence times. *Proceedings of the National Academy of Sciences 113*: 1285–1290.

Smallman, T. L., Exbrayat, J.-F. Mencuccini, M. Bloom A. A., and Williams M. (2017). Assimilation of repeated woody biomass observations constrains decadal ecosystem carbon cycle uncertainty in aggrading forests. *Journal of Geophysical Research Biogeosciences 122*: 528–545.

Compare the CARDAMOM approach with other model-data fusion approaches in the papers presented below. What are the advantages of these other approaches?

Peylin, P., Bacour, C., MacBean, N., Leonard, S., Rayner, P., Kuppel, S., Koffi, E., Kane, A., Maignan, F., Chevallier, F., Ciais, P., and Prunet, P. (2016). A new stepwise carbon cycle data assimilation system using multiple data streams to constrain the simulated land surface carbon cycle. *Geophysical Model Development 9*: 3321–3346.

Kaminski, T., Scholze, M. L. U., Vossbeck, M., Knorr, W. L. U., Buchwitz, M., and Reuter, M. (2017). Constraining a terrestrial biosphere model with remotely sensed atmospheric carbon dioxide. *Remote Sensing of Environment 203*: 109–124.

QUIZ

1. Models aim to be general, realistic, and accurate. Why is it so hard to meet all three goals at once?
2. What are the arguments for and against calibrating and validating global C cycle models at eddy covariance sites alone?
3. How does the CARDAMOM approach to generating C model parameters differ from the typical plant functional type approach? What are the relative advantages of each method?
4. What are the challenges to applying CARDAMOM at very high resolutions (e.g., 1 ha) across the globe? How might these challenges be addressed?

28 Practice 7

Data Assimilation at the SPRUCE Site

Shuang Ma
University of California, Los Angeles, USA

This practice uses the SPRUCE field experiment as a case study to explore how a model's parameter values influence its output via sensitivity analysis, and how parameters may be estimated via assimilation of data from field measurements. Exercise 1 addresses how certain model output variables are sensitive to changes in certain parameter values. Exercise 2 illustrates how observational data change posterior parameter distributions in comparison to the priors. In this practice, we use data from the SPRUCE research site in Northern Minnesota, USA, as a context to show possible studies you could conduct by using data assimilation. The exercises may be performed using the CarboTrain software. Instructions for installing and working with CarboTrain are available in Appendix 3.

PRACTICE DESIGN

Data sets that are used in this practice are listed in Table 28.1.

Exercise 1 Parameter sensitivity of model output

In this exercise we will examine how the parameter values of a model affect its outputs. Launch CarboTrain on your computer (see Appendix 3) and follow these steps:

- a Select Unit 7 ➔ Exercise 1 ➔ Set output directory (e.g., mydir/EX1/default) ➔ Click *Run Exercise*.
- b Output files appear in subdirectory simulation of the output directory you specified in the previous step. CarboTrain plotted daily_cpool.png and daily_fluxes.png with the data in the same folder for a quick look. The full model outputs are also provided as .csv files. The model output can be found in the file: 'your_output_dir/simulation/Simu_dailyflux14001.csv'. The observation data can be found here: '/CarboTrain/Source_code/TECO_2.3/input/SPRUCE_cflux.txt' and '/CarboTrain/Source_code/TECO_2.3/input/SPRUCE_cpool.txt'. Annotations for output data columns is here: your_CarboTrain_dir/CarboTrain/Source_code/TECO_2.3/annotations in TECO modeloutput_file_format.jpg
- c Select Unit 7 ➔ Exercise 1 ➔ Set output directory (e.g., mydir/EX1/increase_vcmax). Click *Set Initial parameters* and increase the parameter 'Vcmax' value by 30%. Click *Run Exercise*.
- d Repeat step (c), set a new output directory (e.g., mydir/EX1/decrease_vcmax). Click *Set Initial parameters* and decrease the parameter 'Vcmax' value by 30%.
- e Repeat steps (c) and (d) for increases and decreases of the following parameters. Remember to create a new name for output directory each time after you changed the parameter value. You should end up with ten different output files:
  - a SLA, the specific leaf area;
  - b Tau_leaf, leaf carbon turnover time;
  - c Tau_root, root carbon turnover time;

TABLE 28.1
The SPRUCE site data used in this practice

Purpose	Data name	Year	Period	Time step
Environmental variables (input) to drive the TECO model, spin up and forward run	Soil temperature at 0, 5, 10, 20, 30, 40, 50, 100, 200cm depth	2011–2018	Whole year	Hourly
	Air temperature at 2m			
	Relative Humidity at 2m			
	Wind speed at 10m			
	Precipitation			
	Photosynthetically Active Radiation (PAR) at 2m			
Water balance calibration	Soil moisture at 0, 20cm	2014–2018	Whole year	Hourly
	Water table depth	2014–2018	Whole year	Hourly
Data streams used in data-model fusion	Leaf, wood, root biomass	2014–2018	End of growing season	Once a year
	Soil C content	2012	August 13–15	One time
	NEE, GPP, ER fluxes	2015–2018	Growing season	1–2 times a month

Note: NEE = Net Ecosystem Exchange, GPP = Gross Primary Production, Reco = Ecosystem Respiration.

DOI: 10.1201/9781032711126-35

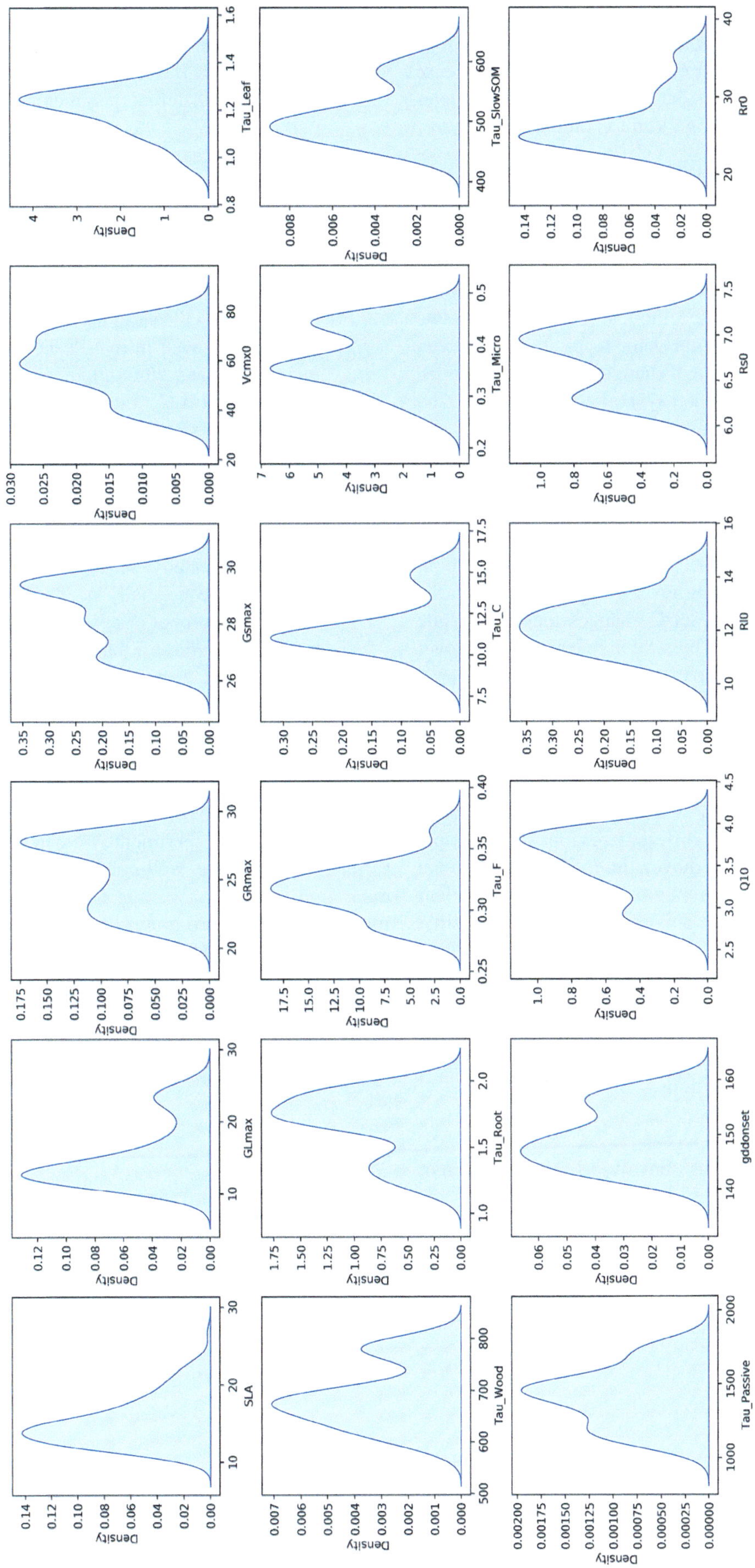

FIGURE 28.1 Parameter posterior distribution from data assimilation based on SPRUCE experimental data using the TECO model.

d Tau_slowSOM, recalcitrant soil C pool turnover time.
f Plot the output data from your ten model simulations and compare the differences. It will be helpful for ease of comparison to plot the results for one parameter with ±30% changes (e.g., Vcmax) on the same figure.

After Step (b), you can inspect the default run results, such as daily leaf, wood, and root pool changes, GPP, NEE, Reco, and other carbon fluxes from 2011 to 2016. When you decrease and increase the Vcmax value by 30% in Steps (c)–(d), you will see substantial changes in GPP and autotrophic respiration. Similarly, when the value of specific leaf area is changed, we can see effects on GPP and autotrophic respiration. However, in both cases, we don't see much change in the carbon pools. From these comparisons, you may realize that Vcmax and SLA have similar effects on model outputs.

By contrast, when you change the turnover time of the leaf pool, you may see that the steady state size of the leaf pool has considerably changed whereas GPP or NEE do not change much. At steady state, the leaf pool is about 350 gC m^{-2} when we decrease the turnover time by 30%, increasing to about 450 gC m^{-2} if we increase the turnover time by 30%. Similarly, tuning the turnover time of root C changes the steady state size of the root carbon pool. Thus, we may learn that turnover time can affect the steady state of carbon biomass.

Exercise 1 shows us that certain output variables are sensitive to changes of certain parameter values. We have also seen that adjustments in different combinations of parameter values can sometimes generate the same changes in output variables. In other words, we might sometimes get right answers but for a different reason. An example is that the value of leaf C turnover time, SLA, and Vcmax can all be responsible for the magnitude and pattern of leaf carbon pool sizes and GPP. This is so-called equifinality, an issue for both manual tuning and Bayesian-guided parameter estimation (Luo et al. 2009). However, manually tuning parameter values can be tricky and subjective. Data assimilation searches for the whole prior ranges of parameter values with millions of iterations to generate posterior probability distributions.

Question:
Which output variables are most sensitive to change in each of the five parameters? Can you explain why?

Exercise 2 Data assimilation with TECO
This exercise will help you learn to perform data assimilation with the comprehensive ecosystem model, TECO. We will use CarboTrain as a toolbox for the exercise.

1 Launch CarboTrain and select Unit 7 ➔ Exercise 2 ➔ Set output directory (e.g., mydir/EX2/default) ➔ Click *Run Exercise*. The model runs iteratively 20,000 times with different parameter values, the accepted parameter values are saved in the file DA/paraest.txt. It takes several hours to finish the full data assimilation process. If you want to get fast but not accurate results, you may let it run for 500 times, reducing the run time to 5–10 mins. To do this, monitor the value for 'isimu' in your terminal window; when it gets to 500, terminate the run by pressing Ctrl+C. Then CarboTrain will randomly select 100 sets of accepted parameters saved in the file DA/paraest.txt, and save model forward run output in the file forecasting/Simu_dailyflux***.txt. '***' could be any of the 3-digit numbers ranging between 1–100. A quick plot of the results is also provided in the file pool_and_flux.png.
2 This exercise is a teaser for you to walk through the whole process of data assimilation. In a formal data assimilation study, tens of thousands to millions of iterations are normally needed to get to a stabilized chain. The number of iterations needed depends on the model structures, the parameter sampling and selection algorithms (e.g., different variants of the Monte Carlo Markov Chain algorithm; see Chapter 22), and the quality and quantity of observations.

Figure 28.1 shows parameter posterior distributions from a full data assimilation run informed by observational data from the SPRUCE field experiment. We can see that SLA and leaf C turnover time are well-constrained by the observational data. In Exercise 2, we see that parameters to which the model is sensitive are more likely to be constrained by data. Instead of one value, the data assimilation study generates probabilistic distributions of parameter values, allowing the range of uncertainty to be displayed and analyzed. With data assimilation, you can quantify uncertainties and partition the uncertainties into different sources of error and process variabilities.

QUESTIONS

1 What is a constrained posterior distribution?
2 Which parameters are most likely to be constrained by the observation data sets, why? (hint: think of equifinality and parameter sensitivity)

Unit Eight

Ecological Forecasting with EcoPAD

29 Introduction to Ecological Forecasting

Yiqi Luo
Cornell University, Ithaca, USA

In this rapidly changing world, improving the capacity to forecast future dynamics of ecological systems and their services is essential for better stewardship of the earth system. This chapter introduces ecological forecasting, the next frontier of research in ecology. Using weather forecasting as an analog, this chapter discusses four elements for ecological forecasting. The four elements are: predictability of the land carbon cycle; observations to constrain forecasting; data assimilation to integrate data with models; and a workflow system to automate ecological forecasting. This chapter also describes applications of an ecological forecasting system to a warming and CO_2 experiment in northern Minnesota and a precipitation mean and variance experiment in New Mexico.

INTRODUCTION

As you have seen in Figure 21.1, to realistically forecast ecosystem responses to environmental change, we need three elements: (1) model structure to represent the real-world processes that control system functions; (2) parameterization to reflect system properties; and (3) external forcing variables that an ecosystem experiences. We have studied the matrix approach to process-based modeling for the carbon cycle in Units 1–5. This chapter will examine the predictability of the land carbon cycle, according to the matrix equation, to understand the expectation of how well carbon forecasting can be achieved. We have also studied data assimilation for parameter estimation in Units 6 and 7. This chapter will explore the availability of observations to achieve accuracy in ecological forecasting with different levels of model complexity. Training in this unit (i.e., this chapter, Chapter 30, and Practice 8 in Chapter 32) is focused on a workflow system, Ecological Platform of Assimilating Data (EcoPAD) into model, to link real-time forcing and automate ecological forecasting.

WEATHER FORECASTING

Before we get into ecological forecasting, let us learn something from weather forecasting. Probably everyone is very familiar with weather forecasting. First, please take a moment to answer a few multiple-choice questions. How frequently do you look at weather forecasting? A. never; B. once every a few days; C. once a day; and D. a few times a day. You may make your own choice. Why do you look at weather forecasting? A. deciding what clothes to wear; B. deciding what kinds of outdoor activities to do; C. deciding whether you will do some field research; or D. doing something else. What do you think are the benefits that weather forecasting brings to society? A. saves lives; B. supports emergent management; C. mitigates the impact of and prevents economic losses from high-impact weather; D. facilitates financial revenue in energy, agriculture, transport, and recreational sectors; or E. all the above. These questions show that weather forecasting has become part of our lives, influences our daily activities, and is relevant to many aspects of our society.

According to a review paper by Peter Bauer et al. (2015) published in *Nature*, weather forecasting skills have been steadily improving. The skill reaches 98% by 2014 for a three-day weather forecast and about 60% for a seven-day weather forecast. I have personal experience on the accuracy of the weather forecast. Many of you may also notice how accurate the weather forecast has become.

Numeric weather prediction as a scientific discipline has been developing for more than 100 years. The major milestones of weather prediction include knowing the laws of physics to make weather forecasting possible in 1901, developing and using super-computing in the 1970s, using satellite and other observations in the 1980s, and using data assimilation in the 1990s. For example, it is relatively well known that the physical processes that determine weather dynamics include energy and water fluxes, momentum dynamics, and land surface conditions, among others.

Numeric weather prediction uses extensive observations from radar and other observations in data assimilation to generate weather patterns. Observations are used to constrain initial values every few hours. The data assimilation methods include 3D-var, 4D-var, and nowadays ensemble Kalman Filter. Data assimilation with complex weather models is computationally expensive. Accuracy and resolution of numerical weather prediction models increase over time as computational power exponentially increases. In short, success in weather forecasting depends on understanding of physical laws to develop models, collecting satellite and other data, using data assimilation to constrain initial values every a few hours, and relying on supercomputing to carry out calculation of the numeric models.

Similar to weather forecasting, ecological forecasting also needs process-based models, observations, data assimilation, and supercomputing. The process-based models offer model structure whereas observations are assimilated into the process-based models for parameterization through data assimilation via supercomputing.

DOI: 10.1201/9781032711126-37

TABLE 29.1
Intrinsic predictability of response patterns of the terrestrial carbon cycle to five classes of external forcing. The predictability of the carbon cycle measures a degree to which the response pattern is predictable given one class of external forcing. The predictability is usually judged by the sensitivity (e.g., diverging vs. converging) of systems behavior in response to various classes of perturbation and external forcing. In general, carbon cycle responses *per se* are more predictable than external forcing, which causes much high uncertainty in predicting carbon cycle responses to climate change

External forcing		Response of the terrestrial carbon cycle		
Class	**Example**	**General pattern**	**Component**	**Intrinsic predictability**
Cyclic environment	Diurnal, seasonal, and interannual	Cyclic	Diurnal and seasonal	High
			Interannual	Less known
Disturbance event	Fire, land use, insect outbreak, and storms etc.	Pulse-recovery	Time of events happening	Low
			Immediate impacts of disturbance events on carbon cycle	Medium
			Recovery	High
			Recovery to original or new attractor	Less known
Climate change	Rising $[CO_2]_a$, climate warming, altered precipitation	Gradual	Direct impacts	High
			Indirect impacts via induced changes in disturbance regimes and ecosystem states	Less known
Shifts in disturbance regimes	Regional, long-term patterns of fire, land use, insect outbreak, and storm etc.	Disequilibrium	Joint probability to describe disturbance regimes and their shifts	Unknown
			Impacts of shifted disturbance regimes on mean carbon storage	High
Ecosystem state change	Forest to cropland, grassland to cropland, reforestation, etc.	Abrupt changes	When and where ecosystem states change	Less known
			Carbon cycle change with ecosystem states	High

Source: Luo et al. (2015).

MODELS AND PREDICTABILITY OF THE TERRESTRIAL CARBON CYCLE

Process-based models of the terrestrial carbon cycle have different levels of complexity but are examined in Units 1–5 for their general properties through the matrix approach. One of the key properties of terrestrial carbon dynamics is the convergence toward some moving attractor states over time, even if external forcing and disturbances often push the carbon cycle to be in disequilibrium. Using this intrinsic property, Luo et al. (2015) examined the predictability of the terrestrial carbon cycle. While the rate of approach to an attractor, and attractor itself, is relatively predictable given knowledge about carbon input rates, loss rates, the initial conditions, and governing environmental constraints, there are three levels of predictability: high, medium, and low, plus two cases: less known and unknown about the predictability for individual processes (see Table 29.1).

For example, some external variables exhibit cyclic changes, typically causing the carbon flux rates, such as photosynthesis and respiration, to vary with the same period as the forcing (Table 29.1). The responses of terrestrial carbon to daily and seasonal cyclic forcing should be highly predictable. However, interannual variability in the terrestrial carbon cycle, as reflected in eddy-flux measurement and variations in the growth rate of atmospheric CO_2, is less known for its underpinning mechanisms, making it difficult at present to evaluate its predictability.

Disturbance events, such as wildfire and climate extremes themselves, however, have an inherent random component (e.g., chances of a hurricane), making the predictability of individual events relatively low. Likewise, the severity of disturbance impacts on the carbon cycle is not very predictable, either. The recovery dynamics following a disturbance, however, appear to be highly predictable given adequate knowledge of the carbon influx rates, the residence times, and the pool sizes following disturbance (Table 29.1). Moreover, there is evidence that some ecosystems may recover to an alternative steady state (or different moving attractors) following disturbance. Our lack of understanding of why this occurs limits our assessment of its consequences for carbon cycle predictability.

Most of the direct effects of climate changes on the terrestrial carbon cycle can be predicted via relatively simple response functions in ESMs. However, climate change also causes indirect effects on the terrestrial carbon cycle, such as changes in plant species composition, microbial priming, and respiratory acclimation. The indirect effects are much less well understood, making it currently unclear just how predictable they are (Table 29.1). Moreover, climate change may also induce shifts in disturbance regimes and changes in ecosystem states, which is less predictable as discussed below.

Disturbance regimes can be quantified by joint probabilistic distributions of disturbance frequency and severity, which, in turn, can be used to generate a probability distribution of ecosystem carbon storage. The mean of the probability distribution determines the realizable carbon storage capacity under a given regime, reflecting the mean carbon storage capacity over a sufficiently long time period or over a sufficiently large area (Luo and Weng 2011). This means carbon storage capacity could thus be predictable. However, we do not have enough knowledge to predict when the disturbance regime changes by direct or indirect anthropogenic forcing.

When ecosystem states change, rates of carbon cycling among the plant, litter, and soil carbon pools also change. Given the change in vegetation structures and corresponding parameters, a consequent change in the carbon cycle is quantifiable. However, while vegetation state changes have been studied, their relationships with those carbon cycle parameters remain poorly understood.

Overall, many processes of the terrestrial carbon cycle are intrinsically predictable. For these processes, forecasting is expected to be highly achievable. However, the indirect effects of climate change on terrestrial carbon cycling become less predictable, especially those which, via changes in species composition and disturbance regimes, lead to ecosystem state changes. In addition, individual disturbance events usually occur stochastically even within a stationary disturbance regime. For such processes, forecasting is expected to be more uncertain.

DATA AVAILABILITY TO CONSTRAIN FORECAST VIA DATA ASSIMILATION

There are plenty of data available to support forecasting the carbon cycle. For example, eddy-flux networks provide half-hourly data streams over hundreds of sites to support near-term forecasting of carbon cycle dynamics over daily, seasonal, and interannual timescales.

Data is also available on long-term processes, such as disturbance events and subsequent recovery. Forecasting disturbance events themselves is not easy at this stage. However, data is available to test forecasting of recovery processes. As disturbance regimes and their impacts on carbon cycle take place over quite long timescales, it is clear how real or near-term forecasting would be useful to research or management. But it is feasible to use available data to test the capability of forecasting. Plenty of data are also available on ecosystem state changes for studying ecological forecasting.

There are many global change experiments ongoing right now. They offer great opportunities to test forecasting capability at some of the experimental sites. We have been forecasting responses of the carbon cycle to experimental treatments at five levels of temperature and two levels of atmosphere CO_2 concentration at the SPRUCE site since 2016 (see Chapter 25). We are also setting up the forecasting system for a drought experiment at Sevilleta Long Term Ecological Research (LTER) site in New Mexico.

Data from observational networks and experiments are integrated into models via data assimilation before ecological forecasting is made. Units 6 and 7 describe basic concepts, procedure, and application cases of data assimilation.

WORKFLOW SYSTEM TO FACILITATE ECOLOGICAL FORECASTING

Ecological forecasting is usually carried out in an automatic fashion. We have developed a workflow system for near-time forecasting. The system is Ecological Platform for Assimilating Data (EcoPAD) into models (Huang et al. 2019) (also see Chapter 30). EcoPAD is a software system that links sensor networks to ecological forecasting. It integrates eco-informatics, web-technology, ecological models, data assimilation techniques, and visualization. EcoPAD was designed to promote interactions among modeling, experimentation, and observations to gain the best science.

Data and models are integrated through a data assimilation system before the trained models are used for forecasting, optimization of measurement plans, and uncertainty analysis. The data to be integrated can come from the real-time sensor networks or from spreadsheets with records of hand measurement (Figure 29.1). EcoPAD offers three options, which are simulation, data assimilation, and forecast, in response to a request from users. Once a user makes a request, EcoPAD can execute the task automatically to generate results. The generated results can be visualized in real-time as well. Thus, EcoPAD is an interactive software system for researchers to automatically execute model simulation, data assimilation, and ecological forecasting in real or near-time.

We have applied EcoPAD to the SPRUCE experimental site located in Northern Minnesota (see Chapter 25). SPRUCE is a whole ecosystem warming and CO_2 enrichment project. It has five levels of temperature treatments and two levels of CO_2 concentrations. The experiment follows a gradient design with five chambers for five levels of temperature treatments at ambient CO_2 concentration and five chambers at elevated CO_2 concentration. The project is very well equipped with lots of real-time sensors and involves more than 100 scientists who perform many kinds of measurements. The real-time sensors send data to data servers. Data servers also store the data from hand measurements. EcoPAD automatically ingests data from data servers through a web app server for data assimilation and forecasting. The forecast results are automatically sent to two sites, one at the SPRUCE site and one at our lab website, for visualization. EcoPAD has done ecological forecasting automatically at midnight on Saturday every week since June 2016.

The forecasting variables include snow cover, soil thermal dynamics, and frozen depth (Huang et al. 2017); many carbon cycle variables, such as gross primary production, net primary production, net ecosystem production, and ecosystem

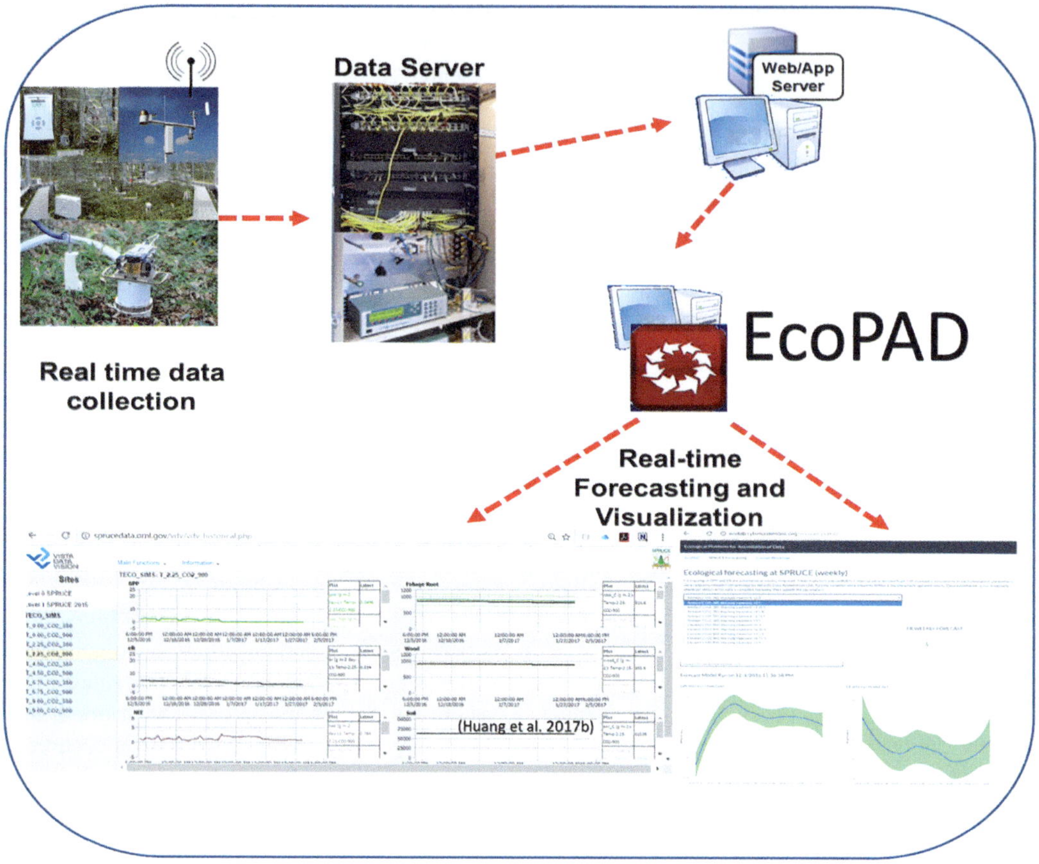

FIGURE 29.1 EcoPAD to streamline data ingestion from sensors and servers, model simulation, data assimilation, forecasting, and visualization. By timely updating of the parameter values, EcoPAD enables close interactions between experimenters and the modelers.

respiration (Jiang et al. 2018); and methane flux and pathways (Ma et al. 2017).

We are applying EcoPAD for data assimilation and ecological forecasting at Sevilleta Long-Term Ecological Research (LTER) site in New Mexico. Sevilleta LTER site currently has a few experiments going on. These experiments are mainly related to precipitation and nitrogen fertilization. In addition, there are several long-term eddy-flux towers, measuring carbon, water, and energy fluxes for years. We are developing the capability to do real-time or near-time data assimilation and ecological forecasting at those experimental and eddy-flux sites.

In fact, EcoPAD can be used as a smart experiment-modeling system. First, the system can predict what ecosystems may respond to treatments once you have selected a site and decided your experimental plan. When we were writing a proposal to continue the LTER study at Sevilleta, New Mexico, we used the TECO model to do pre-experiment analysis on possible ecosystem responses to increasing variability in precipitation. The modeling results were included in the proposal. Once you get funding to do the experiment, you can use EcoPAD to assimilate the data you are collecting to constrain model forecasts. The model forecasts what ecosystem responses may likely be for the remaining period of your experiment. The forecast ecosystem responses can be used as references for you to design your measurement plan. At the SPRUCE project, Shuang Ma's forecast results stimulated discussion about how much methane may be released through bubbling. Discussion on this issue lasted for a few weeks. The extensive discussion led to improvement of Shuang's methane model, new ideas to design additional measurements of methane concentrations along the soil profiles, and more collaborations between experimental and modeling teams. Moreover, the uncertainty analysis with EcoPAD can tell us what those important datasets are and which need more measurement in order to understand the system dynamics. During the course of our study, we can use EcoPAD to periodically update the forecast by repeating steps from data assimilation to forecasting to improvement of measurement and model. We are using EcoPAD to do weekly forecasts at the SPRUCE project. During this process, we improve the models, the experiments, and the data assimilation system.

In summary, carbon cycle forecasting is a new frontier of research in ecology. As many processes are highly predictable, forecasting carbon cycle dynamics is expected to be highly achievable. Plenty of data are available to constrain forecasting with data assimilation. Workflow systems to automate data-model integration and ecological forecasting are becoming available. The challenge remains to identify societally relevant issues that carbon cycle forecasting can address.

SUGGESTED READING

Jiang Jiang, Yuanyuan Huang, Shuang Ma, Mark Stacy, Zheng Shi, Daniel M. Ricciuto, Paul J. Hanson, Yiqi Luo. 2018. Forecasting responses of a northern peatland carbon cycle to elevated CO_2 and a gradient of experimental warming. *Journal of Geophysical Research: Biogeosciences*, *123*: 1057–1071.

QUIZ

1. What are the elements leading to the success in weather forecasting?
2. We need forcing variables to be consistent between models and field sites for realistic forecast because (choose one answer)
   a. the model needs forcing variables to drive simulations.
   b. it is easy to get forcing variables from field sites.
   c. environment variables that drive the model prediction have to represent what ecosystems experience.
   d. we can measure environmental variables at field sites.
3. What is a workflow system?
4. Pre-experiment modeling analysis is useful because (choose as many as correct ones)
   a. it gives some general ideas on what ecosystem responses may look like.
   b. it will give us the precise prediction of ecosystem responses.
   c. we do not have to carry out the experiment anymore.
   d. it helps us adjust the measurement plan.

30 Ecological Platform for Assimilating Data (EcoPAD) for Ecological Forecasting

Yuanyuan Huang
Institute of Geographic Sciences and Natural Resources Research, Chinese Academy of Sciences, Beijing, China

Tremendous scientific endeavors in ecology have been driven by the goal of forecasting future ecological dynamics. This chapter introduces the web-based Ecological Platform for Assimilating Data into model (EcoPAD) to facilitate ecological forecasting. The objectives of this lecture are to understand why we need a platform like EcoPAD, the structure of the platform, and how to use EcoPAD to facilitate ecological forecasting.

WHY DO WE NEED ECOPAD?

Ecological research is driven by the quest of understanding the dynamics of biota and their interactions with the environment. To understand ecological patterns, processes, and functions, we conduct field or manipulative experiments. From these experiments, we combine a wide range of approaches to obtain relevant observational data (e.g., through real-time sensor, laboratory measurements, remote-sensing, and video-based observations). For example, the National Ecological Observatory Network (NEON) of the United States monitors ecosystems across the United States by collecting hundreds of data products including organismal counts and measurements, water and soil quality, energy fluxes, and remotely sensed vegetation indices. Many similar observational networks now provide us with a wide range of ecological relevant datasets for different regions. To name a few, FLUXNET tracks the exchanges of carbon dioxide, water vapor, and energy between the biosphere and atmosphere for a network of flux tower sites around the world, while DroughtNet focuses on the responses of terrestrial ecosystems to drought. Knowledge obtained from data and experiments enables us to make inferences about ecological dynamics under novel situations. The inference could be based on complex mathematical models built upon data as well as simple relationships derived from data. We call inference under novel conditions prediction. Forecasting is a type of prediction in which we make predictions about the future.

Ecological forecasting is not only valuable for contributing to scientific advances but is also practically valuable in guiding resource management and decision-making towards a sustainable future. The practical need for ecological forecasting is particularly urgent for our current rapidly changing world, which is experiencing unprecedented food insecurity, natural resource depletion, biodiversity loss, climate changes, and pollution of air, waters, and soils. This practical need has brought a growing number of forecasting-oriented studies, for example, on fisheries, crop yield, species dynamics, algal blooms, phenology, pollinator performance, and biodiversity. Ecological forecasting is valuable in almost every subdiscipline of ecology. Recent progress especially in available data, improved process understanding, data assimilation techniques, and advanced cyber-infrastructure, is converging to transform ecological research into advanced quantitative forecasting. In practice, however, ecological forecasting remains largely aspirational, with the number of forecasting studies lagging behind demand. One bottleneck is the lack of infrastructure to enable timely integration of data into models. EcoPAD is designed as a solution to widen this bottleneck. It provides a fully interactive infrastructure to facilitate ecological forecasting, especially near-time ecological forecasting based on iterative data-model integration.

EcoPAD (https://ecolab.cals.cornell.edu/?ecopad, last accessed: October 2023) serves to link ecological experiments and data with models and provides easily accessible and reproducible data-model integration with interactive web-based simulation, data assimilation, and forecasting. The system is designed to streamline web request-response, data management, modeling, prediction, and visualization to boost the overall throughput of observational data, promote data-model integration, facilitate communication between modelers and experimenters, inform ecological forecasting, and improve scientific understanding of ecological processes. EcoPAD facilitates estimation of model parameter values, evaluation of model structure, assessment of information content of datasets, and understanding of uncertainties revealed by model-data fusion exercises. Additionally, EcoPAD automates data management, model simulation, data assimilation, ecological forecasting, and result visualization. It provides an open, convenient, transparent, flexible, scalable, traceable, and readily portable platform to systematically conduct data-model integration towards better ecological forecasting. The automated near-time ecological forecasting through EcoPAD updates periodically in a manner similar

DOI: 10.1201/9781032711126-38

to weather forecasting. This design of EcoPAD enables it to function as a smart interactive model-experiment (ModEx) system (Figure 30.1). ModEx forms a feedback loop in which field experiments guide modeling and modeling influences experimental focus. Information is constantly fed back between modelers and experimentalists, and simultaneous efforts from both parties advance and shape understanding towards better forecasts. ModEx can: (1) predict what an ecosystem's response might be to treatments once the experimenter has selected a site and decided the experimental plan; (2) assimilate the data collected during the experiment to constrain model predictions; (3) project the expected ecosystem responses in the rest of the experiment; (4) tell experimenters which priority datasets to collect in order to better understand the system; (5) periodically update the projections; and (6) improve the models, the data assimilation system, and field experiments during the process.

In addition to forecasting and facilitating interaction between modeling and experimental communities, EcoPAD is desirable because of the potential service it can bring to society. Forecasting with carefully quantified uncertainty is helpful in providing support for natural resource managers and policy makers. It is always difficult to bring complex mathematical ecosystem models to end users who do not have training in modeling, which creates a gap between current scientific advances and public awareness. The web-based interface of EcoPAD makes modeling as easy as possible without losing the connection to the mathematics, knowledge, and data behind the models. In this way, infrastructure like EcoPAD has the potential to transform environmental education and encourage citizen science in ecology and climate change.

GENERAL STRUCTURE OF ECOPAD

The essential components brought together by EcoPAD include experiments and data, ecological models, data assimilation techniques, and the scientific workflow (Figure 30.1).

Data are the foundation of ecological modeling and forecasting. We have entered the "big data" age, characterized by the ready availability of different, often extensive datasets across various temporal-spatial scales. These datasets might have high temporal resolution, such as time series from real-time ecological sensors, or extensive spatial coverage from remote sensing sources and data stored in geographic information systems. Data may contain information related to environmental forcing (e.g., precipitation, temperature, incoming radiation), site characteristics (e.g., soil texture and species composition), or biogeochemistry of soils and waters. EcoPAD offers systematic data management to digest diverse data streams. Datasets in EcoPAD are derived from research projects in comma-separated value (csv) files or other loosely structured data formats. These datasets are first described and

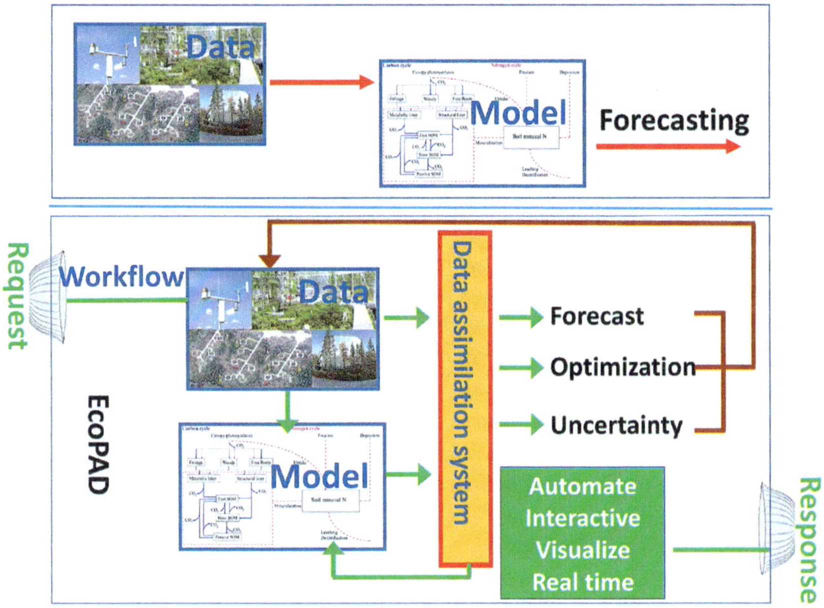

FIGURE 30.1 Schema of approaches to forecast future ecological responses under (a) current practice; and (b) the Ecological Platform for Assimilation of Data (EcoPAD). Current practice makes use of observations to develop and/or calibrate models to make predictions. EcoPAD goes further by linking models to data through a formalized, iterative cycle using a fully interactive platform. EcoPAD consists of four major components: experiment and data, model, data assimilation, and the scientific workflow (green arrows or lines). Data and model are iteratively integrated through its data assimilation systems to improve forecasting. Its near real-time forecasting results are shared among research groups through a web interface to guide new data collections. The scientific workflow enables web-based data transfer from sensors, model simulation, data assimilation, forecasting, result analysis, visualization, and reporting, encouraging user-model interactions, especially for experimentalists and end users having a limited background in modeling.

Adapted from Huang et al., (2019).

stored with appropriate metadata via either manual operation or scheduled automation from sensors. Data are generally separated into two groups. One comprises forcing variables to drive modeling, the other, observations used for data assimilation. Scheduled sensor data are appended to existing data files with prescribed frequency. Attention is given to how the particular dataset varies over space and time. When the spatio-temporal variability is understood, it is then placed in metadata records that allow for query through EcoPAD's scientific workflow.

The workflow and data assimilation system of EcoPAD are relatively independent of any specific ecological model. To illustrate the integration of models with EcoPAD, we take the Terrestrial ECOsystem (TECO) model as a general example. Linkages among the workflow, data assimilation system, and ecological model are based on messaging. For example, the data assimilation system generates parameters that are passed to ecological models. The state variables simulated from ecological models are passed back to the data assimilation system. Models may have different formulations. As long as these models take in the same parameters and simulate the same state variables, they are functionally identical from the point of view of the data assimilation system. TECO simulates ecosystem carbon, nitrogen, water, and energy dynamics. The original TECO model has four major submodules (canopy, soil water, vegetation dynamics, and soil carbon and nitrogen) (Weng and Luo, 2008) and is further extended to incorporate methane biogeochemistry and snow dynamics (Huang et al., 2017; Ma et al., 2017).

Data assimilation (see Chapter 21) provides a framework to combine models with data to estimate model parameters, test alternative ecological hypotheses through different model structures, assess the information content of datasets, quantify uncertainties, derive emergent ecological relationships, identify model errors, and improve ecological predictions. Under the Bayesian paradigm, data assimilation techniques treat the model structure, the initial and parameter values as priors that represent our current understanding of the system (see Chapter 22). As new information from observations or data becomes available, model parameters and state variables can be updated accordingly. The posterior distributions of estimated parameters or state variables are imprinted with information from the model, observations (or data) as the chosen parameters are constrained to reduce mismatches between observations and model simulations. Future predictions benefit from such constrained posterior distributions through forward modeling. As a result, the probability density functions of predicted future states following data assimilation normally have narrower spreads than those without data assimilation. EcoPAD can accommodate different data assimilation techniques since the scientific workflow of EcoPAD is independent of the specific data assimilation algorithm. One example of a data assimilation method is the Markov Chain Monte Carlo (MCMC) introduced in Chapter 22.

The scientific workflow of EcoPAD wraps around one or more user-specified ecological models and data assimilation algorithms and acts to move datasets in and out of structured and cataloged data collections (metadata catalog), while leaving the logic of the ecological models and data assimilation algorithms untouched (Figure 30.2). When a user makes a request through the web browser or command line utilities, the scientific workflow takes charge of triggering and executing corresponding tasks, pulling data from a remote server, running a particular ecological model, automating forecasting, or making the result easily accessible and understandable to users through web based graphic displays (Figure 30.2). The workflow system is portable across operation system and programming language and is built to be scalable to meet the demands of the model and the end-user community. The essential components of the scientific workflow of EcoPAD include the metadata catalog, web application-programming interface (API), the asynchronous task or job queue (Celery), and the container-based virtualization platform (docker) (Figure 30.2). The workflow system of EcoPAD also provides structured result access and visualization. Scientific workflow is a relatively new concept in the ecology literature but is essential to realize real- or near real-time forecasting. Thus, we describe it in detail below. Readers who are not interested in technical details may skip the following paragraphs and jump to the section: "Applications of EcoPAD: The Example of SPRUCE".

Datasets can be placed and queried in EcoPAD via a common metadata catalog, which allows for effective management of diverse data streams. The EcoPAD metadata scheme includes a description of the data product, access pattern, and time stamp of last metadata update. MongoDB (https://www.mongodb.com/, last accessed: November 2020), a NoSQL database technology, is employed to manage heterogeneous datasets to make documentation, query, and storage fast and convenient. Through MongoDB, measured datasets can be easily fed into ecological models for various purposes such as to initialize the model, calibrate model parameters, evaluate model structure, and drive model forecasts. For datasets from real-time ecological sensors that are constantly updating, EcoPAD can be set to automatically fetch new data streams with adjustable frequency according to research needs.

The "gateway" of EcoPAD is the Representational State Transfer (RESTful) application programming interface (API). It can deliver data to a wide variety of applications and enables a wide array of user interfaces and data dissemination activities. Once a user makes a request, such as through clicking on relevant buttons from a web browser, the request is passed through the RESTful API to trigger specific tasks. Thus, the API bridges communication between the client (e.g., a web browser or command line terminal) and the server (Figure 30.2). The API exploits the HyperText Transfer Protocol (HTTP) such that data can be retrieved and ingested from EcoPAD through the use of simple HTTP headers and verbs (e.g., GET, PUT, POST, etc.). Since HTTP is also understood by web servers and clients, a user can incorporate summary data from EcoPAD into a website with a single line of html code. Users are able to access data directly through programming environments like R, Python, and MATLAB. Simplicity, ease of use, and interoperability are among the main advantages of the API, which enables web-based modeling.

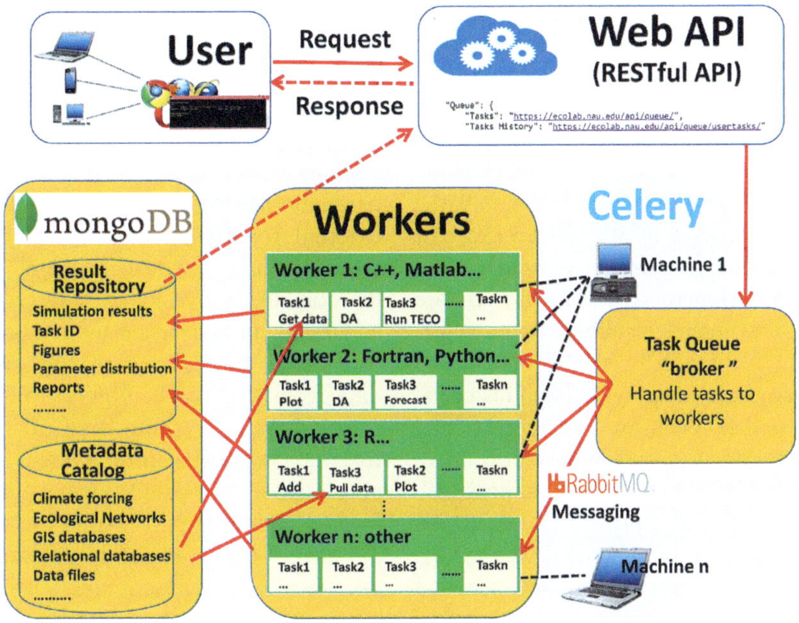

FIGURE 30.2 The scientific workflow of EcoPAD. The workflow wraps ecological models and data assimilation algorithms with the docker containerization platform. Users trigger different tasks through the representational state transfer (RESTful) application-programming interface (API). Tasks are managed through the asynchronous task queue, Celery. Tasks can be executed concurrently on a single or more worker servers across different scalable IT infrastructures. MongoDB is a database software that takes charge of data management in EcoPAD, and RabbitMQ is a message broker.

Adapted from Huang et al., (2019).

The task queue is a mechanism used to distribute work across work units such as threads or machines. EcoPAD uses Celery (https://github.com/celery/celery, last accessed: November 2020) as an asynchronous task or job queue that runs in the background (Figure 30.2). Celery communicates through messages, and EcoPAD takes advantage of RabbitMQ (https://www.rabbitmq.com/, last accessed: November 2020) to manage messaging. After the user submits a command, the request or message is passed to Celery via the RESTful API. These messages may trigger different tasks, which include but are not limited to pulling data from a remote server where original measurements are located, accessing data through a metadata catalog, running model simulations with user specified parameters, conducting data assimilation that recursively updates model parameters, forecasting future ecosystem status, and postprocessing model results for visualization. The broker inside Celery receives task messages and hands out tasks to available Celery "workers" that perform the actual tasks (Figure 30.2). Celery workers are in charge of receiving messages from the broker, executing tasks, and returning task results. The worker can be a local or remote computation resource (e.g., the cloud) that has connectivity to the metadata catalog. Workers can be distributed into different information technology infrastructures, which makes the EcoPAD workflow expandable in accommodating more computational resources. Each worker can perform different tasks depending on the tools installed in each worker. One task can also be distributed to different workers. In such a way, the EcoPAD workflow enables the parallelization and distributed computation of actual modeling tasks across various IT infrastructures and is flexible in implementing additional computational resources by connecting additional workers.

Another key feature that makes EcoPAD easily portable and scalable among different operation systems is the utilization of a container-based virtualization platform, the docker (https://www.docker.com/, last accessed: January 2019). The docker can run many applications that rely on different libraries and environments on a single kernel with its lightweight containerization. Tasks that execute TECO in different ways are wrapped inside different docker containers that can "talk" with each other. Each docker container embeds the ecosystem model into a complete file system that contains everything needed to run an ecosystem model: the source code, model input, run time, system tools, and libraries. Docker containers are both hardware-independent and platform-independent, and they are not confined to a particular language, framework, or packaging system. Docker containers can be run from a laptop, workstation, virtual machine, or any cloud compute instance. This is done to support the widely varied number of ecological models running in various languages (e.g., MATLAB, Python, Fortran, C, and CCC) and environments. In addition to wrapping the ecosystem model into a docker container, software applied in the workflow, such as Celery, RabbitMQ, and MongoDB, are all lightweight and portable encapsulations through docker containers. Therefore, EcoPAD is readily portable to different environments.

EcoPAD enables structured result storage, access, and visualization to track and analyze data-model fusion practice. Upon the completion of the model task, the model wrapper code calls a postprocessing callback function. This callback function allows model-specific data requirements to be added to the model result repository. Each task is associated with a unique task ID and model results are stored within the local repository that can be queried by the unique task ID. The storage and query of model results are realized via the MongoDB and RESTful API (Figure 30.2). Researchers are able to review and download model results and parameters submitted for each model run through a web-accessible URL (link). The EcoPAD web page also displays a list of historical tasks (with URL) performed by each user. All current and historical model inputs and outputs are available to download, including the aggregated results produced for graphical web applications. In addition, EcoPAD also provides a task report that contains an all-inclusive recap of submitted parameters, task status, and model outputs with links to all data and graphical results for each task. Such structured result storage and access make sharing, tracking, and referring to modeling studies instantaneous and clear.

APPLICATIONS OF ECOPAD: THE EXAMPLE OF SPRUCE

The SPRUCE experiments and datasets were introduced in Chapter 25. Here, we demonstrate the use of the EcoPAD infrastructure as a way to assimilate multiple streams of data from the SPRUCE experiment to the TECO model using the MCMC algorithm and forecast ecosystem dynamics in both near time and for the next ten years. A similar example was presented in Chapter 26. The forecasting system for SPRUCE is available at: https://ecolab.nau.edu/ecopad_portal/ (last accessed: November 2020). From the web portal, users can check our current near- and long-term forecasting results, conduct model simulation, data assimilation, and forecasting runs, and analyze and visualize model results. We set up the system to automatically pull new data streams every Sunday from the SPRUCE FTP site that holds observational data and updates the forecasting results based on new data streams. Updated forecasting results for the following week are customized for the different manipulative treatments of the SPRUCE experiments and displayed in the EcoPAD-SPRUCE portal. At the same time, these results are sent back to SPRUCE communities and displayed together with near term observations for experimentalists to study.

In the SPRUCE project, we take advantage of this platform to stimulate interactive communication between modelers and experimentalists, study the acclimation of ecosystem carbon cycling to experimental manipulations, partition uncertainty sources in forecasting, improve the biophysical estimation for better forecasts, and explore how the updated model and data contribute to reliable forecasting. Our case studies confirm that realistic model structure, correct parameterization, and accurate external environmental conditions are critical for forecasting carbon dynamics. The following section describes how the updated model and data contribute to reliable forecasting in the SPRUCE experiment. For other applications, the interested reader is referred to Huang et al. (2019).

In the SPRUCE project, the system automatically conducts data assimilation with the new observational data streams from each week, successively improving the model parameterization. With constantly adjusted model and external forcing and weekly archived model parameters, model structure, external forcing, and forecasting results, the contribution of model and data updates to forecasting accuracy can be tracked by comparing the previous week's forecasted simulations to the current one informed by data from that week. Figure 30.3 illustrates how updated external forcing (compared to stochastically generated forcing) and shifts in ecosystem state variables shape the predictions. "Updated" means the real meteorological forcing monitored from the site's weather station. In the absence of observations, stochastically generated forcing is used as a proxy for future meteorological conditions. Future precipitation and air temperature are generated by vector autoregression using a historical dataset (1961–2014) monitored by the weather station. Photosynthetically active radiation (PAR), relative humidity, and wind speed are randomly sampled from the joint frequency distribution at a given hour each month. Detailed information on weather forcing is available in Jiang et al. (2018). TECO is trained through data assimilation with observations from 2011–2014 and used to forecast GPP and total soil organic carbon content at the beginning of 2015.

For demonstration purposes, Figure 30.3 shows three series of forecasting results instead of updates from every week. Series 1 (S1) records forecasted gross primary production (GPP) and soil carbon with stochastically generated weather forcing from January 2015–December 2024 (Figure 30.3a, b, cyan). Series 2 (S2) records simulated GPP and soil carbon with observed (updated) climate forcing from January 2015–July 2016 and forecasted GPP and soil carbon with stochastically generated forcing from August 2016–December 2024 (Figure 30.3a, b, red). Similarly, the stochastically generated forcing in Series 3 (S3) starts from January 2017 (Figure 30.3a, b, blue). For each series, predictions were conducted with randomly sampled parameters from the posterior distributions and stochastically generated forcing. 100 mean values are displayed (across an ensemble of forecasts with different parameters) corresponding to 100 forecasts with stochastically generated forcing.

GPP is highly sensitive to climate forcing. The differences between the updated (S2, 3) and initial forecasts (S1) reach almost 800 gC m^{-2} yr^{-1} (Figure 30.3c). The discrepancy is strongly dampened in the following 1–2 years. The impact of updated forecasts is close to zero after approximately five years. However, the soil carbon pool shows a different pattern. The soil carbon pool is increased by less than 150 gC m^{-2}, which is relatively small compared to the carbon pool size of ca. 62,000 gC m^{-2}. For soil, the impact of updated forecasts grows with time and is highest at the end of the simulation year 2024. GPP is sensitive to the immediate change in climate forcing, while

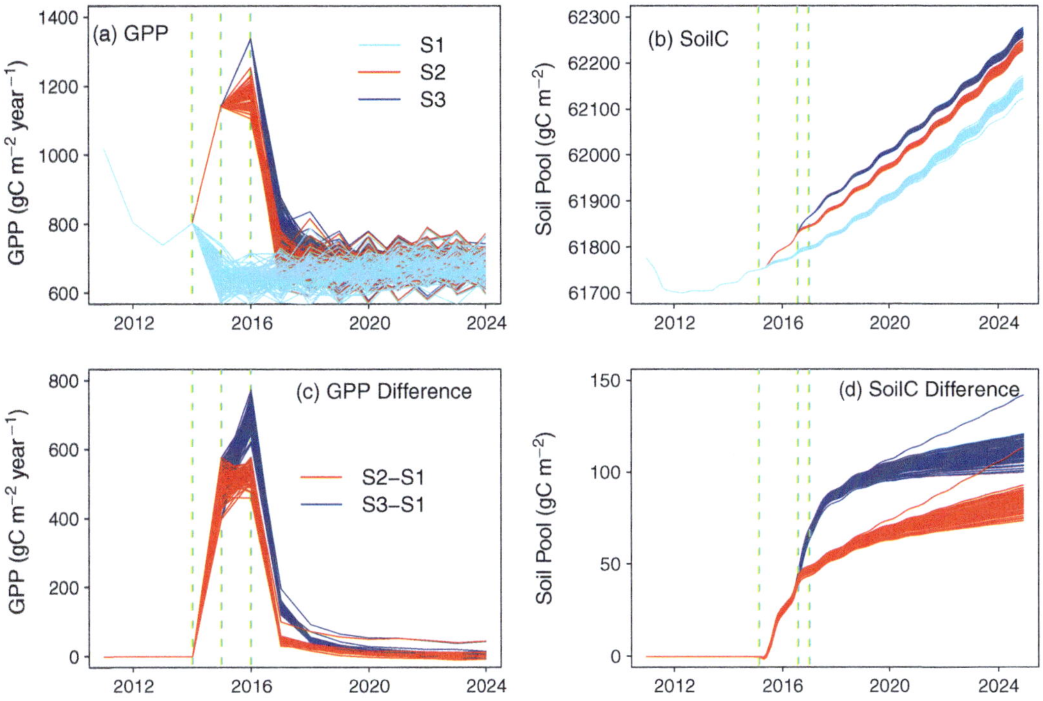

FIGURE 30.3 Updated vs. original forecasting of gross primary production (GPP; panels a, c) and soil organic C content (SoilC; panels b, d). The upper panels show three series of forecasts with updated vs. stochastically generated weather forcing. "Updated" = forced by actual meteorology from field weather stations. Cyan lines indicate forecasting with 100 stochastically generated weather forcing time series from January 2015 to December 2024 (S1); red lines correspond to forecasting updating with measured weather forcing from January 2015 to July 2016, followed by forecasting with 100 stochastically generated weather forcing time series from August 2016 to December 2024 (S2); blue lines show updated forecasting with measured weather forcing from January 2015 to December 2016, followed by forecasting with 100 stochastically generated weather forcing time series from January 2017 to December 2024 (S3). Panels (c) and (d) display mismatches between updated forecasting (S2, 3) and the original forecasting (S1). Red displays the difference between S2 and S1 (S2–S1), and blue shows the discrepancy between S3 and S1 (S3–S1). Dashed green lines indicate the start of forecasting with stochastically generated weather forcing. Note that panels (a) and (c) are plotted on a yearly timescale and panels (b) and (d) show results on a monthly timescale.

Adapted from Huang Use a consistent format for citing a reference in figure legend et al., (2019).

the updated ecosystem state (or initial value) has a minimum impact on the long-term forecast of GPP. The impact of updated climate forcing is relatively small for soil carbon forecasts during our study period. Soil carbon is less sensitive to the immediate change in climate compared to GPP. However, the alteration of system status affects the soil carbon forecast, especially on a longer timescale. Since we are archiving updated forecasts every week, we can track the relative contribution of ecosystem status, forcing uncertainty, and parameter distributions to the overall forecasting patterns of different ecological variables and how these patterns evolve in time. In addition, as more observations of ecological variables (e.g., carbon fluxes and pool sizes) become available, it is feasible to diagnose key factors that promote robust forecasts by comparing the archived forecasts to observations and these to the model parameters, initial values, and climate forcing used.

In addition to scientific capability, the EcoPAD system brings new opportunities to broaden user–model interactions and facilitate forecasting practice. The high complexity and long learning curve of ecological models and data–model fusion techniques frequently discourage researchers and impede progress in forecasting practice. EcoPAD is designed to reduce these hurdles. It can be accessed from a web browser and does not require any coding by the user, which opens the door for non-modelers to work with models. The online storage of results lowers the risk of data loss. The results of each model run can be easily tracked and shared with a unique ID and web address. In addition, the web-based workflow saves time for experts through automated model running, data assimilation, forecasting, structured result access, and instantaneous graphic outputs, allowing the researcher to focus on a thorough exploration of results. At the same time, advanced users have the flexibility to scrutinize or modify model code, embed a different ecological model, change the data assimilation algorithm or add new forecasting properties.

In summary, EcoPAD provides an effective infrastructure with its interactive platform that rigorously integrates merits from models, observations, statistical advance, information technology, and human resources from experimentalists and modelers to practitioners and the general public. This facilitates progress in ecological forecasting and analysis. That being said, ecological forecasting and the EcoPAD platform are both at their early development stage. We need more creative ideas and community efforts to realize the potential of this promising field.

SUGGESTED READING

Huang, Y., Stacy, M., Jiang, J., Sundi, N., Ma, S. et al. 2019. Realized ecological forecast through an interactive Ecological Platform for Assimilating Data (EcoPAD, v1.0) into models. *Geoscientific Model Development 12*: 1119–1137.

QUIZ

1 What is ecological forecasting?
2 What key challenges and barriers does EcoPAD help overcome?
3 What are the four major components of EcoPAD?
4 List four potential applications of EcoPAD?

31 Community Cyberinfrastructure for Ecological Forecasting

Xin Huang
National Center for Atmospheric Research, Boulder, USA

Lifen Jiang
Cornell University, Ithaca, USA

Ecological forecasting uses a workflow system that contains cyberinfrastructures from data collection to visualization of the forecasting results. Over the past decades, some community cyberinfrastructures have been developed for ecological forecasting. This chapter provides an introduction of the structure and key features of community cyberinfrastructures available for ecological forecasting, and the major challenges and solutions in building future community cyberinfrastructures.

INTRODUCTION

In the previous two chapters of the ecological forecasting unit, the three elements that are required for realistic ecological forecasting were introduced. These are: selecting a model structure to represent the real-world ecological processes, optimizing model parameters through data assimilation, and generating external forcing that represents the future climate conditions (Luo et al., 2016).

To realize ecological forecasting, a cyberinfrastructure is needed to automate the workflow involving the three above-mentioned elements. EcoPAD, presented in the previous chapters, is such a cyberinfrastructure. It is a software system that links real-time forcing and observational data to ecological forecasting. Thus, it facilitates integration of field experiments and modeling. A web-based Graphical User Interface (GUI) is also developed to access and operate the EcoPAD system easily and users can request a specific scenario to run through it. Once the request is submitted successfully, EcoPAD will receive the command and execute the tasks such as forward model simulation, data assimilation, and forecasting of a specific climate scenario.

This chapter will introduce the needs for community cyberinfrastructures to support ecological forecasting, the structure and key features of such community cyberinfrastructures, available cyberinfrastructures for ecological forecasting, and the challenges and possible solutions in building future community cyberinfrastructures.

THE NEED FOR COMMUNITY CYBERINFRASTRUCTURES

Once a cyberinfrastructure is developed for ecological forecasting, it can be operated at a specific study site with any incorporated model(s). However, it will be desirable for such a cyberinfrastructure to be easily adapted to other ecosystems or to use other models, that is, to share the research resources across the ecological community. This will not only reduce redundant work repeating the whole process of constructing the cyberinfrastructure, but also leverage collaboration across the research community to foster innovation and promote success.

Presently, it is still challenging to adopt a new model by other users in the traditional cyberinfrastructure due to technical barriers. A critical step to overcome these technical bottlenecks is to design community-wide cyberinfrastructures.

THE STRUCTURE OF A COMMUNITY CYBERINFRASTRUCTURE

Figure 31.1 provides an example structure of a community cyberinfrastructure for data-model integration (Fer et al., 2021). First, a GUI is developed for users to set up a modeling activity. The selected settings for the modeling activity are then translated into a human and machine-readable markup language and read in by the master workflow to execute a sequence of modularized tasks. During this setup step, a unique identifier is assigned to the full workflow to be executed, which includes output of the results and access to the metadata required to repeat this modeling activity. This unique identifier can be published in papers and data products and shared with other collaborators.

Next, the user's selections are queried with the incorporated databases, including data from meta-analysis, for prior distributions of model parameters and for data assimilation to optimize model parameters. Actions to be executed at this step depend on whether the newly requested tasks have already been conducted in an earlier modeling activity. For example, if the tasks in the new request have been conducted previously, the system can retrieve the saved results to be used in this newly requested modeling activity.

DOI: 10.1201/9781032711126-39

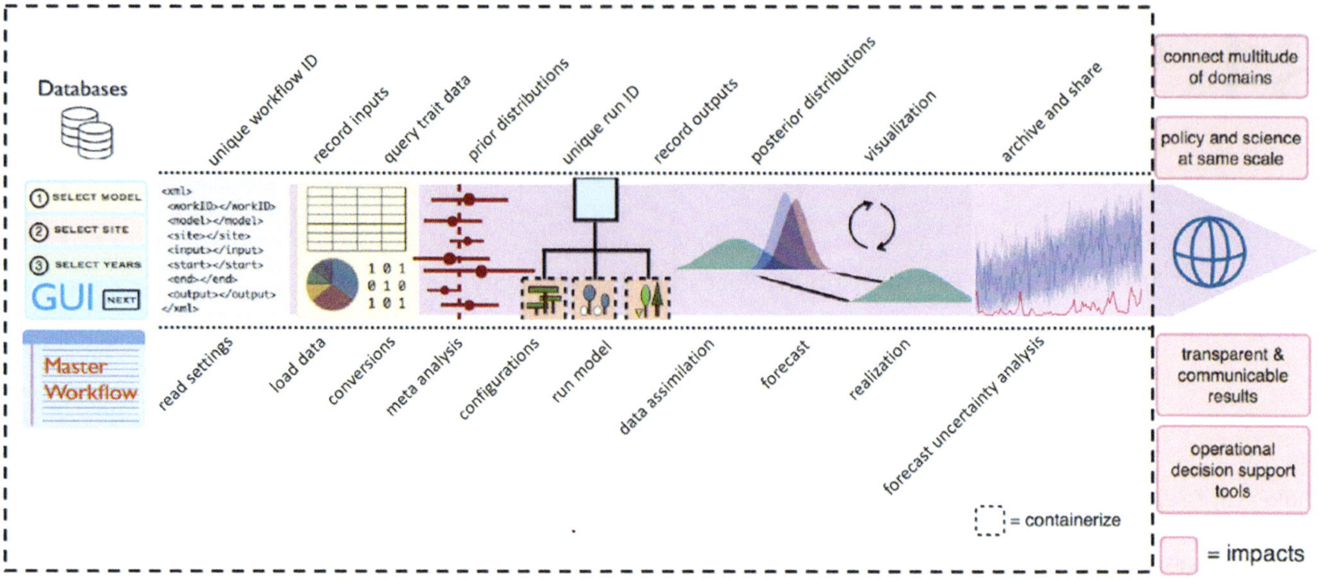

FIGURE 31.1 The schematic of a community cyberinfrastructure for ecological forecasting.

Modified from Fer et al., 2021.

The data assimilation module uses the databases incorporated in the system to find posterior distributions of model parameters after parameter optimization. The optimized parameters are then used to realize forecasting under different scenarios. The forecasting results are visualized for the user to interpret, and forecast uncertainty can be quantified. Key outputs, such as calibration posteriors with a specific database, are stored in a way that can be reused. Crucial information for each step of the workflow is also recorded along with the final modeling results. The forecasting results from such a cyberinfrastructure could be useful to inform policy as well as science.

THE KEY FEATURES OF A COMMUNITY CYBERINFRASTRUCTURE

The key feature of a community cyberinfrastructure is the FAIR essentials, which denote digital assets that are Findable, Accessible, Interoperable, and Reusable (Fer et al., 2021). Both research data and software should follow the FAIR essentials to allow replicability, reproducibility, and reusability to facilitate the ease of research practices. Findability is achieved by assigning a unique and persistent identifier to a research work, so that users are able to access all necessary data to reproduce the work. GitHub is a popular repository for this purpose. Accessibility is related to the functionality offered by a software system to reduce technical barriers in reproducing the results. For example, the workflow should be automated and provide users with an intuitive interface to handle all tasks. To achieve this, the key point is abstraction, that is, the system needs to standardize the tasks and schedule them to work in a specific order while users do not need to deal with all the details on how they are standardized and scheduled. Interoperability means that the cyberinfrastructure works seamlessly across different models. This requires consistency of data format, such as model inputs and outputs. Standardizing data format can also promote knowledge sharing across the community.

All the three essentials mentioned above are to achieve replicability and reproducibility. Reusability refers to the use of the software system or part of it for different purposes. It requires the tasks involved in the cyberinfrastructure to be isolated and modularized. A possible solution for realizing this function is to use virtualization techniques, such as Docker.

AVAILABLE CYBERINFRASTRUCTURES FOR ECOLOGICAL FORECASTING

Currently, there are some open-source cyberinfrastructures available for ecological forecasting or related purposes mentioned previously. Below we introduce such cyberinfrastructures.

DART is the Data Assimilation Research Testbed (https://dart.ucar.edu/). It is developed under the Data Assimilation Initiative at the National Center for Atmospheric Research (NCAR). The original motivation of DART was to develop methodologies, tools, and software to enable data assimilation capabilities within models developed at NCAR by providing a bridge between observationalists and modelers. DART helps model development by revealing model error and enables fundamental discoveries by allowing observationalists and modelers to share their expertise by combining models and data using a rigorous assimilation framework. DART has been growing beyond its original purpose and now supports data assimilation with models from both NCAR and a broader community in the geosciences.

EcoPAD is the Ecological Platform for Assimilating Data and Forecasting in Ecology (https://ecolab.cals.cornell.edu/?ecopad, Huang et al., 2019; see also Chapter 30). It focuses on inverse modeling and forward prediction and is supported by meta-databases of biogeochemical variables, libraries of modules of process models, and a toolbox of inversion techniques. EcoPAD components include: (1) core models that are specifically designed to solve ecological issues; (2) a variety of optimization techniques for data assimilation; (3) various databases that feed into forecasting and other tasks; and (4) functionality for a diversity of applications. The available functionality enables users to (i) estimate model parameters or state variables, (ii) quantify uncertainty of estimated parameters and projected states of ecosystems, (iii) evaluate model structures, (iv) assess sampling strategies, and (v) conduct ecological forecasting. EcoPAD has been successfully applied to ecological forecasting in the SPRUCE project (see Chapter 25) since 2016.

MIDA is a Model-Independent Data Assimilation module (https://zenodo.org/record/4762725, Huang et al., 2021). Accurate predictions of future states of terrestrial ecosystems depend on not only model structures but also parameterizations. Data assimilation is a robust statistical method to optimize model parameters, but its application in ecology is restricted by highly technical requirements such as model-dependent coding. MIDA is developed to alleviate this technical burden. MIDA works in three steps (see also Chapter 24). The first step is to prepare prior ranges of parameters, a defined number of iterations, and directory paths to access files of observations and models. The second step is execution, in which MIDA calibrates parameter values to best fit the observations and generates the posterior distributions of the parameters. The last step is to automatically visualize the calibration performance and posterior distributions. MIDA is model independent, meaning that modelers can use MIDA for data assimilation in a simple, interactive way without modification of their original model code. MIDA has been applied to conduct data assimilation with four types of ecological models: an ecosystem carbon model (DALEC); a surrogate-based energy exascale earth system model: the land component (ELM); nine phenological models; and a stand-alone biome ecological strategy simulator (BiomeE). These applications demonstrate that MIDA can effectively solve data assimilation problems for different ecological models. The easy implementation and model-independent feature of MIDA break the technical barrier of applications of data–model fusion and facilitate the assimilation of various observations into models to reduce uncertainty in ecological modeling and forecasting.

PEcAn is the Predictive Ecosystem Analyzer (https://pecanproject.github.io/), which is open-source and built using open source software (Dietze et al., 2013; LeBauer et al., 2013). The PEcAn project is motivated by the explosion in the amount and types of data in climate change science, which are valuable for addressing questions about the responses of the terrestrial carbon cycle and biodiversity to global change. The PEcAn project is designed to integrate existing data with process-based models to better understand the ecological processes behind patterns and trends in the available data. The PEcAn system includes three components to achieve data-model integration: (1) ecosystem models; (2) a workflow management system to handle the numerous streams of data; and (3) a data assimilation statistical framework in order to synthesize the data with the model. Currently, PEcAn supports over a dozen ecosystem models, including CABLE, CLM5, ED, FATES, JULES, and LPJ-GUESS (https://github.com/PecanProject/pecan/tree/develop/models).

PEST is a software package for Model-Independent Parameter ESTimation and Uncertainty Analysis (https://pesthomepage.org/). It automates calibration and calibration-constrained uncertainty analysis of numerical models, such as a synthetic groundwater model (i.e., the enhanced Freyberg model, https://github.com/usgs/pestpp; White et al., 2020), by interacting with the models through the models' own inputs and outputs. It runs a model many times while estimating or adjusting the model's parameters. Such model runs can be conducted either in serial or in parallel and the details of the tasks that have been done are recorded in output files to aid understanding. The software suite of PEST performs multiple tasks to assist and complement model parameter estimation and uncertainty analysis, which include: setup facilitation, flexible spatial parameterization, objective function definition, linear prior and posterior uncertainty analysis, and nonlinear prior and posterior uncertainty analysis.

LAVENDAR is the Land Variational ENsemble Data Assimilation Framework (https://github.com/pyearthsci/lavendar, Pinnington et al., 2020). LAVENDAR uses a four-dimensional ensemble variational (4D-En-Var) data assimilation method for land surface models. This method negates the costly calculation of a model adjoint required by traditional variational techniques (e.g., 4D-Var) for optimizing parameters or state variables. LAVENDAR has been applied to the Joint UK Land Environment Simulator (JULES) land surface model, recovering seven parameters controlling crop behavior. When applied to a field site with maize where the measured amount of harvestable material from the maize crop was available for data assimilation, LAVENDAR was able to accurately capture observations of leaf area index, canopy height, and gross primary productivity. As LAVENDAR requires no modification to the model used, it can be applied more easily than other data assimilation methods.

LIS refers to the Land Information System framework, which is a software suite for high-performance terrestrial hydrology modeling and data assimilation (https://lis.gsfc.nasa.gov/, Kumar et al., 2006; Peters-Lidard et al., 2007). It has been developed by the Hydrological Sciences Laboratory at NASA's Goddard Space Flight Center, with the goal of integrating satellite and ground-based observational data products and advanced modeling techniques to produce optimal fields of land surface states and fluxes. The software suite consists of three modeling components: (1) Land surface Data Toolkit (LDT), which is a formal environment that handles the data-related requirements of LIS including land surface parameter processing, geospatial transformations, consistency checks, data assimilation preprocessing, and forcing bias correction; (2) Land Information System (LIS), which is the modeling system that encapsulates physical models, data assimilation algorithms, optimization and uncertainty estimation algorithms, and high-performance computing support; and (3) Land surface Verification Toolkit (LVT), which is a formal model verification and benchmarking environment that can be used for enabling rapid prototyping and evaluation of model simulations by comparing against a comprehensive suite of in-situ, remote sensing, and model and reanalysis data products.

CHALLENGES AND SOLUTIONS IN BUILDING FUTURE COMMUNITY CYBERINFRASTRUCTURES

Some of the key challenges in building a community cyberinfrastructure to conduct data-model integration include data ingestion, model calibration, forecasting, and uncertainty analyses.

The first challenge is data ingestion. Data are crucial for ecological forecasting. However, it is usually difficult to find the needed observational data, including forcing data, and to use these data directly due to their diversity of formats. It is also difficult to maintain these data manually because of their massive volume and diverse formats. Therefore, standardization of data to be ingested is important to make data follow the FAIR essentials of community cyberinfrastructures. Standardization of data includes a consistent naming structure. For details about consistent naming structure, please refer to the Climate & Forecast convention (http://cfconventions.org). Using an open file format, such as csv or netCDF, is also an important standard. The netCDF format has many advantages for scientific data management; for example, supporting a multi-dimensional data structure. Standardization of data also confers some requirements for data repositories. For example, a dataset should have an associated version number or code, a citation should be provided, the metadata should be searchable, and an Application Programming Interface (API) should be available to enable automated retrieval of the data by external software, web platforms, and the like. Widely used data repositories include GitHub and DataONE. The bottleneck for big data relies on the efficiency of interacting with high-volume, high-velocity data. Some cutting-edge software, such as cloud computing, allows users to operate big data without downloading to local computers, which saves lots of time. The data ingestion process in EcoPAD is pipelined and modularized. Therefore, it is relatively easy to connect data ingestion with other simulation tasks in EcoPAD. Specifically, model inputs are environmental data measured at the SPRUCE site. Scripts have been written for EcoPAD to automatically download these measurements from a remote server every week, to conduct data filling and cleaning, and then to convert the data into standardized csv files to drive the model.

The second challenge in building a community cyberinfrastructure is model calibration, which is to integrate observational data into models to constrain model parameters, usually via data assimilation. Some parameters are able to be informed directly by observations, such as turnover rates. There are measurements of turnover rates available in open-access databases for constraining these parameters. There are also some meta-analyses to extract turnover rate data from

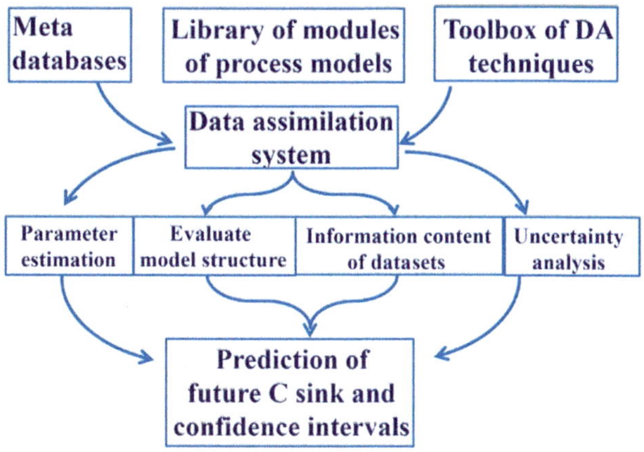

FIGURE 31.2 Data assimilation in EcoPAD.

Huang et al., 2019.

published results in the literature. Other parameters are not directly measurable. For such parameters, a Bayesian approach, such as MCMC, is often used to assimilate observational data into models to generate posterior distributions for these parameters. Traditional Bayesian tools (e.g., JAGS and STAN) cannot work with external black-box models. Users need to modify the source code of the models to incorporate Bayesian algorithms by themselves, creating large technical barriers. In addition, users have to repeat such coding work when a new model is needed or new observations become available. To overcome these challenges, we developed the MIDA module to release users from coding (Huang et al., 2021). MIDA enables seamlessly linking multiple observational datasets with different models. All these benefits facilitate the use of data assimilation techniques in ecology.

A key issue in calibrating models is that a calibration for one study site may not yield the same reliable performance at another site because the environmental conditions are different. Hierarchical Bayesian calibration allows parameters to vary, but not in a completely independent manner, between sites. Hierarchical Bayesian tools are able to quantify ecological variability and inform the direction of model development. Again, standardizing the calibration results will allow a wider pool of users to use the calibration results for subsequent analyses. Moreover, the standardization may also foster improvements of models, or constitute a step towards more advanced calibration methods.

The data assimilation system in EcoPAD is modularized (Figure 31.2), allowing different data assimilation techniques to be readily integrated. Other than estimating parameters, data assimilation in EcoPAD also supports multiple research purposes, such as testing alternative ecological hypotheses through different model structures, assessing information content of individual datasets, quantifying uncertainties, and so on.

Next, we will address the challenge of forecasting. Generally, ecological forecasting starts with the development of a model that is repeatedly executed to generate forecasts on a certain time horizon, for instance weekly. Forecasting relies on data acquisition to obtain real-time observations and to compare these observations with suggested output from the model, given certain settings or parameter values. These settings are then updated with the help of the observations through statistical analyses such as data assimilation, until a reliable forecast is achieved. The forecasting results are archived for future analyses and might be used in communication with policy-makers to inform their decision-making processes. When forecasting uses multiple models that differ in their representations of ecological processes, the relative success or failure of their forecasts helps advance our fundamental understanding of the ecological processes. Furthermore, forecasting performance can guide field experimentalists to refine data collection, prioritizing measurements of the greatest relevance to improving forecasting skill. A community cyberinfrastructure can automate the forecast cycle. Modelers, data collectors, and decision-makers can all benefit from the system in a cost-effective way.

The last challenge of community cyberinfrastructures is forecast uncertainty analysis. Once we have forecasting results, uncertainty in these forecasts needs to be quantified to tell us how confident we can be in acting on the forecast. Decision-makers may also be particularly interested in the 'worst-case' as opposed to the most likely or average property of a forecast, for instance when the focus is on protecting lives, property, or natural values such as biodiversity or carbon stocks. Information on the uncertainty surrounding the forecast can help with this. The uncertainty of the forecast stems from the uncertainty in relation to initial conditions, model structure, parameters, and external forcing (Luo et al., 2016; Dietze 2017; Bonan and Doney 2018). Uncertainty in initial conditions (i.e., initial values of state variables) can be estimated by running models to a steady state, for example, using a Semi-Analytic Spin-Up (SASU) method as described in Chapter 14. Different model structures can lead to divergent results, as often seen in model intercomparison projects. Traceability analysis as described in Chapter 17 provides a way to systematically diagnose model uncertainty propagating from each component or process. Benchmark analysis as described in Chapter 19 can also be used to evaluate model uncertainty in reference to observations. Forecast uncertainty arising from model parameters can be reduced by assimilating observational data to constrain models.

Uncertainty resulting from external forcing can strongly affect forecasting results, and can be the dominant uncertainty source in carbon cycles projections (Ahlström et al., 2013). Future forcing in ecological forecasting is often sampled from historical climate databases or derived from simulations by general circulation models (GCMs). The uncertainty caused by forcing is usually assessed by the spread of model outputs (e.g., C flux) with ensemble forcing, with the spread of model outputs indicating the contribution of forcing uncertainty or sensitivity of the forecast to the variations in forcing, and a reduced spread indicating well-constrained forcing uncertainty. However, such assessments do not compare forecasts with reality and, thus, do not directly evaluate forecast accuracy. Archiving past forecasted results allows them to be compared with realized measurements at the time point targeted by the forecast. 'Hindcasts' can also be undertaken, focusing on past periods for which measurements and forcing data are already available. As hindcast eliminates forcing uncertainty (i.e., it replaces the uncertain forcing from a GCM with real forcing), a comparison of model output between forecast and hindcast is an indicator of the uncertainty due to forcing.

SUGGESTED READING

Fer I, Gardella AK, Shiklomanov AN et al. 2021. Beyond ecosystem modeling: A roadmap to community cyberinfrastructure for ecological data-model integration. *Global Change Biology*, 27: 13–26.

QUIZ

1. What are the FAIR essentials for community cyberinfrastructures?
2. How do you think the available cyberinfrastructures for ecological forecasting can facilitate your research?
3. What are the major challenges in building community cyberinfrastructures for ecological forecasting?

32 Practice 8

Ecological Forecasting at the SPRUCE Site

Jiang Jiang
Nanjing Forestry University, Nanjing, China

This practice session aims to learn how to use EcoPAD for ecological forecasting at the SPRUCE experiment. The web portal of EcoPAD-SPRUCE provides automated ecological forecasting at a weekly time scale. Using CarboTrain, a training version of EcoPAD, this practice illustrates how constrained posterior parameters influence forecast uncertainty, how different forcings influence forecast uncertainty, and investigates the responses of ecosystem forecast to warming and elevated CO_2 with fully specified uncertainties.

INTRODUCTION

To generate a realistic projection of terrestrial carbon dynamics, we need to have three elements perfectly aligned (Luo et al. 2016). First, we need a good model structure, which represents underlying ecosystem processes. Then we need a good parameterization method, such as data assimilation discussed in Units 6 and 7, to estimate parameter values. Finally, we also need a workflow system to link real-time forcings to the model. In this practice session, we will focus on the third step, using CarboTrain to link forcing data to carbon cycle models that usually need climatic forcing to drive the projections.

For this purpose, we developed an Ecological Platform for Assimilating Data into the model (EcoPAD). EcoPAD is described in more detail in Chapter 30. Traditionally, modelers usually tune a model, validate the model with data, and then generate model output as prediction or forecast. This is called forward modeling. In addition to forward modeling, EcoPAD can perform data assimilation and ecological forecasting. EcoPAD links data and model via data assimilation to optimize parameter estimation. The system can update parameter estimation when new data becomes available to generate real-time forecasting. From an ensemble of parameters, EcoPAD can also produce an ensemble of forecasts, instead of just one projection. This method allows us to quantify the uncertainty of forecasting. The insights we get from this sort of analyses can provide feedback to experimenters on which data sets are needed to further improve model predictions; and feedback to modelers on which parts of a model need to be improved.

EcoPAD requires two categories of datasets, observation data and forcing data, to be able to do forecasting. Observation data is used to optimize the model parameters through data assimilation. The observations might comprise inventory data, laboratory measurements, or high temporal resolution data such as flux tower measurements from the FLUXNET database. The other category of data sets is used as forcing to drive a model. The temporal resolution of forcing data should be same as the time steps of the model used for projections, and the time span should also be same as the future projections.

In this practice, EcoPAD is applied to the SPRUCE experiment located at the USDA Forest Service Marcell Experimental Forest in Northern Minnesota. The SPRUCE project was introduced in Chapter 25. It is an ongoing project that focuses on long-term responses of northern peatland to climate warming and increased atmospheric CO_2 concentration. The SPRUCE project generates a large variety of observational datasets that reflect ecosystem dynamics from different scales. These datasets are available from the project web page and file transfer protocol (FTP) site, which provides easy integration with EcoPAD.

DATASET PREPARATION FOR ECOPAD

In EcoPAD-SPRUCE, observational datasets were pulled from SPRUCE archives and stored in the EcoPAD metadata catalog for running the TECO model and conducting data assimilation. The forcing datasets to drive TECO in EcoPAD are hourly climate data, which can be separated into two parts: one is past climate and the other is projection of future climate. We pulled past climate data from the SPRUCE FTP site weekly, and generated an ensemble of future climate using vector auto-regressive modeling. Observational data could be updated any time when available. In this training version, however, we use pre-treatment datasets from 2011 to 2014 to investigate how constrained posterior parameters influence forecasting uncertainty. We then go on to partition the uncertainty sources.

Pre-treatment observations include three data sets of community-scale flux measurements (GPP, NEE, and ecosystem respiration) in 1.2-m internal diameter chambers, six data sets of plant biomass growth and carbon content (foliage, wood, and root), one data set of carbon in peat soil, and leaf phenological data. During 2011–2014, CO_2 flux observations were collected monthly during the growing season at ambient plots in the experimental site. A total of 30 data measurements were collected in August, September, and October 2011; May through November 2012; July, September, and October 2013; and June and July 2014. Three annual data points from 2012 to 2014 for plant foliage, woody biomass, and aboveground net primary production were estimated from inventory data. Biomass data were compiled by combining

DOI: 10.1201/9781032711126-40

allometric data for shrubs, all ground layer species, and trees. Only one data point was collected each for fine-root and peat soil C. We collected leaf-out dates, which corresponded to the date growing degree-days above a threshold in the TECO model. The standard deviations reported in these data sets were also compiled to estimate uncertainties for each data stream. In CarboTrain, the observational data files can be seen in the '\CarboTrain\Source_code\TECO_2.3\input\' folder.

External forcing of the TECO model includes hourly climate data of photosynthetically active radiation (PAR), air temperature, soil temperature, precipitation, relative humidity, vapor pressure deficit, and wind speed. We took records of 2011–2014 from the weather station of the experimental site, and then generated an ensemble of 300 trajectories of 10 year forcing variables from 2015 to 2024 as inputs for the forecasting period. Precipitation and air temperature were generated by vector auto-regression (VAR) using the 1961–2014 data set using the R package RMAWGEN.

Three of the meteorological drivers: PAR, relative humidity, and wind speed, were coupled and randomly drawn from frequency distributions at a given hour each month. Specifically, we constructed a 24 (hours) by 12 (months) matrix field. Within each field, a coupled pool of the three drivers was obtained from the weather station data during 2011–2014. We then generated PAR, relative humidity, and wind speed for 2015–2024 by resampling the set of drivers from the pools. Soil temperature at 20-cm depth was scaled from the generated air temperature based on a linear regression between soil temperature and air temperature from 2011 to 2014. Vapor pressure deficit was calculated from the difference between the saturation vapor pressure and actual partial pressure of the water vapor in the air, which was computed by multiplying relative humidity with the saturation vapor pressure. The generated climate data were archived in the EcoPAD metadata catalog. For each operational forecast, the system updates climate data streams from the SPRUCE FTP site to replace stochastically generated forcing.

PRACTICE WITH CARBOTRAIN AT THE SPRUCE SITE

Exercise 1: How Constrained Posterior Parameters Influence Forecasting Uncertainty?

Open the user interface of 'CarboTrain', and select the Task 'Unit 8 DA'. It will allow you to set initial parameters, and parameter ranges for data assimilation. Once you click 'Run Exercise', the system will run data assimilation to constrain parameters and then undertake forecasting. It may take a long time to run data assimilation depending on your computer performance. You can find all the results in the user defined output folder.

Here are the steps to generate different sets of constrained parameters:

a. Select Unit 8 → DA → Set DA output directory (e.g., mydir/unit8/DAoutputs1) → Click 'Run Exercise' (It may take a few hours depending on your computer performance and number of iterations).
b. Find your DA output files in 'mydir/unit8/DAoutputs1/DA'. You can find the constrained parameter values in 'Paraest.txt'.
c. Select Unit 8 → DA → Set DA output directory (e.g., 'mydir/unit8/DAoutputs2') → click 'select DA pars' and set the parameters that participated in data assimilation ('1' signifies that the parameter is selected for data assimilation).
d. Repeat steps a to c for different combinations of DA parameters being selected.

Here are the steps to do forecasting with different DA outputs:

a. Select Unit 8 → Forecast → Set DA folder → Set Forecast output folder (e.g., 'mydir/unit8/forecast1') → Click 'Run Exercise'
b. Find your output files in 'mydir/unit8/forecast1/forecasting'. You can find a repetition of 100 times daily output data in the same folder.
c. Repeat step a with other DA outputs.
d. Plot the output data from your forecasting runs, and compare the differences.

Questions:

- Do the trajectories of forecasting results match the realized data? What's the difference between forecasting results among different variables?
- Do the forecasting results depend on parameter posterior distributions? How do the forecasting results change with constrained parameters?
- What is the difference between forecasting results when using different constrained parameters, comparing pool-based parameters and flux-based parameters?

A unique feature of the data assimilation portal is that users can pick whatever parameters they wish to be constrained among the pool of 18 parameters. Users can change the range of parameters they are interested in and modify the initial values of parameters to be used for MCMC. Figure 32.1 shows an example of forecasting using posterior parameters when all the 18 parameters were chosen to do data assimilation.

Exercise 2: How Does Different Forcing Influence Forecasting Uncertainty?

External forcing variables such as temperature, precipitation, and light regulate various aspects of carbon cycle processes (e.g., plant photosynthesis, water use, and soil carbon decomposition), and therefore influence carbon stocks in different compartments of the ecosystem. The challenge of precisely predicting the future state of an ecosystem is partly due to low predictability of future trajectories of forcing variables. To make the task simple, CarboTrain provides

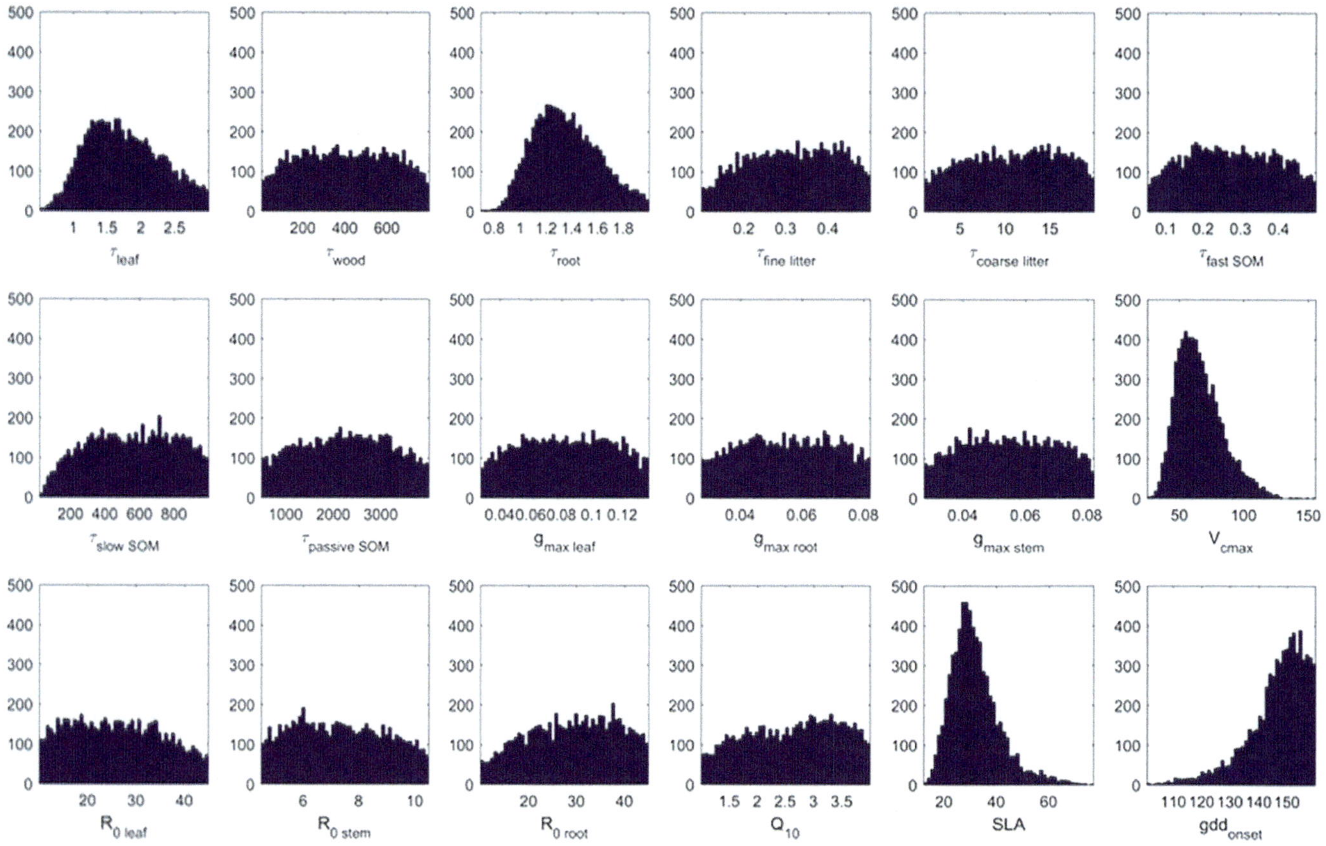

FIGURE 32.1 Histogram of the posterior distribution of each parameter that produced by data assimilation.

two sets of forcing variables: one is fixed forcing, the other, random forcing chosen from a forcing pool.

a. Select Unit 8 → Forecast → Set DA folder (e.g., 'mydir/unit8/DAoutputs1/DA') → Set Forecast output folder (e.g., 'mydir/unit8/forecast5') → change 'Times' to 1 → click 'Run Exercise'
b. Repeat step a with 'Times' set to 100.
c. Plot the output data from your forecasting runs, and compare the differences.

Questions:

- What is the difference between forecasting results when using different forcing files, comparing fixed forcing and random forcing?

Exercise 3: Specifying Uncertainty in Forecasting Ecosystem Responses to Warming and Elevated CO_2

Forecasting is not very informative without fully specified uncertainties. It is especially useful if we try to anticipate when carbon pool changes are expected to be significantly different between temperature or CO_2 treatments after the SPRUCE experiment started. It is because it takes time for the carbon pools to accumulate over time under different treatments. In general, if the magnitude of uncertainty in the forecast of one pool is small, the statistical power to detect the treatment effects on the pool will be large. And the time points at which statistical difference can be observed among treatments is short after the treatment, and vice versa.

a. Select Unit 8 → Forecast → Set DA folder (e.g., 'mydir/unit8/DAoutputs1/DA') → Set Forecast output folder (e.g., mydir/unit8/forecast6) → click 'Run Exercise'
b. Repeat step a for different combination of warming treatment (+0, +2.25, +4.5, +6.75, +9.0 °C) and CO_2 fertilization treatment (380 and 900 ppm)
c. Plot the output data from your forecasting runs, and compare the differences.

Questions:

- How do the ecological carbon dynamics respond to warming and elevated CO_2?
- How do the uncertainties of the forecasting vary with treatments?

Figure 32.2 illustrates that both warming and elevated CO_2 increased carbon stock in foliage, wood, root, and soil. Plant compartments had stronger responses to either warming or elevated CO_2 than soil carbon. Differences of response between ambient and elevated CO_2 were more likely significant at higher temperature treatments. Each scenario contains 300 simulations with sets of coupled parameters randomly chosen from accepted parameters by data assimilation and the forcing trajectories from the stochastic weather generator.

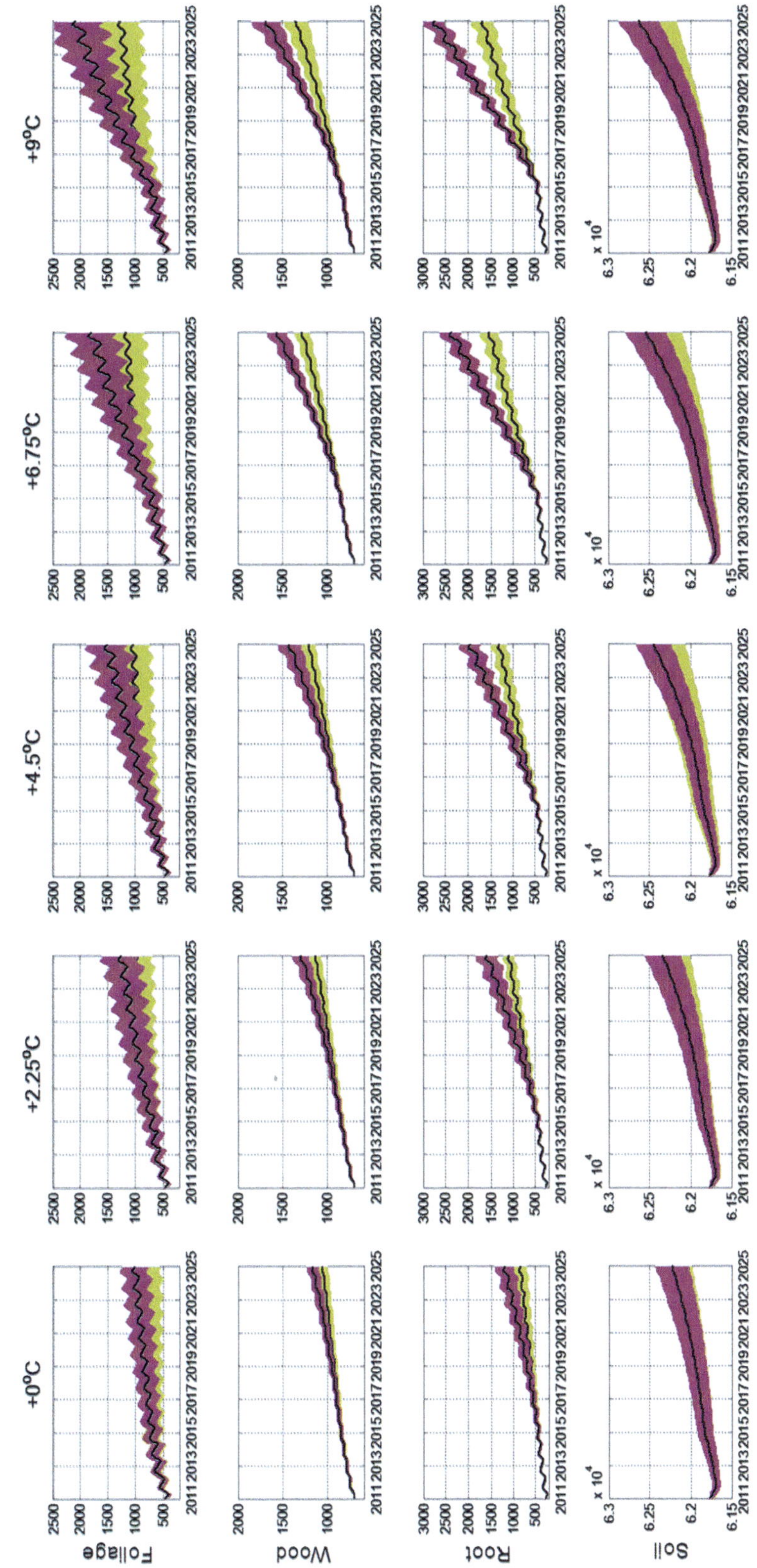

FIGURE 32.2 An example illustrates effects of warming and elevated CO_2 on carbon stock in foliage, wood, root, and soil.

Jiang et al. 2018.

Unit Nine

Machine Learning and its Applications to Carbon Cycle Research

33 Introduction to Machine Learning and its Application to Carbon Cycle Research

Yuanyuan Huang
Institute of Geographic Sciences and Natural Resources Research, Chinese Academy of Sciences, Beijing, China

Machine learning is increasingly being applied in multiple disciplines. This chapter serves as an introduction to some basic concepts related to machine learning. The goal of this chapter is to lead you to understand what machine learning is, the different types of machine learning algorithms, and the applications of machine learning in studies of the carbon cycle or mitigating greenhouse gas emissions.

WHAT IS MACHINE LEARNING?

Machine learning (ML) is a sub-class of artificial intelligence in which we enable computers to learn from data, to develop pattern recognition, and to continuously learn and make predictions or decisions based on data. Data is essential for ML, but data alone is not ML. ML algorithms normally have strong mathematical and statistical bases. We provide the computer with a large amount of data and a way (through algorithms or models) to process and learn from that data. The computer then uses this data to identify patterns, relationships, and interpretations, which can be used for predictions or decisions on future states of a system based on states revealed within the realm of training data. A general introduction to ML methods is presented in Chapter 37.

TYPES OF ML ALGORITHMS

There are generally two categories of machine leaning: unsupervised learning and supervised learning. The difference is whether the labeled output variable is involved or not. Unsupervised learning corresponds to inferring underlying patterns from an unlabeled dataset without any reference to labeled outcomes or predictions. Clustering is a typical unsupervised machine learning method. When we plot a simple set of (x,y) data on a two-dimensional plot, data points with similar characteristics are automatically grouped together. Likewise, through ML we can identify different clusters, but in many more than two dimensions if we wish. We can use clustering, for example, to detect anomalies (e.g., in weather patterns or in time series of carbon fluxes relative to average conditions), and to study plant traits or genetic similarities between species populations.

Supervised learning is the machine learning process of learning a function that maps an input to an output based on examples of input-output pairs. It infers a function (or relationship) from labeled training data consisting of a set of training examples. For example, if we make predictions of house price based on data of historical house price and influencing factors, we call this supervised learning. If the training or target samples use category values, we call this type of task classification. If the target has continuous numerical values, we refer to this task as regression.

EXAMPLES OF ML ALGORITHMS

Machine learning is evolving very quickly, with a wide range of methods designed to solve different practical issues. This chapter will briefly introduce two commonly used methodologies, decision tree-based and neural network-based methods.

A **decision tree** is a hierarchical modeling technique that uses a tree-like structure to represent decisions and their possible outcomes. A decision tree consists of three components: root node, decision node, and leaf node. A decision tree algorithm divides a training dataset into branches. The branches can be further separated into child branches and so on until a leaf node that cannot be separated further. Information gain metrics like entropy (or Gini Impurity) can identify gaps or separations in the data to split branches. A high information gain means that a high degree of uncertainty (information entropy) has been removed. The idea is to assign branches in such a way that the information contained in the data of each branch is as high as possible. To illustrate how this works, take the case of predicting if a customer will purchase a house (Figure 33.1). The features or characteristics of the purchase opportunity – such as the price and size of the house, and how safe the neighborhood is – form the basis of the decision. The root node and decision nodes of the decision tree represent these features. The leaf node represents the final output, either buying or not buying (Figure 33.1a).

Now that you know what a decision tree is, we will introduce the Random Forest (RF), a commonly used ML technique in the geosciences. RF is made up by multiple decision trees (Figure 33.1b). Conducting RF involves the following steps.

1. We typically create bagged samples of size n from the original dataset, with n being smaller than the original

DOI: 10.1201/9781032711126-42

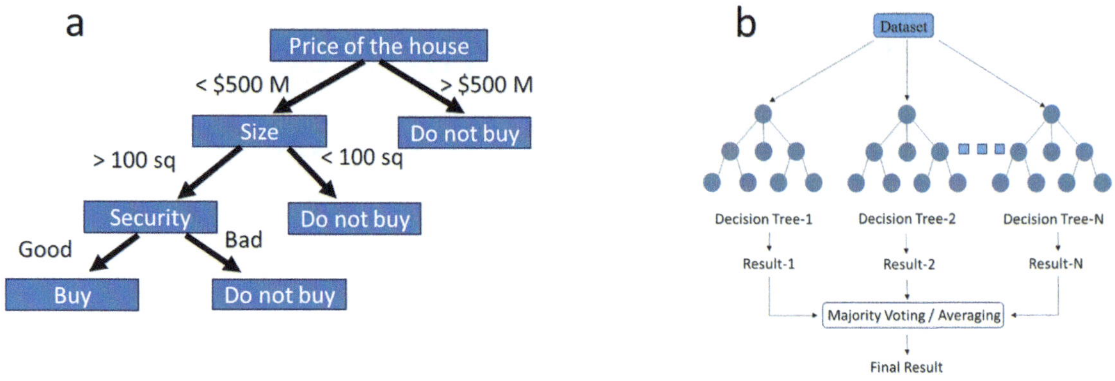

FIGURE 33.1 Illustration of a decision tree (a) and Random Forest (b).

Carolina Bento and Chirag Goyal.

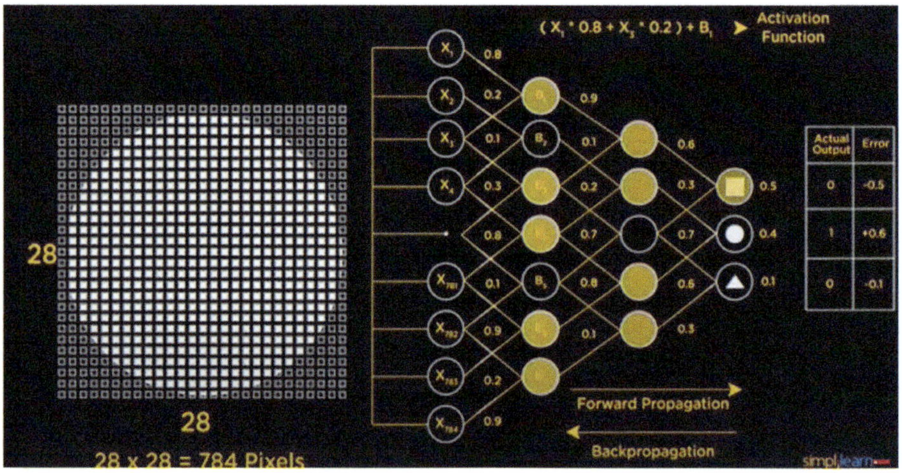

FIGURE 33.2 Illustration of a neural network.

Adapted from Simplilearn (https://www.simplilearn.com/).

size of the dataset. Bagging involves using different samples of training data rather than just one sample.

2. Afterwards we train a decision tree with each of the bagged datasets as inputs. But, when we do a node split, we do not use all features in the dataset. Instead, we randomly select a smaller number from all features in the training set. Then we pick the best split using an information gain measure, like Gini Impurity or entropy.
3. We aggregate the results of the individual decision trees into a single output. If we are working on a regression task, we average the values for each observation produced by each tree. If we are working on a classification task, we do a majority vote across all trees for each observation.

These steps help to avoid overfitting issues that could arise if instead we directly developed one decision tree using all the data.

An **artificial neural network (ANN)** is a series of algorithms that mimics the way the human brain operates to recognize underlying relationships in a dataset. Neural networks refer to systems of neurons. A simple neural network includes an input layer, an output (or target) layer and, in-between, a hidden layer which performs most of the computations required by our network. The layers are connected via nodes, and these connections form a "network" – the neural network – of interconnected nodes. Neural networks with more than three layers are known as "deep" networks and are used for deep learning algorithms. "Deep" in deep learning refers to the presence of multiple hidden layers in neural networks. Deep learning has more complex network architecture, which enables hierarchical feature learning, increased model capacity and performance. It also brings challenges such as the requirements for a large dataset and high computational resources.

To explain how neural networks work, we will take an example from Simplilearn (Figure 33.2). Suppose we are going to predict the shape of an image (a square, circle, or triangle). The image is made up of 28 × 28 = 784 pixels (Figure 33.2). The pixel values (either 0 or 1) are used as the input to the first layer of neurons. Neurons of the first layer are interconnected with neurons of the second layer through channels. Each channel is assigned a numerical value known as the weight. The input is then multiplied by the weight and the sum is then used as input to the neurons of the hidden layer. In addition, each of the neurons is associated with a numerical value, called the bias (B), which is then added to the input sum. This value is then passed to a threshold function, called the activation function. The result of the activation function determines if the neurons will be activated or not. The activated neurons pass data to the neurons of the next layer over the channels. In this way, data is propagated through the network, which is called forward propagation. In the output layer, the neuron with the highest value determines the output. The value in this example is probability. For example, the value of square has the highest probability in the illustration. So, square is the predicted output. In the example case, however, our input is a circle. Here comes the training processes we have not done yet. In the training, our predicted output is compared against the real output to quantify the error in prediction. The computed errors give the direction and magnitude of change to reduce the error. The error information is then transferred backward through the network. This process is called back-propagation. The weights are adjusted accordingly. The forward and backward propagations are conducted iteratively until weights are determined with which the network can predict the shape correctly in most tests. Depending on the complexity of the data, neural networks may take hours or even months to train. With the quick advancement of this area, different variations of ANN, such as convolutional neural network and recurrent neural networks, have been designed to fit different computation requirements.

APPLICATIONS OF ML TO CARBON CYCLE RESEARCH

Applications of ML to carbon cycle research or studies of greenhouse gas (GHG) emission mitigation are widespread and advancing quickly. You can find ML studies in almost every corner of carbon-related science as long as abundant data are available. For example, Rolnick et al. (2023) summarized how ML can be used for understanding the land carbon cycle, reducing GHG emissions and helping society adapt to a changing climate. ML is used in precision agriculture, remote sensing, quantifications of carbon stocks and fluxes, monitoring peatlands, and forest management among many other application areas. In Industry sectors, ML is used for reducing GHG emissions through, for example, optimizing supply chains, adaptive control, detecting GHG emissions, optimizing shipping routes, and preventing overstocking. To help reduce transportation emissions, ML can for example be applied to analyzing and understanding transportation patterns and drivers, forecasting, interpreting remote sensing data, detecting loading inefficiency, automating vehicles, understanding and scheduling charging patterns, or managing congestion.

Table 33.1 synthesizes typical applications of ML to GHG emission mitigations. These applications tackle issues related to monitoring, budget quantifications, upscaling, mapping, optimization, scheduling, management, forecast, etc. Topics involve causal inference, computer vision, interpretable models, natural language processing, reinforcement learning, time series analysis, transfer learning, uncertainty quantification, and unsupervised learning. Computer vision is by far where ML is most widely used at present.

After this broad overview, let's dive into some specific case studies. The following sections discuss how ML can be applied to quantify carbon fluxes and stocks with examples of mapping and predictions. We will also examine how ML can help identify drivers of carbon-cycle relevant processes and improve the predictive skill of process-based carbon models.

In the land carbon cycle literature, you can find various maps generated from ML models trained with site level samples, such as gross primary production (GPP), aboveground biomass, belowground biomass, and soil respiration.

To do mapping or upscaling, typically, we first need to spend a considerable amount of time preparing our samples, standardizing target variables (e.g., GPP) and covariates. In ML terminology, a covariate is also called a feature. Covariates are variables that share an association with our target variable, affecting (or determining) its value in the ML prediction. For GPP, covariates might include climate variables, soil conditions, vegetation traits, and so forth. Depending on the understanding of our target, we select predictors, features, or, in other words, covariates that are relevant to our target. We apply ML to understand the relationships, especially non-linear relationships among our predictors and targets. This is called training the machine learning model. After training, and if model evaluation reveals satisfactory performance, we can then use the model to do predictions or mapping. The spatio-temporal extent of our prediction depends on the predictors. A number of global gridded datasets can be useful as predictors in carbon cycle work. These include FluxCom (Jung et al., 2020), SoilGrid (Hengl et al., 2017), MODIS land products, digital elevation models (DEM), and climate datasets. When these global predictors are used as inputs for the trained ML models, we can generate global estimates of GPP. To account for uncertainty of the predictions or mappings, multiple ML algorithms and different predictor datasets can be used to generate an ensemble of predictions.

To evaluate the trained ML model, we conduct cross validation, for example, K-fold cross validation, or leave-one-out cross validation when sample size is small. In K-fold cross validation, for example, suppose $K = 5$. We divide all our samples into five folds (five blocks, ideally equal in the size of samples). We use the first four folds of samples to train our ML model. The fifth fold of samples is used to evaluate the performance of the trained model. In a second round, we use the first, second, third, and fifth folds of samples to train the model, and the fourth fold to evaluate the model performance. We repeat these steps five times in total. In this way, we use all the samples that are not used to train the model to evaluate the model to avoid overfitting. ML models that perform well on the training samples may not be good at predicting samples that are not part of the training samples. An ML model that is good at predicting new samples is what we need. Cross-validation provides an approach to verify the performance of the ML model, increasing our confidence in its ability to make reliable predictions for new samples. Leave-one-out cross validation can be viewed as a special case of K-fold validation (K being the total number of samples) and takes similar procedures as the K-fold validation.

TABLE 33.1
Typical applications of ML to greenhouse gas mitigation

		Causal interference	Computer vision	Interpretable models	NLP	RL & control	Time-series analytics	Transfer learning	Uncertainty quantification	Unsupervised learning
Mitigation	**Electricity systems**									
	Enabling low-carbon electricity		•	•		•	•		•	•
	Reducing current-system impacts		•				•		•	•
	Ensuring global impact		•					•		•
	Transportation									
	Reducing transport activity		•				•		•	•
	Improving vehicle efficiency		•			•				
	Alternative fuels and electrification					•				•
	Modal shift	•	•				•		•	
	Buildings and cities									
	Optimizing buildings	•				•	•	•		
	Urban planning		•				•	•		•
	The future of cities				•			•	•	•
	Industry									
	Optimizing supply chains		•			•	•			
	Improving materials									•
	Production and energy		•	•		•				
	Farms & forests									
	Remote sensing of emissions		•							
	Precision agriculture		•			•	•			
	Monitoring peatlands		•							
	Managing forests		•			•	•			
	Carbon dioxide removal									
	Direct air capture									•
	Sequestering CO_2		•						•	•

Source: Adapted from Table 1 of Rolnick, et al. (2023).

As mentioned above, one typical issue in ML is overfitting, especially when sample size is not big enough. An overfit model performs well on the training data but poorly on the new data (e.g., test data). For some variables related to the land carbon cycle that rely on manual measurements, small sample size is typical. Cross-validation and sometimes careful feature selections (i.e., not using too many different predictors) can help to avoid overfitting issues. Autocorrelation, as discussed in Ploton et al. (2020), and uncertainty estimation are also issues that remain to be adequately addressed in many studies. Whether and how autocorrelation should be handled in ML-based spatial mapping are widely debated.

Another important application of ML within land carbon cycle research is to aid in understanding drivers of different land carbon cycling processes. We normally plot partial dependence plots or individual conditional expectation plots to make inferences on the relationship between drivers and our predicted variables (see also Chapter 35). This is illustrated in the study on the response and recovery of tropical forest biomass to the 2015/16 drought by Yang, et al. (2022), depicted in Figure 33.3. The red lines show the response of AGB changes during the drought to the most important feature variables, listed in order of importance in the bar chart on the left. If only the red lines are plotted, we call this partial dependence plots, which show the marginal effect one feature has on the predicted outcome of a machine learning model across samples. In addition to the red lines, many individual black lines are shown. Now the plots become individual conditional expectation plots. Each black line refers to one sample, or one site. For each sample, covariates other than soil moisture are held constant at the site-level value. The black line shows, for each sample, the response of the target variable we are predicting (here ΔAGB) to variations in the feature variable it is plotted against.

Shapley values also help to move ML from a black box to explainable ML. Shapley values are a concept borrowed from the cooperative game theory. They are very powerful and commonly used in ML. Let's say we would like to predict GPP from four features: air temperature, precipitation, photosynthetically active radiation (PAR), and soil nitrogen content. Say that in one of our samples with air temperature 20°C, precipitation 500 mm, PAR 2000 MJ m^{-2} yr^{-1}, and soil nitrogen content 30 mg N kg^{-1}, our model predicted a GPP of 2 g C m^{-2} day^{-1}. Shapley values answer how much of each of the features contribute to the difference between the prediction of 2 g C m^{-2} day^{-1} and the average of the predictions across all samples. Let us suppose the average is 5 g C m^{-2} day^{-1}. Below are steps on how to calculate Shapley values,

1. Create the set of all possible feature combinations (called coalitions).
2. Calculate the average model prediction using all coalitions.
3. For each coalition, calculate the difference between the model's prediction without a specific feature (we call it F here) and the average prediction.
4. For each coalition, calculate the difference between the model's prediction with F and the average prediction.
5. For each coalition, calculate how much F changed the model's prediction from the average (i.e., step 4 – step 3) – this is the marginal contribution of F.
6. Calculate the Shapley value as the average of all the values calculated in step 5 (i.e., the average of F's marginal contributions).

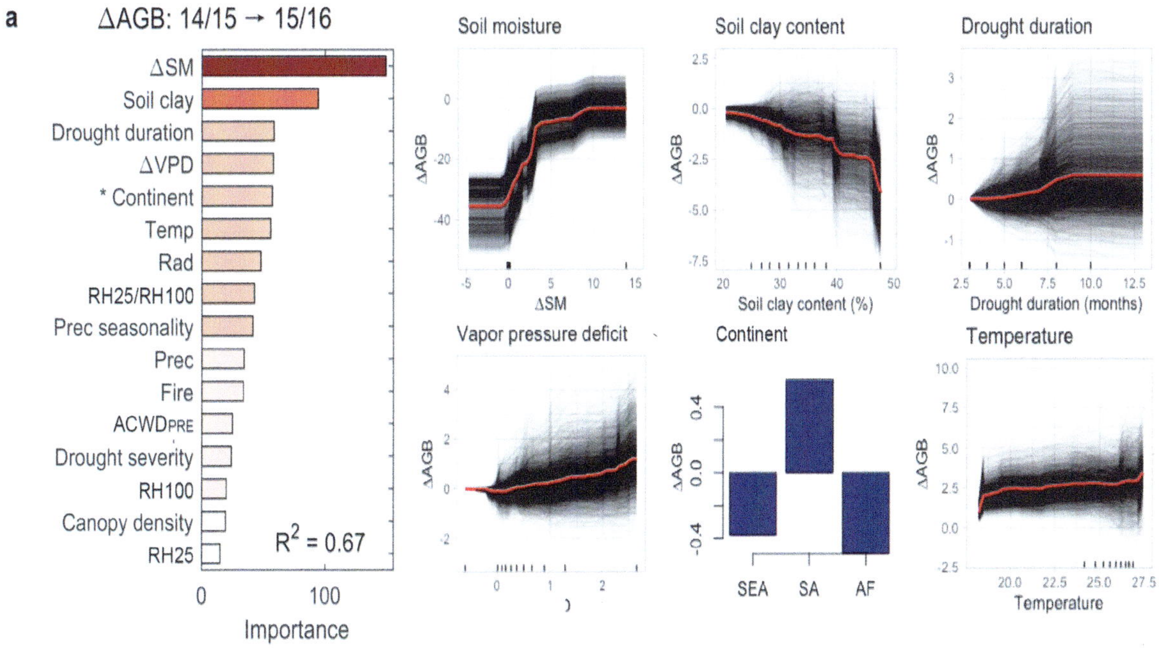

FIGURE 33.3 Illustration of feature importance, partial dependence plot and individual conditional expectation plot.

Adapted from Yang et al., (2020).

There are different ways to plot the Shapley values, which can give detailed information on the contributions of each feature for each sample, or rank the relative importance of features across all samples, or the response curve of the target to a specific feature.

ML can be used to improve process-based carbon models in different ways. Let us consider two examples. In the first example, ML is combined with data assimilation, process-based models, and big data to improve process-based predictions in soil carbon (Tao et al., 2020). Chapter 38 gives a more detailed explanation. The second example uses ML to accelerate the spin-up of a large-scale process-based model. The idea is, instead of running the model for spin-up across the whole global grid, we train a machine learning model to predict the steady state based on selected representative samples. Such a study undertaken with ORCHIDEE model showed that the performance of ML is very good at predicting the steady state, with a cross-validation R^2 of more than 0.9 (Sun et al., 2023). So, instead of running computationally expensive global samples, we can achieve a reliable spin-up state using much smaller grids with the help of ML. In the future, there will be more creative ways to combine ML algorithms to advance land carbon studies with your contribution.

SUGGESTED READING

Rolnick, D. *et al.* (2023). Tackling Climate Change with Machine Learning. *Acm Computing Surveys* **55**. doi:10.1145/3485128

QUIZ

1 What are two types of machine learning according to whether or not labeled training data is used?
2 What is cross-validation? Could you give any examples of cross-validation?
3 Could you give an example of applications of machine learning in carbon cycle research?

You can read more about neural networks and how they function in Chapter 37.

34 Estimation of Terrestrial Gross Primary Productivity Using Long Short-Term Memory Network

Yao Zhang and Jinghao Qiu
College of Urban and Environmental Science, Peking University, Beijing, China
Institute of Carbon Neutrality, Peking University, Beijing, China

This chapter introduces an innovative technique for upscaling Gross Primary Productivity (GPP) from eddy covariance measurements using the Long Short-Term Memory (LSTM) network. Leveraging solely meteorological forcings as model inputs, LSTM demonstrates superior performance on spatial extrapolation compared to artificial neural network (ANN) approaches. This is achieved by using climate information of the past to predict hidden variables of ecosystem states.

INTRODUCTION

Over the past decades, various machine learning (ML) architectures have been applied to upscaling in land carbon studies, such as artificial neural networks (ANN) (Papale & Valentini, 2003), support vector regression (Ueyama et al., 2013), model tree ensemble (Jung et al., 2009), random forest (Bodesheim et al., 2018), Long Short-Term Memory (LSTM) (Liu et al., 2023) and attention-based temporal fusion transformer (Nakagawa et al., 2023). While all these ML methods are powerful for predicting carbon fluxes (Tramontana et al., 2016), most of them process input features at a specific time to generate output for the concurrent time. For example, GPP can be predicted as a product of photosynthetically active radiation absorbed by green leaves and a light use efficiency factor which is modulated by environmental stress (Zhang et al., 2017). These relationships can be easily captured by machine learning models. However, when satellite greenness is not used as an input, the performance of the model declines considerably as current vegetation greenness is partly a result of past environment which determines plant phenology and green-up speed. Similarly, soil water stress depends on precipitation and potential evapotranspiration over a period of time before present. Plant resistance to stress may also depend on the history of past stress, affecting plant carbohydrate status, acclimation processes, etc. Some of these variables may not be directly observable or easily obtained, but may be predicted based on past information (Thornton et al., 2002). As such, models that do not consider the temporal dependencies may inadvertently overlook critical relationships inherent in the data, thereby potentially undermining the models' performance.

To achieve a more accurate depiction of the terrestrial carbon cycle process, it is crucial to incorporate time-aware (informed by temporal sequences) models to unravel the latent temporal linkages embedded within the training data. Recurrent neural network (RNN) is a class of neural networks designed to handle dataset of temporal sequences (Elman, 1990). RNN derives hidden state information from one time step and allows it to be passed to the next, such that the model output at each step is influenced by inputs of previous steps. As a special type of RNN, LSTM addresses the issues of vanishing or exploding gradients by incorporating the "gate" structure. These "gates" control the information flow into and out of the hidden state, i.e., a vector that stores information and is updated at each time step, so that the hidden state information can better represent the dynamics of the system.

This chapter introduces the LSTM as an approach to upscale site-level GPP to global scales. We will briefly explain basic LSTM network structure, describe LSTM workflow for GPP upscaling, use two examples to show the performance of LSTM, and make concluding remarks.

LSTM is a time-aware model that leverages timewise dynamic input features to estimate target features (Hochreiter & Schmidhuber, 1997). By simultaneously utilizing the long-term and short-term information inherent in time series data, LSTM has the potential to be an effective tool for carbon flux upscaling.

THE STRUCTURE OF LSTM

The fundamental structure of LSTM is characterized by its internal memory cell (c), hidden state (h), and three gates: the input gate (i), forget gate (f), and output gate (o). These elements are key to the operation of the LSTM (Figure 34.1a), as they enable it to selectively store, update, or export the information flow. LSTM has two elements (memory cell and hidden states) to store the latent information flow through time. The difference between these two is that the cell state is less variant and can store long-term memory; while hidden state is strongly influenced by short-term fluctuation of the input data, so it mostly stores the short-term memory. By integrating the current input (x[t]) with the preceding hidden state and subsequently applying a sigmoid activation function, the gates are like valves which generate a value ranging between 0 and 1 to control the information flow in LSTM.

DOI: 10.1201/9781032711126-43

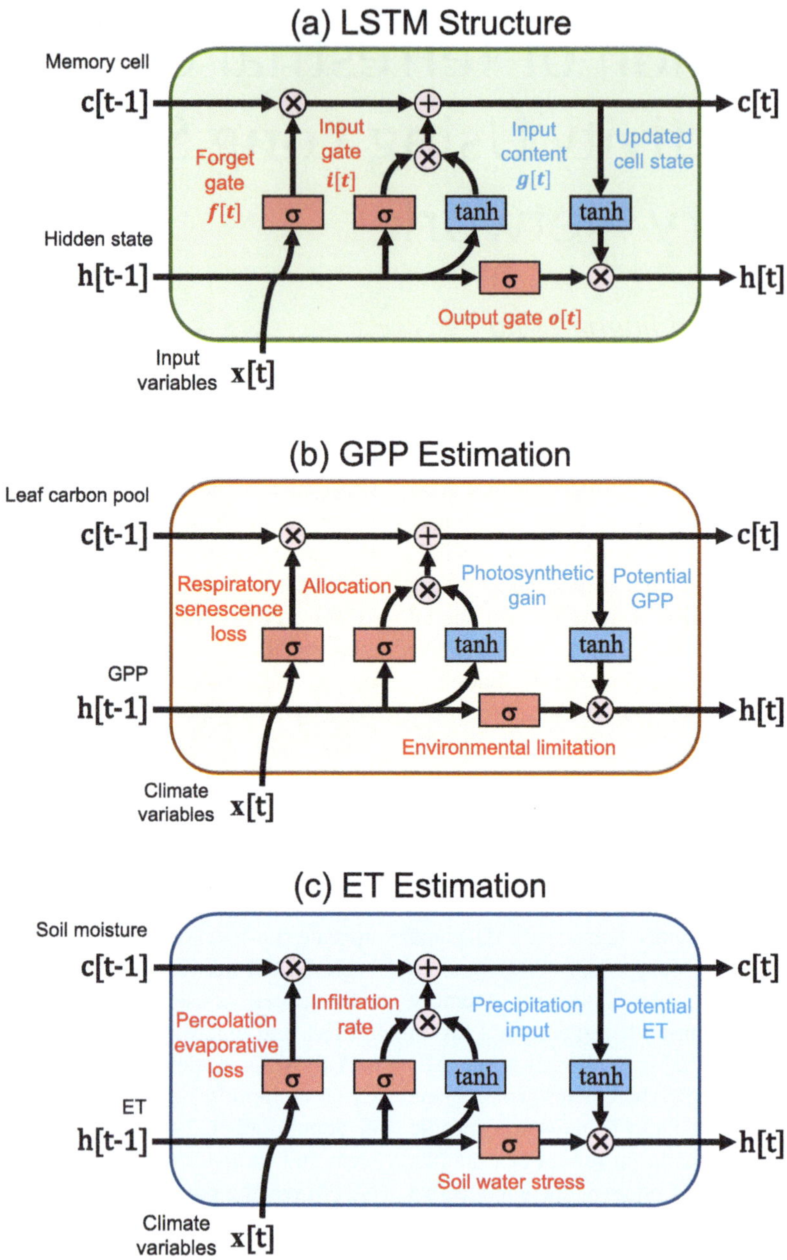

FIGURE 34.1 (a) The Long Short-Term Memory (LSTM) network architecture. (b) Similarity between LSTM and terrestrial biome models (TBM) with regard to GPP estimation. (c) Similarity between LSTM and TBM with regard to evapotranspiration estimation.

This unique structure is designed to alleviate the vanishing or exploding gradients problem that is often encountered in traditional RNNs.

THE INFORMATION FLOW IN LSTM

In the framework of LSTM, the flow of information is regulated by different gates at each time step (t). As for the memory cell, its value is first modified by the forget gate. The forget gate has a value ranging from 0 to 1, which determines how much information from the previous memory cell ($c[t-1]$) will be kept or discarded. After forget, cell state will be updated by adding new content predicted from the input gate and input content. The input gate and input content describe how much and what new information should be passed to the memory cell, respectively. A new memory cell ($c[t]$) is then obtained after the forget and update process.

As for the hidden state, its value is determined by the new memory cell ($c[t]$) and the output gate. While the memory cell stores the long-term memory, the output gate determines how much information should flow from the memory cell to the hidden state. Therefore, the new hidden state ($h[t]$) is expressed as a product of the tanh transformed memory cell ($tanh(c[t])$ and the output gate. It should be noted that

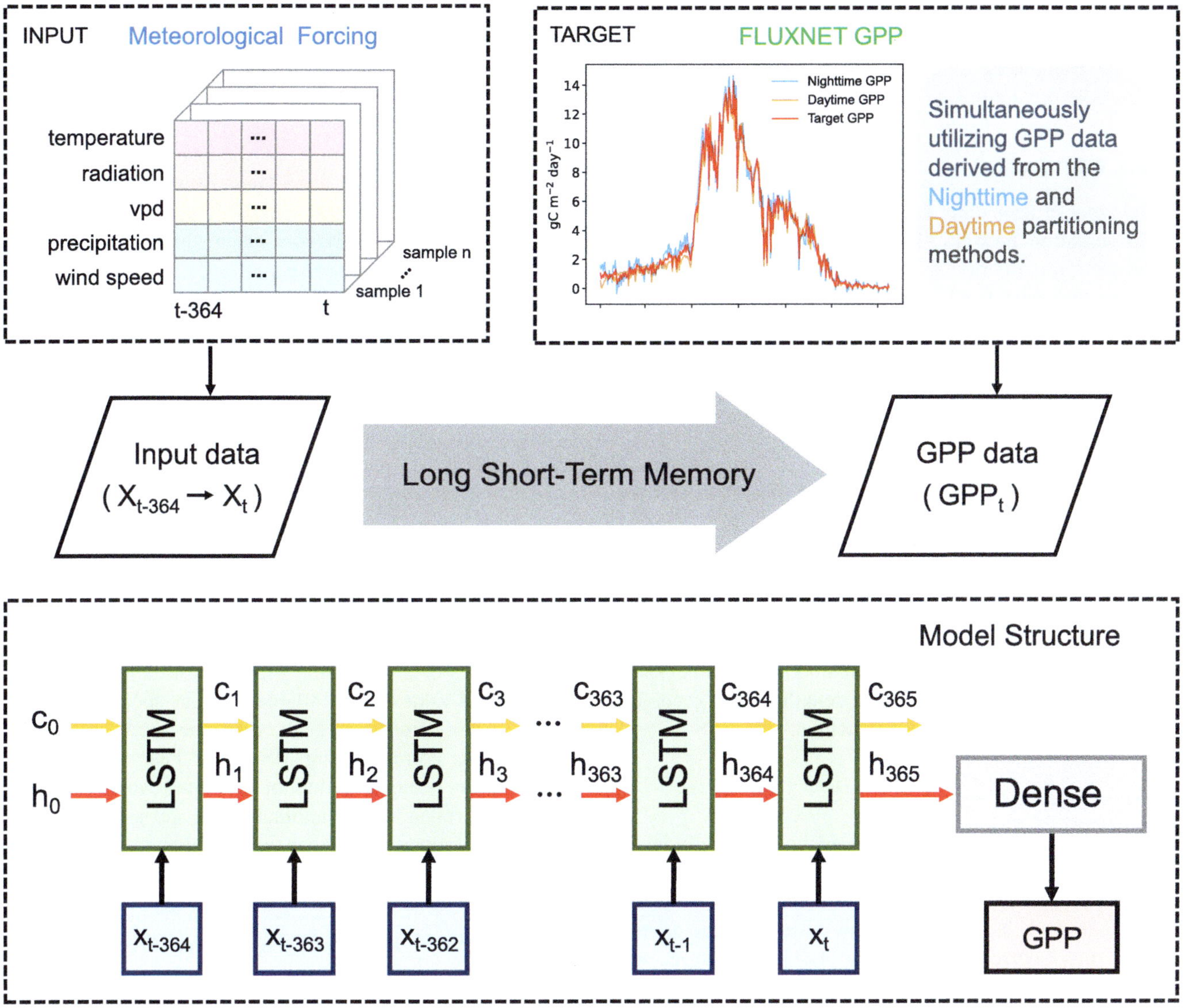

FIGURE 34.2 Schematic of a data-driven approach to upscale in situ GPP estimates using LSTM. The LSTM model is trained using meteorological data from day *t*-364 to day *t* to emulate the target GPP at day *t*. To more accurately represent the target GPP, the model incorporates the mean of GPP data obtained from both nighttime and daytime partitioning methods.

all gates are different transformations of input and hidden states, thus, they are affected by both concurrent and past information.

The unique structure of LSTM allows it to represent complex processes of Earth systems in a simplified manner (Figure 34.1b,c). For example, to mimic the terrestrial carbon cycle, the memory cell can represent the dynamics of the foliar carbon pool, which determines the photosynthetic capacity. Its dynamics are influenced by two processes: the loss from respiration and senescence, and the gain from carbon allocation, both of which are determined by environmental factors and GPP in the previous time step. GPP itself is affected by the photosynthetic potential predicted by the memory cell, as well as the climate factors which contain both radiation and environmental limitations. Water dynamics can also be represented by the LSTM framework in a similar manner (Figure 34.1c).

THE WORKFLOW OF LSTM IN GPP UPSCALING

The upscaling framework comprises two core components, as illustrated in Figure 34.2: the LSTM layer and the dense layers. The LSTM layer, due to its inherent recurrent nature and intrinsic ability to learn long-term dependencies, is utilized to analyze the meteorological forcing time series spanning from day *t*-364 to day *t*. By recurrently processing meteorological features, the LSTM layer is able to function as a dynamic temporal feature extractor, encapsulating both historical and current climate data into high-dimensional hidden states. Following the LSTM layer, the dense layers play a crucial role in handling the high-dimensional hidden states generated by the LSTM layer. By remapping the high-dimensional information into a lower-dimensional space via non-linear activation functions, the dense layers output the final GPP prediction. This prediction combines the temporal dependencies learnt by the LSTM layer and the non-linear dependencies revealed by the dense layers, enabling the upscaling framework to effectively harness meteorological data for GPP predictions.

In Situ GPP Data

This chapter utilizes daily GPP data sourced from the global flux observation network, FLUXNET (https://fluxnet.org/). The observational period spans variously from 1991 to 2014, encompassing data from 200 sites across 13 types of land cover. The in situ mean GPP estimates that are generated via the nighttime partitioning method and the daytime partitioning method (Pastorello et al., 2020) are selected as the target variables in our model. In total, 353,433 site-days of in situ GPP estimates are utilized in this chapter.

Meteorological Data

Meteorological forcings used in the analysis include air temperature, shortwave radiation, vapor pressure deficit (VPD), precipitation, and wind speed derived from ERA5-Land as input features from 1950 to the present (Muñoz-Sabater et al., 2021). For each target feature, a five-dimensional input meteorological time series of length 365 is generated, serving for model training and prediction.

Machine Learning Models

We will utilize LSTM to develop an upscaling model for GPP. To demonstrate the potential of LSTM, a non-time-aware ANN will also be trained for comparative purposes. Given that ANN is devoid of time-aware characteristics, it can only correlate each target GPP feature with meteorological forcing features specific to the same day, resulting in the ANN's input being a five-dimensional vector with a length of 1.

We will adopt a unified model architecture for both the LSTM model and the ANN model: (1) an input layer (an LSTM layer in the LSTM model, a dense layer in the ANN model) that receives meteorological features; (2) a sequence of three hidden layers, utilizing the rectified linear unit (ReLU) as the activation function; (3) an output layer designed to estimate the logarithmic values of GPP, thereby ensuring non-negative results.

In our procedure for training and validating both models, we execute a spatially random split of FLUXNET sites into three distinct subsets: 160 sites are designated for training purposes to optimize the model parameters, 20 sites are allocated for validation to fine-tune the model configurations and mitigate the risk of overfitting, and the remaining 20 sites are set aside for testing to evaluate the overall performance of the final model. To reduce overfitting, we will adopt an early stopping strategy, i.e., training will be terminated when the model performance on validation sites ceases to improve for a duration exceeding 30 epochs, i.e., after the entire dataset is fed to the model for more than 30 times. Our approach of random spatial split provides an objective and unbiased assessment of the model's capacity for extrapolation, which is essential in the realm of upscaling.

Model Evaluation

To evaluate the performance of the two machine learning models comprehensively, three fundamental metrics to gauge model accuracy on the test sites will be calculated: Nash-Sutcliffe efficiency (NSE), root mean square error (RMSE), and mean absolute error (MAE).

NSE quantifies the relative magnitude of residual variance to the variance of measured data. The value of NSE can range from negative infinity to 1, where an optimal value of 1 represents perfect predictions, and values can be negative for less accurate models. RMSE provides the square root of averaged squared prediction discrepancies, emphasizing larger errors. MAE quantifies average absolute differences between predicted and actual values, providing a straightforward measure of average model prediction error.

In addition to examining the predictive accuracy of the models, we also want to explore their capability to elucidate the mean seasonal cycle (MSC) and anomalies. The MSC at each site is computed by averaging the values for each day over the span of all available years, given the condition that at least two values are available for that specific day. The anomalies are determined as the difference between the GPP value and its corresponding MSC. The Pearson correlation coefficient (r) is introduced as an additional evaluation metric specifically for assessing anomalies. To get a more reliable evaluation of model performance, a K-fold cross-validation is often used, see Chapter 37 for details.

MODEL PERFORMANCE IN ESTIMATING GPP

For overall predictive capability, the LSTM model surpasses the ANN model. The LSTM model yields an NSE of 0.72, suggesting LSTM can effectively incorporate historical meteorological information in predicting GPP (Figure 34.3a). This is further supported by the model's lower RMSE (1.77 gC $m^{-2}d^{-1}$) and MAE (1.15 gC $m^{-2}d^{-1}$), both of which emphasize LSTM's improved prediction precision compared to the ANN model (RMSE=2.18 gC $m^{-2}d^{-1}$, MAE=1.48 gC $m^{-2}d^{-1}$) (Figure 34.3d).

With regard to MSC predictions, the LSTM model also demonstrates a better performance with an NSE of 0.78. LSTM efficiently models the MSC of GPP, which is largely dictated by cyclical meteorological patterns. The lower RMSE and MAE scores further validate LSTM's superior handling of temporal seasonality (Figure 34.3b). By contrast, the non-time-aware ANN model, lacking the ability of learning information from temporal sequences, shows less proficiency in capturing this essential seasonality with lower NSE (0.68) and higher RMSE and MAE values (Figure 34.3e).

In capturing GPP interannual anomalies, both models encounter difficulties. The performance of the LSTM model exhibits a relatively low NSE of 0.10, an RMSE of 1.17 gC $m^{-2}d^{-1}$, an MAE of 0.73 gC $m^{-2}d^{-1}$, and a correlation

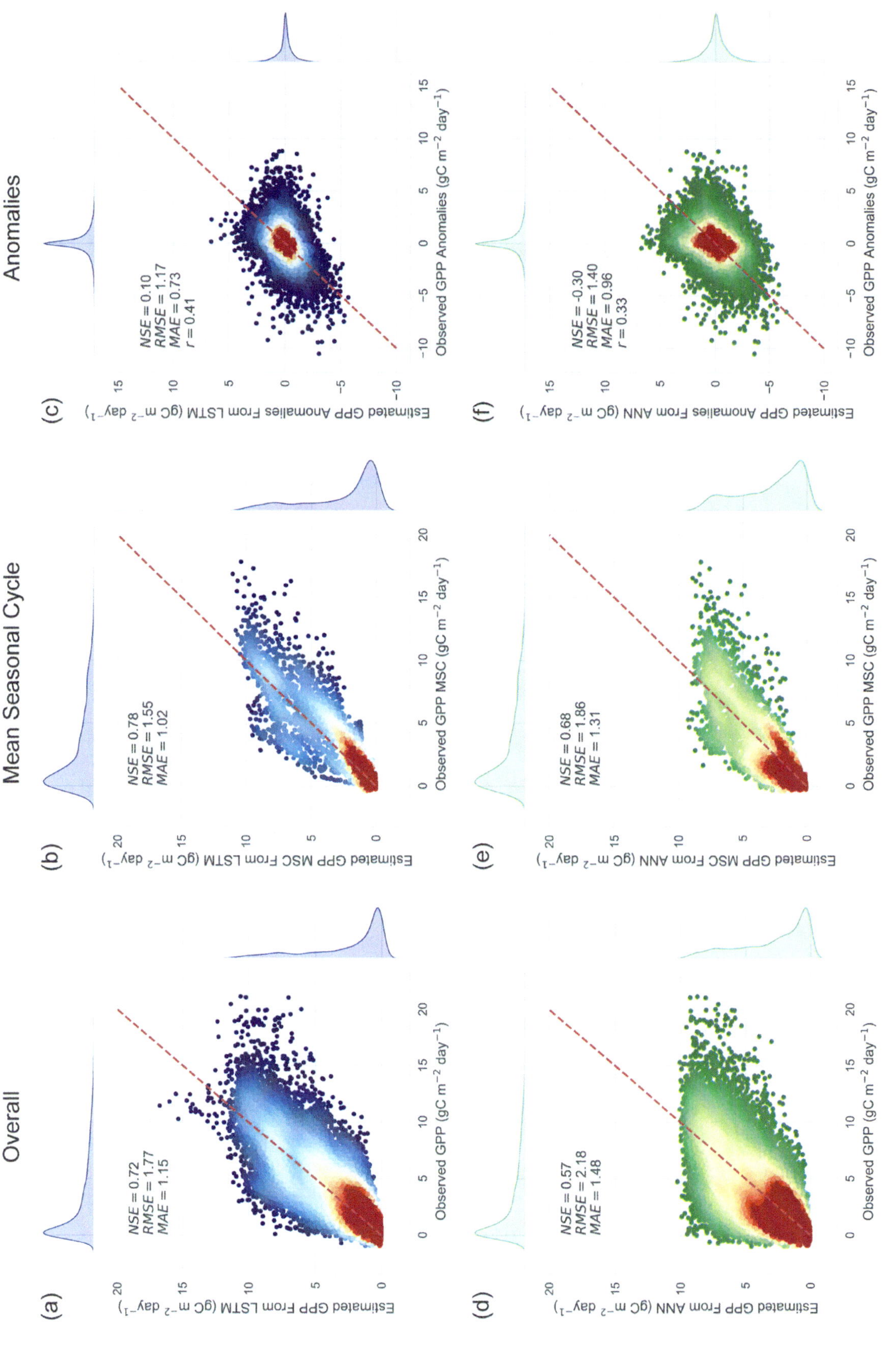

FIGURE 34.3 Comparison between LSTM and ANN on overall GPP predictive ability, mean seasonal cycle of GPP and GPP anomalies detection. The top and right sides of the scatter plots correspond to the kernel density estimation of the observed values and predicted values, respectively.

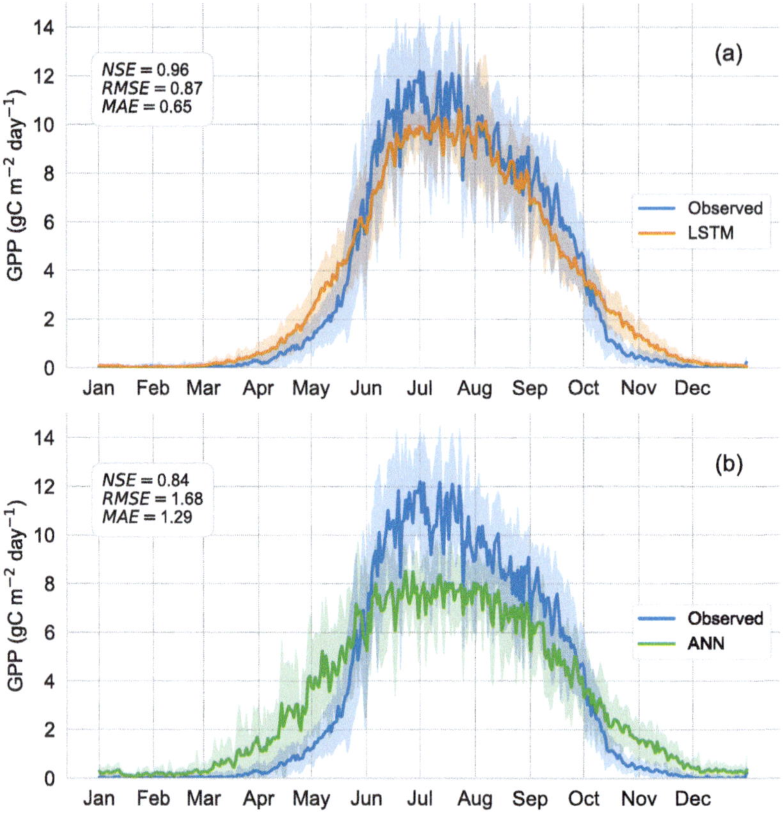

FIGURE 34.4 The mean seasonal cycle of the University of Michigan Biological Station site (Longitude: -84°42'49.68", Latitude: 45°33'35.28"). The solid lines in the graph represent the mean seasonal cycle values, while the shaded areas indicate the corresponding range of the standard deviations, signifying the variability around the MSC.

coefficient of 0.41. The performance of the ANN model is even worse, as indicated by a negative NSE of −0.30, and larger error (RMSE = 1.40 gC $m^{-2}d^{-1}$, MAE = 0.96 gC $m^{-2}d^{-1}$) along with a correlation coefficient of 0.33 (Figure 34.3f). The results indicate that predicting interannual anomalies is still challenging for both models when only meteorological data are employed. However, with the latent state information learnt from past time series, LSTM considerably improves the model performance.

To explicitly illustrate the distinct capabilities of LSTM and ANN models in forecasting the MSC of GPP, an in-depth analysis is conducted, using a deciduous forest site in the US Great Lakes region as a representative case (Figure 34.4).

The LSTM model exhibits a robust performance, with an NSE of 0.96, an RMSE of 0.87 gC $m^{-2}d^{-1}$, and an MAE of gC $m^{-2}d^{-1}$ (Figure 34.4a). These metrics corroborate the LSTM's adeptness in accurately representing the MSC of GPP, especially concerning large and small GPP values. Notably, the sequential nature of the LSTM, characterized by its ability to process and remember past information, mirrors the non-linear dynamics between the accumulation of plant leaf area and climatic variables. This feature allows the LSTM to effectively account for both high values during peak growth phases and low values during periods of plant dormancy or stress.

The ANN model displayed a comparatively weaker performance. With an NSE of 0.84, the ANN model exhibits higher error metrics (RMSE = 1.68 gC $m^{-2}d^{-1}$, MAE = 1.29 gC $m^{-2}d^{-1}$; Figure 34.4b). Specifically, the ANN model demonstrates deficient control over large and small GPP values, tending to underestimate the GPP during peak growth phases and overestimate during periods of plant dormancy or stress. This discrepancy can potentially be attributed to the structure of the ANN model. Being a non-time-aware model, it only takes meteorological forcing features specific to the same day and thus struggles to accurately portray the temporal accumulation process of plant leaf area and other ecosystem state variables.

In summary, this comparison illustrates the LSTM model's superior ability in capturing the nuances of the MSC of GPP, particularly in reflecting the non-linear relationship between the accumulation of plant leaf area and climatic variables. This is an aspect where non-time-aware models fall short due to their limited sequence learning abilities and inherent inability to capture long-term dependencies.

CONCLUDING REMARKS

This chapter delineates the distinct advantage of Long Short-Term Memory network over traditional artificial neural networks. By effectively capturing the intricacies of time-series data, as indicated by a robust Nash-Sutcliffe efficiency of 0.72, and lower root mean square error and mean absolute error, LSTM showed a distinct potential to simulate the vegetation growth process through unique recurrent mechanisms and memory effects.

LSTM model performance could potentially be further improved from more encompassing input features. Using meteorological information over a longer temporal range LSTM may capture the temporal dynamics of GPP more effectively. In addition, the incorporation of static ecological information such as vegetation and climate types could provide additional depth to LSTM's learning process.

SUGGESTED READINGS

Besnard, Simon, Nuno Carvalhais, M. Altaf Arain, Andrew Black, Benjamin Brede, Nina Buchmann, and Jiquan Chen, et al. 2019. "Memory Effects of Climate and Vegetation Affecting Net Ecosystem CO2 Fluxes in Global Forests." *Plos One* 14 (2): e0211510. https://doi.org/10.1371/journal.pone.0211510

Liu, Weihua, Honglin He, Xiaojing Wu, Xiaoli Ren, Li Zhang, Liang shi, Lili Feng, Yangang Wang, and Yan Lv. 2023. "Importance of the Memory Effect for Assessing Interannual Variation in Net Ecosystem Exchange." *Agricultural and Forest Meteorology* 341 (October): 109691. https://doi.org/10.1016/j.agrformet.2023.109691

QUIZ

1. What are the advantages of LSTM as compared to an ANN?
2. What types of data can be used as inputs by LSTM?
3. What is the difference between the cell states and hidden state?
4. Do you think LSTM will still perform better than ANN when satellite observed vegetation greenness is used as input to predict GPP? Why?

35 Machine Learning to Predict and Explain Complex Carbon Cycle Interactions

Julia K. Green
University of Arizona, Tucson, USA

Machine learning models are often considered as being black boxes. This chapter shows ways to gain in-depth understanding of the interactions between predictor and response variables in these algorithms. Specifically, a recent study (see the first suggested reading) is used to illustrate some examples of the utility of machine learning models in the carbon science field, and outline several methods available for machine learning model interpretability.

INTRODUCTION

The Amazon rainforest stores the most aboveground biomass of all ecosystems (Saatchi et al., 2011). However, over the last several decades, the climate has rapidly been changing, leading to moisture conditions that might be less favorable to vegetation growth. Temperature and air dryness have been increasing globally, while precipitation patterns have also been shifting in less predictable ways (Byrne and O'Gorman, 2016; Greve et al., 2014). With these climatic changes, the ability of the Amazon rainforest to continue sequestering these vast amounts of carbon dioxide is unclear, because photosynthesis and carbon storage can be limited by water supply and demand. If rates of photosynthesis across the Amazon rainforest were to reduce in response to these changing moisture conditions, this would lead to the region storing less biomass, and leaving more CO_2 in the atmosphere. Thus, there is an urgent need to better understand the sensitivity of photosynthesis in the Amazon rainforest to changing moisture conditions, and to evaluate whether the earth system models (ESMs) being used for future climate projections are accurately capturing those sensitivities.

A recent study by Green et al. (2020) tackled these issues. It had the following three research objectives:

Obj 1. To determine the photosynthetic sensitivity of the tropical Americas to air and soil dryness as represented in ESMs from the Coupled Model Intercomparison Project Phase 5 (CMIP5).

Obj 2. To use observational data to evaluate the ESM representation (Obj 1) of photosynthetic sensitivity of the tropical Americas to air and soil dryness.

Obj 3. To understand the mechanisms behind the photosynthetic sensitivity of the tropical Americas that were demonstrated using observational data.

Although the published study addressed these research objectives for both the dry and wet seasons, here the focus is on Objectives 1 and 2 during the dry season, which both utilized machine learning algorithms including a K-means clustering analysis, followed by the application of artificial neural networks (ANNs). For more information on Objective 3, please see the recommended readings at the end of the chapter.

PHOTOSYNTHETIC SENSITIVITY IN THE TROPICAL AMERICAS AS REVEALED BY MACHINE LEARNING

Soil moisture and air dryness (i.e., vapor pressure deficit; VPD) are frequently anticorrelated as atmospheric dryness tends to be high when soil moisture is low. Because soil and air dryness are changing differently with climate change (Byrne and O'Gorman, 2016; Greve et al., 2014), it is necessary to understand how they impact photosynthesis independently to better predict future changes in vegetation carbon uptake. This strong anticorrelation between soil and air dryness makes it difficult to decouple their effects on photosynthesis (i.e., gross primary production; GPP). For instance, using a pixel-wise multivariate linear regression model to predict GPP in an ESM would not lead to interpretable results, because the more highly correlated the linear regression model predictors, the lower the precision of the regression coefficients (Harrell, 2015). In fact, with highly correlated predictors, the p-values for each linear regression model coefficient, representing statistical significance, may not even be reliable (Harrell, 2015). Additionally, a multivariate linear regression approach would not work in this circumstance because the interactions between GPP, VPD, and soil moisture are not linear. To separate the effects of VPD and soil moisture on GPP in ESMs, these issues of collinearity had to be dealt with: a method was needed to reduce the correlation between VPD and soil moisture.

K-MEANS CLUSTERING

Despite the strong coupling between VPD and soil moisture in time, the correlation between these two variables can become less pronounced when aggregated spatially. Thus, it is not necessary to analyze the photosynthetic sensitivity to soil moisture and VPD per every 1 × 1 degree pixel location in each ESM separately. Rather, analysis can be performed over

DOI: 10.1201/9781032711126-44

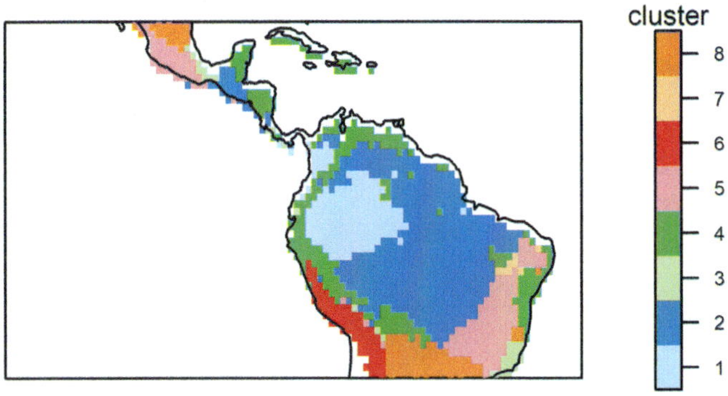

FIGURE 35.1 K-means clustering results (see Green et al., 2020; Supplementary Figure 1).

data from larger regions to reduce the correlation between VPD and soil moisture. The robustness of such analysis usually increases with the amount of data. Thus, a K-means clustering approach was applied to define these regions.

K-means clustering is an unsupervised machine learning algorithm based on vector quantization. It uses input data to group the data into non-overlapping clusters based on minimizing the difference between each data point to the centroid of the cluster (Wu, 2012). A user must first choose the information that they want their data to be grouped by and supply those datasets as input variables to the algorithm, as well as specify the number of clusters desired. To assist in choosing the optimum number of clusters, there are many approaches available, such as the silhouette method, which is included in software packages such as 'cluster' in R (Maechler et al., 2022). The algorithm then works by selecting initial centroid values based on the seed value provided by the user, and the data is grouped to the centroids closest to each point. The centroids are then updated, and the process is performed iteratively to minimize the distance between each data point and its cluster centroid (Wu, 2012). The model output can be quite sensitive to the initial centroids chosen, and thus it is recommended to try multiple seed-values for K-means clustering before selecting the final clusters.

For this analysis, predictors were chosen that related to both climate and vegetation activity, so that the study area could be grouped into regions that were functionally and climatically consistent in space and time, and where the relationship between VPD, soil moisture, and GPP would be expected to be relatively consistent. Input datasets used were remote-sensing based maps of the mean and standard deviation of solar induced fluorescence (SIF; used as a proxy for GPP; Guanter et al., 2012), precipitation, surface shortwave radiation down, 2-meter near surface air temperature, and vapor pressure deficit calculated from near-surface temperature and relative humidity data. In this scenario, eight clusters were used, based on both a silhouette analysis and visual inspection (Figure 35.1).

Because soil moisture data from satellite sources only reflects the top layers of the soil and have high uncertainty associated with their measurements in the Amazon rainforest, precipitation data at various monthly lags were used in their place for this study, to represent the 'memory' of the system. Thus, to evaluate the effectiveness of K-means clustering for reducing the correlation between soil moisture and VPD, the correlation between remotely sensed current and lagged precipitation data and VPD was computed for each cluster and was compared to the average correlation between current and lagged precipitation and VPD for all pixels within a cluster (Figure 35.2). This process reduced the correlation across all clusters where the correlation could be reliably calculated, demonstrating the success of this approach for reducing the collinearity between precipitation and VPD. Because clusters 3 and 7 consist of only a limited number of pixels, a robust correlation could not be computed, and are therefore not included in Figure 35.2.

ARTIFICIAL NEURAL NETWORKS

Because the impact of moisture supply and demand on photosynthesis are not linear (Green et al., 2022), using linear regression models to examine the drivers of photosynthesis was not possible for this application, even after clustering the data. Thus, to address the nonlinear impact of water supply and demand on GPP, artificial neural networks were used.

ANNs are modeled on the functioning of the brain, and comprise an input layer, usually one or more hidden layers, and an output layer, which are known as node layers. ANNs have nonlinear activation functions, and if they are run with multiple hidden layers, can disentangle nonlinear relationships between predictor and response variables (Bell, 2014; see also Chapter 37), making them an effective choice of model for this application. They need to be trained prior to making predictions, and this is done by using a portion of the data for training the model (for this study, 60%) before validating and applying the ANNs. It is important for a user to follow this process of using a portion of the data for training the models and a different portion for testing the models to avoid issues of overfitting (if the R^2 value of the model largely decreases when testing the model on new data, the model is likely overfit) (See Chapter 37 for more discussion on cross-validation to avoid under- and over-fitting). Disadvantages of ANNs are that they require a large amount of data to train, a user must be careful not to overfit the model, and they can be computationally intensive.

For this case study, predictor datasets in each ANN were used related both to climate and vegetation structure. In the ESMs, these included monthly data of precipitation at different monthly lags (to account for the memory of soil moisture), vapor pressure deficit calculated from near surface temperature and humidity data, surface shortwave radiation down, and leaf area index (representative of vegetation structure). These datasets covered the time period of 1976–2005 and were all brought to a spatial resolution of 1 × 1 degree before running the analysis. In the observational analysis (used to evaluate the ESM photosynthetic sensitivity), predictor datasets used

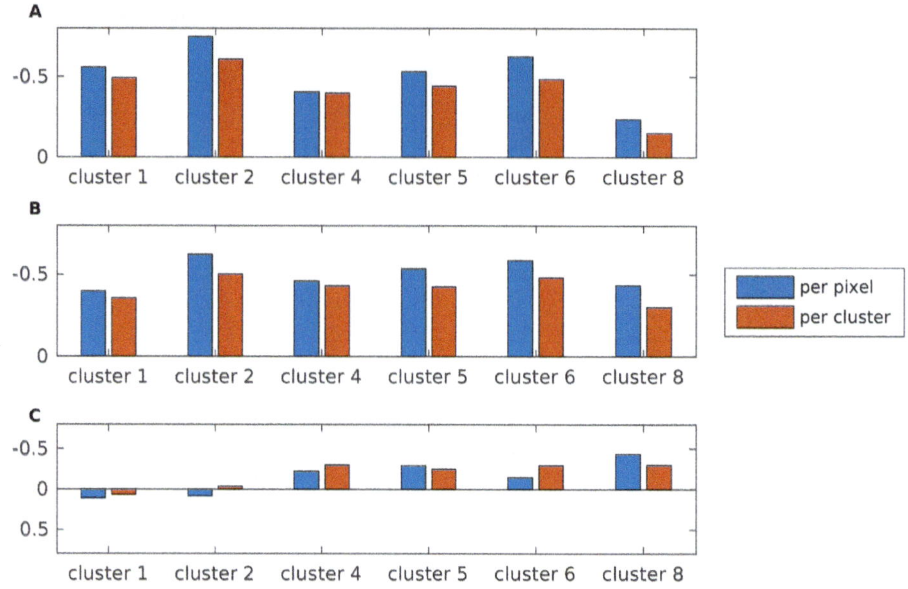

FIGURE 35.2 Correlations between VPD and current precipitation (A), 2-month lagged precipitation (B) and 4-month lagged precipitation (C) (see Green et al., 2020; Supplementary Figure 2).

were monthly precipitation at different monthly lags, vapor pressure deficit calculated from near surface temperature and humidity data, photosynthetic active radiation (similar to shortwave radiation down), and then canopy height and the fraction of absorbed photosynthetically active radiation to represent the canopy structure. These datasets covered the time period of 2007–2016 based on data availability. In the ESMs, GPP was used as a response variable, while SIF data in the observational analysis was used as a response variable as a proxy for GPP. Prior to running the ANNs, all data were normalized by the cluster mean and standard deviation for each dataset.

To increase robustness, an ensemble of ten ANNs were run per cluster across the observational data. In total, ten ESMs from CMIP5 were used in the analysis, and the final figures depict the median ESM behavior (Figure 35.3).

ARTIFICIAL NEURAL NETWORK PERTURBATION ANALYSIS

After developing the ANNs per cluster, the trained models could be used to make predictions with new data, and to evaluate the model performance using the R^2 values. However, without further analysis, it is unclear how or if each input variable is influencing the final ANN prediction. Therefore, a sensitivity analysis, known as a perturbation analysis, was performed in this study to disentangle the impacts of VPD and precipitation at different monthly lags on photosynthetic activity per cluster, for both ESMs and observational data separately.

This process consisted of four steps. First, each ESM (or observational) ANN per cluster was used to predict GPP (or SIF) both temporally and spatially ($GPP_{ANN\ all\ var}$). Second, one predictor variable, which will be called variable X, was perturbed by one standard deviation, and the same ANNs were used to predict the GPP (or SIF) data again ($GPP_{ANN\ X+absval(stdev(X))}$). Third, these predictions from steps 1 and 2 were transformed back to their pixel locations, to create two time-varying map arrays of GPP (or SIF) ('GPP predicted unperturbed' and 'GPP predicted perturbed'). Fourth, using Equation (35.1), the percent change in GPP (or SIF) due to a one standard deviation change in a predictor variable X could be determined per pixel-location, by taking the temporal mean of the difference between the 'GPP (or SIF) predicted perturbed' data array and the 'GPP (or SIF) predicted unperturbed' data array, and dividing that by the temporal standard deviation of the GPP data (or SIF observations):

$$\frac{100 \times mean\left(GPP_{ANN\ X+absval(stdev(X))} - GPP_{ANN\ all\ var}\right)}{stdev(GPP)} \tag{35.1}$$

These results could also be separated into dry and wet seasons as defined by the climatology of precipitation, by applying Equation 35.1 to only dry season months, or only wet season months.

In doing so, the percentage change in GPP (or SIF) due to perturbing each predictor variable by 1 standard deviation could be determined for each predictor variable in the model (Figure 35.3). If the perturbed model resulted in higher GPP (or SIF) values on average, this demonstrated that the predictor variable was having a predominantly increasing impact on the response variable, and that sensitivity was depicted as positive (teal in Figure 35.3). Likewise, if the perturbed model resulted in lower GPP (or SIF) values on average, this demonstrated

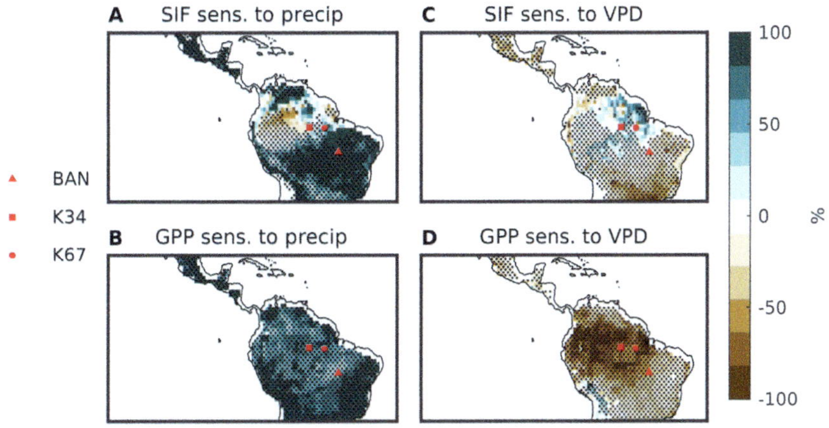

FIGURE 35.3 ANN sensitivity results: Dry season. Remote sensing results for the sensitivity of SIF to precipitation (A), and VPD (C). Model results for the sensitivity of GPP to precipitation (B) and VPD (D) (see Green et al., 2020; Figure 1).

that increases in a particular predictor variable tended to have a reducing impact on GPP (or SIF) values and was depicted as negative (brown in Figure 35.3).

CASE STUDY: DRY SEASON RESULTS

During the dry season, observational data results from the perturbation analysis showed a positive photosynthetic sensitivity to precipitation in drier regions of the study area, but a negative photosynthetic sensitivity to precipitation in the wettest regions of the Amazon rainforest (Figure 35.3A). The observational data also showed a negative photosynthetic sensitivity to VPD across the study area, except in the Amazon rainforest, where the sensitivity was either neutral or slightly positive (Figure 35.3C). Meanwhile, the ESMs showed a systematic increase in GPP in response to increased precipitation, and a systematic decrease in GPP in response to increased VPD throughout the study area (Figure 35.3B and 35.3D).

These results demonstrate that the CMIP5 ESMs used for future climate projections are overestimating the positive influence of soil moisture on photosynthetic activity in the Amazon rainforest, while also overestimating the negative photosynthetic sensitivity to VPD. While this ESM response is well characterized (both low soil moisture and high atmospheric demand frequently cause vegetation to reduce photosynthesis; Ball et al., 1987), it neglects the fact that in nature there can be shifts in vegetation structure and biogeochemistry that can compensate for these reductions in photosynthesis from a leaf-level understanding. CMIP5 ESMs assume static biogeochemistry regardless of vegetation water stress, which would prevent them from capturing these other vegetation responses which occur in nature.

OTHER METHODS OF MACHINE LEARNING INTERPRETABILITY

In the case study presented here, a perturbation analysis was applied to increase the interpretability of the ANN models used. However, there are various methods in addition to this that can be used to understand the importance of machine learning model predictors, and to reveal the nature of variable interactions (see also Chapter 33).

PARTIAL DEPENDENCE PLOTS

One such method is partial dependence plots, which are used to understand the influence of a predictor variable on a model prediction. They depict the prediction function between each predictor variable (or a set of predictors) and the response variable, in a straight-forward and visual way (Molnar, 2022; see Figure 35.4A). They have been used to study variable interactions related to understanding the capacity of mineral-associated organic carbon (Georgiou et al., 2022), as well as soil respiration (Warner et al., 2019).

Some shortcomings of partial dependence plots are that they are limited to depicting the interactions between two predictor variables and one response variable at most, and they also make assumptions about variable independence. Therefore, if there are interactions between variables within the model, this could lead to spurious results (Molnar, 2022).

SHAPLEY VALUES

Shapley values are another method for disentangling the impacts of predictor variables on model predictions based on the output of a machine learning model. Shapley values are based on game theory, and decompose the anomalies of a machine learning response variable, to measure the contributions of each predictor variable to that anomaly (Hart, 1989). They can be used to calculate overall variable importance, and can also be used with categorical response variables, in which their values represent how an individual predictor influences the probability of a particular model outcome (see Figure 35.4B). They have been used in carbon cycle science to explore the interactions between water stress and GPP (Wang et al., 2022) as well as to determine the drivers of changes in aboveground biomass (Winkler et al., 2023). While they can be used with

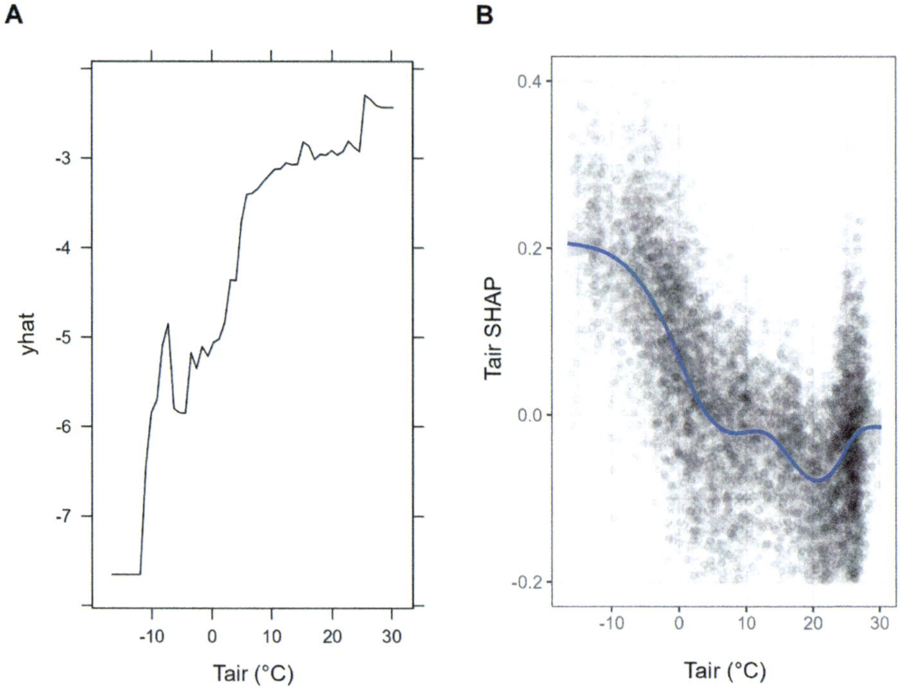

FIGURE 35.4 An example partial dependence plot showing the relationship between near surface air temperature (Tair) and a random forest model prediction of a response variable yhat (**A**), and an example Shapley plot depicting the influence of Tair on a random forest model prediction of a binary response variable (**B**). Panel A shows that as Tair increases, model predictions of yhat tend to increase, but that the relationship is nonlinear. Panel B demonstrates that at Tair values < ~2°C, Tair increases the probability that the model prediction is 1, while at Tair > 5°C and Tair < 25°C, Tair decreases the probability that the model prediction is 1.

many machine learning model outputs, they can be computed most efficiently for the output of tree-based models (such as random forest models). A disadvantage of Shapley values compared to partial dependence plots is that they can be very computationally expensive. However, an advantage of Shapley values over partial dependence plots is that they are better at disentangling feature interactions.

CONCLUSIONS

Machine learning models have opened the door to both predict and explain complex interactions in environmental systems. In the case study presented in this chapter, K-means clustering was used to reduce the correlation between VPD and lagged precipitation data, to more easily disentangle their effects on photosynthetic activity across the tropical Americas. This was followed by using artificial neural networks, that allowed for the decomposition of non-linear effects of lagged precipitation and VPD on photosynthetic activity. Lastly, a perturbation analysis was performed to be able to quantify the sensitivity of photosynthesis to a standard deviation of change in each predictor variable. The observational results could then be used as a benchmark for the CMIP5 ESMs, leading to an investigation to understand why these ESMs were not capturing the photosynthetic sensitivity that was being detected in nature. Although a perturbation analysis was used here, other methods of interpreting machine learning output exist, and two others presented here are partial dependence plots and Shapley values. These are just several examples of how machine learning algorithms can be applied in carbon cycle science, with this case study's objective being to improve understanding of vegetation response to water supply and demand across the tropical Americas.

SUGGESTED READINGS

Green, J. K., Berry, J., Ciais, P., Zhang, Y. & Gentine, P. Amazon rainforest photosynthesis increases in response to atmospheric dryness. *Science Advances* **6**, eabb7232, (2020).

Molnar, C. *Interpretable Machine Learning: A Guide for Making Black Box Models Explainable*. Second Edition. (2022).

QUIZ

1. Explain why it was necessary to use a K-means clustering analysis in the case study presented here rather than disentangling the effects of VPD and precipitation on GPP per pixel location.
2. Explain some of the pros and cons of using artificial neural network models.
3. You would like to examine the output of a machine learning model. Give three examples of sample problems you could be working with where it would be appropriate to use: i) a perturbation analysis, ii) partial dependence plots, and iii) Shapley values, to disentangle the variable interactions.

36 Practice 9

Applications of Machine Learning to Predict Soil Organic Carbon Content

Feng Tao
Cornell University, Ithaca, USA

This practice focuses on one of the simplest yet effective machine learning models, the random forest model. We will first introduce basic concepts in random forest modeling and then apply this model to an example of predicting soil organic carbon (SOC) distributions across the ecosystems of North Macedonia, a country in Southeast Europe. We will compare the predicting results from the random forest with linear regression models. At the end of this practice, we will discuss the interpretability of the random forest model.

BASIC CONCEPTS OF RANDOM FOREST AS A MACHINE LEARNING METHOD

This popular machine learning technique in the geosciences was introduced in Chapter 33. A random forest is an ensemble learning method for classification, regression, and other possible tasks that operate by constructing a multitude of decision trees at training time. Specifically, in a classification study, the final outputs of a random forest would be the class selected by most decision trees. For a regression problem, the final results would be the mean value of predictions by all the individual trees.

To further illustrate this idea, let us imagine a task where we will identify the names of different fruits by different decision trees in a random forest (Figure 36.1). Through its internal processing algorithm, the first decision tree tells us the fruit we need to identify is an apple. However, the second decision tree concludes differently that the fruit is a lemon. Meanwhile, the third decision tree says that the fruit should be an apple. In this process, all these three decision trees compose the random forest. In making the final decision about the name of the fruit, the random forest ensembles all the outputs from different decision trees together and chooses the most popular suggested decision (known as majority voting) as the final prediction. According to this rule, the random forest predicts the fruit to be an apple in this case.

As illustrated in our example, the decision trees are the fundamental units for a random forest. A decision tree is a tree-shaped diagram used to choose among a spruce of actions based on some criteria or questions. Analogous to a real tree, decision trees in a random forest also have branches and leaves. Each branch of a decision tree represents a possible decision, occurrence, or action for specific problems. Each branching point or node corresponds to a question we can ask in relation to the choices available to us.

Back to the problem of identifying different fruits. If we have a bowl of different kinds of fruits, how are we supposed to differentiate them from one another? Intuitively, we may ask different questions that are related to the typical features of different fruits. For a bowl with apples, lemons, and cherries (Figure 36.2), probably we can first ask a question about the diameter of those fruits. If we ask the question: "Is the diameter of the fruits larger than 3 cm?" then we can pick cherries from apples and lemons. We can ask a second question, for example: "Is the color red?" to further differentiate apples from lemons. After two questions, we can separate all these three kinds of fruits from the bowl. In this process, the two questions split different fruits

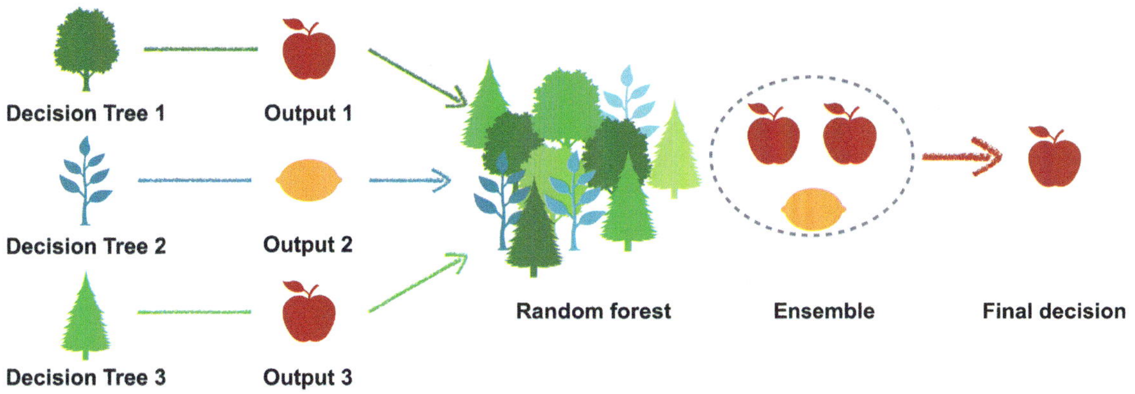

FIGURE 36.1 The basic working algorithm in a random forest.

Adapted from https://builtin.com/data-science/random-forest-algorithm.

DOI: 10.1201/9781032711126-45

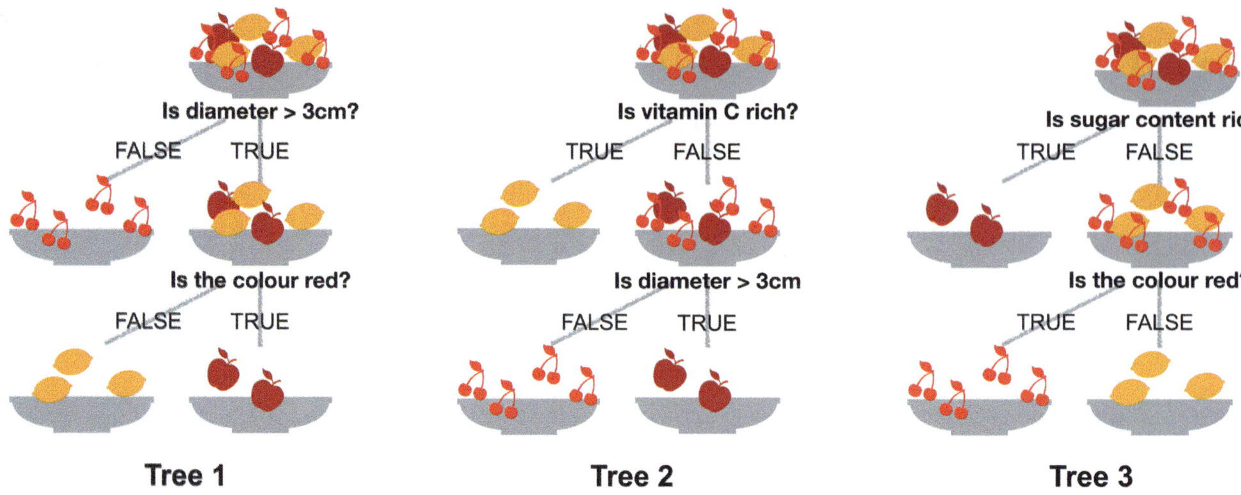

FIGURE 36.2 Branches of different decision trees.

Adapted from https://builtin.com/data-science/random-forest-algorithm.

and we call them the branches of the decision tree. Meanwhile, right before the questions, we have the so-called decision nodes where the fruits are waiting to be separated.

In the real world, we have lots of information about the features of different fruits. Therefore, we can take an array of available information into our predictions, which makes multiple decision trees possible. Figure 36.3 shows three possible decision trees to separate different fruits in a bowl. All of these three decision trees eventually successfully separate different kinds of fruits from one another. If all the possible observations in the real world followed the rules entailed by these three decision trees, each of the decision trees would make 100% correct predictions. However, such a successful prediction does not happen most of the time. For example, red may be the most common color for apples. But we can always buy some green apples which are as fruitful as the red ones from supermarkets. When we have an observation that is out of the samples which we used in building those decision trees, how will different decision trees perform?

Suppose, for example, we find a fruit that has a diameter of 7 cm with the color of green, medium vitamin C, and rich sugar content. According to Figure 36.2, the first decision tree tells us that this fruit should be a lemon, because its diameter is larger than 3 cm, and its color is not red. However, the second decision tree tells us that this should be an apple because it does not have very high vitamin C, and at the same time, its diameter is larger than 3 cm. Meanwhile, when we look at the third decision tree, it indicates our observation should be an apple because it has a pretty high sugar content. Two of our three trees voted for the observation being an apple, while one voted for it being a lemon. According to the majority voting rule, the final decision of the random forest will be for the fruit being an apple. From this simple demonstration, we will notice that predictions based on a single or few decision trees are not necessarily reliable. Yet after building a bunch of decision trees in the random forest, although some of the trees may not give the correct answer, the ensemble of the random forest can still make useful predictions. The larger the ensemble, the more likely the predictions will be correct.

To bring the random forest from abstract concepts to real practice, we will have three exercises for you to get more familiar with the random forest. Soil organic carbon (SOC) is one of the most important parts of the land carbon cycle, yet it is not predicted especially well by conventional regression methods or process models. Here we will try to use a random forest to map the SOC distributions in North Macedonia (hereafter Macedonia), a mountainous country in Southeast Europe. In this case, SOC is the prediction target, and we have 11 different environmental variables that can serve as predictors in the random forest. The abbreviations and full names of these variables are listed in Table 36.1.

In Exercise 1, we will first screen the available soil data to gain a preliminary impression of SOC and related environmental variables.

Exercise 1: Screening soil data in Macedonia. Follow the instructions in CarboTrain:

a. Select **Unit 9**
b. Select **Exercise 1**
c. Select environmental variable (i.e., the "**Select Var**" button)
d. Select **Output Folder**
e. **Run Exercise**
f. Check results in your Output Folder. Two figures will be generated (Figure 3 and Figure 4). The figure in soc_histogram.png describes the distribution of SOC content in Macedonia. soc_var_dist.png describes the spatial distributions of SOC content and the selected environmental variable.

Questions:

1. What is the distribution shape of SOC content in Macedonia? Which part of the distribution (lower values with abundant observations or higher values with fewer observations) is easier to predict?

TABLE 36.1
SOC and environmental variables used in the random forest model

Variable	Full Name
SOC	Soil organic carbon content (%)
B04CHE3	Temperature seasonality (°C)
PRSCHE3	Total annual precipitation (mm)
TMDMOD3	Mean annual land surface temperature (daytime) from MODIS (°C)
DEMENV5	Land surface elevation (m)
B07CHE3	Temperature annual range (°C)
HIST	Histosol probability
B13CHE3	Precipitation of wettest month (mm)
B14CHE3	Precipitation of driest month (mm)
REDL00	Landsat band 3 (red)
TWIMRG5	Topographic wetness index
LandCover	Land cover types

Source: Data originates from the Soil Information System of North Macedonia: http://www.maksoil.ukim.mk/masis/index.html?lang=en& It is openly available through the GitHub data repository of the Soil Organic Carbon Mapping Cookbook: https://github.com/FAO-GSP/SOC-Mapping-Cookbook/tree/master/data

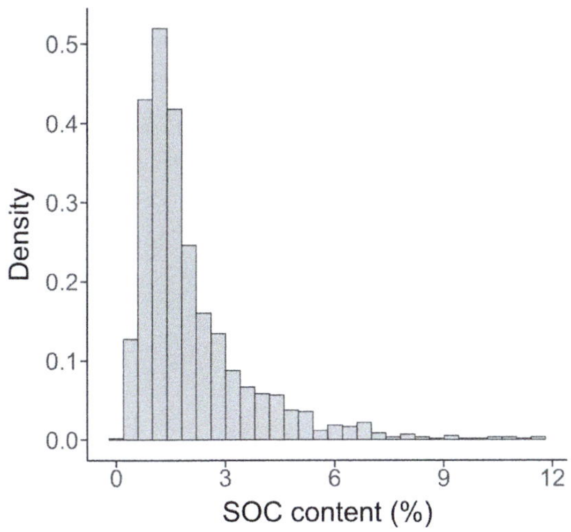

FIGURE 36.3 SOC distribution in Macedonia.

2. Describe the spatial patterns of SOC content across Macedonia. Can you understand how the spatial patterns of SOC correlate with the environmental variables you selected?

RANDOM FOREST IN R LANGUAGE

After looking at the data, Exercise 2 will help you to build a simple random forest model and make predictions on SOC content. The R language provides the package "randomForest" to construct random forest models. Specifically, we will use the "randomForest" function:

```
randomForest(formula,
             x,
             y,
             ntree=500,
             mtry=2,
             ...)
```

This function has many hyper-parameters, but we will pay attention to the most important four of them. The hyper-parameter "formula" describes the structure of the random forest model, which includes the target you want to predict and the predictors you want to use in your prediction. In our case, the prediction target is the SOC content, and you can use any combination of the 11 environmental variables in Table 36.1 as predictors. An example formula goes: "SOC ~ Var1 + Var2 + Var3".

In addition to the formula, you may specify the data of both your predictors (i.e., x in the function) and prediction target (i.e., y in the function). Meanwhile, as we have discussed, a successful random forest needs lots of decision trees. You may decide how many trees you want to plant in the random forest (i.e., ntree in the function). For each of the decision trees, you may also need to decide how many variables you want to use in building the branches (i.e., mtry in the function).

Exercise 2: Building up a random forest model. Follow the instructions in CarboTrain:

a. Select **Unit 9**
b. Select **Exercise 2**
c. Select mtry number
d. Select ntree number
e. Specify the formula of the random forest in the box of **Model Formula**
f. Set output folder
g. **Run Exercise**
h. **Try different combinations of the above hyper-parameters**
i. Check results in your Output Folder. One figure will be generated (Figure 5). The figure in soc_rf_predict.

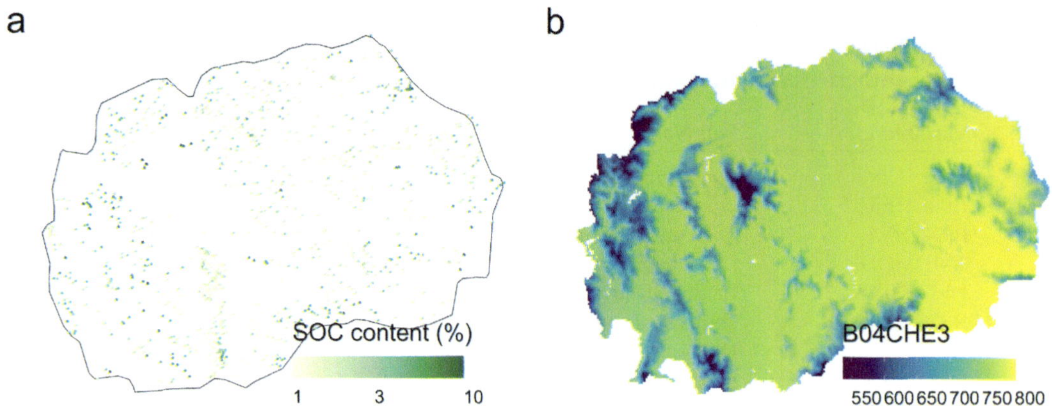

FIGURE 36.4 Spatial distributions of SOC content and the selected environmental variables in Macedonia.

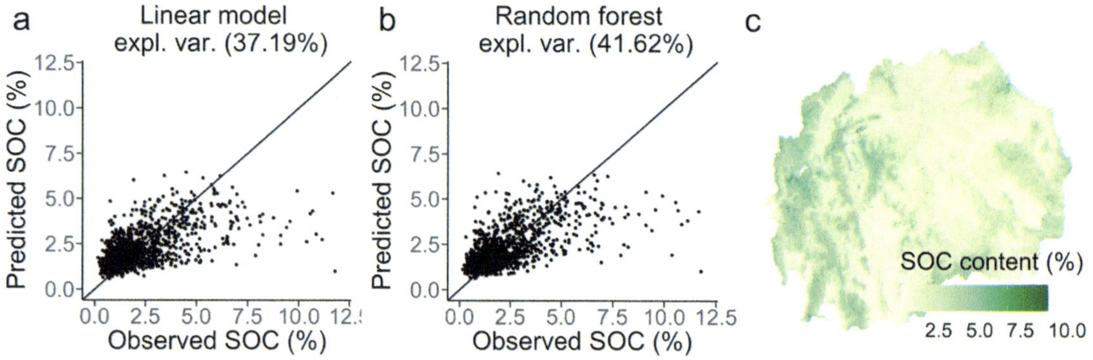

FIGURE 36.5 Model performance using the (a) linear and (b) random forest models. (c) Predicted SOC content map using the trained random forest model.

png is composed of three panels. Panel a and b show the model performance of the linear and random forest models in predicting SOC content in Macedonia using the same formula you specified. Panel c shows the predicted SOC content map of Macedonia using the trained random forest model.

Questions:

1. How well did your random forest model outperform the linear model?
2. Among all the sets of hyper-parameters you have tried, which set generated the best performance of the random forest model?
3. Can you propose other ways that we can further improve the performance of the random forest model, other than changing the hyper-parameters?

INTERPRETABLE RANDOM FOREST RESULTS

After we generate the predicted maps of SOC content in Macedonia, a natural question arising is how to interpret the results in terms of the relative influence of the multiple environmental variables we used to train the model. We might also wonder, what happened inside the random forest that made it better than a normal linear model in predicting SOC? People often take the machine learning model as a black box which is excellent in solving specific problems yet can be hard to interpret compared to simple models such as a linear model, in which the coefficients tell us how important each predictor is in terms of its influence on the values of the response variable. However, if we learn to apply a few tools, machine learning models are not that difficult to interpret.

We will introduce two basic metrics for understanding a machine learning model. The first one is the feature importance, which quantifies the importance of predictors to the final prediction target in a machine learning model. In practice, we have different ways to express the feature importance. A permutation experiment replaces the original predictor values with some random values and thus breaks the relationship between the investigated features and the outcomes. We measure the increase in the prediction error (e.g., the mean squared error) as the permutation importance. A larger permutation importance value of one feature (i.e., variable) indicates model performance is more reliant on the information contained in that variable. Alternatively, Gini importance goes through all the splits for which the feature was used and then measures how much that feature reduced the variance or the Gini index in comparison with the parent

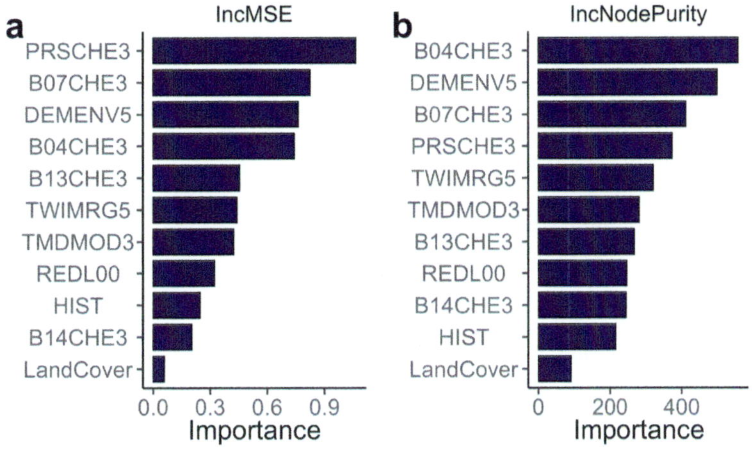

FIGURE 36.6 Importance of environmental variables in predicting SOC. Panel a shows the results measured by permutation importance, and panel b shows the results measured by Gini importance.

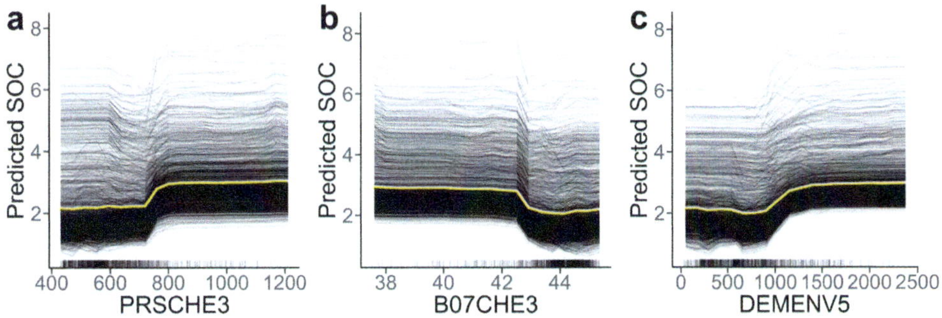

FIGURE 36.7 Partial dependence plots between environmental variables and SOC. Environmental variables are the first three most important ones measured by permutation importance. Black lines indicate the individual conditional expectations, and yellow lines show the global partial dependence plots.

node. The larger the Gini importance value, the more efficient the investigated feature in reducing the prediction variance of the random forest model.

After identifying the importance of different variables, we can further look into the specific relationships between those variables and the prediction target that was learned by the random forest. The partial dependence plot (PDP) shows the marginal effects of one variable in predicting the target outcomes. Through the PDP, we can visually show whether the relationship between the feature and target outcomes is a linear, monotonic or more complex nonlinear relationship. When we look at the PDP at an individual site, it is also called the individual conditional expectation (ICE). The ensembled mean values of all ICE compose the global partial dependence plot.

In our practice of predicting SOC, after you have trained the random forest model, you can also identify the most important variables in predicting SOC and explore the relationships between those variables and SOC. We will try this in Exercise 3, using the same model configuration you tried in Exercise 2.

Exercise 3: Interpret random forest model results. Follow the instructions in CarboTrain:

a. Select **Unit 9**
b. Select **Exercise 2**
c. Select mtry number
d. Select ntree number
e. Specify the formula of the random forest in the box of **Model Formula**
f. Set output folder
g. **Run Exercise**
h. Check results in your Output Folder. Two figures will be generated (Figure 6 and Figure 7). The figure rf_importance.png shows the importance of different variables in predicting SOC content measured by both permutation importance and Gini importance. The figure soc_rf_var_pdp.png shows the partial dependence plots between the first three most important variables (ranked by the permutation importance) and SOC.

QUESTIONS:

1. Which variable in your random forest model is the most important one in predicting SOC in Macedonia?
2. Is the rank of the variable importance the same when measured by permutation importance and Gini importance? Why?
3. Individual conditional expectation (ICE) lines are often different from each other. Can you explain what caused such a difference?
4. Do you think the relationship between the most important environmental variable and SOC retrieved from the random forest is reasonable, based on your knowledge of soil science? If they are reasonable, can you use your empirical knowledge to explain the emerged relationship?

In this practice, we learned basic concepts of the random forest as a machine learning approach and built up a simple random forest model. In the end, we tried to interpret the results learned from the random forest. Yet the story of the machine learning is still to be continued. For machine learning, it is never enough to only build up a model without testing its robustness to avoid overfitting. Meanwhile, it is important to quantify the uncertainty of predictions made by machine learning models. Furthermore, as excellent as machine learning has proven to be in pure data mining, in ecological studies, we are also interested in mechanistic understanding from big data. How to fuse machine learning with process models to retrieve mechanistic understanding from big data is an emerging field to be explored. These issues will be further addressed in Unit 10. Chapter 37 will discuss how to use cross-validation to avoid under and over-fitting and address prediction uncertainties in machine learning. Chapters 38 and 39 will introduce how to integrate process principles into machine learning to better understand key processes in the land carbon cycle from big data.

Unit Ten

Process-based Machine Learning

37 Introduction to Machine Learning and Neural Networks

Toby Dylan Hocking
Northern Arizona University, Flagstaff, USA

In this chapter we introduce basic concepts and algorithms from machine learning. We explain how neural networks can be used for regression and classification problems, and how cross-validation can be used for training and testing machine learning algorithms.

INTRODUCTION AND APPLICATIONS OF MACHINE LEARNING

Machine learning is the domain of computer science which is concerned with efficient algorithms for making predictions in all kinds of big data sets. A defining characteristic of supervised machine learning algorithms is that they require a data set for training. The machine learning algorithm then memorizes the patterns present in those training data, with the goal of accurately predicting similar patterns in new test data. Many machine learning algorithms are domain-agnostic, which means they have been shown to provide highly accurate predictions in a wide variety of application domains (computer vision, speech recognition, automatic translation, biology, medicine, climate science, chemistry, geology, etc.).

For example, consider the problem of image classification from the application domain of computer vision. In this problem, we would like a function that can input an image, and output an integer which indicates class membership. More precisely, let us consider the MNIST and Fashion-MNIST data sets (Figure 37.1), in which each input is a grayscale image with height and width of 28 pixels, represented as a matrix of real numbers $\mathbf{x} \in \mathrm{R}^{28\times28}$ (LeCun et al., 1998, Xiao et al., 2017). In both the MNIST and Fashion-MNIST data sets each image has a corresponding label which is an integer $y \in \{0, 1, \dots, 9\}$. In the MNIST data set each image/label represents a digit, whereas in Fashion-MNIST each image/label represents a category of clothing (0 for T-shirt/top, 1 for Trouser, 2 for Pullover, etc.). In both data sets the goal is to learn a function $f\colon \mathrm{R}^{28\times28} \rightarrow \{0, 1, \dots, 9\}$ which inputs an image x and outputs a predicted class $f(\mathbf{x})$ which should ideally be the same as the corresponding label y.

As mentioned above, a big advantage of supervised learning algorithms is that they are typically domain-agnostic, meaning that they can learn accurate prediction functions f using data sets with different kinds of patterns. That means we can use a single learning algorithm LEARN on either the MNIST or Fashion-MNIST data sets (Figure 37.1, left). For the MNIST data set the learning algorithm will output a function for predicting the class of digit images, and for Fashion-MNIST the learning algorithm will output a function for predicting the class of a clothing image (Figure 37.1, right). The advantage of this supervised machine learning approach to image classification is that the programmer does not need any domain-specific knowledge about the expected pattern (e.g., shape of each digit, appearance of each clothing type). Instead, we assume there is a data set with enough labels for the learning algorithm to accurately infer the domain-specific pattern and prediction function. This means that the machine learning approach is only appropriate when it is possible/inexpensive to create a large, labeled data set that accurately represents the pattern/function to be learned.

How do we know if the learning algorithm is working properly? The goal of supervised learning is **generalization**, which means the learned prediction function f should accurately predict $f(\mathbf{x}) = y$ for any inputs/outputs $(\mathbf{x}, y)$ that will be seen in a desired application (including new data that were not seen during learning). To formalize this idea, and to compute quantitative evaluation metrics (accuracy/error rates), we need a test data set, as explained in the next section.

K-FOLD CROSS-VALIDATION FOR EVALUATING PREDICTION/TEST ACCURACY

Each input $\mathbf{x}$ in a data set is typically represented as one of N rows in a "design matrix" with D columns (one for each dimension or feature). Each output y is represented as an element of a label vector of size N, which can be visualized as another column alongside the design matrix (Figure 37.2, left). For example, in the image data sets discussed above we have $N = 60{,}000$ labeled

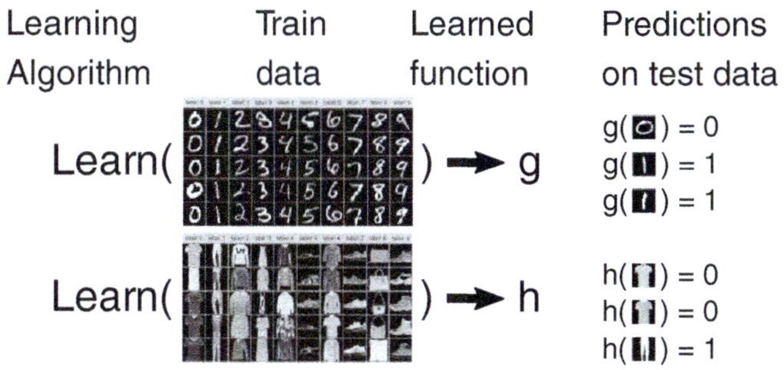

FIGURE 37.1 A learning algorithm inputs a train data set, and outputs a prediction function, g or h. Both g and h input a grayscale image and output a class (integer from 0 to 9), but g is for digits and h is for fashion.

DOI: 10.1201/9781032711126-47

images/rows, each with $D = 784$ dimensions/features (one for each of the 28×28 pixels in the image).

The goal of supervised learning is to find a prediction function f such that $f(\mathbf{x}) = y$ for all inputs/outputs $(\mathbf{x}, y)$ in a test data set (which is not available for learning f). So how do we learn f for accurate prediction on a test data set, if that test set is not available? We must assume that we have access to a train data set with the same statistical distribution as the test data. The train data set is used to learn f, and the test data can only be used for evaluating the prediction accuracy/error of f.

Some benchmark data sets which are used for machine learning research, like MNIST and Fashion-MNIST, have designated train/test sets. However, in most applications of machine learning to real data sets, train/test sets must be created. One approach is to create a single train/test split by randomly assigning a set to each of the N rows/observations, say 50% train rows and 50% test rows. The advantage of that approach is simplicity, but the drawback is that we can only report accuracy/error metrics with respect to one test set (e.g., the algorithm learned a function which accurately predicted 91.3% of observations/labels in the test set, meaning 8.7% error rate).

In addition to estimating the accuracy/error rate, it is important to have some estimate of variance in order to make statements about whether the prediction accuracy/error of the learned function f is significantly larger/smaller than other prediction functions. The other functions to compare against may be from other supervised learning algorithms, or some other method that does not use machine learning (e.g., a domain-specific physical/mechanistic model). A common

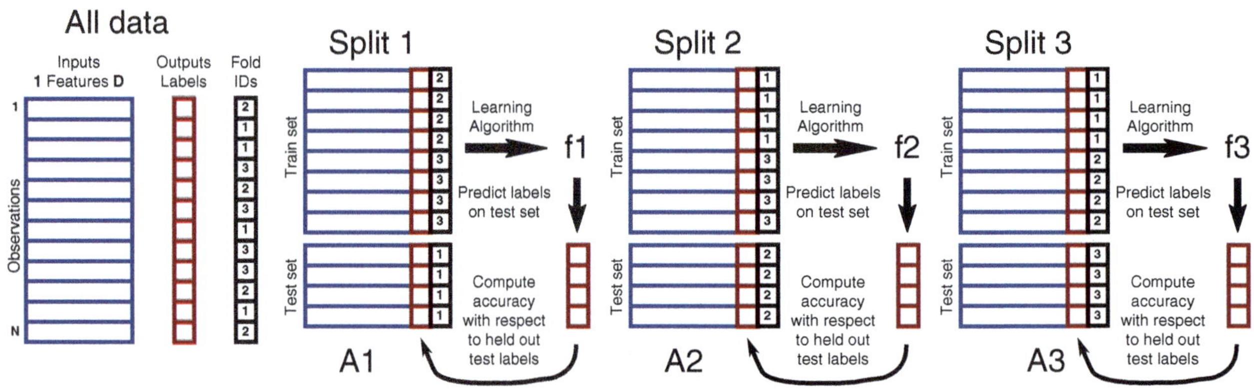

FIGURE 37.2 K = 3-fold cross-validation. Left: the first step is to randomly assign a fold ID from 1 to K to each of the observations/rows. Right: in each of the k ∈ {1, . . ., K} splits, the observations with fold ID k are set aside as a test set, and the other observations are used as a train set to learn a prediction function (f1–f3), which is used to predict for the test set, and to compute accuracy metrics (A1–A3).

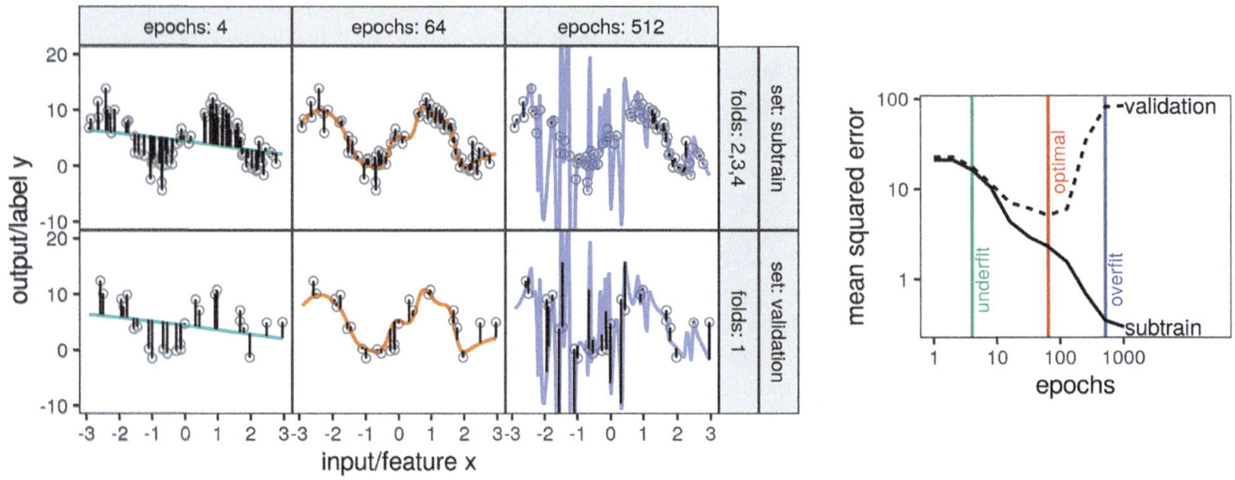

FIGURE 37.3 Illustration of underfitting and overfitting in a neural network regression model (single hidden layer, 50 hidden units). Left: noisy data with a nonlinear sine wave pattern (grey circles), learned functions (colored curves), and residuals/errors (black line segments) are shown for three values of epochs (panels from left to right) and two data subsets (panels from top to bottom). Right: in each epoch the model parameters are updated using gradient descent with respect to the subtrain loss, which decreases with more epochs. The optimal/minimum loss with respect to the validation set occurs at 64 epochs, indicating underfitting for smaller epochs (green function, too regular/linear for both subtrain/validation sets) and overfitting for larger epochs (purple function, very irregular/nonlinear so good fit for subtrain but not validation set).

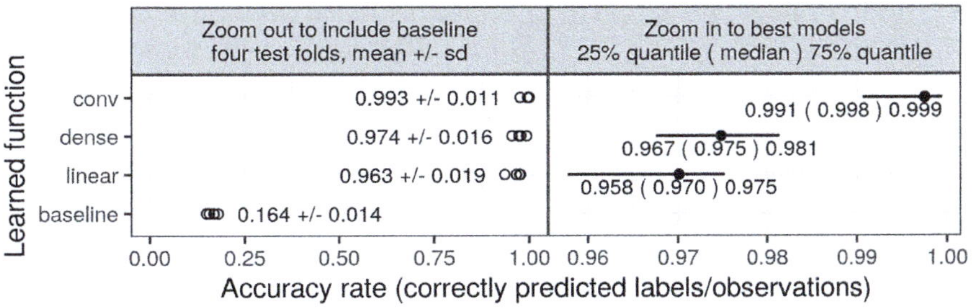

FIGURE 37.4 Prediction accuracy of functions learned for image classification of handwritten digits. The baseline function always predicts the most frequent class in the train set; other three learned functions are neural networks with different numbers of hidden layers (linear = 0, conv = 2, dense = 8).

baseline is the constant function $f(\mathbf{x}) = y_0$ where y_0 is the average or most frequent label in the train data. This baseline ignores all of the inputs/features $\mathbf{x}$, and can be used to show that the algorithm is learning some non-trivial predictive relationship between inputs and outputs (e.g., see Figure 37.4).

The K-fold cross-validation procedure generates K splits, and can therefore be used to estimate both mean and variance of prediction accuracy/error. The number of folds/splits K is a user-defined integer parameter which must be at least 2, and at most N. Typical choices range from $K = 3$ to 10, and usually the value of K does not have a large effect on the final estimated mean/variance of prediction accuracy/error. The algorithm begins by randomly assigning a fold ID number (integer from 1 to K) to each observation (Figure 37.2, left). Then for each unique fold value from 1 to K, we hold out the corresponding observations/rows as a test set, and use data from all other folds as a train set (Figure 37.2, right). Each train set is used to learn a corresponding prediction function, which is then used to predict on the held-out test data. Finally, accuracy/error metrics are computed in order to quantify how well the predictions fit the labels for the test data. Overall, for each data set and learning algorithm the K-fold cross-validation procedure results in K splits, K learned functions, and K test accuracy/error metrics, which are typically combined by taking the mean and standard deviation (or median and quartiles). Other algorithms may be used with the same fold assignments, in order to compare algorithms in terms of accuracy/error rates in particular data sets.

For example, Figure 37.4 uses K = 4-fold cross-validation to compare four learned functions on an image classification problem. The accuracy rates of the "dense" and "linear" functions, 97.4 ± 1.6% and 96.3 ± 1.9% (mean ± standard deviation) are not significantly different. Both rates are significantly larger than the accuracy of the "baseline" constant function, 16.4 ± 1.4%, and smaller than the accuracy of the "conv" function, 99.3 ± 1.1%. We can therefore conclude that the most accurate learning algorithm for this problem, among these four candidates, is the "conv" method (which uses a convolutional neural network, explained later). It is important to note that statements about which algorithm is most accurate can only be made for a particular data set, after having performed K-fold cross-validation to estimate prediction accuracy/error rates.

OTHER APPLICATIONS

So far we have only discussed machine learning algorithms in the context of a single prediction problem, image classification. In this section we briefly discuss other applications of machine learning. In each application the set of possible inputs $\mathbf{x}$ and outputs y are different, but machine learning algorithms can always be used to learn a prediction function $f(x) \approx y$. Jones et al. (2009) proposed to use interactive machine learning for cell image classification in the CellProfiler Analyst system. This application is similar to the previously discussed digit/fashion classification problem, but with only two classes (binary classification). In this context the input is a multi-color image of cell $\mathbf{x} \in \mathbb{R}^{h \times w \times c}$ where h, w are the height and width of the image in pixels, and $c = 3$ is the number of channels used to represent a color image (red, green, blue). The output $y \in \{0, 1\}$ is a binary label which indicates whether or not the image contains the cell phenotype of interest.

Some email programs use machine learning for spam filtering, which is another example of a binary classification problem. When you click the "spam" button in the email program you are labeling that email as spam ($y = 1$), and when you respond to an email you are labeling that email as not spam ($y = 0$). The input $\mathbf{x}$ is an email message, which can be represented using a "bag-of-words" vector (each element is the number of times a specific word occurs in that email message).

Russell et al. (2008) proposed the LabelMe tool for creating data sets for image segmentation, which is more complex than the previously discussed image classification problems. In this context the input $\mathbf{x} \in \mathbb{R}^{h \times w \times c}$ is typically a multi-color image, and the output $\mathbf{y} \in \{0, 1\}^{h \times w}$ is a binary mask (one element for every pixel in the image) indicating whether or not that pixel contains an object of interest.

Machine learning can be used for automatic translation between languages. In this context the input is a text in one language (e.g., French) and the output is the text translated to another language (e.g., English). The desired prediction function f inputs a French text and outputs the English translation.

Machine learning can be used for medical diagnosis. For example, Poplin et al. (2017) showed that retinal photographs can be used to predict blood pressure or risk of heart attack. Since the output y is a real number (e.g., blood pressure of 120 mm mercury), we refer to this as a regression problem.

AVOIDING UNDER/OVERFITTING IN A NEURAL NETWORK FOR REGRESSION

In this section we begin by explaining the prediction function and learning algorithm for a simple neural network. We then demonstrate how the number of iterations of the learning algorithm can be selected using a validation set, in order to avoid underfitting and overfitting.

We consider a simple regression problem for which the input $x \in R$ is a single real number ($D = 1$ feature/column in the design matrix), and the output $y \in R$ is as well. Using a neural network with a single hidden layer of U units, two unknown **parameter** vectors are apparent which need to be learned using the training data, $\mathbf{w} \in R^U$ and $\mathbf{v} \in R^U$. The prediction function f is then defined as:

$$f(x) = \mathbf{w}^T \sigma(x\mathbf{v}) = \mathbf{w}^r \mathbf{z}, \quad 37.1$$

where σ: $R^U \to R^U$ is a non-linear activation function, and $z \in R^U$ is the vector of hidden units. Typical activation functions include the logistic sigmoid $\sigma(t) = 1/(1 + \exp(-t))$ and the rectifier (or rectified linear units, ReLU) $\sigma(t) = \max(0, t)$. The prediction function is learned using gradient descent, which is an algorithm that attempts to find parameters w, v which minimize the mean squared error between the predictions and the corresponding labels in the N train data:

$$L(w,v) = \frac{1}{N}\sum_{i=1}^{N}\left[W^T\sigma(x_i V) - y_i\right] \quad 37.2$$

Gradient descent begins using uninformative parameters $\mathbf{w}_0$, $\mathbf{v}_0$ (typically random numbers close to zero), then at each iteration $t \in \{1, \ldots, T\}$ the parameters are improved by taking a step of size $\alpha > 0$ in the negative gradient direction,

$$\mathbf{w}_t = \mathbf{w}_{t-1} - \alpha\nabla_{\mathbf{w}} L(\mathbf{w}_{t-1}, \mathbf{v}_{t-1}) \quad 37.3$$

$$\mathbf{v}_t = \mathbf{v}_{t-1} - \alpha\nabla_{\mathbf{v}} L(\mathbf{w}_{t-1}, \mathbf{v}_{t-1}) \quad 37.4$$

The algorithm described above is referred to as "full gradient" because the gradient descent direction is defined using the full set of N samples in the train set. Other common variants include "stochastic gradient" (gradient uses one sample) and "minibatch" (gradient uses several samples). When doing gradient descent on a neural network model, one "epoch" includes computing gradients once for each sample (e.g., 1 epoch = 1 iteration of full gradient, 1 epoch = N iterations of stochastic gradient).

In the algorithm above, the number of hidden units U, the number of iterations T, and the step size α must be fixed before running the learning algorithm. These **hyper-parameters** affect the learning capacity of the neural network. An important consideration when using any machine learning algorithm is that you most likely need to tune the hyper-parameters of the algorithm in order to avoid underfitting and overfitting. **Underfitting** occurs when the learned function f neither provides accurate predictions for the train data, nor the test data. **Overfitting** occurs when the learned function f only provides accurate predictions for the train data (and not for the test data). Both underfitting and overfitting are bad, and need to be avoided, because the goal of any learning algorithm is to find a prediction function f which provides accurate predictions in test data.

How can we select hyper-parameters which avoid overfitting? Note that the choice of hyper-parameters such as number of hidden units U and iterations T affect the learned function f, so we cannot use the test data to learn these hyper-parameters (by assumption that the test data are not available at train time). Then, how do we know which hyper-parameters will result in learned functions which best generalize to new data?

A general method which can be used with any learning algorithm is splitting the train set into subtrain and validation sets, then using grid search over hyper-parameter values. The subtrain set is used for parameter learning, and the validation set is used for hyper-parameter selection. In detail, we first fix a set of hyper-parameters, say $U = 50$ hidden units and $T = 100$ iterations. Then the subtrain set is used with these hyper-parameters as input to the learning algorithm, which outputs the learned parameter vectors **w**, **v**. Finally, the learned parameters are used to compute predictions $f(x)$ for all inputs x in the validation set, and the corresponding labels y are used to evaluate the accuracy/error of those predictions. The procedure is then repeated for another hyper-parameter set, say $U = 10$ hidden units with $T = 500$ iterations. In the end we select the hyper-parameter set with minimal validation error, and then retrain using the learning algorithm on the full train set with those hyper-parameters. A variant of this method is to use K-fold cross-validation to generate K subtrain/validation splits, then compute mean validation error over the K splits, which typically yields hyper-parameters that result in more accurate/generalizable predictions (when compared to hyper-parameters selected using a single subtrain/validation split). Note that this K-fold cross-validation for hyper-parameter learning is essentially the same procedure as shown in Figure 37.2, but we split the train set into subtrain/validation sets (instead of splitting all data into train/test sets as shown in the figure).

For example, we simulated some data with a sine wave pattern (Figure 37.3), and used the R package nnet to fit a neural network with one hidden layer of $U = 50$ units (Venables and Ripley, 2013). We demonstrate the effects of under/overfitting by varying the number of iterations/epochs from $T = 1$ to 1000. In this example $K = 4$-fold cross-validation was used, so each data point was randomly assigned a fold ID integer from 1 to 4. The result for only the first split is shown, so observations assigned fold ID = 1 are considered the validation set, and other observations (folds 2–4) are considered the subtrain set (which is used at input to the nnet R function which implements the gradient descent learning algorithm). We then used the predict function in R to compute predictions for subtrain and validation data, and analyzed how the prediction error changes as a function of the number of iterations/epochs T of gradient descent. The data exhibit a nonlinear sine wave pattern, but the learned function for $T = 4$ iterations/epochs is mostly linear (underfitting, large error on both subtrain/validation sets). For $T = 512$ iterations/epochs the learned function is highly non-linear (overfitting, small error for the subtrain set but large error for the validation set). When the error rates are plotted as a function of a model complexity hyper-parameter such as T (Figure 37.3, right), we see the characteristic U shape for the validation error, and the monotonic decreasing train error. The hyper-parameter with minimal validation error is $T = 64$ iterations/epochs; smaller T values underfit or are overly regularized, and larger T values overfit or are under-regularized.

Overall, in this section we have seen how a neural network for regression can be trained using gradient descent (for learning parameter vectors, given fixed hyper-parameters) and subtrain/validation splits (for learning hyper-parameter values to avoid under/overfitting).

COMPARING NEURAL NETWORKS FOR IMAGE CLASSIFICATION

In this section we provide a comparison of several other neural networks for image classification. In general, in a neural network with $L-1$ hidden layers we can represent the prediction function as the composition of L intermediate f_l functions, for all layers $l \in \{1, \ldots, L\}$:

$$f(\mathbf{x}) = f_L\left[\cdots f_1[\mathbf{x}]\right] \tag{37.5}$$

Each of the intermediate functions has the same form:

$$lf_l(t) = A_l\left(\mathbf{W}_1^t\right), \tag{37.6}$$

where A_l is an activation function and $\mathbf{W}_l \in \mathrm{R}^{ul \times ul-1}$ is a weight matrix with elements that must be learned based on the data. This model includes several hyper-parameters which must be fixed prior to learning the neural network weights:

- The number of layers L.
- The activation functions A_1.
- The number of units per layer u_1.
- The sparsity pattern in the weight matrices $\mathbf{W}_l$.

The number of units in the input layer is fixed, $u_0 = D$, based on the dimension of the inputs $\mathbf{x} \in \mathrm{R}^D$. The number of units in the output layer u_L is also fixed based on the outputs/labels y. The numbers of units in the hidden layers ($u_1, \ldots, u_{L-1}$) are hyper-parameters which control under/overfitting. Increasing the numbers of hidden units u_l results in larger weight matrices $\mathbf{W}_l$, which in general means more parameters to learn, and larger capacity for fitting complex patterns in the data. The sparsity pattern of $\mathbf{W}_l$ means which entries are forced to be zero; this technique is used in "convolutional" neural networks for avoiding overfitting and reducing training/prediction time. When the matrix is not sparse (all entries non-zero), we refer to the layer as dense or fully connected.

For example, in the previous section we used a neural network for regression with one hidden layer, which in this more general notation means using $L = 2$ intermediate functions; the input dimension is $u_0 = D = 1$, the number of hidden units is $u_1 = U = 50$, and there is a single output $u_2 = 1$ to predict. The weight matrices are dense/fully connected (no convolution/sparsity), of dimension $\mathbf{W}_1 \in \mathrm{R}^{50\times 1}$, $\mathbf{W}_2 \in \mathrm{R}^{1\times 50}$. The hidden layer activation function A_1 used by the R nnet package is the logistic sigmoid, $\sigma(t) = 1/(1 + \exp(-t))$, and the output activation for regression (real-valued outputs) is the identity, $A_2(t) = t$.

In this section we implement three other neural networks for image classification. Using the "zip.train" data set of $N = 7291$ handwritten digits (Hastie and Tibshirani, 2009), each input is a greyscale image of 16 × 16 pixels which means that number of input units is $u_0 = 256$. As in Figure 37.1 (top) there are ten output classes, one for each digit. For the activation function A_L in the output layer we use the "softmax" function which results in a score/probability for each of the ten possible output classes, so the number of output units is $u_L = 10$.

The three neural networks that we consider are:

linear $L = 1$ intermediate function with 2,570 parameters to learn (linear model, inputs fully connected to outputs, no hidden units/layers).

dense $L = 9$ intermediate functions with 97,410 parameters to learn (nonlinear model, each hidden layer dense/fully connected with 100 units).

sparse $L = 3$ intermediate functions with 99,310 parameters to learn (nonlinear model, one convolutional/sparse layer followed by two dense/fully connected layers).

We defined and trained each neural network using the keras R package (Allaire and Chollet, 2020). We used the fit function with argument validation_split = 0.2, which creates a single split (80% subtrain, 20% validation). We selected the number of epochs hyper-parameter by minimizing the validation loss, and we used the selected number of epochs to retrain the neural network on the entire train set (no subtrain/validation split).

We did this entire procedure $K = 4$ times, once for each fold/split in K-fold cross-validation. Note that even though these data have a pre-defined split into "zip.train" and "zip.test" files, we used K-fold cross-validation on the "zip.train" file, yielding K train/test splits that we used to estimate mean and variance of prediction accuracy for these models (the "zip.test" file was ignored). In each split we used the test set to quantify the prediction accuracy of the learned models. It is clear that the test accuracy of all three neural networks is significantly larger than the baseline model which always predicts the most frequent class in the train set (Figure 37.4, left); they are clearly learning some nontrivial predictive relationship between inputs and outputs. Furthermore, it is clear from Figure 37.4 (right) that the dense neural network is slightly more accurate than the linear model ($p = 0.032$ in

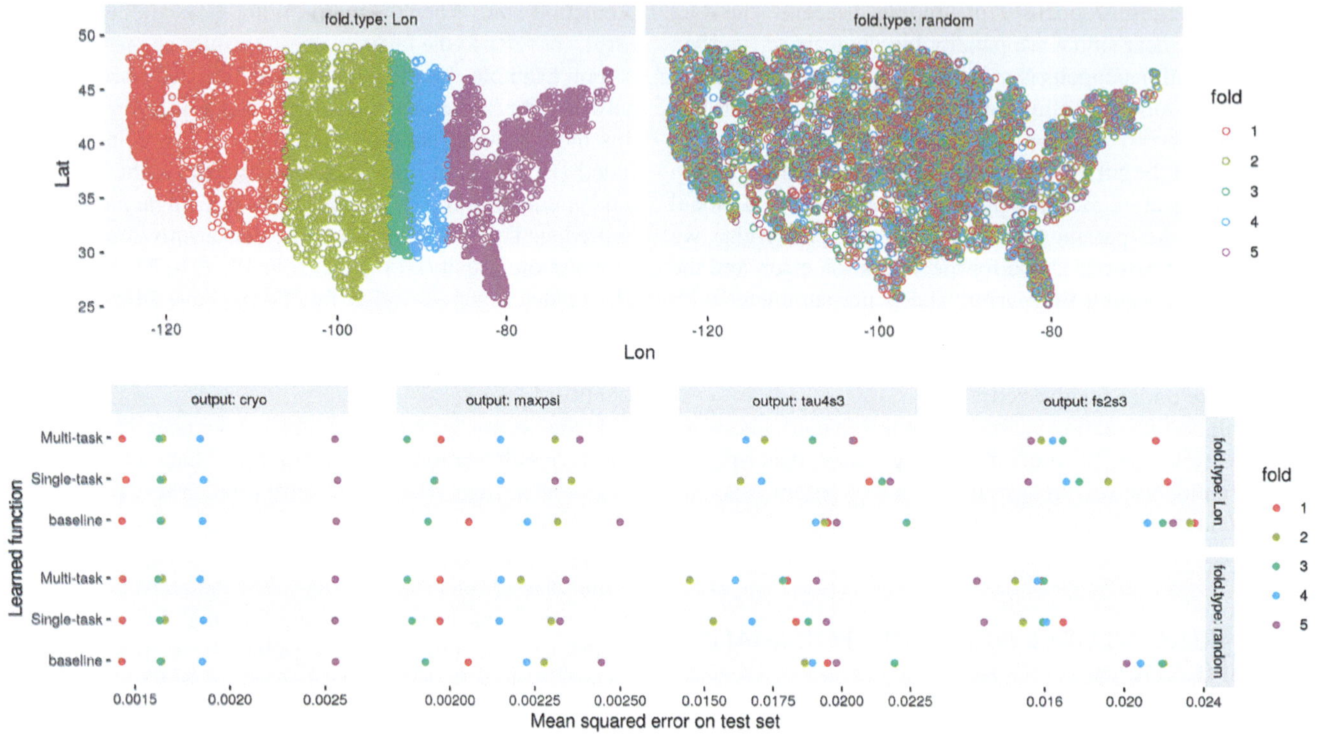

FIGURE 37.5 Cross-validation for estimating error rates of machine learning algorithms that predict earth system model parameters. **Top:** fold IDs were assigned to each observation using longitude (left) or randomly (right). **Bottom:** prediction error for four of the 25 outputs. Please see (Tao et al., 2020) for meanings of abbreviations (cryo, maxpsi, tau4s3, fs2s3).

paired one-sided t_3-test), and the sparse/convolutional neural network is significantly more accurate than the dense model ($p = 0.009$).

In summary, from this comparison it is clear that among these three neural networks, the sparse model should be preferred for most accurate predictions in this particular "zip" data set. However, we must be careful not to generalize these conclusions to other data sets — even for some other image classification data sets such as MNIST (Figure 37.1), the most accurate algorithm may be different. For very difficult data sets, it may even be the case that these three neural networks are no more accurate than the baseline model which always predicts the most frequent class in the train set. In general, we always need to use computational cross-validation experiments to determine which machine learning algorithm is most accurate in any given data set. To learn a predictive model with maximum prediction accuracy, machine learning algorithms other than neural networks should be additionally considered (e.g., regularized linear models, decision trees, random forests, boosting, support vector machines).

CROSS-VALIDATION FOR EVALUATING PREDICTIONS OF EARTH SYSTEM MODEL PARAMETERS

As a final example application, we consider using cross-validation to evaluate a neural network that predicts carbon cycle model parameters (Tao et al., 2020). In this context there is a data set with $N = 26{,}158$ observations, each one a soil sample with $D = 60$ input features. There are 25 real-valued output variables to predict; each is the value of an earth system model parameter at the location of the soil sample. We want a neural network that will be able to predict the values of these earth system parameters at new locations. Tao et al. (2020) proposed using a neural network with $L = 4$ fully connected layers and dropout regularization for this task (see paper for details). In this section the "multi-task" model uses the same number of layers/units as described in that paper; the term multi-task means that the neural network outputs a prediction for all 25 outputs/tasks. For comparison, we additionally consider "single-task" models with the same number of hidden layers/units, but only one output unit. We expect the multi-task model to sometimes be more accurate, because of the expected correlation between outputs (earth system model parameters). To see whether or not these neural networks learn any nontrivial predictive relationship between inputs and output, we consider a baseline model which always predicts the mean of the train set label/output values (and does not use the inputs at all).

Here we show how $K = 5$-fold cross-validation can be used to evaluate how well these neural networks predict each of the outputs at new locations. We first assign a fold ID from 1 to 5 to each observation/row, either systematically using the longitude coordinate, or randomly (Figure 37.5, top). We can define a cross-validation procedure using both sets of fold IDs, in order to answer the question, "is it more difficult to predict at new longitudes, or new random locations?" We expect that predicting at new longitudes should be more difficult, because that involves more extrapolation (predicting outside the range of observed data values). In detail, for each fold ID from 1 to 5, we define the test set as the data points which have been assigned that fold ID using both methods (longitude and random). For these data with $N = 26,158$ observations total, each fold has approximately 5000 observations, so each resulting test set has approximately 1000 observations. As described in the last section on image classification, we used the R keras package to compute the neural network parameters and predictions (using a maximum of 100 epochs, and a single 80% subtrain 20% validation split to choose the optimal number of epochs for retraining on the entire train set). For each fold/model/output we computed mean squared error with respect to the test set, and we plot these values for four of the 25 outputs (Figure 37.5, bottom). It is clear that some outputs are more difficult to predict than others; for cryo and maxpsi outputs the neural networks show little or no improvement over baselines, whereas for tau4s3 and fs2s3 outputs we observed substantial improvements over baselines. As expected, there is a difference in test error between fold assignment methods (random has lower error rates than Lon for several outputs), indicating that it is indeed easier to predict at new random locations, and harder to predict at new longitudes. Finally, the multi-task models are slightly more accurate than the single-task models, indicating that the neural network is learning to exploit the correlations between outputs. Overall this comparison has shown how cross-validation can be used to quantitatively evaluate and compare machine learning algorithms for predicting earth system model parameters.

In comparison to the neural network practice in Unit 10, the main difference is that here we discussed how held-out test sets can be used to estimate prediction accuracy/error rates of learning algorithms. Chapter 38 discusses how a validation set can be used to avoid overfitting, as we have done in this chapter as well. We have additionally discussed how $K = 5$-fold cross-validation can be used to generate several train/test splits, which can be used to estimate prediction error rates for each fold/data/algorithm combination (e.g., Figure 37.5, bottom). This technique is useful since it allows us to see which algorithms are significantly more/less accurate than others on given data sets.

SUGGESTED READINGS

Machine learning is a large field of research with many algorithms, and there are several useful textbooks that provide overviews from various perspectives:

C. M. Bishop. 2006. *Pattern Recognition and Machine Learning*. Springer, New York.

I. J. Goodfellow, Y. Bengio, and A. Courville. 2016. *Deep Learning*. MIT Press, Cambridge, MA, USA.

T. Hastie, R. Tibshirani, and J. Friedman. 2009. *The Elements of Statistical Learning. Springer Series in Statistics*, Second edition. Springer, Springer Science+Business Media, New York NY.

K. P. Murphy. 2013. *Machine Learning: A Probabilistic Perspective*. 2013. MIT Press, Cambridge, MA.

L. Wasserman. 2010. *All of Statistics: A Concise Course in Statistical Inference*. Springer, New York.

Reproducibility statement. Code for figures in this chapter can be freely downloaded from https://github.com/tdhock/2020-yiqi-summer-school

QUIZ

1. When using a design matrix to represent machine learning inputs, what does each row and column represent? What other data/options does a supervised learning algorithm such as gradient descent need as input, and what does it yield as output?
2. When splitting data into train/test sets, what is the purpose of each set? When splitting a train set into subtrain/validation sets, what is the purpose of each set? What is the advantage of using K-fold cross-validation, relative to a single split?
3. In order to determine if any non-trivial predictive relationship between inputs and output has been learned, a comparison with a baseline that ignores the inputs must be used. How do you compute the baseline predictions, for regression and classification problems?
4. How can you tell if machine learning model predictions are underfitting or overfitting?
5. When using the nnet function in R to learn a neural network with a single hidden layer, do large or small values of the number of iterations hyper-parameter result in overfitting? Why?
6. When using the nnet function in R learn a neural network with a single hidden layer, and you do not yet know how many iterations to use, what data set should you use as input to nnet? How should you learn the number of iterations to avoid underfitting and overfitting? After having computed the number of iterations to use, what data set should you then use as input to nnet to learn your final model? Hint: possible choices for set to use are all, train, test, subtrain, validation.

38 PROcess-Guided Deep Learning and DAta-Driven Modeling (PRODA)

Feng Tao
Cornell University, Ithaca, USA

Yiqi Luo
Cornell University, Ithaca, USA

This chapter describes a PROcess-guided deep learning and DAta-driven modeling (PRODA) approach to optimize parameterization of Earth system models (ESMs) using spatio-temporal datasets. PRODA involves both data assimilation to estimate parameter values and deep learning to predict spatial and temporal distributions of parameter values so as to optimize ESM prediction. An application to the Community Land Model version 5 (CLM5) using soil organic carbon (SOC) distributions in the conterminous United States illustrates the potential and utility of the PRODA approach.

THE NEED FOR OPTIMIZING PARAMETERIZATION OF EARTH SYSTEM MODELS

Earth system models (ESMs) are used to simulate historical and potential future states of climate and ecosystems. However, simulations often deviate substantially from observations. For example, soil carbon dynamics simulated by ESMs vary widely among models and often fit poorly with observations. Modeled global soil carbon storage differs by up to six-fold among 11 models of the Coupled Model Intercomparison Project Phase 5 (CMIP5) ensemble (Todd-Brown et al. 2013). None of the models reproduces the spatial distribution of SOC stocks presented in the Harmonized World Soil Database (HWSD) (Luo et al. 2015).

Uncertainty in simulating SOC dynamics with ESMs could stem from poor parameterization, incorrect model structure, or biased external forcing (Luo and Schuur 2020; see Chapter 33). While model structure represents ecological processes (e.g., decomposition of soil organic matter), parameters in ESMs characterize properties of the processes, such as baseline decomposition rate at reference temperature and moisture content, or sensitivity to these drivers. The choice of parameter values can strongly influence model projections of SOC dynamics. Parameter values in the current generation of ESMs, however, are mostly determined on an *ad hoc* basis. They may be derived from the results of field experiments, other models, or informed from scientific or grey literature (Luo et al. 2001), but rarely take into account the range of possible values encompassed by such sources.

Data assimilation techniques to estimate parameter values from observations were discussed and illustrated in earlier chapters (Units 6, 7, 8). Parameter values constrained by data assimilation can improve SOC simulation in ESMs compared to the default parameter values. For instance, the global representation of SOC distribution in the Community Land Model version 3.5 (CLM3.5) was improved from explaining 27% to 41% of variation in the HWSD database by constraining model parameters with a Bayesian Markov Chain Monte Carlo (MCMC) data assimilation method (Hararuk et al. 2014). The large unexplained variation in observed SOC with ESMs is partly due to a textbook concept that parameter values of a simulation model must be constant in contrast to variables that can vary over the time course of simulation (Forrester 1961). In reality, ecosystem properties, which parameters characterize in models, constantly evolve via acclimation and adaptation. In addition, a model, no matter how complex it is, can never represent all the processes of a system at resolved scales (Luo and Schuur 2020). Interactions of processes at unresolved scales with those at resolved scales should be reflected in model parameters. Therefore, Luo and Schuur (2020) argue that parameter values in ESMs may have to vary over space and time (i.e., heterogeneous parameter values) to represent changing properties of evolving ecosystems and unresolved processes.

The advent of big ecological data provides a golden opportunity to reconcile model representations with observations and quantify the spatial and temporal features of key parameters in soil carbon cycle simulation. Meanwhile, new techniques such as deep learning have been proposed to improve performance of ESMs (Reichstein et al. 2019). By constructing computational models with multiple processing layers and allowing the models to learn representations of data from multiple levels of abstraction (LeCun et al. 2015), deep learning techniques have promising applications in Earth system science, such as pattern classification, anomaly detection, regression, and space- or time-dependent state prediction (Reichstein et al. 2019). Exploration is warranted on how to properly employ deep learning techniques in reducing uncertainties of simulated carbon dynamics in ESMs.

Here, we propose the PROcess-guided deep learning and DAta-driven modeling (PRODA) approach to estimate

DOI: 10.1201/9781032711126-48

spatially and temporally heterogeneous parameter values for ESMs from extensive spatio-temporal datasets ('big data') at regional or global scales. The PRODA approach estimates parameter values at individual sites via data assimilation and builds a deep learning model to upscale the site-level estimates of parameters to predict spatially heterogeneous parameters at regional and global scales so that modeled and observed SOC are maximally matched.

In this chapter, we introduce the PRODA approach by using an extensive dataset of vertical soil profiles across the conterminous United States to optimize SOC representation by CLM5. We discuss the PRODA-optimized model performance in representing SOC stock and its vertical and spatial distributions, and compare it with results of the default model simulation and after the data assimilation optimization. In particular, we highlight that the PRODA approach helps the process model to achieve the most precise SOC distribution ever represented in ESMs. An accurate SOC representation in ESMs is critical to fully understand soil carbon feedbacks to future climate change.

THE WORKFLOW OF PRODA

Three fundamental components together formulate the PRODA approach (Figure 38.1a), namely the process-based model, the site-level data assimilation, and the deep learning model. Process-based models with their predefined structure and default parameter values simulate SOC distributions using meteorological forcing data. Data assimilation is used to estimate parameter values of a process-based model with soil carbon data at sites where the observations were made. The deep learning model is used to predict optimized site-level parameter values with their associated environmental variables. Eventually, the process-based model will apply the optimized parameter values upscaled by the deep learning model to simulate SOC distributions at regional or global scales.

Process-based model: We use the matrix representation of the Community Land Model version 5 (CLM5) to facilitate data assimilation and model simulation in the PRODA approach (Figure 38.1b). CLM5 is the latest version of CLM models (Lawrence et al. 2019). Its soil carbon module is similar to that in CLM4.5 (Koven et al. 2013), except that it has an option to change the number of soil layers from a default of 20. In this example, we use ten soil layers with a vertical transformation among carbon pools from the surface to a maximum depth of 3.8 m as in CLM4.5. The soil carbon component of CLM5 includes carbon transfer among four litter pools (coarse woody debris, metabolic litter, cellulose litter, and lignin litter) and three soil organic carbon pools (fast, slow, and passive SOC) in each of ten layers, totaling 70 pools. The thickness of soil layers increases exponentially from the surface layer (1.75 cm) to deep layers (151 cm), with a total depth of 3.8 m over the ten layers. Vertical carbon transfer between soil

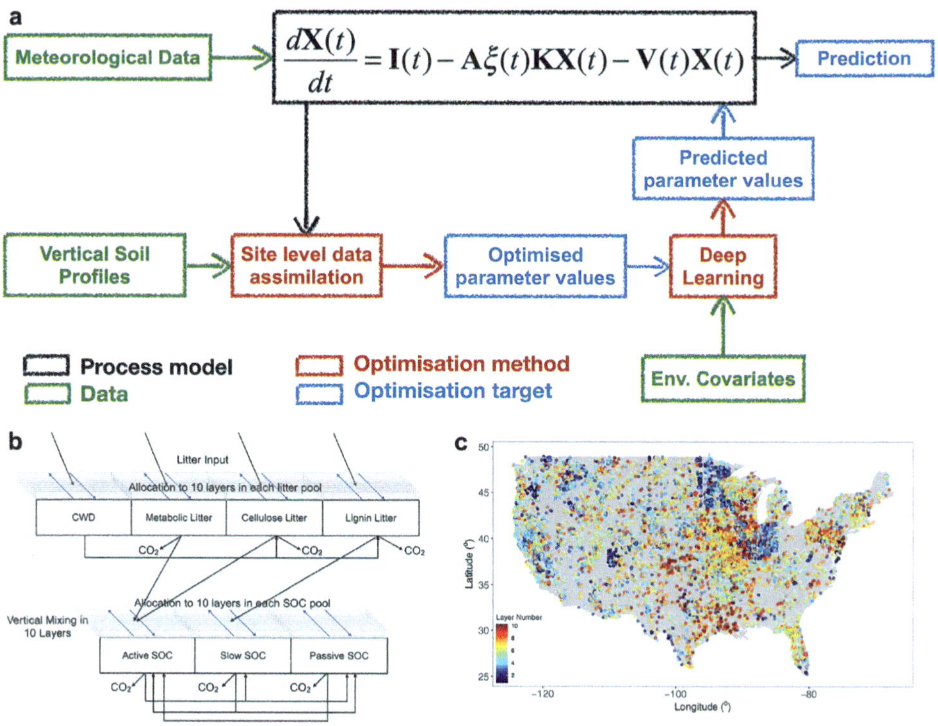

FIGURE 38.1 Workflow of the PRODA approach. (a) PRODA optimally matches CLM5 as the process-based model (b) with vertical SOC profiles on the conterminous United States (c). We first assimilate data at each site into CLM5 to estimate its parameters through the Markov Chain Monte Carlo method (MCMC). We further assemble the estimated site-level parameter values (i.e., the mean value of the posterior distribution after MCMC) as targets to be predicted by a multilayer neural network with environmental covariates in a deep learning model. The predicted parameters by the deep learning model are applied to CLM5 to optimize model representation of SOC distribution.

layers only occurs among the adjacent layers and represents both diffusive and advective carbon flux transportation caused by bioturbation and cryoturbation. The baseline advective rate of carbon flux is set to zero in CLM5 as a default, and this is assumed in our example as well.

We have discussed in Units 1–5 that carbon balance equations in land carbon models can be unified to a matrix form. For CLM5, we use the matrix equation to describe carbon transfer among the 70 pools with state variables $\boldsymbol{X}(t)$ as:

$$\frac{d\boldsymbol{X}(t)}{dt} = Bu(t) + A\xi(t)KX(t) - V(t)X(t) \tag{38.1}$$

where $\boldsymbol{B}$ is a vector (70×1) of partitioning coefficients from C input to each of the pools (unitless), and u(t) is C input rate (gC m^{-3} day^{-1}). $\boldsymbol{A}$ represents the transfer coefficients among litter and soil pools (unitless), including the transfer coefficients from four litter pools to three soil carbon pools as well as the transfer coefficients of SOC among soil carbon pools in the same layer. $\boldsymbol{\xi}$ (t) represents effects of environmental variables on decomposition of litter and soil (unitless). It includes scalars of temperature (ξ_T), soil water (ξ_W), oxygen (ξ_O), nitrogen (ξ_N), and depth (ξ_D). $\boldsymbol{K}$ indicates the decomposition rate of SOC in different litter and soil carbon pools (day^{-1}). $\boldsymbol{V}(t)$ represents SOC mixing among vertical soil layers through cryoturbation or bioturbation (day^{-1}). The t in parentheses indicates that the corresponding element is time-dependent. At a steady-state of the carbon cycle $\left(\frac{dX(t)}{dt} = 0\right)$, the SOC content of each carbon pool at each layer can be calculated as:

$$X(t) = \left[A\xi(t)K - V(t)\right]^{-1}\left(-Bu(t)\right) \tag{38.2}$$

Soil carbon data and site-level data assimilation: We use vertical SOC profiles in the conterminous U.S. from the World Soil Information Service (WoSIS) dataset (www.isric.org) for the site-level data assimilation (Figure 38.1c). The depth of recorded SOC layers ranges from the surface to more than 3 m. A total of 26,509 soil profiles with a total of 240,148 layers at different depths in the conterminous U.S. are available in this study.

In addition, we use the mean values of global net primary productivity (NPP) from 2000 to 2014 as carbon input (DAAC 2018). After running the CLM4.5 model to a steady state by the pre-industrial climate forcing (version code of forcing database: I1850CRUCLM45BGC), ten-year records of soil temperature and soil water potential of the conterminous U.S. were obtained from the model outputs.

The site-level data assimilation constrains parameter values of CLM5 with one data set of a vertical SOC profile at each site with the Markov Chain Monte Carlo (MCMC) method (as described in Chapter 22). Three parallel chains are generated each containing a test run of 20,000 iterations and a formal run of 30,000 iterations. To effectively capture the vertical distribution pattern of soil content along the depths, we put weights to observations at different depths in calculating the discrepancy between modeled and observed SOC content (i.e., cost function). These weights decrease exponentially with the depth (i.e., $weight_i = e^{-|depth_i|}$, where i refers to the layer's soil depth in observations) except for the top layer and the bottom layer, where a weight of ten is assigned to accelerate calibrating the upper and lower bounds of the SOC distribution curve. To monitor the efficiency of the MCMC process, an acceptance rate threshold is set. For Markov chains whose acceptance rate is higher than 50% or lower than 15%, the corresponding data assimilation results are rejected. After the MCMC process, the first half of the accepted parameter values in the formal run are discarded as burn-in. The Gelman-Rubin statistics of each parameter are then calculated for each soil profile to ensure the convergence of these three independent MCMC results. We randomly select one chain after eliminating the burn-in period to generate the posterior distributions of parameters. The mean value of the parameter's posterior distribution is calculated and chosen to serve as the training target in the deep learning model.

We evaluate the effectiveness of the site-level data assimilation by the coefficient of efficiency:

$$E = 1 - \frac{\sum\left(obs_i - mod_i\right)^2}{\sum\left(obs_i - \overline{obs}\right)^2} \tag{38.3}$$

where obs_i and mod_i are the observed and modeled SOC content at ith soil layer of one soil profile; $\overline{obs}$ is the mean value of observed SOC content of the soil profile. In this study, we take profiles having negative E values as invalid and discard the results from the corresponding deep learning model. Moreover, at those sites where an observation is available at only one soil depth, we do not apply the data assimilation to the data point. After those data sets are excluded, 25,444 out of 26,905 soil profiles, or 94.6% of the entire dataset, are used in the PRODA approach.

Deep learning model: We design a deep learning model with multiple processing layers to predict optimized parameter values with environmental covariates. A total of 60 environmental variables that describe the climatic, edaphic, and vegetation features at the observational sites is used. We used 80% of the total dataset to train and validate the neural network. After model training, we use the remaining 20% of the dataset to quantify the prediction accuracy of the deep learning model. The predicted parameter values are first compared with those retrieved in site-level data assimilation and then applied to the matrix CLM5 model to simulate soil organic carbon stock at each observational site. Meanwhile, we used the trained deep learning model to generate parameter maps across the United States based on gridded environmental covariates. The parameter maps are then applied to the matrix

CLM5 to simulate the SOC distributions across the United States at a resolution of 0.5 degrees.

SOC distributions optimized by data assimilation: To analyze the significance of the spatially-explicit parameter estimation of PRODA compared with a traditional approach, we perform a batch data assimilation using all the observational dataset as one batch in the MCMC method to estimate parameter values of CLM5 with data assimilation. The estimated parameter values from this method are spatially homogeneous, in contrast with the site-level data assimilation, which is a middle step of the PRODA approach to estimate spatially heterogeneous parameter values. SOC distributions simulated by CLM5, trained by the batch data assimilation versus the results by PRODA, can then be compared.

The batch data assimilation runs three parallel MCMC chains, each containing 50,000 iterations as test run and 200,000 iterations as formal run. Weights at different depth in calculating the cost function and acceptance control are the same as those in the site-level data assimilation. After the MCMC method, we first discard the first half of the accepted parameter values of the formal run as burn-in. The Gelman-Rubin statistics for each parameter are then calculated to ensure the convergence of these three independent MCMC results. We randomly select one Markov chain after eliminating the burn-in period to generate the posterior distribution for each parameter. We then randomly sample parameter values from the posterior distributions 1,000 times and apply the sampled parameter values to the CLM5 matrix model. We estimate SOC content distributions at different sites by calculating the average of the results. The same sampled parameter values are further assigned in CLM5 to estimate SOC content distributions at each grid cell on the map of the conterminous U.S. at a resolution of 0.5 degrees.

Reference SOC data products: We use two sets of SOC data, WISE30sec and SoilGrids250m (Hengl et al. 2017), as references to compare with spatial and vertical distributions of SOC obtained from our study over the United States. WISE30sec is an updated version of the dataset HWSD, generated by using traditional mapping methods at a resolution of 30 × 30 arc sec. SoilGrids250m is a global gridded soil information dataset generated by using machine learning techniques at 250 m resolution. We took data of SOC content over three depth intervals from these two datasets, 0–30 cm, 0–100 cm and 0–200 cm. All the original data with high resolution were resampled to a resolution of 0.5 × 0.5 degrees.

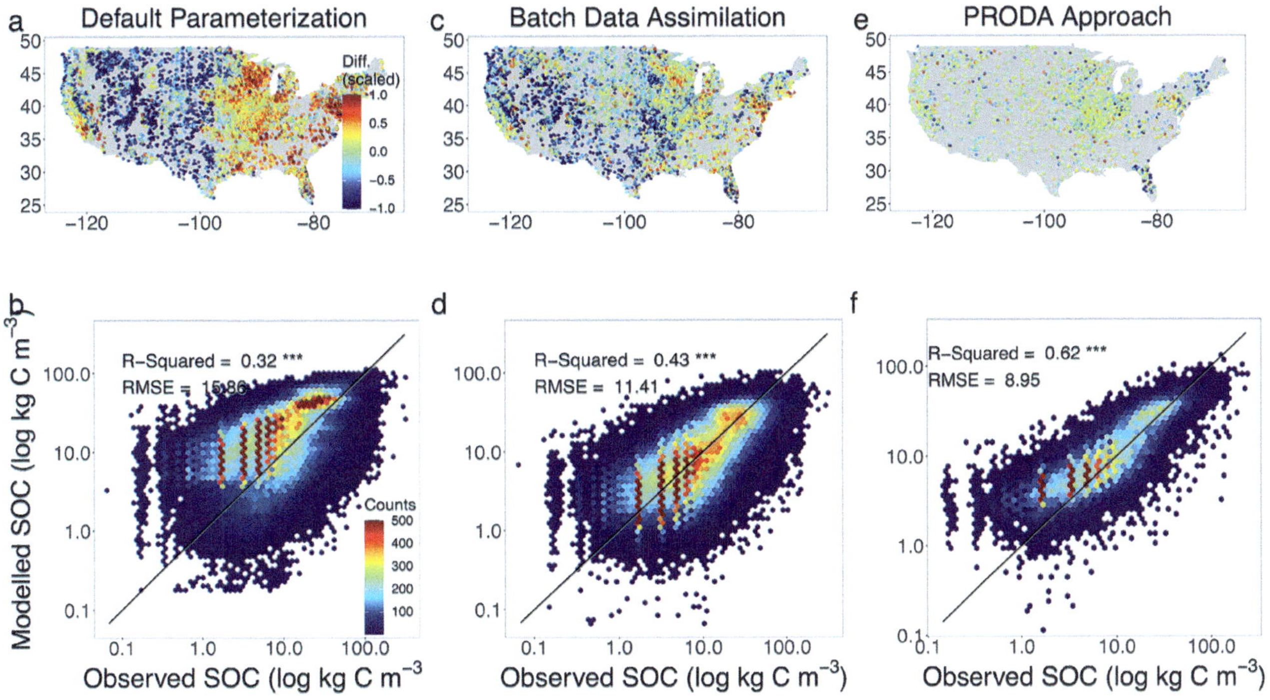

FIGURE 38.2 The agreement between observed and modeled SOC content with different approaches. SOC estimates modeled by CLM5 were extrapolated to the depths of observations to evaluate model performance. The upper panel indicates the deviation of the modeled SOC storage from the observation of the whole profile for each site. The lower panel shows the results of linear regression between observed and modeled vertical SOC content at different depths in different methods. In calculating the deviation of modeled SOC storage from observations, for better presentation, the positive (overestimation) and negative (underestimation) discrepancy between the observed and modeled SOC content were scaled based on the 95% quantile of the positive discrepancy and 5% quantile of the negative discrepancy, respectively. Meanwhile, only the results of the testing set were presented in PRODA approach.

TABLE 38.1
Performance of CLM5 in representing SOC distribution under different approaches

	Model Performance	
Method	**R^2**	**RMSE (kg C/m^3)**
Default CLM5	0.32	15.86
Batch Data Assimilation	0.43	11.41
PRODA Approach	0.62	8.95

Note: R^2 is the coefficient of determination from linear regression between the observed and modeled SOC content. RMSE is the root mean square error.

MODEL REPRESENTATION OF SOC CONTENT ACROSS OBSERVATION SITES

The original CLM5 model with default parameterization presents significant geographical biases on the estimation of SOC content in comparison with observations. Modeled SOC in the grid cell in which the site of observation was located is compared with observations (Figure 38.2a). SOC storage is systematically overestimated by the original model near the east and west coasts of the U.S. but underestimated in the Midwest. The consistency between observed and modeled SOC content is low, with $R^2 = 0.32$ and RMSE = 15.9 kgC m^{-3} (Figure 38.2b and Table 38.1).

The batch data assimilation method generates the distribution of SOC from continentally homogeneous posterior distributions of parameters estimated from all the observation data at once in data assimilation. With the batch data assimilation, the mismatch between observed and modeled SOC content in the CLM5 model is moderately reduced in the northern and eastern parts of the U.S. (Figure 38.2c). However, geographical biases in model representation of SOC are not eliminated. CLM5 optimized by the batch data assimilation still underestimates SOC storage in the Intermontane Plateaus and southern Great Plains. Meanwhile, overestimation still exists in the Great Lakes areas and the Northeast. Overall, CLM5 after optimization by the batch data assimilation explains 43% variation in the observed SOC content with RMSE = 11.4 kg C m^{-3} (Figure 38.2d and Table 38.1).

Through the deep learning model, the PRODA approach predicts the optimized parameter values at each site across the conterminous U.S. by its environmental variables. PRODA-optimized CLM5 achieves a better representation of SOC distribution compared to the batch data assimilation. Little systematic geographical biases in estimating SOC storage are observed across the study domain (Figure 38.2e). The modeled and observed SOC content are highly correlated with $R^2 = 0.62$ and RMSE = 9.0 kg C m^{-3} (Figure 38.2f and Table 38.1).

SPATIAL DISTRIBUTION OF SOC ACROSS THE CONTERMINOUS U.S.

We take point observations (Figure 38.3a–c) and estimations from WISE30sec (Figure 38.3d–f) and SoilGrids250m (Figure 38.3g–i) as references to compare the SOC estimations by CLM5 with default parameterization, optimized parameterization after the batch data assimilation, and the PRODA approach. At the continental scale, the reference data suggest large volumes of SOC in the northeast and northwest of the conterminous U.S. The magnitude of SOC content in these regions can be as high as 30 kg C m^{-2} for the 0–200 cm depth interval. Meanwhile, a decreasing gradient of SOC from the northeast to the southwest is observed. High SOC exists in areas across the Great Plains, extending from Texas to the Great Lakes.

The default CLM5 model (Figure 38.3j–l) captures the continental SOC content gradient from the northeast to the southwest but fails to reproduce sub-regional features of SOC distribution in the Great Plains. Meanwhile, SOC content in the east and northwest estimated by the original CLM5 is significantly higher than that indicated by the reference data. After optimization by the batch data assimilation, CLM5 reproduces the continental SOC gradient from the northeast to the southwest with reasonable values (Figure 38.3m–o). However, high SOC content in the Great Plains is still not well represented. The PRODA approach performs best overall, helping achieve the most realistic spatial SOC distribution (Figure 38.3p–r) in comparison with observations (Figure 38.3a–c) and data products (Figure 38.3d–i). In addition to capturing the continental SOC distribution pattern, the PRODA-optimized CLM5 presents more accurate subregional SOC distribution patterns in the Great Plains.

VERTICAL DISTRIBUTION OF SOC ACROSS THE CONTERMINOUS U.S.

We take results from WISE30sec and SoilGrids250m as references in estimated SOC stocks at different depth intervals (Figure 38.4). For the first 2-meter soil, WISE30sec suggests 243 PgC and SoilGrids250m estimates 269 PgC stored as SOC. Along the soil depth, WISE30sec suggests 98 PgC at 0–30 cm depth, 81 PgC at 30–100 cm, and 64 PgC at 100–200 cm. SoilGrids250m estimates 102, 86, and 81 PgC at the same three depth intervals, respectively.

The original CLM5 model with default parameterization substantially overestimates SOC stocks in comparison with the references at all three soil depths (Figure 38.4). Compared with the references, the overestimation becomes

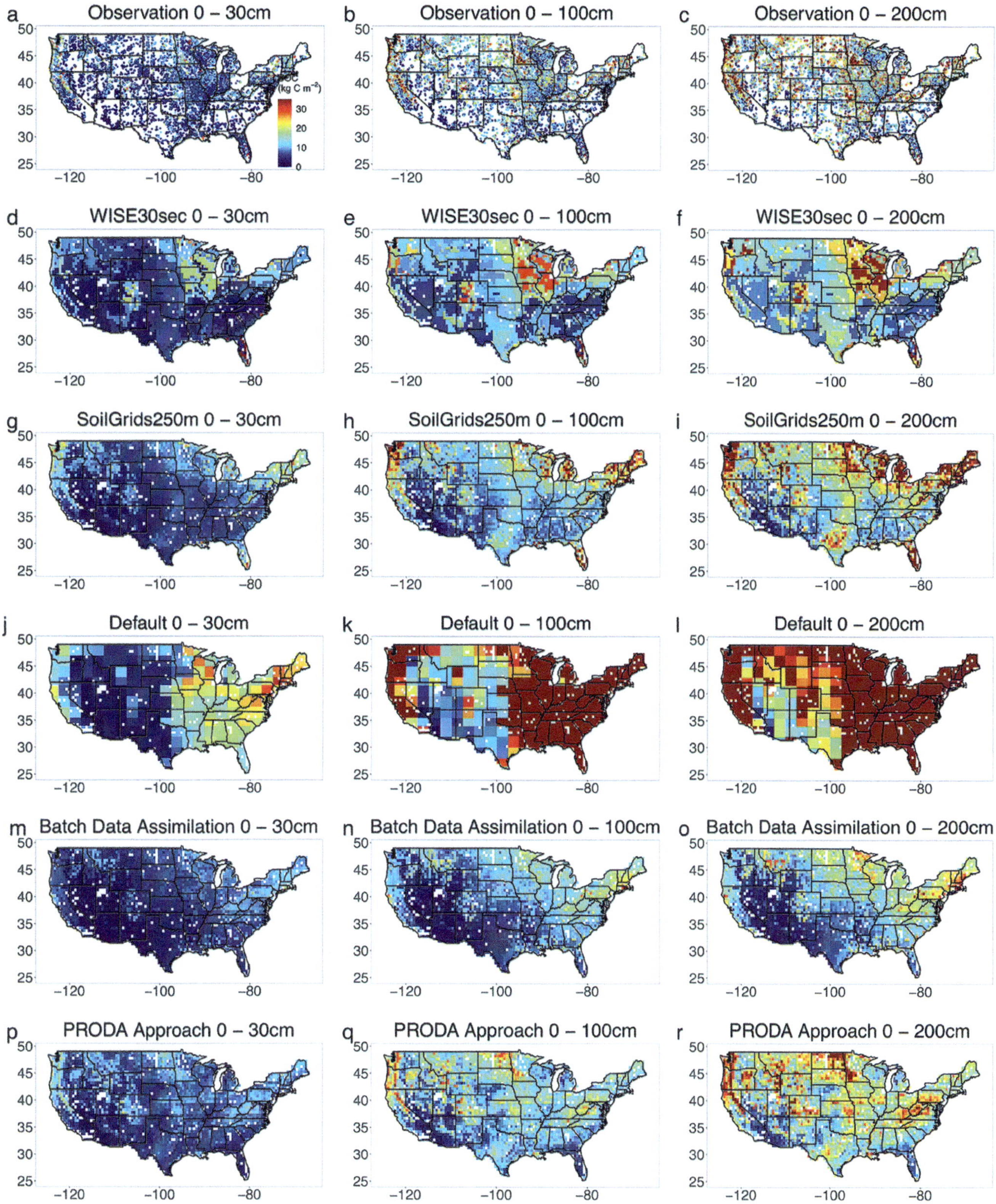

FIGURE 38.3 Modeled spatial SOC distributions in three depth intervals across the conterminous U.S. by different approaches and datasets.

stronger with increasing soil depth. Both the batch data assimilation and the PRODA approach help CLM5 estimate more reasonable SOC storage compared with the original CLM5 model. We estimate 165 PgC using the batch data assimilation and 246 PgC for the first 2-m soils using the PRODA approach.

For different vegetation types, the PRODA approach presents more accurate estimations of the vertical SOC distribution than the batch data assimilation (Tao et al. 2020).

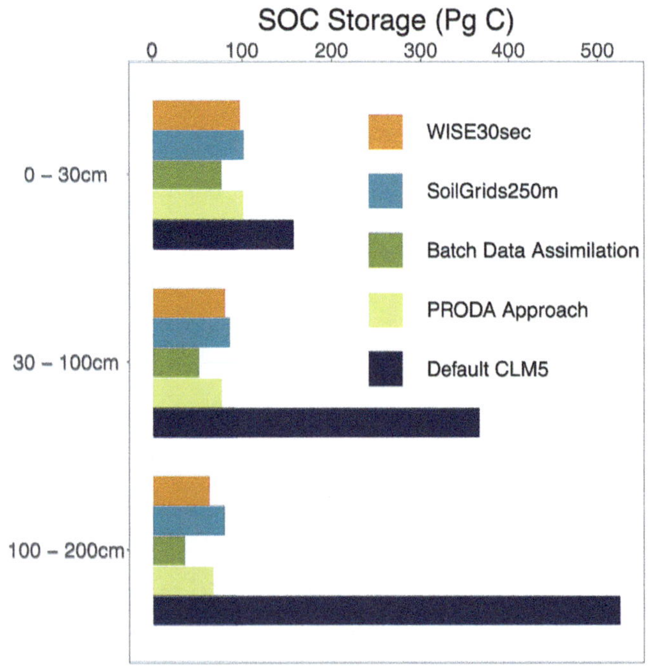

FIGURE 38.4 SOC storage across the conterminous U.S. at different depths estimated by different approaches and data sources.

CLM5 underestimates the SOC content in the evergreen forest, shrubland, savanna, grassland, and wetland regions after the optimization by the batch data assimilation. The PRODA-optimized CLM5, in contrast, presents the least biased estimations in comparison with observations at all depth intervals in the aforementioned regions.

TOWARD MORE REALISTIC REPRESENTATIONS OF SOC DISTRIBUTION

This chapter systematically explored the significance of spatially heterogeneous parameterization for the adequate prediction of SOC distribution in ESM, with CLM5 as a representative case. The results support the PROcess-guided deep learning and DAta-driven modeling (PRODA) as a promising approach to optimize model representation of SOC, utilizing the explanatory power implicit in immense observational data. PRODA considers biogeochemical processes in the soil carbon cycle while preserving strong big data analysis ability to integrate soil data into complex models. We compared the PRODA-optimized SOC representation by CLM5 with the default model simulation and the results optimized by batch data assimilation and conclude that PRODA helped CLM5 achieve the most accurate SOC representation. Indeed, no better fit to reference data on SOC has ever been simulated by process-based models.

In the past decades, different approaches have been developed for representation of SOC distribution (Figure 38.5). Soil scientists collect soil data and develop mechanistic understanding of soil carbon cycling from field observations or experiments. The simulation modeling approach conceptualizes those mechanisms into mathematical equations and strives to simulate SOC according to process understanding. Notwithstanding the detailed description of carbon cycle processes, the models struggle to realistically

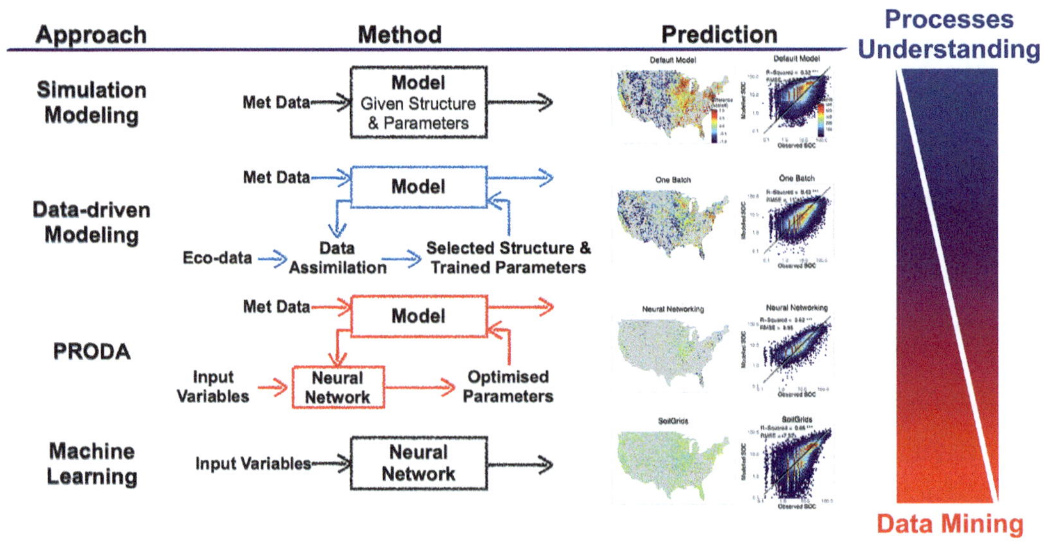

FIGURE 38.5 Schema of different approaches to represent SOC distributions. The PRODA approach benefits from both process understanding (as featured by simulation modeling) with the real-world information brought out by big data analysis from machine learning. The latter is primarily to obtain accurate representations of the spatial distribution of SOC and its underlying mechanisms.

simulate SOC distribution. Such unrealistic model simulations mainly arise from inadequate parameterization. Parameters that represent critical processes of the soil carbon cycle in the real world are not sufficiently constrained with widely distributed observational data. Therefore, it is difficult for process-based models to accurately represent SOC distributions. In our example, CLM5 with default parameter values substantially overestimates the total SOC storage of the conterminous U.S. and presents strong geographical biases in the representation of SOC distribution.

Batch data assimilation provides a way of incorporating observational data information into the process model to improve SOC simulation. Such data-driven optimization harmonizes site-level data information as a whole to adjust the parameter values for better representation of the SOC. We have shown in the example study that the optimized CLM5 with data assimilation successfully corrects the considerable overestimation of total carbon storage across our study domain.

In terms of representing the spatial variability of SOC, however, batch data assimilation fails to capture the spatial variability of observed SOC. The spatially invariant parameter values optimized from the batch data assimilation approach are insufficient in describing the heterogeneity of SOC distribution at large scales. In our example study, geographical bias still exists after the optimization by the batch data assimilation.

The PRODA approach solves the issue of geographical bias by using a deep learning model to first fully estimate parameters at the site level using the data assimilation and then upscales the site-level estimates of parameters to the whole U.S. continent. The spatially varying parameter values retrieved from the PRODA approach contribute to a more accurate model representation of SOC across the range of ecosystem types (vegetation class, soil type, geology, etc.) across the continent. PRODA-optimized CLM5 simulates the most realistic SOC distribution ever simulated by process models. The high agreement between observed and modeled SOC content (R^2 = 0.623 across the conterminous U.S.) achieved by the PRODA approach is comparable with that for harmonization mapping in SoilGrids250m by machine learning (R^2 = 0.635 across the globe) (Hengl et al. 2017), and greater than the agreement between separate gridded empirical data products (Wu et al. 2019).

More importantly, the PRODA approach paves the way for mechanistic understanding of the soil carbon cycle from big data analysis with machine learning. A recent study integrated 57,267 globally distributed vertical SOC profiles and a microbial-process explicit model (note that it is a different model from CLM5) using the PRODA approach (Tao et al. 2023). The study found microbial carbon use efficiency (CUE, the ratio of organic carbon accrued as microbial growth to the quantity of carbon utilized in microbial metabolism) to be a dominant process in determining global SOC storage and its spatial distribution (Figure 38.6). Results after fusing the global-scale database and the process-based model by PRODA suggested microbial CUE is at least four times as important as any other evaluated components (e.g., plant carbon input, decomposition, and vertical transport) in determining SOC storage and its spatial variation across the globe. Moreover, the emerging positive CUE-SOC relationship reflected the critical role of organic carbon partitioning by microbes at the global scale: a high CUE means more allocation to biomass and by-products, which leads to SOC accumulation, whereas a low CUE value indicates the partitioning of more carbon

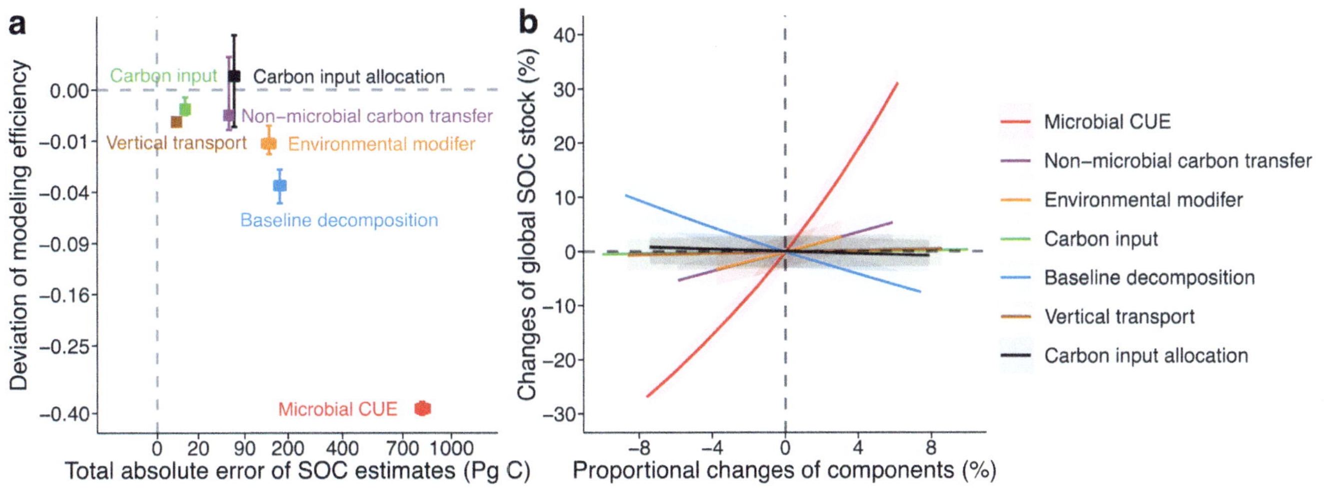

FIGURE 38.6 Microbial CUE as the primary regulator of global SOC storage. Spatially explicit model components obtained through the PRODA approach with the microbial model were substituted by their spatially invariant counterparts to assess their influence on SOC stocks (i.e., the sum of absolute deviations from PRODA estimates over the globe) and distributions (i.e., deviation of explained variation in observations defined by Equation 38.3) (a). We further proportionally changed the retrieved parameter values of different components and found that global SOC stock is most sensitive to changes of CUE (b). Error bars and shaded areas show the two-sigma confidence interval in the 200-time bootstrapping. Note that axes in panel a were scaled by a signed square root function.

towards cellular respiration, which drives SOC loss. The findings by Tao et al. (2023) helped prioritize future research on microbial processes in addition to SOC decomposition and organic carbon input for improving prediction of SOC dynamics. Understanding microbial processes underlying CUE and their environmental dependence is critical for predicting SOC feedback to a changing climate.

To summarize, a better understanding of SOC formation and its persistence at the global scale lies in leveraging the potential of both big data mining and process-based modeling. Machine learning alone is good at accurately describing SOC distribution, yet previous applications used in digital soil mapping (e.g., Chapter 36) focus only on the complex statistical relationship between environmental variables and SOC. Conversely, while process-based models summarize the understanding of complex mechanisms in the soil carbon cycle from empirical studies, a model calibrated at one site is usually not applicable at larger scales due to the high spatial heterogeneity of soil processes. In this chapter, we showed that the PRODA approach provides a common tool for reconciling extensive observations in fields and theoretical reasoning in modeling. It not only precisely maps SOC distributions but also provides the spatial patterns of different mechanisms (as represented by different parameters) of the soil carbon cycle at continental and global scales. By disentangling how these mechanisms vary with environments and quantifying their importance to SOC storage, new findings and relationships emerging from the PRODA approach will further stimulate both new empirical studies in laboratory and field, and improvement of process-based models.

SUGGESTED READINGS

Tao, F., Z. Zhou, Y. Huang, Q. Li, X. Lu, S. Ma, X. Huang, Y. Liang, G. Hugelius, L. Jiang, R. Doughty, Z. Ren, and Y. Luo. 2020. Deep Learning Optimizes Data-Driven Representation of Soil Organic Carbon in Earth System Model over the Conterminous United States. *Frontiers in Big Data*, *3*, 17.

Tao, Feng, Yuanyuan Huang, Bruce A Hungate, Stefano Manzoni, Serita D Frey, Michael WI Schmidt, and Markus Reichstein, et al. 2023. "Microbial Carbon Use Efficiency Promotes Global Soil Carbon Storage." *Nature*, 1–5.

QUIZ

1. What is the main difference, in terms of parameterization scheme, between the batch data assimilation and the PRODA approach as described in this chapter?
2. Describe the input and output of the deep learning model in the PRODA approach?
3. What is the advantage of the PRODA approach in comparison to conventional machine learning methods in representing SOC distributions?

39 Hybrid Modeling in Earth System Science

Yu Zhou
Cornell University, Ithaca, USA

Hybrid modeling harnesses data-driven methods to enhance earth system models for higher accuracy and deeper understanding, which brings about a new era of innovation and insights in earth system sciences. This chapter uses the estimation of latent heat as an example to illustrate different strategies of hybrid modeling.

INTRODUCTION

Hybrid modeling in earth system science combines two distinct yet complementary approaches: data-driven methods, such as machine learning (ML) and artificial intelligence (AI), and process-based earth system models (ESMs). This combination has the potential not only to improve the accuracy of the model, but also to deepen our understanding of ecological processes, allowing for more accurate predictions of system dynamics.

ML/AI models have demonstrated impressive abilities in capturing complex, non-linear relationships between target variables and their driving factors, enhancing data analysis in different fields, such as image classification and natural language translation. As noted in Chapters 33–35, such models are increasingly applied in the earth system sciences as well. The primary limitations of utilizing such methodologies in science is their nature as "black boxes" that present difficulties in interpretation, their lack of physical significance and scientific justification, and the risk of generating imprecise results when applied beyond their training domain, i.e., out-of-sample scenarios, and used for prediction.

Conversely, an ESM functions as a "glass box" constructed upon physical-based principles and empirical knowledge. It describes systemic changes using empirical equations and offers process-based understanding about changes in climate, ecosystems, and their interactions across space and time. ESMs have the added benefit of predicting future changes, both over long (decades to centuries) and short (hours to weeks) time frames. Nonetheless, ESMs face the challenge of accurately representing the intricate dynamics of the terrestrial ecosystem, resulting from inherent limitations associated with simplified equations and empirical parameters. The equations and parameters adopted by most such models are derived from limited experiments and observations in specific geographic contexts.

Nowadays, there is an opportunity for parameters and equations of ESMs to be refined, since data availability has been increasing dramatically, a trend commonly referred to as "big data". Big data encompasses measurement data collected from laboratory experiments, field measurements, unmanned aerial vehicle (UAV) and flight campaigns, and satellite observations. One important explanation for the greatly increased availability of such data compared to the earlier era in which most current ESM frameworks were originally developed is the open data policies adopted by many academic journals in recent years. We can take advantage of current data availability to improve ESMs in terms of parameterization and process-based optimization by the cautious incorporation of data-driven methods, aiming to resolve accuracy and interpretability issues and ultimately enhance model performance. Reichstein et al. (2019) listed several aspects that characterize hybrid modeling, including (1) improving parametrization; (2) replacing a 'physical' sub-model with an ML/AI model; (3) analysis of model-observation mismatch; (4) constraining sub-models; and (5) surrogate modeling or emulation. For example, parameters in ESMs may vary spatially and temporally (Luo and Schuur, 2020), but both traditional hand tuning and data assimilation are computationally infeasible to get the optimized parameters. Tao et al. (2020) proposed a hybrid modeling approach, i.e., PRODA (Tao and Luo, 2022), which integrates data assimilation at the plot scale and machine learning to extrapolate optimized parameters to the global scale. This hybrid method has improved our understanding of changes in soil organic carbon (see Chapter 38 and Tao et al., 2023). This chapter focuses on a hybrid modeling technique which involves replacing a physical sub-model with an ML/AI model and constraining sub-models, using an example of latent heat flux simulation to illustrate the concept. A sub-model in the ESM can be replaced directly by an ML/AI model or by a knowledge-constrained ML/AI model.

AN EMPIRICAL SUB-MODEL IN AN ESM

Here we take a common sub-model of surface latent heat flux (LE) as an example. In most ESMs, LE is calculated using the Penman–Monteith (PM) equation:

$$LE = \frac{\Delta(R_n - G) + \frac{\rho_a c_p (e_s - e_a)}{r_a}}{\Delta + \gamma\left(1 + \frac{r_s}{r_a}\right)} \tag{39.1}$$

DOI: 10.1201/9781032711126-49

where R_n is the net radiation, G is the soil heat flux, $(e_s - e_a)$ represents the vapor pressure deficit of the air, ρ_a is the mean air density at constant pressure, c_p is the specific heat of the air, Δ represents the slope of the saturation vapor pressure temperature relationship, γ is the psychrometric constant, and r_s and r_a are the bulk surface and aerodynamic resistances. The surface resistance describes the resistance of vapor flow through stomatal openings, total leaf area and soil surface. Aerodynamic resistance describes the resistance from the vegetation upward and involves friction from air flowing over vegetative surfaces.

r_s and r_a are two critical parameters in estimating LE, but in most ESMs, they are calculated based on empirical knowledge. The transfer of heat and water vapor from the evaporating surface into the air above the canopy is determined by the aerodynamic resistance:

$$r_a = \frac{\ln\left(\frac{z_m - d}{z_{om}}\right)\ln\left(\frac{z_h - d}{z_{oh}}\right)}{k^2 u_z} \tag{39.2}$$

where z_m is the height of wind velocity measurements, z_h is the height of air temperature and humidity measurements, d is the zero-plane displacement height, z_{om} is the roughness length relative to momentum transfer, z_{oh} is the roughness length relative to heat and vapor transfer, u_z is the wind velocity at height z_m, and k is the von Karman constant (0.41). The surface resistance for full-cover canopies is often expressed by:

$$r_s = \frac{r_l}{LAI_{eff}} \tag{39.3}$$

where r_l is the bulk stomatal resistance of a well-illuminated leaf, and LAI_{eff} is the effective leaf area index, usually taken as 0.5 LAI for a full cover canopy; or more generally:

$$r_s = r_a \frac{\Delta}{\gamma}\beta - 1 + 1 + \beta\frac{\rho c_p VPD}{\gamma} \tag{39.4}$$

where β is the Bowen ratio (the ratio between the sensible and latent heat fluxes).

Since values of parameters used in calculating r_s and r_a are determined by the vegetation coverage, canopy type, and can also differ across space, it is not possible to get these values for global simulation.

The PM equation has several limitations. It relies upon several assumptions, including assuming a uniform vegetation canopy, constant aerodynamic resistance, and a uniformly mixed boundary layer. The empirical parameters, r_s and r_a, have limited ability to adapt to changing vegetation composition or long-term climate conditions. As a result, the PM equation does not explicitly account for the state of vegetation and soil moisture, both of which can impact transpiration rates.

REPLACING THE ESM SUB-MODEL WITH AN ML/AI SURROGATE

ElGhawi et al. (2023) proposed using a ML/AI surrogate model to estimate LE directly from environmental conditions, rather than relying solely on a physics-based PM model that includes multiple uncertain parameters. In the simplest case, the surrogate model could be entirely data-driven (see Figure 39.1a). The ML model is a feed-forward neural network (FNN) that directly predicts LE based on environmental drivers including water availability index (*WAI*), vapor pressure deficit (*VPD*), air temperature (*TA*), R_n, incoming shortwave radiation (SW_{in}), mean incoming shortwave potential (SW_{pot}), G, and wind speed (*WS*). The loss function (***L***) is defined as the sum of difference between observed and modeled LE at each time step across the simulation period. The neural network (NN) usually consists of multiple layers of nodes. The activation value in each node per layer is the weighted sum of the activations from prior layers in addition to bias. The bias is modified by non-linear activation functions. The nodes are connected between each layer resulting in an end-to-end fully connected NN. The use of hidden layers creates deeper neural networks, but presents a drawback: the vanishing gradient problem. This issue can be bypassed by implementing the rectified linear units (ReLU) activation function. The non-linear functions utilized in this work include ReLU and Softplus, which enable the network to perform non-linear calculations. Hyperparameters are predetermined before training and affect the performance of a model. Tuning hyperparameters involves algorithms and tools to search for optimal combinations of values for epochs, batch size, learning rate, number of hidden layers, hidden size, and weight decay (Table 39.1). Various methods can be employed to perform hyperparameter tuning on neural networks, such as grid search, random search, and Bayesian optimization. Hyperparameter optimization can enhance both the accuracy and efficiency of a neural network. This is an essential step in achieving optimal performance.

For other cases in Figure 39.1, the hybrid models consist of two FNNs that separately surrogate estimations of latent variables r_s and r_a instead of using previous empirical equations (Equations 39.2–39.4). The physically based PM equation is retained for mechanism understanding and integrated into the network architecture where the estimated latent variables are plugged into the equation along with other variables and constants need to calculate LE. Therefore, the hybrid models learn and predict the variations of the r_s and r_a that exist between the initial input variables and the resulting target variable LE in the following step using the PM equation. The difference among theory-driven models (Figure 39.1b–d) is the loss function. In the unconstrained theory-driven model, similar to a purely data-driven approach, ***L*** is the difference between modeled and observed LE (Figure 39.1b). However, this theory-driven model suffers from an equifinality problem due to its design, i.e., different combinations of r_s and r_a can result in the same LE prediction. Equifinality, or non-uniqueness, arises when dissimilar model parametrization or structures generate equivalent representations of the system.

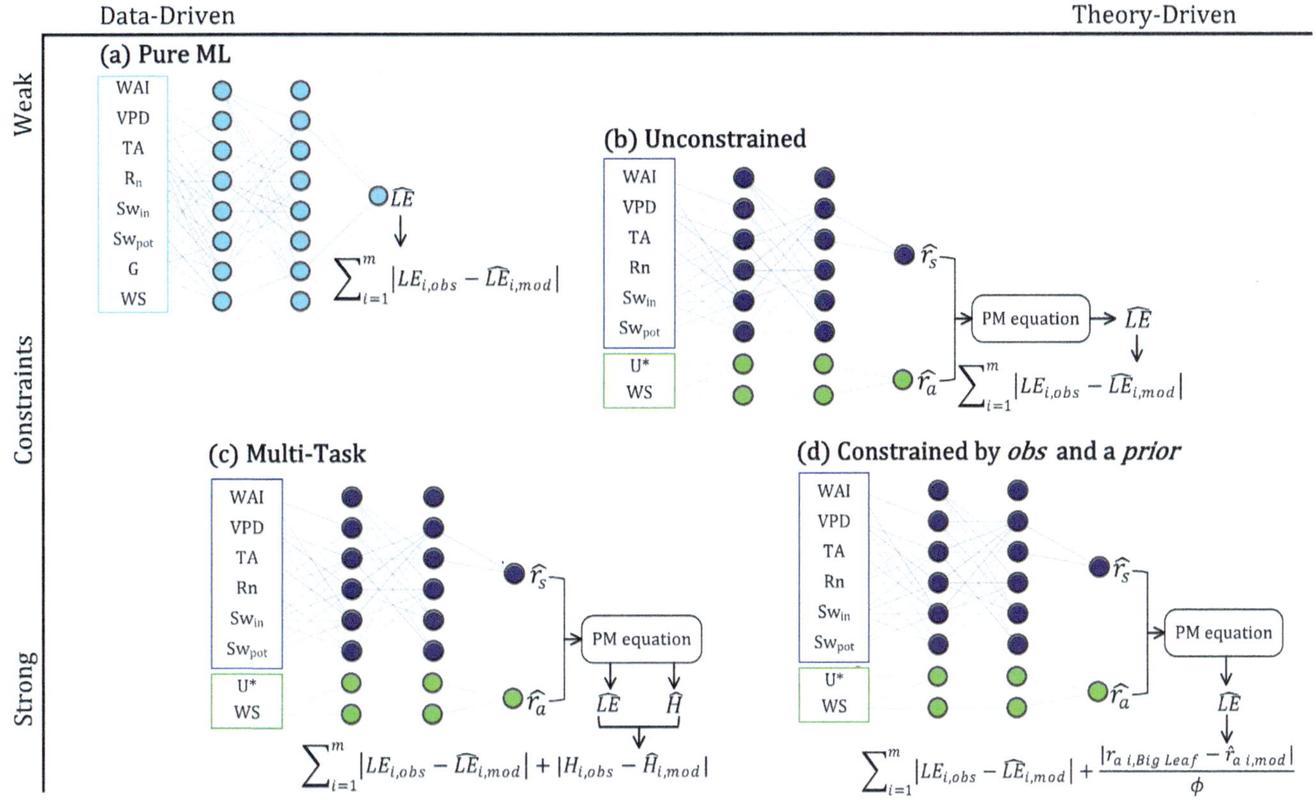

FIGURE 39.1 Schematic overview and classification of all models with respect to being more theory- and/or data-driven, as well as the strengths of the constraints on the loss function. Multiple surrogate models to simulate latent heat flux, including (a) purely data-driven only and (b–d) physics-informed learning by adjusting the constraints on the loss function. Cyan, dark blue, and green indicate the input variables and neural network to simulate latent heat flux, surface (r_s), and aerodynamic (r_a) resistances, respectively. The variables and hyperparameters of each model are included in Table 39.1.

Adopted from ElGhawi et al. (2023).

TABLE 39.1
Variables and hyperparameters used for the different models

	Variables		Hyperparameters					
Models	**Latent variable**	**Target variable**	**Batch size**	**Hidden size**	**Hidden layers**	**Learning rate**	**Weight decay**	**Epochs**
Pure ML LE model	NA	*LE*	1000	32	3	0.0005	0	2000
Under-constrained hybrid model	r_a	LE	1000	32	2	0.005	0.0001	2000
	r_s				4			
a priori constrained hybrid model	r_a	*LE*	1000	32	2	0.005	0.0001	2000
	r_s				4			
Multi-task learning hybrid model	r_a	*H*	1000	32	2	0.0005	0	2000
	r_s	*LE*			4			

Source: Adopted from ElGhawi et al. (2023), supplement.

CONSTRAINING A SUB-MODEL

To identify and reduce equifinality of the theory-driven model, one way is to restrict the parameter space through model regularization. This can be achieved through two approaches, including either additional theory or data via additional loss terms. In the two constrained cases (Figure 39.1c–d), the additional loss term adds constraining ability in modeling r_s and r_a. The multitask case adds sensible heat (H) as the other target variable calculated as:

$$H = \frac{\rho c_p (T_s - T_A)}{r_a} \tag{39.5}$$

where T_s and T_A are surface and air temperature respectively. The T_s is estimated using the Stefan–Boltzmann equation:

$$T_s = \sqrt[4]{\frac{LW_{out}}{\sigma\epsilon}} \tag{39.6}$$

where LW_{out} is the outgoing longwave radiation, σ is the Stefan–Boltzmann constant (5.789×10^{-8} $Wm^{-2}K^{-4}$), and ϵ is emissivity (dimensionless). The $\boldsymbol{L}$ includes the difference between modeled and observed H because r_a is used in H calculation, therefore, adding H in the loss function can constrain the estimation of r_a.

Similarly, in case (d) of Figure 39.1, the alternative model "Big Leaf" is introduced to estimate r_a as a *prior*, and the difference between two modeled r_a values is the additional term in loss function. The *prior* is not considered to be fully accurate, and the relative importance of r_a in the new loss is regulated by ϕ, which is varied between the high influence and low influence of the constraint.

EVALUATING PERFORMANCE OF DIFFERENT SURROGATE MODELS

ElGhawi et al. (2023) evaluated the hybrid models' consistency with respect to the interannual variability of *LE* for different sites (Figure 39.2). The R^2 values range between 0.47 and 0.49 for the interannual *LE* anomalies for forests, and thus exhibit a comparable performance to predictions at half-hourly frequency (R^2 between 0.46 and 0.49). Overall, the evaluation of the models across various temporal scales indicates their ability to learn not only the dominant structure of the diurnal and seasonal cycle, but also the more nuanced year-to-year anomalies. This consistency demonstrates that the models capture

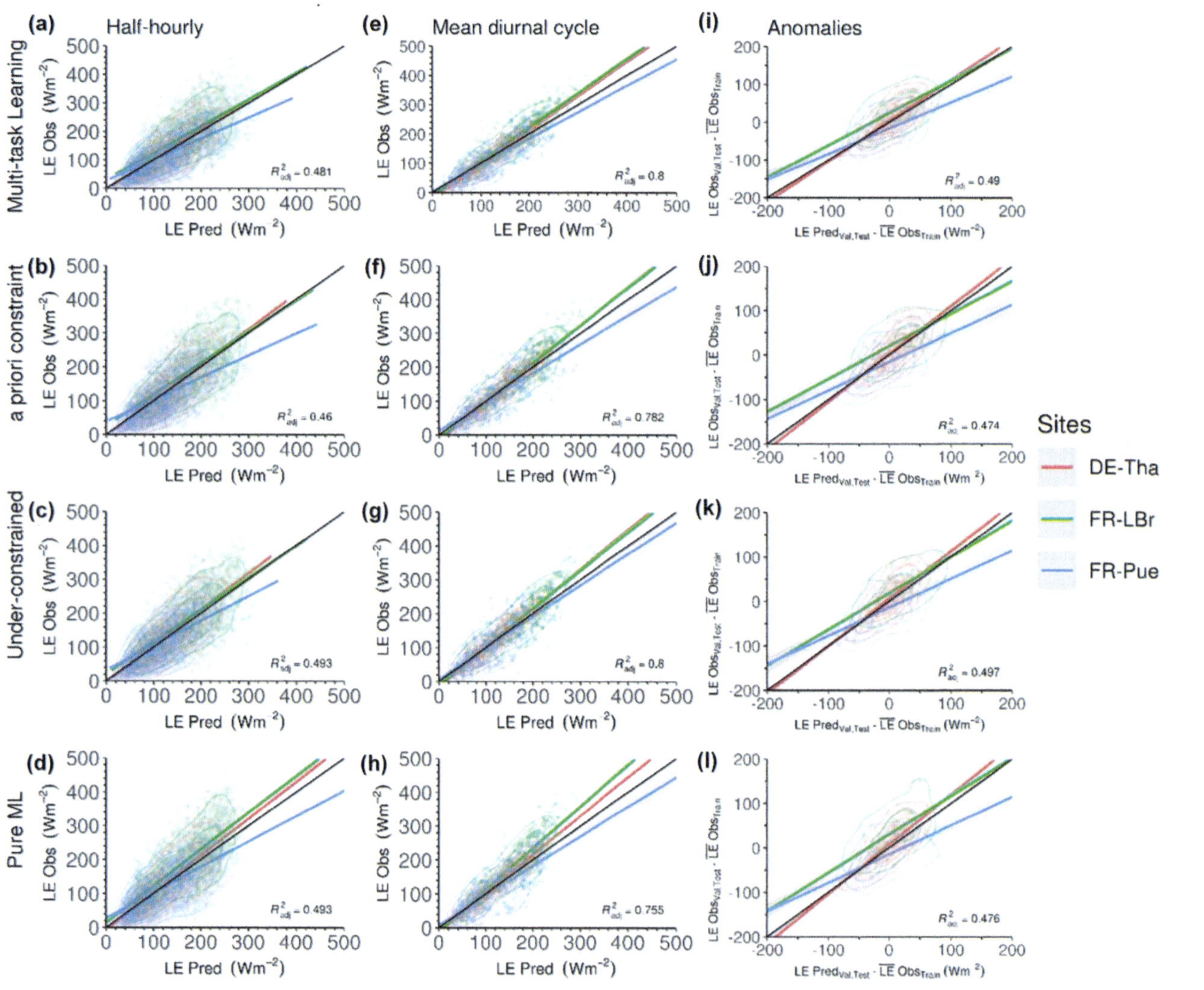

FIGURE 39.2 Evaluation of *LE* predictions at different temporal scales for forests. (a–d) show predictions against observations at a half-hourly scale for different models; (e–h) show predictions against observations at mean diurnal scale; (i–l) show *LE* anomalies at an interannual scale for the different models. The colored lines represent the linear regression lines that fit linear models for each site. The contour lines represent 2D kernel density estimate.

Adopted from ElGhawi et al. (2023).

the physically accurate dependence of the meteorological predictor variables that control *LE*.

CHALLENGES AND FINAL REMARKS

The success of ML/AI has inspired scientists to combine data-driven and process-based modeling approaches for hybrid modeling to discover intrinsic features from observational data (Reichstein et al., 2019). Hybrid modeling helps gain a better understanding of earth system science problems (Shen et al., 2023), such as the memory effect on ecosystem dynamics (Kraft et al., 2019). ESMs provide information about the physical mechanisms of real-world processes, which offer a check against data-driven approaches to avoid false discoveries and inconsistencies with established biophysical processes and interactions. To further advance hybrid modeling, a current frontier uses knowledge-guided ML/AI models (such as physics-informed natural network, PINN) that incorporate physics principles and human knowledge into the model construction and training process to optimize learning efficiency. Consequently, fostering collaboration between ML/AI experts and domain scientists becomes essential in the realm of AI for earth system science.

SUGGESTED READINGS

ElGhawi, R., Kraft, B., Reimers, C., Reichstein, M., Körner, M., Gentine, P. and Winkler, A.J., 2023. Hybrid modeling of evapotranspiration: Inferring stomatal and aerodynamic resistances using combined physics-based and machine learning. *Environmental Research Letters*, 18(3), pp. 034039.

Reichstein, M., Camps-Valls, G., Stevens, B., Jung, M., Denzler, J., Carvalhais, N. and Prabhat, F., 2019. Deep learning and process understanding for data-driven Earth system science. *Nature*, 566(7743), pp. 195–204.

QUIZ

1. List the advantages and limitations of earth system models and the machine learning/artificial intelligence models, respectively.
2. How can current ESMs be improved by machine learning/artificial intelligence?
3. Can you give an example in your field of work of an empirical model that can be improved by ML/AI algorithms? What would be the design and constraints of the hybrid model?

40 Practice 10

Deep Learning to Optimize Parameterization of CLM5

Feng Tao
Cornell University, Ithaca, USA

This practice offers guidance on how to use the PROcess-guided deep learning and DAta-driven modeling (PRODA) approach to integrate observations with the biogeochemical module of the Community Land Model version 5 (CLM5) to best represent regional soil organic carbon distributions. Over three exercises, we focus on how to build, train, and tune a deep learning model in the PRODA approach to predict parameters estimated from site-level data assimilation. Readers can use the CarboTrain platform to explore different deep learning options and to understand and modify the optimization methods.

RATIONALE OF ESTIMATING PARAMETER VALUES BY A DEEP LEARNING MODEL

In Chapter 38, we discussed the performance of different approaches that assimilate observations into CLM5 to best represent soil organic carbon (SOC) stocks across the conterminous United States (Tao et al. 2020). We concluded that the PRODA approach outperforms data assimilation alone by fully interpreting the spatial heterogeneity of parameters. In the PRODA approach, parameter values are first retrieved where the observation resides through the Markov Chain Monte Carlo (MCMC) method and then upscaled to the region by a deep learning model. Eventually, we apply the parameter values predicted by the trained deep learning model to CLM5 to simulate SOC stock and distributions (see Chapter 38 for details of the PRODA workflow).

We use various environmental variables to predict the parameter values by a neural network in the deep learning model. The rationale behind this procedure is that parameters in the process-based model can be expected to vary with environmental conditions over space and time in order to account for changing properties of evolving ecosystems and unresolved processes (Luo and Schuur 2020). The relationships between parameter values and environmental variables, however, are often not easily identified so as to be explicitly represented in model structure. Therefore, we introduce a deep learning model to explore such complex relationships. Specifically, we set the local environmental variables as the input and the estimated parameter values in the site-level data assimilation as the prediction target in a deep learning model. Then, the deep learning model is trained to best predict the parameter values based on input environmental variables (Figure 40.1).

WHAT IS A NEURAL NETWORK?

Chapter 37 introduced the basic concepts of machine learning/deep learning. In this practice, we focus on how to build and optimize a deep learning model that is structured by a multilayer neural network. Four elements together construct the skeleton of a typical multilayer neural network, which are: **inputs**, **outputs**, **neurons**, and **neuron weights** (Figure 40.2). In this practice, the inputs are environmental variables that are relevant to the parameters in CLM5, such as climatic, edaphic, and vegetation variables. The outputs are the parameter values we predict using the trained neural network, which will be further applied in CLM5 to simulate SOC distributions. The neurons are the basic unit of a neural network. They distribute at each hidden layer of the neural network (Figure 40.2), receive information from the inputs or the previous layer, and generate all possible predictions for the next layer or as the final outputs. Finally, the neuron weights are the weights assigned to the predictions of each neuron. We get the final neural network predictions by combining all the predictions by different neurons with their weights.

In addition to the four basic elements of a neural network, we use the **activation function** to generate the neuron weights. We generally use nonlinear activation functions to enable the neural network to explore the nonlinearities between the inputs and the final outputs.

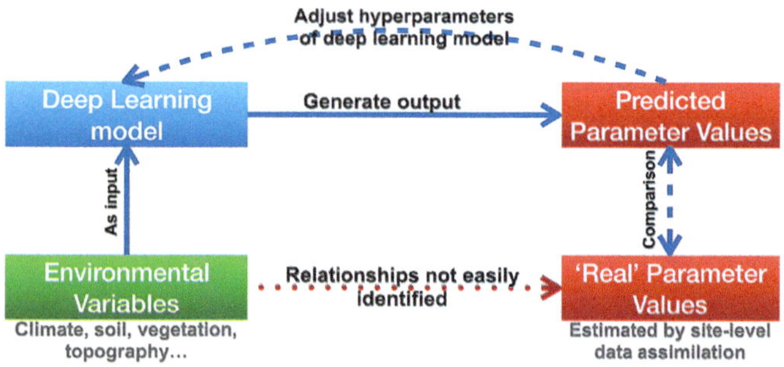

FIGURE 40.1 Schematic diagram of using a deep learning model to predict parameter values. The deep learning model is used to interpret relationships between environmental variables and parameter values retrieved from the site-level data assimilation.

DOI: 10.1201/9781032711126-50

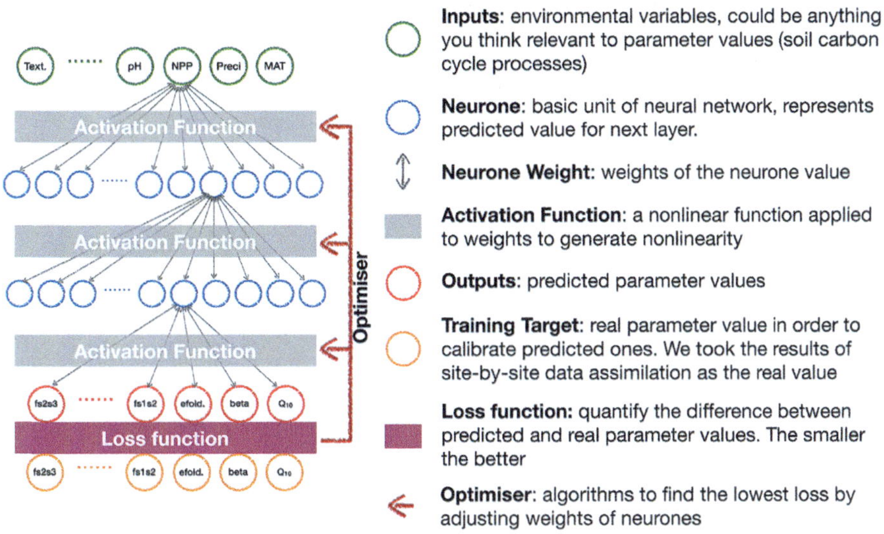

FIGURE 40.2 Basic elements in a typical multilayer neural network.

We train the neural network to best predict our **training target** (i.e., the parameter values retrieved from the site-level data assimilation) via a certain optimization algorithm. A **loss function** is used to quantify the difference between the predicted parameter values from the "real" parameter values retrieved from the site-level data assimilation. The lower the loss function value is, the closer the predictions are from the "real" parameter values, and the better the model performance is.

The loss function value alone, however, cannot initiate the optimization. We need an algorithm to determine how to adjust the neural network according to the results of the loss function so that the final predictions can best fit the training target. In the neural network, we have the **optimizers** to adjust the weights of neurons according to the results of the loss function. Ideally, after sufficient training, the optimizer will eventually lead the neural network to reach a point where the loss function reserves the lowest value it can pursue. We regard the neural network at this point as reaching its global optimum. Predictions by the neural network will consequently be the closest to the training targets.

HYPERPARAMETERS IN THE NEURAL NETWORK

Hyperparameters are the parameters whose values control the training process in the neural network. Hyperparameters that control the shape of a neural network include the number of hidden layers and the neuron numbers of each hidden layer. Hidden layer numbers determine the depth of a neural network. Neuron numbers of each hidden layer, at the same time, control the width of the neural network. Choices of hidden layer numbers and neuron numbers are largely empirical. You can try neural networks with different shapes and choose the one that can best predict your training target.

The epoch number determines how many times the deep learning algorithm will go through the whole dataset for training. During each epoch, the neural network can propose a set of neuron weights and adjust them by the optimizer according to the loss function results. The number of epochs ranges from hundreds to thousands in different deep learning applications. You may try different numbers to find the best epoch number that allows the loss function value to be minimized, so that the neural network can accurately predict the training target.

In addition, the batch size defines how many training data you want to use as one batch when working through the whole training dataset in each epoch. For example, for a training set with 10,000 samples, if we set the batch size as 50, then it will take 200 iterations $\left(iteration\,number = \frac{sample\,size}{batch\,size}\right)$ to go through the whole training set in each epoch of optimization. Possible choices of batch size vary from 1 to the size of the whole training set. Setting batch size as 16, 32, or 64 would be a plausible start.

We also need to decide several hyperparameters that control the optimization process in the neural network before initiating the training. You may need to specify the loss function, activation function, and the optimizer. We use the Keras package in Python to build and train the neural network in this practice. Keras provides multiple choices for these hyperparameters. The loss function can be the mean squared error (expressed as "mean_squared_error" in Keras) or other functions that can quantify the difference between predicted and target values. The activation function can be the Rectified Linear Unit (expressed as "ReLU" in Keras), the hyperbolic tangent function (expressed as "tanh" in Keras), the sigmoid function (expressed as "sigmoid" in Keras), other activation functions that are pre-defined by Keras, or a function that you define yourself. Keras provides several options for the optimizer. Choosing "Adam", "Adadelta", or "RMSprop" would generally be a good start. You can refer to the website https://keras.io/api/ for more possibilities of the hyperparameters you can use in Keras.

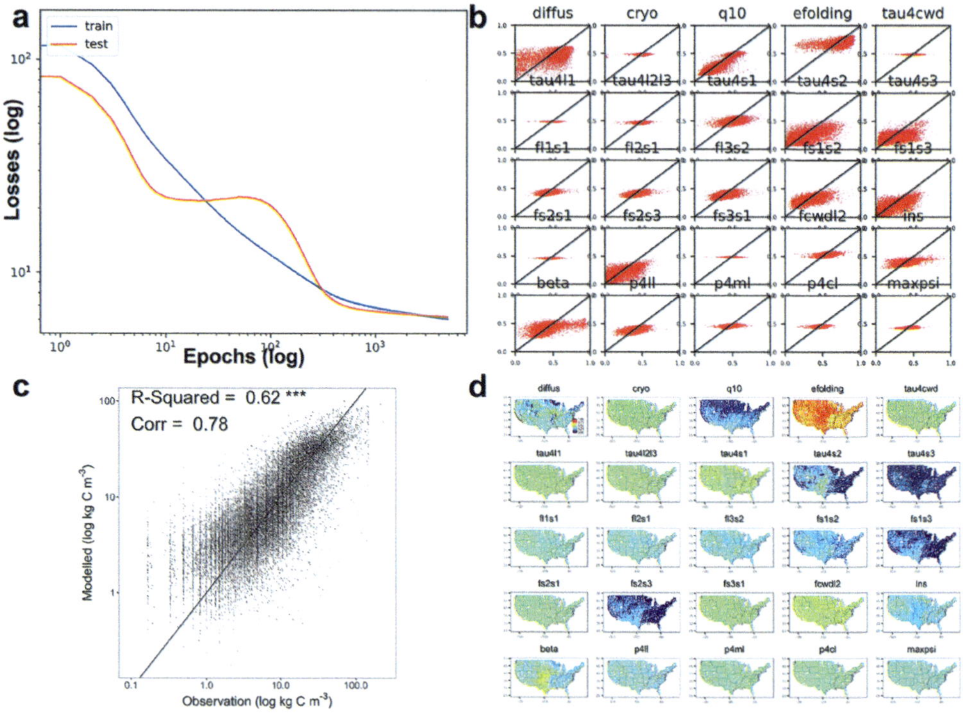

FIGURE 40.3 Output figures in Exercise 1.

Exercise 1

Building and training a neural network that uses environmental variables to predict parameter values in CLM5. Follow the instructions in CarboTrain:

a Select **Unit 10**
b Select **Exercise 1**
c Select **Output Folder**
d **Open Source Code**
e Read section "Setting NN Structure" to get familiar with hyperparameters that we used in this neural network.
f **Run Exercise**
g Check results in your Output Folder. Four figures will be generated (Figure 40.3). The figure in loss_nn.png describes the changes of the loss function value in the training and validation set with increased epochs. para_nn.png describes how well the trained neural network predicts different parameters. nn_obs_vs_mod.jpeg indicates how well the CLM5 model can simulate SOC after applying the predicted parameter values from the trained neural network. map_para_us.jpeg describes the predicted parameter maps from the trained neural network.

Questions:

1 How many layers are set in the default neural network? How many neurons are distributed at each layer?
2 What is the activation function used in the default neural network?
3 How many epochs will the default neural network run in training? What is the batch size?
4 What is the loss function used in the default neural network? Can you justify the reason why we need a loss function?
5 In Figure 40.3a, which two hyperparameters in the neural network control the calculation of loss value and changes to the loss value?
6 In Figure 40.3b, which parameters are predicted well by the trained neural network, and which not? Why do you think the neural network predicts some parameters well, and others not that well?

According to what you have learnt in Chapter 38, what will the results presented in Figure 40.3d be used for?

TUNING THE NEURAL NETWORK FOR BETTER PERFORMANCE

Setting up basic neural network structures and hyperparameters does not guarantee satisfactory model performance in prediction. We need to tune the neural network for better performance. We will briefly introduce some common procedures in tuning the neural network (Figure 40.4).

Suppose you find the predicted parameter values cannot fit well with the targets after training the neural network. In this case, you are recommended to first try a new model structure with more hidden layers (deeper neural network) and/or more neurons for each hidden layer (wider neural network). Expanding the depth and/or width of the neural network will generally increase its ability to interpret more complex

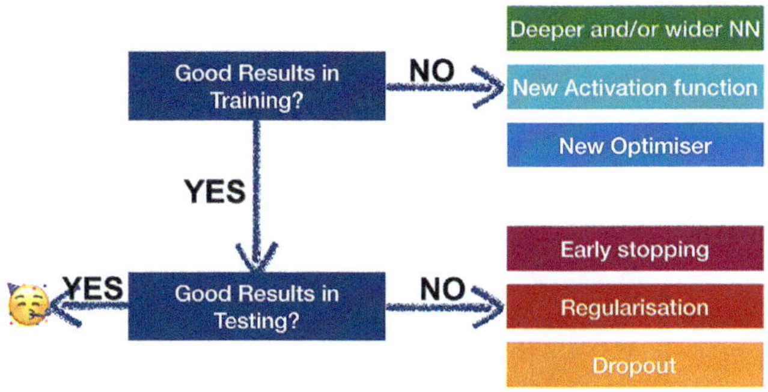

FIGURE 40.4 Tips on tuning the neural network for better model prediction performance.

Reproduced from Lee (2016).

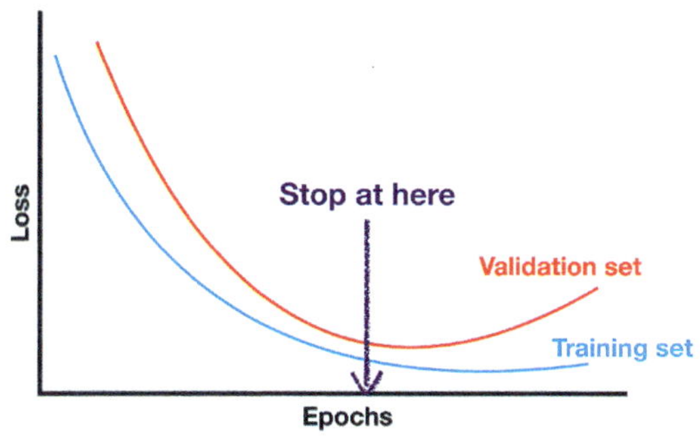

FIGURE 40.5 Overfitting in neural network training and early stopping option.

relationships between the input environmental variables and the target parameter values. Meanwhile, you can also consider applying different activation functions or optimizers.

We may sometimes encounter the overfitting problem. Overfitting happens during training of the neural network when the loss value of the validation set stops decreasing with the loss value of the training set but begins to increase (Figure 40.5). In such a case, even though the final prediction of the training set may fit well with the target, the trained neural network cannot make precise predictions in the validation set (see a detailed discussion about the overfitting in Chapter 37). To avoid overfitting, we can adopt early stopping to stop training the neural network at reasonable epochs. After early stopping, the loss values of both the validation and training sets stay low and thus ensure the robustness of neural network predictions in both datasets.

Regularization is another option to prevent overfitting in training the neural network. Regularization applies penalties to the weights of neurons to avoid the predictions of the trained neural network relying too much on the performance of one or some small group of neurons. Regularization is a more advanced option in tuning the neural network. We may not touch on it too much in this book. If you are interested, you can refer to the website at https://machinelearningmastery.com/how-to-reduce-overfitting-in-deep-learning-with-weight-regularization/.

Dropout offers a further option if we do not want the performance of a small group of neurons to matter too much in the final prediction of the trained neural network. The dropout option allows us to randomly permute some certain percent of neurons in each epoch of optimization. The neural network will then be trained to not depend too much on any specific neurons in prediction, thereby improving its robustness. If you check the default setting in Exercise 1, you will find we used the dropout option in the neural network training.

Exercise 2

Tuning the neural network used in Exercise 1. Follow the instructions in CarboTrain:

a Select **Unit 10**
b Select **Exercise 2**
c Select **Output Folder**
d Change one or more hyperparameter values, e.g., select and change the Optimizer to "adam".
e **Run Exercise**
f Check results in your Output Folder. Four figures as described in Exercise 1 will be generated.

Questions:

How do the output figures change compared to those generated in Exercise 1? Can you explain the reasons for these changes?

PRODA VERSUS DATA ASSIMILATION ALONE FOR OPTIMIZED SOC DISTRIBUTIONS IN CLM5

In Exercises 1 and 2, we have introduced how to build, train, and tune a neural network to best predict parameter values by environmental variables. The deep learning model is the core part of the PRODA approach. We optimize SOC representation in a process model by fully interpreting the environmental dependencies of its parameters through a deep learning model.

Data assimilation alone can also utilize multisite observations to retrieve parameter values so that the biogeochemical model can better represent SOC distributions. Instead of optimizing parameter values at each observational site as in the PRODA approach, the retrieved parameter values with the data assimilation alone are spatially invariant. Exercise 3 will compare results of CLM5 in terms of the SOC representation when using PRODA to set parameter values, compared with data assimilation alone.

Exercise 3

Comparing SOC representations in CLM5 between PRODA and data assimilation alone. Follow the instructions in CarboTrain:

a Select **Unit 10**
b Select **Exercise 3**
c Select **Neural Network Task Folder**

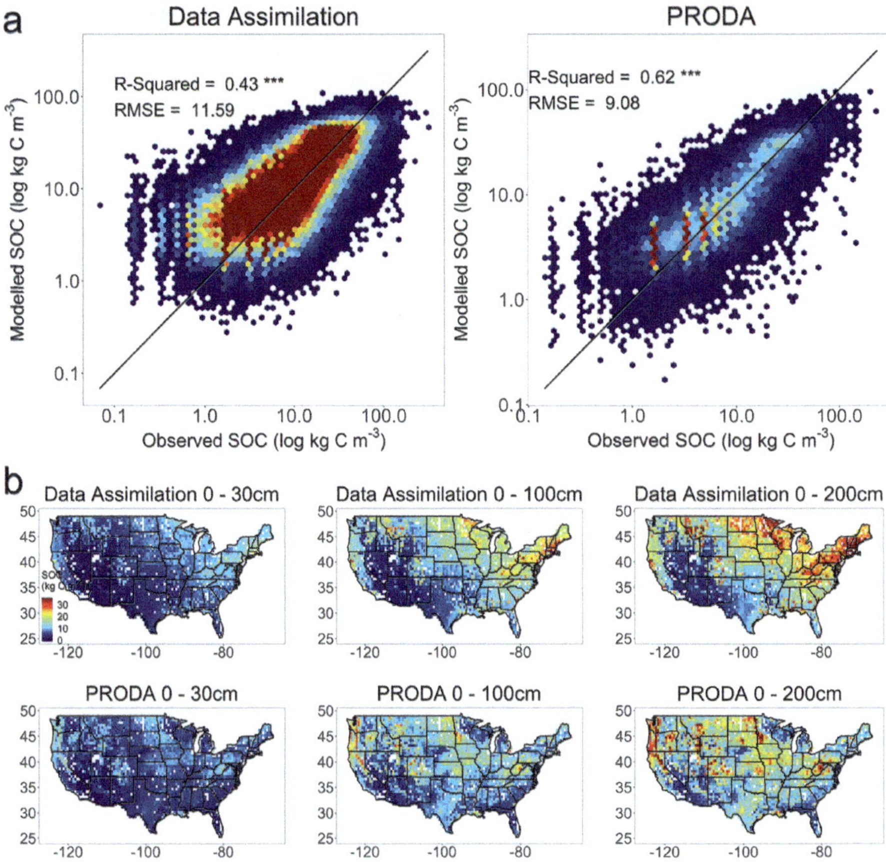

FIGURE 40.6 Output figures in Exercise 3. (a) Agreement between model simulations and observations; (b) SOC distributions at different depth across the conterminous United States simulated by CLM5 with parameter estimation by data assimilation alone vs. PRODA.

d Select **DA alone Task Folder**
e Select **Output Folder**
f **Run Exercise**
g Check results in your Output Folder. Three figures will be generated (Figure 40.6). Figure mod_vs_obs_nn.jpeg shows the agreement between the PRODA-optimized and observed SOC. Figure mod_vs_obs_ob.jpeg describes the agreement between retrieved and observed SOC using data assimilation alone. Figure soc_map.jpeg shows the SOC distributions simulated by CLM5 at different depths across the conterminous United States after optimization by the PRODA approach, and following parameter estimation by data assimilation alone.

Question:
Which approach performs better in representing SOC distribution? Why?

To summarize, this practice has explored the technical details of the PRODA approach in optimizing parameter values and retrieving SOC distributions. The deep learning model is the core part of the PRODA approach. The first two exercises illustrate the basic components in a neutral network (Exercise 1) and how to tune a configured neural network for better performance (Exercise 2). In Exercise 3, we applied CLM5 to simulate the SOC distribution across the conterminous United States after parameter estimation by data assimilation alone and further parameter optimization by the PRODA approach. We compared the agreement between simulations and observations after the two approaches. The results show the advantage of using the PRODA approach to optimize SOC distribution in biogeochemical models.

Appendix 1
Matrix Algebra in Land Carbon Cycle Modeling

Ye Chen
Northern Arizona University, Flagstaff, USA

The purpose of this appendix is to deliver the necessary matrix algebra foundation you will need to understand the matrix model of the land carbon cycle, introduced in Chapter 1. If you have taken the undergraduate level of matrix algebra or above, you may skip this appendix. Please note that some important topics in the following are presented in a simple way and they could be easily extended into longer sections. Professor David Austin's free online textbook (see Suggested Reading) is a good source for extra learning material and self-study.

MOTIVATIONS

Many processes can be described using a dynamical system. In one type of dynamical system, the state $\vec{x}_{n+1}$ of a system at time $n + 1$ can be derived from the state $\vec{x}_n$ at time n using a transition matrix A:

$$\vec{x}_{n+1} = A\vec{x}_n$$

See the following problem for more details.

Problem 1
Consider a simplified carbon transfer model in which 3% of the carbon in the fast soil pool moves to the slow soil pool each year while 95% of the fast soil pool stays the same, and 1% of the slow soil pool moves to the fast soil pool while 97% of the slow soil pool stays the same with no other influences on the two pools. What are the pool sizes after 10 years? 50 years? 100 years?

This question will be fully answered at the end of this appendix. To see how linear algebra can help us, we will do some basic analysis in this section.

Note that there is 2% lost in each pool when transfer occurs every year. To simplify the problem, we are not going to track what happens with that carbon once it exits the two pools.

Let $\vec{x}_n = \begin{bmatrix} f_n \\ s_n \end{bmatrix}$ be the status of the two pools at nth year. That is, f_n is the size of the fast soil pool at nth year, and s_n is the size of the slow soil pool at nth year. Then

$$f_{n+1} = 0.95 f_n + 0.01 s_n$$

$$s_{n+1} = 0.03 f_n + 0.97 s_n$$

First of all, this linear system can be written down in the matrix form using the knowledge from sections 2 and 4 below,

$$\begin{bmatrix} f_{n+1} \\ s_{n+1} \end{bmatrix} = \begin{bmatrix} 0.95 & 0.01 \\ 0.03 & 0.97 \end{bmatrix} \begin{bmatrix} f_n \\ s_n \end{bmatrix}$$

or equivalently, $\vec{x}_{n+1} = A\vec{x}_n$, where A is the 2×2 coefficient matrix.

Given the pool size of the first year, $\vec{x}_1 = \begin{bmatrix} f_1 \\ s_1 \end{bmatrix}$, one interesting question is to find out the pool size a long time after, i.e., $\vec{x}_n$ for a large n. In this problem, $n = 10, 50, 100$.

Note that

$$\vec{x}_{n+1} = A\vec{x}_n = A\left(A\vec{x}_{n-1}\right) = A^2\vec{x}_{n-1} = \ldots = A^n\vec{x}_1,$$

and finding $A^n\vec{x}_1$ for large n can be computationally expensive when the dimension of A is large. However, if we can find the eigenvalues of A as introduced in section 5, then we may be able to turn the problem of finding $A^n\vec{x}_1$ with large n into finding $\lambda^n\vec{x}_1$, where λ is a number, which is called the eigenvalue of A.

MATRIX OPERATIONS

Basic Operations

Definition 1
A **matrix** is a rectangular array of numbers. We say that a matrix A is an m × n matrix when it has m rows and n columns, and the **dimension** of A is m × n.

For example, this is a 2 × 3 matrix:

$$A = \begin{bmatrix} 2 & 3 & 5 \\ 4 & 1 & -9 \end{bmatrix} \tag{A1.1}$$

Definition 2
A matrix is ***square*** *if it has the same number of rows and columns.*

We can use subscript notation to refer to particular entries in a matrix: the notation A_{ij} refers to the entry of matrix A in row i and column j.

Problem 2
Let A be the matrix in equation A1.1. What is A_{13}?

Answer: The number in row 1 and column 3 is 5.

Definition 3
The **transpose** of a matrix A, denoted A^T, is the matrix A with the rows and columns switched. That is, $(A^T)_{ij} = A_{ji}$.

For example, for the matrix in Equation A1.1

$$A^T = \begin{bmatrix} 2 & 4 \\ 3 & 1 \\ 5 & -9 \end{bmatrix}$$

We can add two matrices if they have the same dimensions. We do this by adding corresponding entries. For example,

$$\begin{bmatrix} 2 & 3 \\ 4 & -1 \end{bmatrix} + \begin{bmatrix} 5 & 0 \\ -6 & 3 \end{bmatrix} = \begin{bmatrix} 2+5 & 3+0 \\ 4-6 & -1+3 \end{bmatrix} = \begin{bmatrix} 7 & 3 \\ -2 & 2 \end{bmatrix}$$

Problem 3
For two matrices A and B, is $A + B$ always the same as $B + A$?

Answer: Yes.

Matrix Multiplication

To multiply a scalar number k by a matrix, simply multiply k with every element of the matrix. For example,

$$2\begin{bmatrix} 0 & 1 & 5 \\ 2 & 1 & -1 \end{bmatrix} = \begin{bmatrix} 0 & 2 & 10 \\ 4 & 2 & -2 \end{bmatrix}.$$

To multiply two matrices, the number of *columns* of the first matrix must be the same as the number of *rows* of the second matrix. Let's say that we have two matrices, X, which is $m \times k$, and Y, which is $k \times n$. Then their product, denoted XY, will be an $m \times n$ matrix. Here is how to determine the elements of the matrix product XY: to get $(XY)_{ij}$ (the entry in the ith row and jth column), take the ith row of X and the jth column of Y, multiply their corresponding elements, and add the results.

Suppose we have the matrices

$$X = \begin{bmatrix} 1 & -2 & 3 \\ 0 & -1 & 6 \end{bmatrix} \quad Y = \begin{bmatrix} 0 & -2 & 1 & 0 \\ 5 & 4 & 2 & 6 \\ 1 & -3 & 0 & -1 \end{bmatrix}$$

X has dimensions 2×3, and Y has dimensions 3×4. Since the number of columns of X equals the number of rows of Y, we can multiply them, and the result will be a 2×4 matrix.

Now, to determine the entry in the first row and first column of the product XY, we look at the first row of X and the first column of Y, here shown highlighted in blue and red:

$$X = \begin{bmatrix} 1 & -2 & 3 \\ 0 & -1 & 6 \end{bmatrix} \quad Y = \begin{bmatrix} 0 & -2 & 1 & 0 \\ 5 & 4 & 2 & 6 \\ 1 & -3 & 0 & -1 \end{bmatrix}.$$

We now take these two lists of three numbers, multiply them element-by-element, and add the results:

$$1 \cdot 0 + -2 \cdot 5 + 3 \cdot 1 = -7$$

So far, we know that the matrix product XY looks like this:

$$XY = \begin{bmatrix} -7 & ? & ? & ? \\ ? & ? & ? & ? \end{bmatrix}$$

Now let's compute $(XY)_{12}$ (highlighted in red above). Since we are trying to compute the entry of XY in the *first row* and *second column*, we take the *first row* of X, and the *second column* of Y:

$$X = \begin{bmatrix} 1 & -2 & 3 \\ 0 & -1 & 6 \end{bmatrix} \quad Y = \begin{bmatrix} 0 & -2 & 1 & 0 \\ 5 & 4 & 2 & 6 \\ 1 & -3 & 0 & -1 \end{bmatrix}$$

Multiplying them pairwise and then adding yields

$$1 \cdot -2 + -2 \cdot 4 + 3 \cdot -3 = -19$$

Now XY looks like

$$XY = \begin{bmatrix} -7 & -19 & ? & ? \\ ? & ? & ? & ? \end{bmatrix}$$

Problem 4
Finish computing the matrix product XY.

Answer: $XY = \begin{bmatrix} -7 & -19 & -3 & -15 \\ 1 & -22 & -2 & -12 \end{bmatrix}$

Problem 5
Use the matrices X, Y defined in the previous example, and also

$$A = \begin{bmatrix} 2 & 4 \\ -1 & 1 \end{bmatrix}, \quad B = \begin{bmatrix} 1 & 3 \\ 2 & 2 \end{bmatrix}$$

Compute each of the following if it is defined, or say undefined otherwise.

1.YX 2.XA 3.AB 4.BA

Answer: 1. YX is undefined. 2. XA is undefined as the dimension of X is 2×3, and the dimension of A is 2×3: the inner dimension does not match. 3. $AB = \begin{bmatrix} 10 & 14 \\ 1 & -1 \end{bmatrix}$. 4. $BA = \begin{bmatrix} -1 & 7 \\ 2 & 10 \end{bmatrix}$.

Problem 6

Based on the results of Problem 5, is AB always the same as BA? Explain why.

Answer: No. What you found in Problem 5 is that matrix multiplication is not commutative. This is one way in which matrices are different from real numbers. With real numbers, you are able to switch around the order of operands being multiplied. But with matrices, you have to be very careful: you cannot do this with matrices!

It turns out, however, that matrix multiplication is associative: for any matrices X, Y, and Z, as long as they have dimensions that match up properly, it is always true that $(XY)Z = X(YZ)$. That is, when doing more than one matrix multiplication, it doesn't matter which multiplication we do first, as long as we keep them in the right order. This means that we can write things like $ABCDE$ instead of $((AB)C)(DE)$ or $A(B(C(DE)))$ or $((AB)(CD))E$ since they are all the same.

Quiz 1

a If the dimension of A is 10×20, the dimension of B has only one column, and AB is well defined. How many columns does the matrix product AB have? (Answer: The matrix AB has one column).

b If the dimension of the matrix A is 10×25, the dimension of the matrix C is 10×20, and $AB = C$. What is the dimension of B? (Answer: The dimension of B is 25×20.)

MATRIX EQUATIONS

Identity Matrix, Inverse Matrix

Problem 7

Can you find a 2×2 matrix I which, when multiplied by any other 2×2 matrix X, yields X? That is, $IX = XI = X$ for any 2×2 matrix X. I is called the 2×2 identity matrix (sometimes also written I_2).

Answer: $I_2 = \begin{bmatrix} 1 & 0 \\ 0 & 1 \end{bmatrix}$

Problem 8

What is the 3×3 identity matrix, I_3? In general, what does the $n \times n$ identity matrix I_n look like?

Answer: $I_3 = \begin{bmatrix} 1 & 0 & 0 \\ 0 & 1 & 0 \\ 0 & 0 & 1 \end{bmatrix}$. I_n is an $n \times n$ matrix with the main diagonal elements being 1, and all other elements being 0.

Problem 9

There is no such thing as a 2×3 identity matrix. Why not?

Answer: To match the dimension such that $IX = XI = X$ I has to be a square matrix.

Problem 10

Multiply the following two matrices:

$$A = \begin{bmatrix} 2 & 3 \\ 3 & 5 \end{bmatrix}, \quad B = \begin{bmatrix} 5 & -3 \\ -3 & 2 \end{bmatrix}$$

What do you get? Why is this interesting?

Answer: $AB = I_2$.

Definition 4

The **inverse** of a square matrix A, written A^{-1}, is a matrix which multiplied by A results in the identity matrix: $AA^{-1} = A^{-1}A = I$.

Problem 11

As it turns out, not all square matrices have an inverse. But this should not be too surprising. Why not? (Hint: think about the inverse of 0).

Solving Matrix Equations

Matrices which have an inverse are called **invertible** matrices, and matrices which do *not* have an inverse are called **singular** matrices. Why do we care whether a matrix is invertible? Well, remember what you did in algebra to solve an equation like $3x = 12$: you multiply both sides by 1/3, which is the multiplicative *inverse* of 3. In the same way, inverting matrices allows us to solve *matrix equations* like $AX = Y$ (where A, X, and Y are matrices). If A is invertible, we can left multiply both sides of the equation by A^{-1} to get $X = A^{-1}Y$. Note that YA^{-1} is not the solution of X by Problem 6.

Problem 12

If A, B are invertible, solve this matrix equation for X:

$$AXB = Y$$

Answer: $X = A^{-1}YB^{-1}$

Example 1

Solve this matrix equation for X:

$$Y = BC + ADKX$$

If the matrix product ADK is invertible,

$$Y = BC + ADKX$$

$$\Rightarrow ADKX = -BC + Y$$

$$\Rightarrow X = (ADK)^{-1}(-BC+Y)$$

$$\Rightarrow X = (ADK)^{-1}(-BC)+(ADK)^{-1}Y$$

Following the steps of this example, you can solve the next problem.

Problem 13

The matrix equation below is presented in Chapter 1, Equation 1.6:

$$X'(t) = B\mu(t) + A\xi(t)KX(t)$$

Can you solve it to obtain an expression for $X(t)$ if the matrix $A\xi(t)K$ is invertible, $X'(t)$, B, A, $\xi(t)$, K, and $X(t)$ are matrices, and $\mu(t)$ is a scalar?

Answer: You may have noticed this equation is similar to the matrix equation in Example 1, except the notations are different. The computation would still follow the rule of matrix operations. And to solve this equation for X, just follow the steps in Example 1. The solution is

$$X(t) = (A\xi(t)K)^{-1}(-B\mu(t)) + (A\xi(t)K)^{-1}X'(t)$$

Quiz 2

Let A, C be invertible matrices. Solve $AXC + BD = Y$ for X. (Answer: $X = A^{-1}(Y - BD)C^{-1}$.)

LINEAR SYSTEM

Definition 5

A linear equation in the variables $x_1, x_2, \cdots, x_n$ is an equation that can be written in the form:

$$a_1x_1 + a_2x_2 + \cdots + a_nx_n = b$$

where $a_1, a_2, \ldots, a_n$ and b are real numbers. Thus, for example,

$$2x_1 + 3x_2 + x_3 - 11x_4 = 6$$

is a linear equation in the four variables x_1, x_2, x_3, x_4. But the following equations are not linear:

$$2x_1 + 2x_2 + x_3x_4 = 6$$

$$2x_1^2 + 3x_2 = 6$$

Definition 6

A **system of linear equations** (or **linear system**) is a set of linear equations.

The following is a simple linear system.

Example 2

Solve the linear system:

$$(1) - x_1 + \frac{1}{3}x_2 = 6$$
$$(2) x_1 + x_2 = 10$$

This system can be easily solved by hand using forward and backward substitution. The solution is

$$x_1 = -2, \quad x_2 = 12.$$

Note that with the matrix multiplication, the system is equivalent to

$$x_1\begin{bmatrix}-1\\1\end{bmatrix} + x_2\begin{bmatrix}\frac{1}{3}\\1\end{bmatrix} = \begin{bmatrix}6\\10\end{bmatrix}$$

Therefore, by substituting the solution back to the linear system, you can verify that the following equation is true

$$-2\begin{bmatrix}-1\\1\end{bmatrix} + 12\begin{bmatrix}\frac{1}{3}\\1\end{bmatrix} = \begin{bmatrix}6\\10\end{bmatrix}$$

Here is another approach using the matrix inverse. The linear system can be written down in the matrix form:

$$\begin{bmatrix}-1 & 1/3\\1 & 1\end{bmatrix}\begin{bmatrix}x_1\\x_2\end{bmatrix} = \begin{bmatrix}6\\10\end{bmatrix}$$

We denote this matrix form as

$$Ax = b$$

To find the solution of this linear system for x, we left multiply this equation by A^{-1} if it exists, then we can find the solution

$$x = A^{-1}b$$

EIGENVALUES AND EIGENVECTORS

Definition 7

Let A be a square matrix.

A vector $\vec{x}$ is an **eigenvector** of A if $\vec{x} \neq 0$ and there is a scalar λ such that $A\vec{x} = \lambda\vec{x}$.

A scalar λ is an **eigenvalue** of A if there is a vector $\vec{x} \neq \vec{0}$ such that $A\vec{x} = \lambda\vec{x}$.

Remark 1 By this definition, if λ is an eigenvalue of A and $\vec{x}$ is a corresponding eigenvector, then we must have the following:

$$A^n\vec{x} = A^{n-1}\left(A\vec{x}\right) = A^{n-1}\left(\lambda\vec{x}\right) = \lambda A^{n-1}\vec{x} = \lambda A^{n-2}\left(A\vec{x}\right) = \ldots = \lambda^n\vec{x}$$

Next, we are going to apply Definition 7 to verify the eigenvalues and eigenvectors of a matrix.

Example 3

Confirm that both $\vec{v}_1 = \begin{bmatrix} -1 \\ 1 \end{bmatrix}$ and $\vec{v}_2 = \begin{bmatrix} 1/3 \\ 1 \end{bmatrix}$ are eigenvectors for the matrix $A = \begin{bmatrix} 0.95 & 0.01 \\ 0.03 & 0.97 \end{bmatrix}$, and find the corresponding eigenvalues.

Solution: As

$$A\vec{v}_1 = \begin{bmatrix} 0.95 & 0.01 \\ 0.03 & 0.97 \end{bmatrix}\begin{bmatrix} -1 \\ 1 \end{bmatrix} = \begin{bmatrix} -0.94 \\ 0.94 \end{bmatrix} = 0.94\vec{v}_1$$

by the definition of eigenvalue, $\vec{v}_1$ is an eigenvector for A with corresponding eigenvalue $\lambda_1 = 0.94$.

Similarly, it is easy to verify that $\vec{v}_2$ is an eigenvector for A with corresponding eigenvalue $\lambda_2 = 0.98$.

In this example, the eigenvectors are already given, so it is relatively easy to determine the eigenvalues. In general, to calculate the inverse matrix, the eigenvalues and eigenvectors by hand could be very hard or even impossible when the dimension of A is large. Luckily, modern programming environments like MATLAB and Python provide functions for this purpose.

Example 4

Let us continue finding the solution for Problem 1. Let $\vec{x}_1 = \begin{bmatrix} 6 \\ 10 \end{bmatrix}$ and compute the pool sizes after 50 years.

Solution: By Example 2, we have $\begin{bmatrix} 6 \\ 10 \end{bmatrix} = -2\begin{bmatrix} -1 \\ 1 \end{bmatrix} + 12\begin{bmatrix} 1/3 \\ 1 \end{bmatrix}$, which is equivalent to $\vec{x}_1 = -2\vec{v}_1 + 12\vec{v}_2$. So,

$$A^n\vec{x}_1 = A^n\left(-2\vec{v}_1 + 12\vec{v}_2\right) = -2A^n\vec{v}_1 + 12A^n\vec{v}_2$$

By Remark 1 and Example 3, $A^n\vec{v}_1 = \lambda_1^n\vec{v}_1$, $A^n\vec{v}_2 = \lambda_2^n\vec{v}_2$. We then have

$$A^n\vec{x}_1 = -2\cdot 0.94^n\vec{v}_1 + 12\cdot 0.98^n\vec{v}_2$$

By Problem 1, to determine the pool sizes after 50 years, it is equivalent to find $\vec{x}_{51}$, where $\vec{x}_{51} = A^{50}\vec{x}_1$. By the analysis we just did, we know that this can be computed easily:

$$\vec{x}_{51} = A^{50}\vec{x}_1 = -2\cdot 0.94^{50}\vec{v}_1 + 12\cdot 0.98^{50}\vec{v}_2 \approx \begin{bmatrix} 1.55 \\ 4.28 \end{bmatrix}$$

Therefore, after 50 years, the size of the fast pool is 1.55, and the size of the slow pool is 4.28.

Quiz 3

Continue with Problem 1. Let $\vec{x}_1 = \begin{bmatrix} 2 \\ 6 \end{bmatrix}$. Compute the pool sizes after 20 years. (Answer: As $\vec{x}_1 = 6\vec{v}_2$, we then have $A^n\vec{x}_1 = 6\cdot 0.98^n\vec{v}_2$. So, $\vec{x}_{21} = 6\cdot 0.98^{20}\vec{v}_2 \approx \begin{bmatrix} 1.34 \\ 4.01 \end{bmatrix}$. After 20 years, the size of the fast pool is 1.34, and the size of the slow pool is 4.01.

SUGGESTED READING

David Austin, *Understanding Linear Algebra*, 2023. Grand Valley State University, https://understandinglinearalgebra.org/home.html.

Appendix 2
Introduction to Programming in Python

Xin Huang
National Center for Atmospheric Research, Boulder, USA

This appendix is intended to equip readers with little or no programming experience with basic skills to write and read programs in the Python language. Python is a powerful programming language widely used in many applications. This appendix introduces basic programming knowledge in Python including variables, operators, function, class, module, etc. All code examples in this appendix come from Python files (test_p2.py, GeneralModel.py, and model.py), which will be used in practice chapters 8, 12, and 16. Python 3.7 is preferred for the practice chapters of this book. Readers are not expected to be an expert in programming but to acquire the basic ability to read/write Python codes for the practice chapters.

WHAT IS PYTHON AND HOW DOES IT RUN?

Python is a programming language used for general-purpose software engineering. It is one of the most popular languages among scientists, engineers, and mathematicians for the following reasons. First, Python has simple syntax similar to the English language. It is easy to read, write, learn, and maintain Python code. Second, Python can work seamlessly on different platforms such as Windows, Linux, and Mac. Third, Python is an interpreted language, which means that it executes the code line-by-line. If any error occurs, it will stop further execution until programmers are able to locate the errors and fix them. Fourth, Python comes with many great standard libraries to provide users with a vast choice of functions needed for their tasks.

Similar to English, Python is a language composed of vocabulary and syntax. The vocabulary in English comprises different words. The vocabulary in Python comprises operators, variables/operands, and keywords. As in common languages, the right syntax must be applied when linking different parts of the vocabulary into meaningful statements. In English, the sentence ‘learn python I’ is not syntactically valid. In Python, the expression “hello”+9 is not syntactically valid. Similar to learning English, readers will learn the most common vocabulary and syntax of Python from this appendix.

Before reading further, please refer to Appendix 3 to install Python on your computer.

There are two ways to run a Python program. One way is to use an interactive shell window. The shell will prompt >>> and wait for the user to type in a line of Python code. Given the code input, the shell will execute this code, display the results and wait for the next code input. For all but the simplest operations, we normally wish to gather consecutive lines of code in a Python source file (xxx.py). We may execute a Python source file with a command like **python xxx.py**. All code lines in this source file will be executed one after the other until the end of the program is reached. In the practice chapters, we will use this second way.

THE FIRST PYTHON PROGRAM

A Python program is a collection of code that manipulates data. Figure A2.1 is a code segment in the test_p2.py program. Each code line except for the annotation lines is called an expression. These expressions will be executed by the Python interpreter. There are two ways to denote an annotation. One is a single-line annotation starting with # as shown in Figure A2.1. The other is multiple-line annotations using ’ or ” symbols. The expression **if __name__==’__main__’:** in line 6 of Figure A2.1 indicates the start of a Python program. If one line contains multiple code expressions, such as line 12–13, it is recommended to use a semicolon (;) to separate different expressions.

One of the most common expressions is to assign a value to a variable like lines 8–26 in the sample code of Figure A2.1. Taking line 12 as an example, the variable *f31* will store the value of 0.72. This value can be retrieved by this variable name after this code line. A variable must be assigned a value before any expressions that use it. Readers will learn more about variable types in the next section.

Another common expression is to print out on the screen. Figure A2.2 show a display of the last elements in the *res* variable (a two-dimensional array) on the screen. Programmers often use the print expression to check values stored in variables, which is a useful approach for debugging the Python program.

Python uses colons to indicate an indented code block, which often appears in while-loop, for-loop, if-else condition procedure, and function definitions. Readers will learn more about these procedure definitions in the following section. The code groupings in these procedures are indicated by white space or indentation (Figure A2.3). The right level of

```
if __name__ == '__main__':

    output_folder = sys.argv[1]

    B = np.array([0.45, 0.55, 0, 0, 0, 0, 0]).reshape([7,1])  # allocation

    f31 = 0.72; f41 = 0.28; f42 = 1; f53 = 0.45; f54 = 0.275; f64 = 0.275;
    f65 = 0.296; f75 = 0.004; f56 = 0.42; f76 = 0.03; f57 = 0.45;

    A = np.array([-1, 0, 0, 0, 0, 0, 0,
                  0, -1, 0, 0, 0, 0, 0,
                  f31, 0, -1, 0, 0, 0, 0,
                  f41, f42, 0, -1, 0, 0, 0,
                  0, 0, f53, f54, -1, f56, f57,
                  0, 0, 0, f64, f65, -1, 0,
                  0, 0, 0, 0, f75, f76, -1]).reshape([7,7])  # tranfer


    #turnover rate per day of pools: foliage, wood, metabolic litter, structural
    #litter, soil microbial,slow soil, passive soil
    temp = [0.00176, 0.000100104, 0.021468, 0.000845, 0.008534, 8.976e-005, 0.00000154782]
```

Annotations will not be executed

FIGURE A2.1 Code example of defining an annotation with # . This code segment is from the test_p2.py program.

```
    print(res[:,nyear-1])   # print result of the last year
```

FIGURE A2.2 Code example of print expression.

```
    for i in range(1, 4):
        for j in range(1, 4):
            if ((i-1) * 3 + j) > 7:
                break
            ax = plt.subplot(3, 3, (i-1) * 3 + j)
            ax.plot(x, res[(i-1) * 3 + j - 1,:])
            plt.xlabel("year", fontsize = 12)
            plt.ylabel(pool_names[(i-1) * 3 + (j-1)] + " pool ($g C m^{-2}$)", fontsize = 12)
```

The first for-loop code block
The second for-loop code block
If-else code block

FIGURE A2.3 A Python program uses organized spaces to indicate code grouping. This code example includes two for-loop code blocks (red and green arrows and texts) and an if-else code block (blue arrow and text).

indentation is important – too much or too little space will induce an error.

VARIABLES AND OPERATORS

As we learned in the previous section, variables are one of the primary vocabulary elements in the Python language. Similar to words, variables are containers to store information such as numbers and string values. The information type decides the variable type. Generally, the variable types in Python are integers (e.g., 1, 2, 3, ..), rational numbers (e.g., 3.1415926), complex numbers (e.g., 12 + 0.2*i*), strings (e.g., "helloworld"), Boolean values (i.e., TRUE or FALSE), and NaN. The final variable type is special, and it only has one value that is none. Other programming languages such as Java require you to declare a variable type before using it. Python, however, does not require variable declarations. Variables with the same name can be used in different places in a Python program to store different types of values.

Operators are another primary vocabulary element in Python. Table A2.1 shows a collection of binary operators

TABLE A2.1
A collection of binary operators in python

Operator	Description
a+b	sum
a–b	difference
a*b	product
a/b	division
a//b	The integral part of the quotient when a is divided by b
a%b	The remainder when a is divided by b
a**b	a to the power of b
a|b	True when either a or b is True
a&b	True when both a and b is True
not a	True if a is False
a==b	True when a equals to b
a!=b	True when a doesn't equal to b
a+=b	a=a+b
a/=b	a=a/b

(a)

```
if <BooleanExpr>:
    <ExpressionT1>
    ...
    <ExpressionTk>
else:
    <ExpressionF1>
    ...
    <ExpressionFk>
```

(b)

```
if type(self.input_fluxes) == np.ndarray:
    self.tmp_input_fluxes = self.input_fluxes[idx]
else:
    self.tmp_input_fluxes = self.input_fluxes
```

FIGURE A2.4 The statement form of conditional control flow (a), and its code example in GeneralModel.py (b).

(a)

```
for <var> in <listExpr>:
    <Expression1>
    ...
    <ExpressionN>
```

(b)

```
for t in range(10):
    print(t)
```

FIGURE A2.5 The statement form of for-loop control flow (a), and its code example (b).

(a)

```
while <BooleanExpr>:
    <Expression1>
    ...
    <ExpressionN>
```

(b)

```
t=0
while t<10:
    print(t)
    t=t+1
```

FIGURE A2.6 The statement form of while-loop control flow (a), and its code example (b).

performing calculations on two variables. They may work on numbers or Boolean values. The final two operators are special. The numerical calculation and assignment are performed at the same time. The expression **a/=b**, for example, first calculates the value of *a* divided by *b* and then assigns this as a new value for the variable *a*.

Another operator collection is about control flow. Control flow is to decide which and how expressions are to be executed. We will use examples in English to introduce two common control flows in Python. The first one is conditional control flow. The example in English is: *'If tomorrow is sunny, I will go hiking, otherwise I will stay at home'*. In the conditional control flow, only one set of statements will be executed, either *'go hiking'* or *'stay at home'*. The syntax of conditionals in Python is illustrated in Figure A2.4a. It starts with a <BooleanExpr> expression whose value is either TRUE or FALSE. If the value of this expression is TRUE, then <ExpressionT1>, …, <ExpressionTk> will be executed; If it is FALSE, then another set of expressions <ExpressionF1>, …, <ExpressionFk> will be executed. A code example of if-else control flow is shown in Figure A2.4b. If the variable type of *self.input_fluxes* is an array, the variable *self.tmp_input_fluxes* will be assigned a specific element in this array variable. Otherwise, *self.tmp_input_fluxes* saves the whole values in *self.input_fluxes*.

The second type of control flow is a loop, including a for-loop (Figure A2.5) or a while-loop (Figure A2.6). A loop repeats a set of statements over and over until a termination status is reached. An example in English is *'repeat taking medicine until you feel better'*. The syntax of for-loop control flow is shown in Figure A2.5a. It starts from retrieving the first value from a <listExpr> to a variable <var>, then executing <Expression1>, …, <ExpressionN>. This set of statements will be repeatedly executed when every value in <listExpr> is retrieved. At the end, the <var> will store the last element in <listExpr>. The code example of a for-loop (Figure A2.5b) uses a useful procedure, *range(10)*, as the <listExpr>. Generally, *range(n)* returns integers from 0 up to $n-1$. The for-loop iterates this numerical list from 0 to 9, assigns each element in the list to a variable *t* and prints the value of *t* to the screen.

Generally, any for-loop control flow can be rewritten into a while-loop flow. The code example in Figure A2.6b generates the same results as that in Figure A2.5b. The special characteristic of a while-loop is that programmers do not need to know in advance how many times the set of expressions needs to be repeated. The syntax of a while-loop control flow is shown in Figure A2.6a. It starts by evaluating <BooleanExpr> whose value is either TRUE or FALSE. If the value of this expression is TRUE, the set of expressions will be executed and the <BooleanExpr> will be evaluated again. If the expression of <BooleanExpr> gets FALSE, the execution of this loop will be terminated. Generally, programmers initialize a variable before the while loop, such as variable *t* in the code example (Figure A2.6b). The value of this variable is changed each time by the set of expressions in the loop block to be repeated and the new value is be tested in the <BooleanExpr> to decide whether the loop is to be terminated or not.

The **break** keyword offers another mechanism to exit a loop which is currently being executed. It works both with a for-loop or a while-loop. To make the program more readable, it is suggested to avoid the **break** keyword. As mentioned above, colons are used to indicate an indented code block in the control flow. Expressions after the colon have to be indented to one level (Figure A2.3). For example, in an *if-else* control flow, expressions after **if** and **else** keywords are at the same level of indentation and this indentation level is below the **if** and **else**.

ADVANCED VARIABLES AND OPERATORS

The variables and operators we have seen so far are known as primitive variables and operators. They are capable of coping

```
L=[2020,"training",[3,4]] #define a list with 3 elements. The third element is another list
L[0]=2021 # modify the first element to 2021
L[0:2] # return the first and second elements, [2020,"training"]
L[0:1] # return a new list, [2020]
L[0] # return a number, 2020
len(L) # return the length of the list, 3
L.extend([1,2]) # add more elements to L, now L is [2020, "training",[3,4],1,2]
L.remove(2020) # remove the element 2020, now L is ["training",[3,4]]
```

FIGURE A2.7 Common operators of *list* variable type.

with simple programming tasks. However, if we only use the primitive variables and operators, programs can quickly become long and messy. For longer programs, a modular design is preferred to make code clear and readable, and easier to modify or debug. Modularity entails aggregating primitive variables and operators into more advanced variables or operators. Advanced variables are centered around the organization of data. These advanced variables include *list, array, dictionary*, and *set*. Because the practice chapters of this book mainly use the *list* type, this appendix will only introduce *list*. For other advanced variables, please refer to suggested reading materials at the end of this appendix. Advanced operators are collections of primitive operations such as basic arithmetic, conditional evaluation, and recursion. This appendix will introduce three advanced operators: *function, class*, and *module*.

THE *LIST* VARIABLE

A *list* is a collection of ordered and mutable variables such as numbers, strings, or Boolean values. A *list* is written using square brackets [ ], with elements separated by commas. Figure A2.7 shows an example of defining a list and some common operations on the list. Each element in the list is numbered starting from 0. Therefore, the first element is at index 0, the second element is at index 1, and the final index is one less than the size (number of elements) of the list. We can retrieve the final element of a list called *varList* using the expression *varList[len(varList)−1]*. We can also select a continuous part of the list through slicing. For example, varList[0:3] returns the first three elements (i.e., *varList[0], varList[1], varList[2]*) of *varList*. Remember that the index after the colon in the slicing operator (i.e., 3) will not be included in the result. The slicing operator always returns a new list. Therefore, *varList[0]* returns a variable value while *varList[0:1]* returns a list which only includes one variable *varList[0]*. Some other common operations are illustrated in Figure A2.7, such as getting the length of a list, inserting and removing elements.

THE *FUNCTION* OPERATOR

As an advanced operator, function can be viewed as a collection of primitive variables and operators. Specifically, one function *f1* can call another function *f2*. As a result, all basic operators in *f2* are included in *f1*. One powerful characteristic of a function is that it can be used as a black

(a)

```
if 103%2==0:
    print('even')
else:
    print('odd')
if 1100%2==0:
    print('even')
else:
    print('odd')
if 530%2==0:
    print('even')
else:
    print('odd')
if 79%2==0:
    print('even')
else:
    print('odd')
```

(b)

```
def Evenodd(x):
    if x%2==0:
        print('even')
    else:
        print('odd')

Evenodd(103)
Evenodd(1100)
Evenodd(530)
Evenodd(79)
```

FIGURE A2.8 Code example to decide whether each of the numbers 103, 1100, 530, and 79, is even or odd, and display the answer on the screen. (a) Version using simple code operators; (b) version using a function to keep the code clean and efficient.

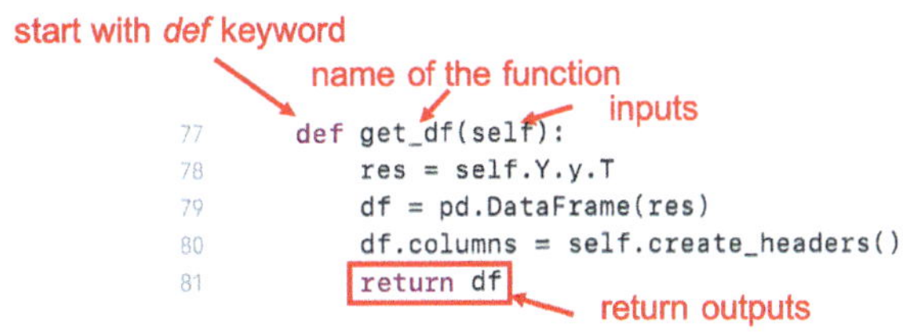

FIGURE A2.9 The syntax of function definition from General Model.py.

box and we only care about the argument inputs and outputs returned. Another feature of function is reusability of code. For example, if we want to decide whether a numerical value is even or odd, we need to write four lines of code for each value. The number of code lines is four times the number of values (Figure A2.8a). Gathering these four lines of code in a function helps keep the code clean and efficient. In Figure A2.8b, we put all expressions used in determining if a number is even or odd into a function called *Evenodd*. We can invoke the function by typing its name, followed by an argument in parentheses. Each time the function is called, it determines whether the argument is even or odd and prints out the result on the screen.

Figure A2.9 shows the syntax of a function definition. We write a sequence of expressions inside the function and give

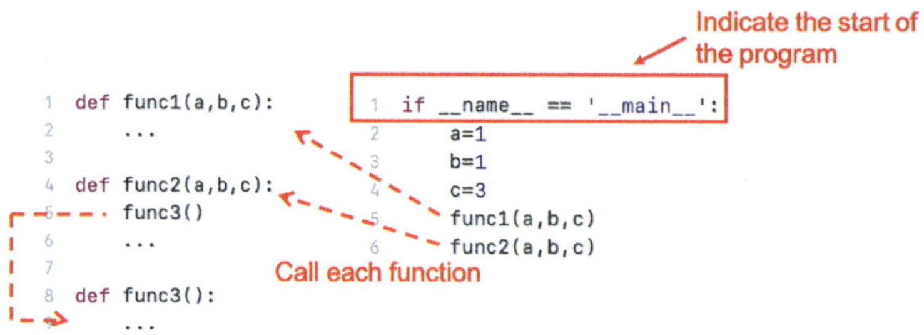

FIGURE A2.10 A workflow of function calls.

that function a name. The sequence of expressions can be executed at any point in the Python program by calling the function name. In the example of Figure A2.9, the function name is *get_df*. In function definition, variables in parentheses ( ) are the input and become available for use by expressions inside the function body. These input variables are also called parameters or arguments. When calling a function, we need to pass a variable storing specific values to the function. In the example of Figure A2.8, the *Evenodd* function requires a numerical value. If we want a value to be returned after a function is called, we need to add a *return* expression at the end of the function definition like line 81 in Figure A2.9. Remember, the indentation in a function definition is important. In the function, all expressions after the **def** keyword and comma have to be indented one level below.

A function can call another function. With more and more functions defined, which one is the highest-level function to call others? As we learned in the first section, a code line with the expression **if __main__ == '__main__'** indicates the start of a Python program no matter where it occurs in a source code file. The code block with the start of program execution is called the main program. In Figure A2.10, three functions (*func1, func2, func3*) are defined. The main program calls *func1* and *func2* and *func2* further calls *func3*. The variables in *func1*, *func2*, and the main program have the same names: *a, b, c*. How to distinguish them? Python uses namespace to do so. Namespace is an isolated scope where variables are valid. So *func1* and *func2* have local namespaces and the main program has a global namespace as well. The *a, b*, and *c* variables can have different values in different namespaces. If we want to change a variable created in the global namespace, it is required to clarify the global property of the variable by using the **global** keyword before the name of the variable.

The *Class* Operator

A class is a mixture of variables and functions, which are called attributes. Just like a function, once defined, can be used many times, a class may be defined, and then multiple instances can be made. Each class instance is called a class object and maintains its own attributes, which provides a way to reuse code to keep the program clean and efficient. This programming style is called object-oriented programming.

Figure A2.11a shows the syntax of a class definition. The *BaseClassName* is an abstract class to support inheritance. A more detailed introduction to inheritance is available in the recommended reading materials. In practice, most expressions inside a class definition are function definitions. Figure A2.11b lists all functions defined in the *GeneralModel* class from GeneralModel.py. A special function is for instantiation and its name is *__init__* (spelled with double underscores on both sides of *init*). The instantiation function is first executed whenever a class object is created and used for the first time within a Python program. The instantiation function in the *GeneralModel* class (Figure A2.11b) requires six parameters and **self** keyword represents the object itself. Figure A2.11c shows an example to initialize a *GeneralModel* object (*mod*). Typically, the expression is the class name followed by a list in parentheses of the parameters required by the instantiation function. The syntax to access attributes of the class object (i.e., variables and functions) is *obj.name* where *obj* is the class object and *name* is the variable name or function name. Figure A2.11c shows some examples of calling functions defined in *GeneralModel*.

The *Module* Operator

A module is a file containing reusable variables and functions. Unlike a class, which enables the instantiation of multiple objects and modification of attributes after creation, a module is a static storage of the reusable attributes. Similar to namespaces, attributes in one module file are not visible to other files. To make them visible, we need to load the module file first before using the variables or functions in the file. Figure A2.12 shows three ways to load a module file using the **import** keyword. One way is to load it directly without assigning a simple name to this module. We can retrieve the variables or functions in the module via its default name. The other way is to load the module and assign a new name to it. In the example of Figure A2.12, we import the *Numpy* module and rename it as *np*. Then we can call functions via *np.funcName* such as *np.zeros(49)*. The third way is to load specific variables or functions from a module file, which saves memory in runtime. Unlike the first two ways, these variables or functions loaded can be used without the module name.

(a)

```
class ClassName(BaseClassName):
    <Expression1>
    ...
    <ExpressionN>
```

(b)

```
class GeneralModel(Model):
    def __init__(self, times, B, A, K, iv_list, input_fluxes, xi=1):
        ...
    def ode_solver(self):
        ...
    def get_input(self, t, y):
        ...
    def right_hand_equation(self, t, y):
        ...
    def get_x(self):
        ...
    def get_df(self):
        ...
    def write_output(self, filename):
        ...
    def get_diagnostic_variables(self):
        ...
    def sasu_spinup(self):
        ...
    def create_headers(self):
        ...
    def get_x_df(self):
        ...
    def get_pool_n(self):
        ...
    def get_c_input(self):
        ...
```

(c)

```
mod = GeneralModel(times, B, A, K, iv_list, input_fluxes)
res = mod.get_x()
```

FIGURE A2.11 (a) Syntax of class definition; (b) functions defined in *GeneralModel* class; (c) code expressions to initialize a *GeneralModel* object and to call a function attribute of the object.

```
from GeneralModel import GeneralModel
import numpy as np
import matplotlib.pyplot as plt
import sys
```

The third way to import a module (line 1)

The second way to import a module (lines 2–3)

The first way to import a module (line 4)

FIGURE A2.12 Three ways to import a module. Codes are from test_p2.py.

In the practice chapters of this book, we will use *Numpy, Pandas, Scipy,* and *Matplotlib* modules. *Numpy* offers comprehensive mathematical functions working on arrays. *Pandas* provides functionality to manipulate tabular data as a two-dimensional data structure. *Scipy* is based on *Numpy* and solves scientific and mathematical problems. *Matplotlib* contains functionality for plotting data. Figure A2.13 shows some common functions from these modules that are useful for the practice chapters of this book.

SUMMARY

If you have read this appendix carefully, you should now have a basic knowledge of Python programming including variables, if-else conditional control, for-loop, while-loop, list, function definition, class definition, and loading module. This knowledge is sufficient to perform the programming tasks of the practice chapters of this book. If you wish to learn more about Python programming, you can refer to the learning resources referenced below.

(a)

```
import numpy as np
K=np.zeros(49).reshape([7, 7]) # create an array with 49 zeros and convert it to 7*7 matrix
np.multiply(K,1/864) # multiple each element in K and 1/864
np.matmul(A,B) # return the multiplied matrix product of two arrays A and B
np.linspace(0,10*365,num = 10) # return 10 evenly spaced numbers over the interval [0,10*365]
np.abs(-18) # return the absolute value, 18
np.ndarray # an array object
np.linalg.inv(matrix_AK) # return the inverse of matrix_AK
np.sum(A) # return the sum of elements in matrix A
```

(b)

```
import pandas as pd
df=pd.DataFrame(Xc) # return a tabular data (df) from Xc
df.Columns=['X1','X2','X3'] # assign labeled column axis
df['X1'] = carbon_storage # arithmetic operations align on both row and column
    labels.
df.to_csv(filename, index=False) # write to a csv file without row names
```

(c)

```
from scipy.integrate import solve_ivp
solve_ivp(func,t,y0,t_eval=times,vectorized = True)
# Solve a system of ordinary differential equations given an initial value, e.g.
# dy/dt=func(t,y) and y(t0)=y0
# t is the interval (t0,tf) where the solver starts with t=to and ends t=tf
# t_eval is times at which to store the computed solution according to t
# vectorized indicates whether func is implemented in a vectorized fashion or not
```

(d)

```
import matplotlib.pyplot as plt
fig = plt.figure(12, figsize=(14, 7))
# create a figure.12 is its identifier.figsize sets its width and height in inches
plt.subplots_adjust() # specify the subplot layout
ax = plt.subplot(nrow, ncols, index)
# add subplot to a current figure at the specified grid
ax.plot(x,y) # plot y versus x as lines and/or markers
plt.xlabel('year') # add x labels
plt.ylabel('pool') # add y labels
plt.savefig(fileName) # save the figure to a file
```

FIGURE A2.13 Code examples in GeneralModel.py and test_p2.py using (a) *Numpy* module, (b) *Pandas* module, (c) *Scipy* module, and (d) *Matplotlib* module.

SUGGESTED READINGS

- https://wiki.python.org/moin/BeginnersGuide
- http://pythontutor.com
- https://thepythonguru.com
- https://pymbook.readthedocs.io/en/latest/
- https://docs.python-guide.org
- https://www.w3schools.com/python/

QUIZ

1 What is an annotation?
2 What is an operator?
3 Can one function call another function?
4 How would you express an if-then sentence in Python?

Appendix 3
CarboTrain User Guide

Jian Zhou
Cornell University, Ithaca, USA

This appendix provides a guide for *CarboTrain*, a *Carbo*n cycle modeling *Train*ing software system tailored for use in the training course *New Advances in Land Carbon Cycle Modeling*. The main goal of the training course is for trainees to learn new theory and skills of modeling land carbon dynamics. *CarboTrain* is designed to help trainees to reach their learning goals without getting bogged down in programming. The software implements all the exercises in Units 2–10 of the training course as described in this book.

INTRODUCTION

CarboTrain has a user interface as shown in Figure A3.1, and it can run on computers running the Windows or macOS operating system.

This software was mainly developed with Python and PyQt. In order to run the software properly, some other software systems have to be pre-installed. The following section provides a step-by-step guide on how to install and use *CarboTrain*.

DOWNLOAD CARBOTRAIN

Due to the diversity of trainees' personal computers, we offer two ways to install the *CarboTrain*: one is to install it by docker image, and another is a non-docker approach, allowing you to directly install the *CarboTrain* software on your computer.

A docker image is available on docker hub (https://hub.docker.com/r/carbotrain/carbotrain). If you have docker installed in your computer, you are ready to go. We recommend using the docker image because all the essential packages that *CarboTrain* needs are included. This means you don't need to create any virtual environments or struggle installing the packages and *CarboTrain* software on your computer. If you choose to use the docker image (more details are available on the referenced docker hub website above), you may skip this part and jump to Uses of *CarboTrain* further down in this appendix. Please be aware that the docker image of the current *CarboTrain* is not compatible with your CPU if you have an Apple Silicon computer.

If you want to run *CarboTrain* without the docker image, you may download and install it with essential software on your

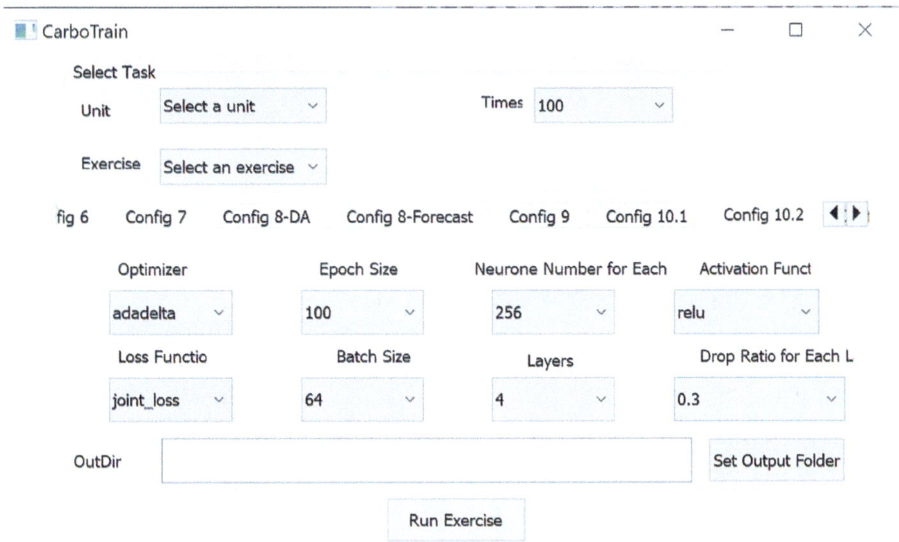

FIGURE A3.1 The Graphical User Interface (GUI) of CarboTrain.

computer. First, please download the software from: https://ecolab.cals.cornell.edu/download/CarboTrain.zip.

Then, there are three pre-installed software packages that are required to run *CarboTrain*:

1. Python environment and relevant packages
2. Fortran compiler
3. R environment and relevant packages

Since different operating systems may have different ways to install all the software needed, we will show how to install these software systems with different operating systems.

INSTALLATION ON WINDOWS

Install Python Environment

Download Python 3.7.9 from https://www.python.org/downloads/release/python-379/ ("**Windows x86-64 executable installer**") and install it on your computer. When installing Python, check "Add Python 3.7 to PATH" as shown in Figure A3.2. After installing Python, open the Command Prompt (CMD) window to see whether it is installed correctly. To open the CMD window, type in "CMD" in the search bar of your computer, as shown in Figure A3.3. Follow the steps in Figure A3.4 to check whether you have installed Python successfully.

Fortran Compiler

Go to the website https://sourceforge.net/projects/mingw/ and download the software as shown in step 1 in Figure A3.3. Then, step 2 is to run the downloaded file by right-clicking it and clicking "Run as administrator" from the menu. Next, you need to click the "Install" button to start the installation. Finally, click "Continue" several times to finish the installation.

Once the installation is finished, another window will pop out automatically as shown in Figure A3.6.

Check each box in Figure A3.6 and then select "Mark for Installation". Then go to "Installation" and click "Update Catalogue" as shown in Figure A3.7 to confirm the changes. You then need to click "Review Changes" in the pop-up window and then click "Apply" in the following pop-up window to install all the packages needed.

When you have finished the installation of all the packages, you need to set the environment variables in your

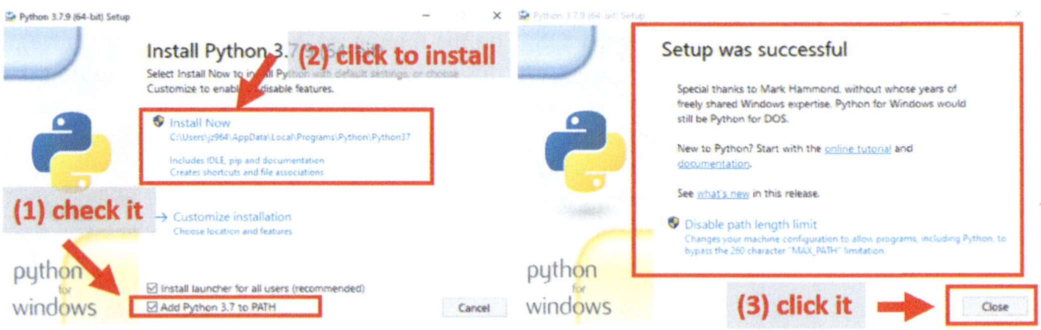

FIGURE A3.2 Steps to install Python on your computer.

FIGURE A3.3 Steps to open a CMD window.

FIGURE A3.4 Steps to check whether Python is installed.

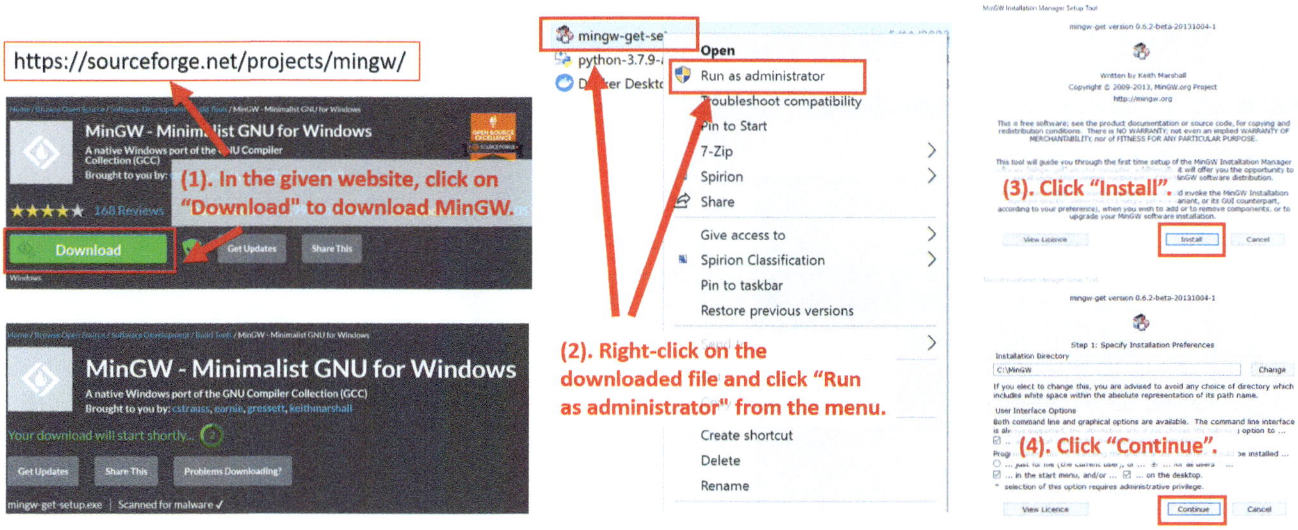

FIGURE A3.5 Steps to install MinGW.

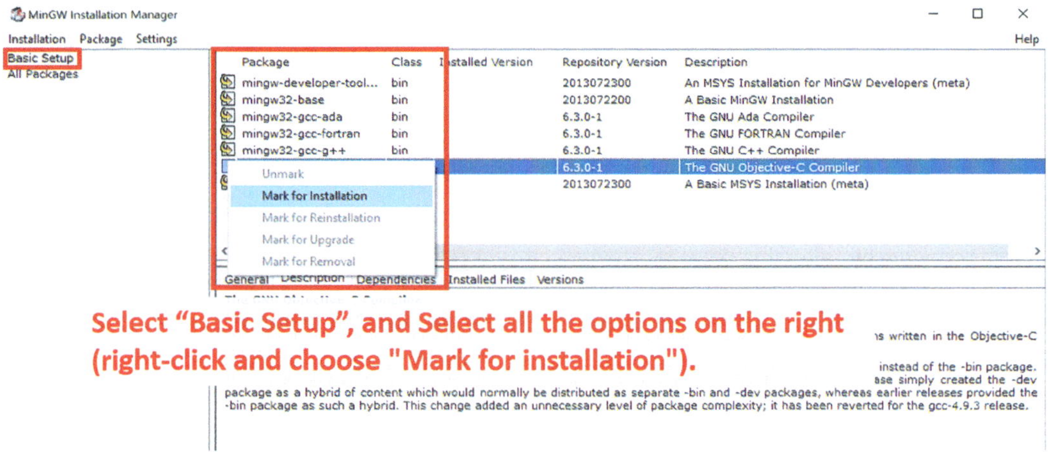

FIGURE A3.6 Mark the packages for installation in MinGW.

computer. To do this, search for "edit the system environment variables" in the search bar of your computer, and then click "Edit the system environment variables" and the "System Properties" window will show up. Go to the "Advanced" menu and then click "Environment Variables... " as shown in Figure A3.8.

Follow the steps in Figure A3.8 to open the environment variable setting window as in Figure A3.9. Follow the steps in the figure to add a new path. First select "path" and then click "Edit... ", steps 1 and 2 in the left panel of Figure A3.9. In the right panel of the dialog shown in Figure A3.9, click "New" and then add "C:\MinGW\bin" to finish the setting of the path.

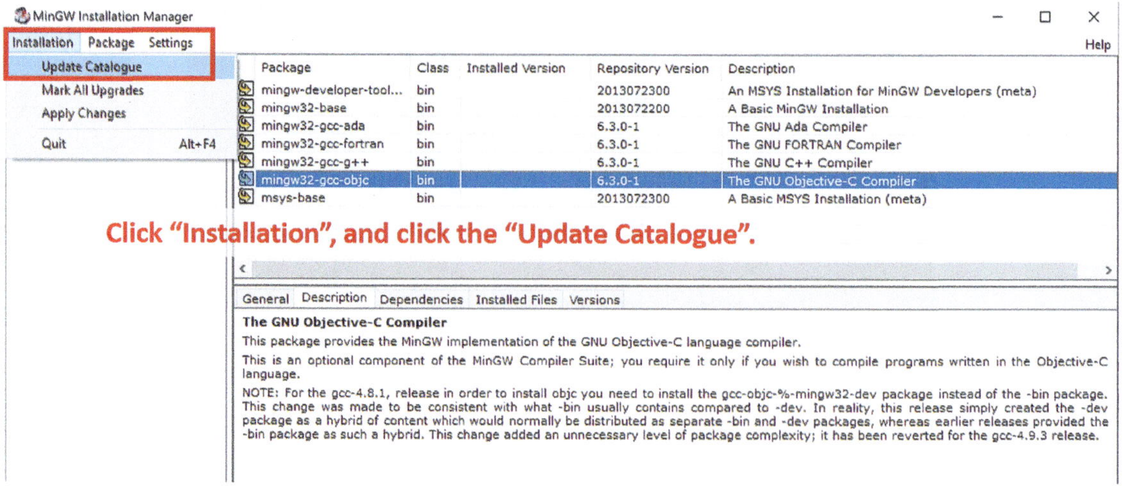

FIGURE A3.7 Update the catalogue in the MinGW.

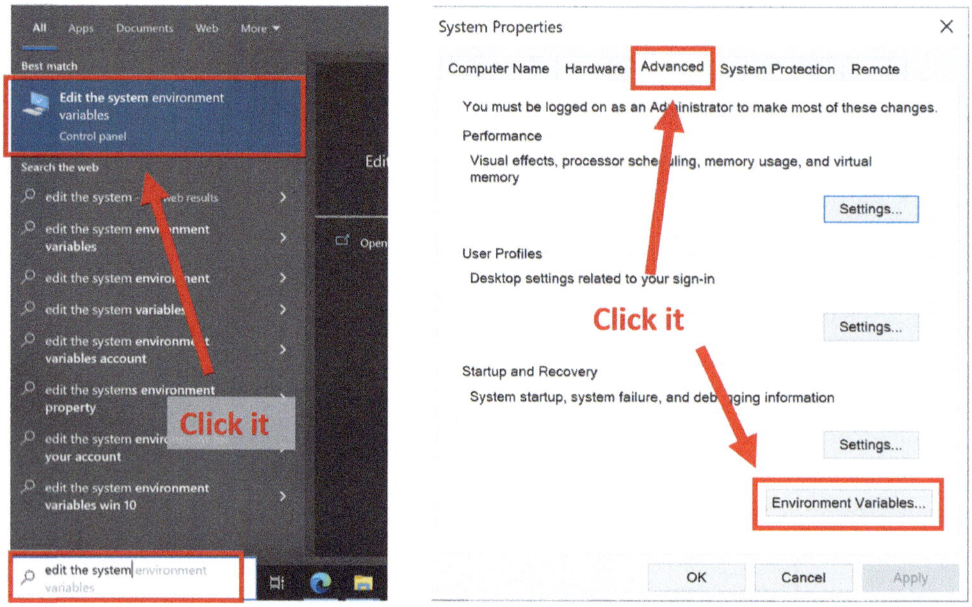

FIGURE A3.8 Steps to set the environment variables under "System Properties".

Finally, open a CMD window by following the steps shown in Figure A3.3 to check whether the Fortran compiler has been installed correctly by following the steps in Figure A3.10. When you type "gfortran --version", it will show the version information of installed gfortran. The Fortran compiler has been successfully installed.

Install R Environment

Download R 3.6.3 from https://cran.r-project.org/bin/windows/base/old/3.6.3/, and install it in your computer as shown in Figure A3.11. Once finished, set the environment variables for R following the two steps in Figure A3.12.

After setting the environment variables, you can now open a CMD window by following the steps shown in Figure A3.3 and follow the steps in Figure A3.12 to see whether you have installed R 3.6.3 correctly.

After the installation, you need to install the Python packages and R packages needed in the training course and then compile the Fortran code. Locate the folder of *CarboTrain*, and copy the path of that folder as shown in Figure A3.13. Once you have copied the path, open a CMD window by following the steps shown in Figure A3.3 and type in the following commands (followed by the Enter key) to install the Python packages and R packages:

1. cd path_your_copied
2. pip3 install -r requirements.txt
3. Rscript Rinstall_packages_win.R

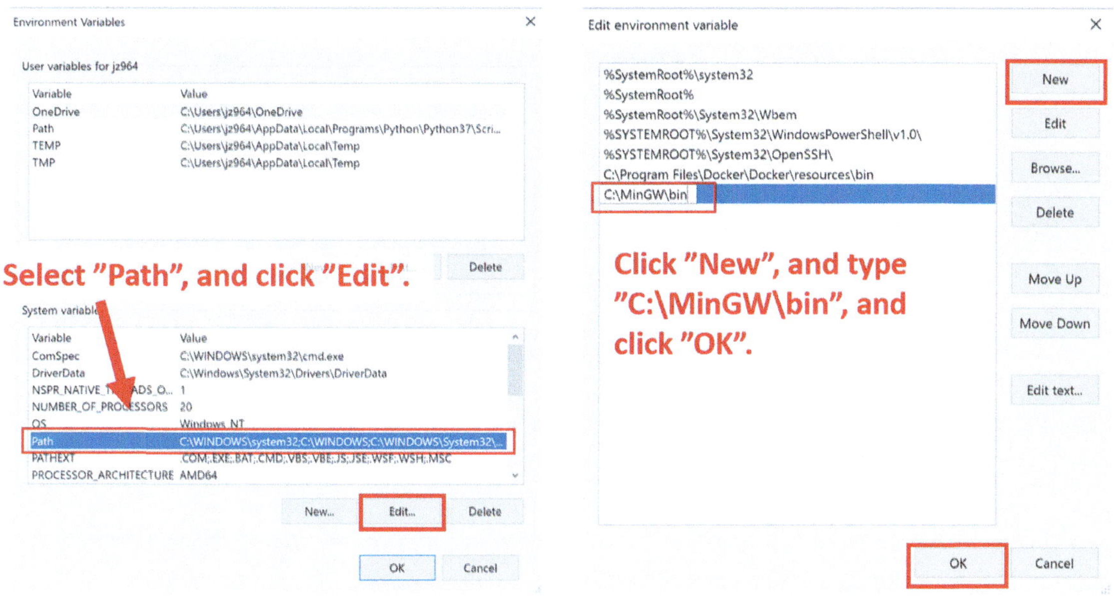

FIGURE A3.9 Steps to add a new path for MinGW.

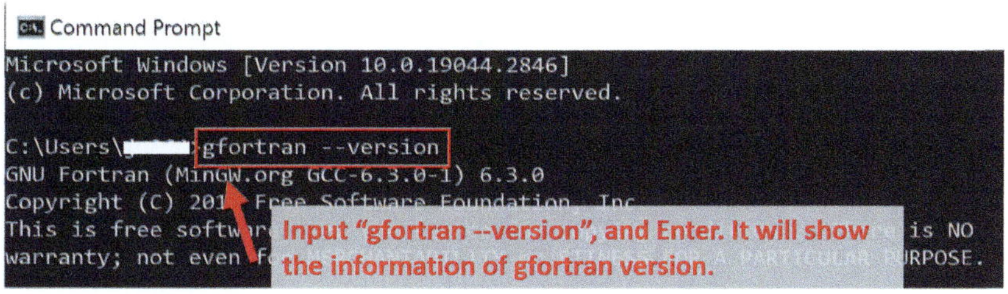

FIGURE A3.10 Check fortran compiler installation.

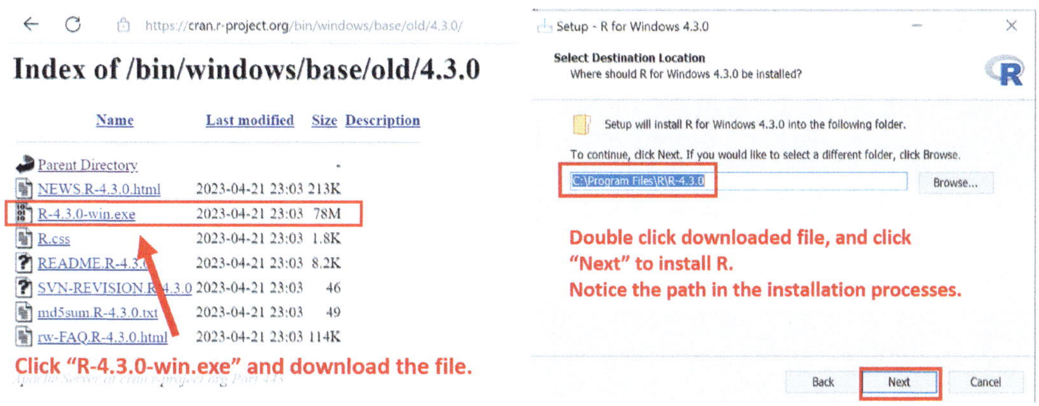

FIGURE A3.11 Installation of the R environment. Most processes are clicking the "Next" button. Please note the installation path during the process.

In order to run the TECO model properly, you also need to compile the source code. Go to the TECO source code folder under "*CarboTrain* → Source_code → TECO_2.3" and copy the full path as we just did in the last step. Once you have copied the path, open a CMD window by following the steps shown in Figure A3.3 and type in the following command to compile the source code as shown in Figure A3.14. You may ignore any warnings that come up.

gfortran -o TECO_2.3.exe TECO_2.3.f90

Windows users are now ready to run the practice sessions for each unit, and may skip to the section Uses of *CarboTrain*, below.

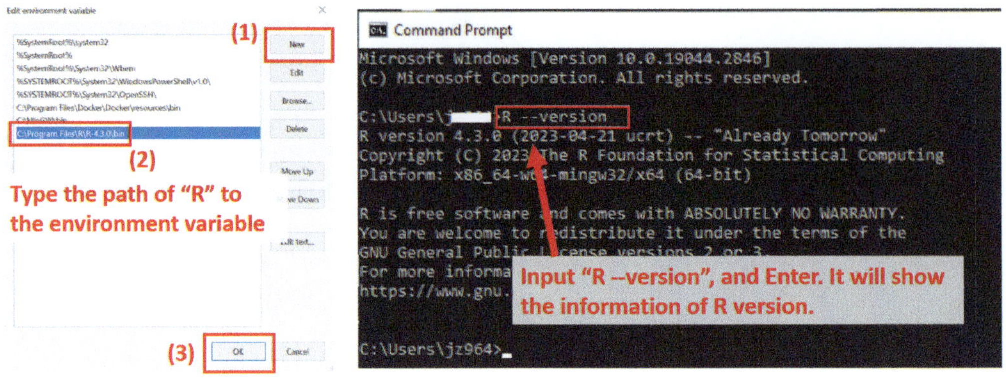

FIGURE A3.12 The main step of adding the R to environment variable, and the step of checking whether R environment is installed.

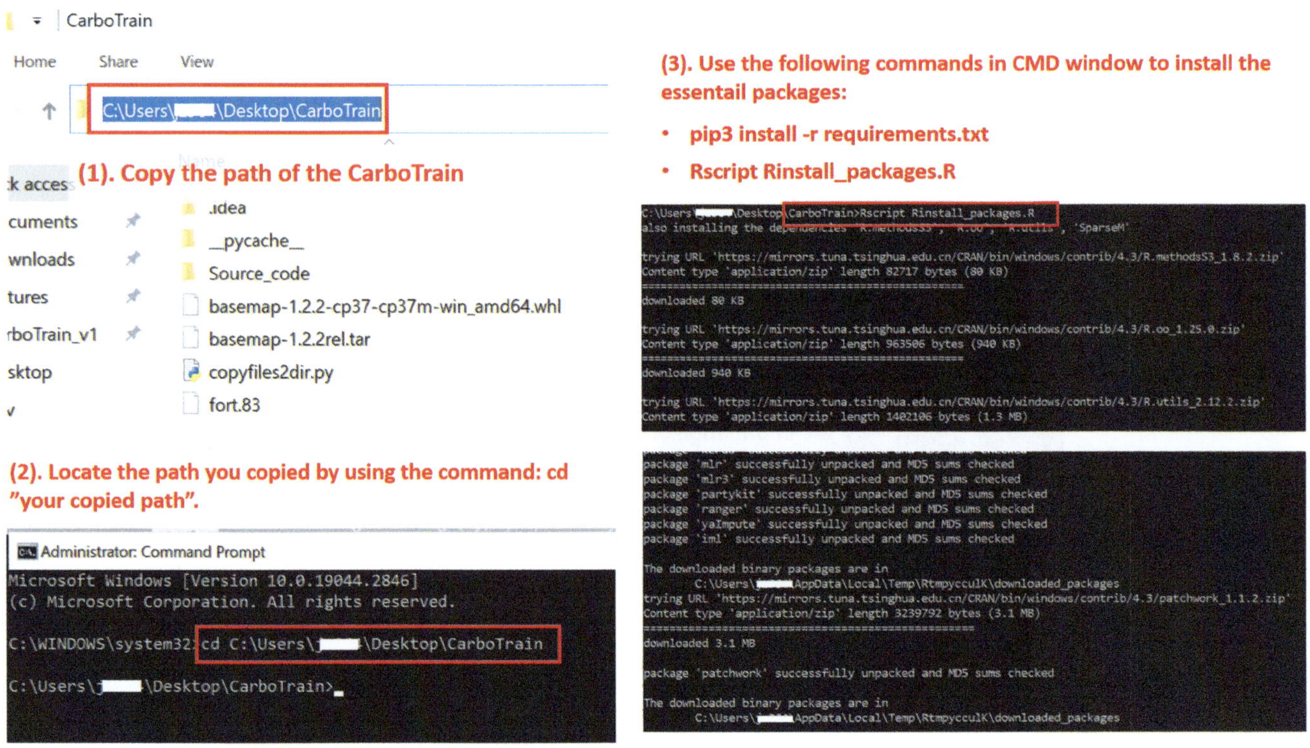

FIGURE A3.13 Locating the path of your *CarboTrain* software folder, and installing the relevant packages of Python and R.

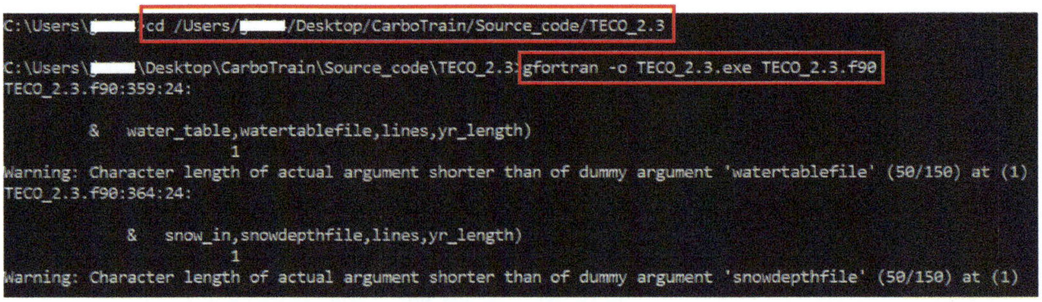

FIGURE A3.14 Compiling the TECO model by gfortran compiler.

FIGURE A3.15 The step of downloading the *Conda* installation file.

We will now provide instructions on how to install *CarboTrain* on a macOS computer.

INSTALLATION ON MACOS

In order to facilitate the installation of the required environment for *CarboTrain* on the Mac system, we are using *Conda* to manage and install *Fortran, Python, R,* and their respective environment packages.

Install Conda and Python

Conda is an open-source package management and environment management system that simplifies the installation and management of software packages, dependencies, and environments across different platforms. It is widely used in the data science and programming communities to create isolated environments for different projects and manage dependencies seamlessly. To install *Conda*, we first need to download the installation file from the following websites, as shown in Figure A3.15:

1. If your Mac is intel chip: https://repo.anaconda.com/miniconda/Miniconda3-latest-MacOSX-x86_64.sh
2. If your Mac is Apple chip (e.g., M1 or M2): https://repo.anaconda.com/miniconda/Miniconda3-latest-MacOSX-arm64.sh

After downloading the *Conda* installation file, move it to the desktop folder. To install *Conda*, we need to first open a terminal window and navigate to the desktop folder in the terminal (where the *Conda* installation file is located). Then, grant execution permission to the *Conda* installation file and run it, following the steps below to complete the *Conda* environment installation (as shown in Figure A3.16):

1. cd & cd Desktop
2. chmod u+x *your_downloaded_Conda_filename*
3. ./*your_downloaded_Conda_filename*

After installing *Conda*, the next step is to create a virtual environment to manage and install the required environment packages for the *CarboTrain* software. We create a virtual environment named "carbotrain" based on Python 3.8 as shown in Figure A3.17. Once you have finished, open a terminal to test the installation following the two steps in Figure A3.17.

After installing the *Conda* and relevant Python environment, use the following commands to install the gfortran, R, and relevant packages of Python and R. (Please make sure you have downloaded the *CarboTrain* software and extracted it to the Desktop. The folder should be named "CarboTrain".)

1. cd & cd Desktop/CarboTrain
2. conda install -c conda-forge --file requirements_macos_intel.txt (if your CPU is intel chip) conda install -c conda-forge --file requirements_macos_Mchip.txt (if your CPU is Apple chip)
3. pip install -r requirements_macos_intel.txt (if your CPU is intel chip)
   pip install -r requirements_macos_Mchip.txt (if your CPU is Apple chip)
4. Rscript Rinstall_packages_mac.R

In order to install these packages successfully, we need to make sure you have activated your created virtual environment (carbotrain) by entering conda activate carbotrain, as shown in Figure A3.17. After installing these packages, we need check the gfortran, Python, and R environment by typing gfortran --version, python --version and R –version, as shown in Figure A3.18.

Once you have installed these required software systems, you need to compile the TECO model. To do this, make sure you are in the *CarboTrain* folder on your Desktop, and then open a Terminal and type in the following commands to compile TECO, as shown in Figure A3.19:

1. cd && cd Desktop/CarboTrain/Source_code/TECO_2.3
2. gfortran -o TECO_2.3.exe TECO_2.3.f90

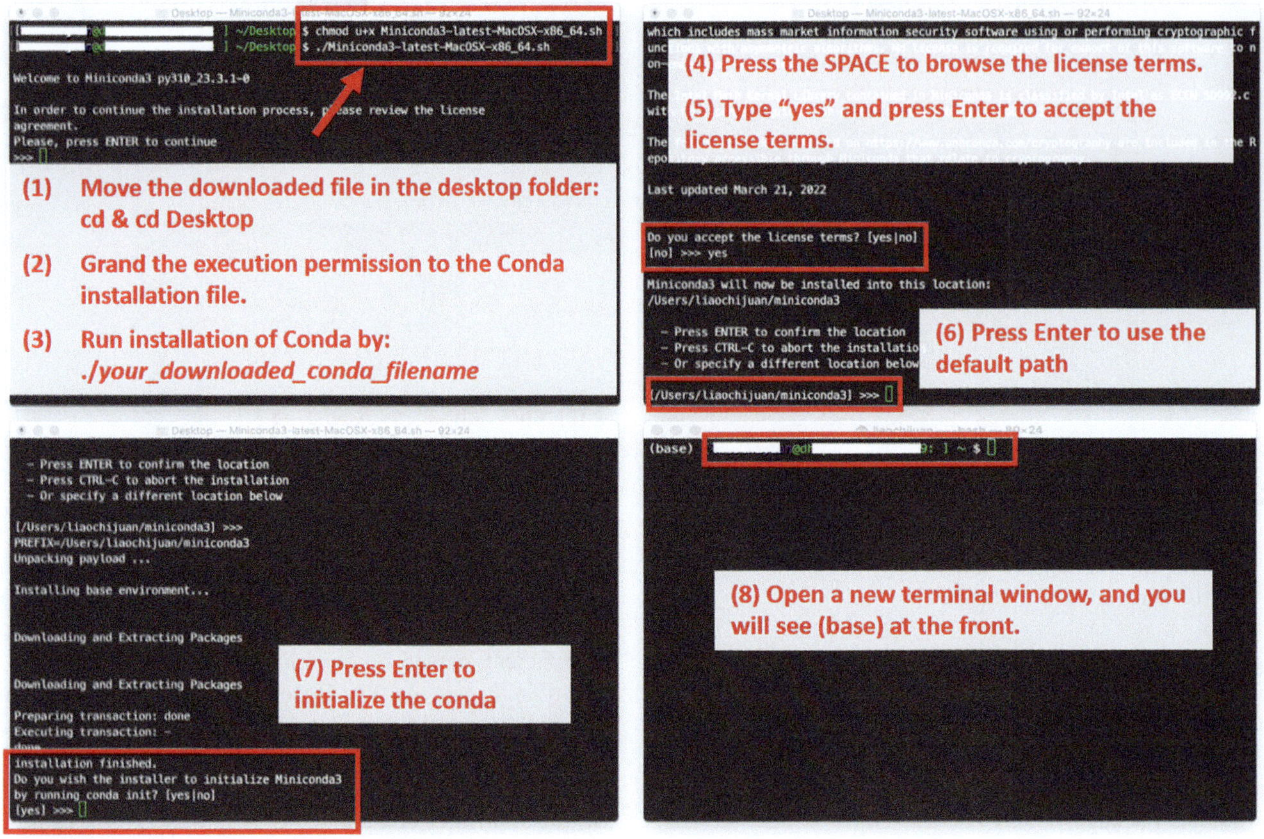

FIGURE A3.16 The main steps to install *Conda*.

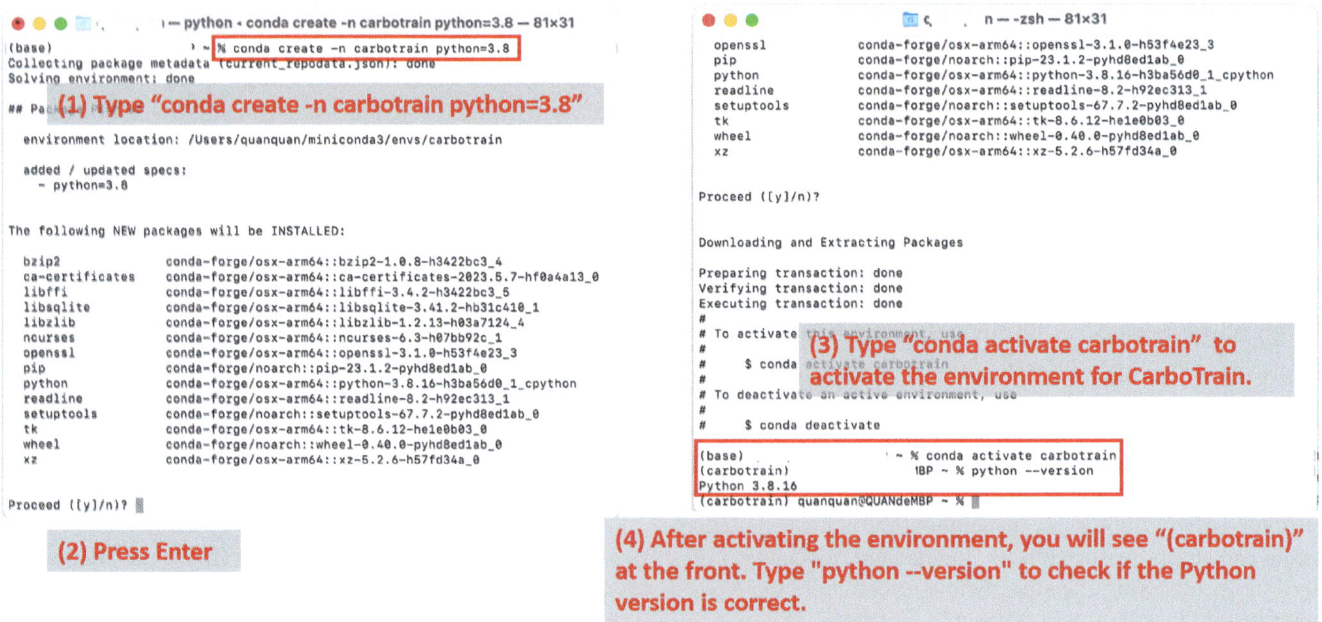

FIGURE A3.17 The steps to create the virtual environment of CarboTrain and install Python 3.8.

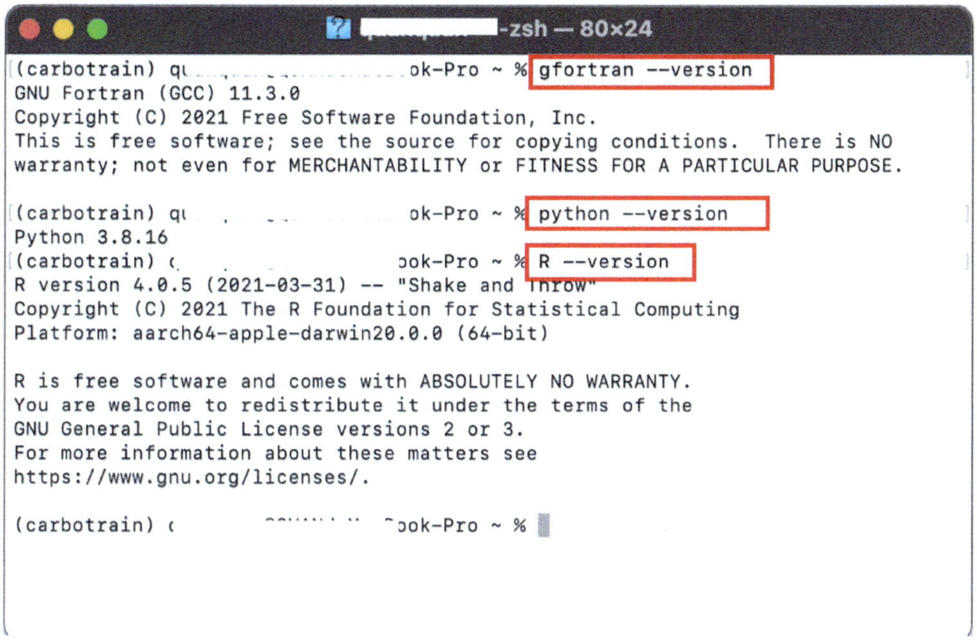

FIGURE A3.18 The steps to check whether the environments of gfortran, python, and R are installed correctly.

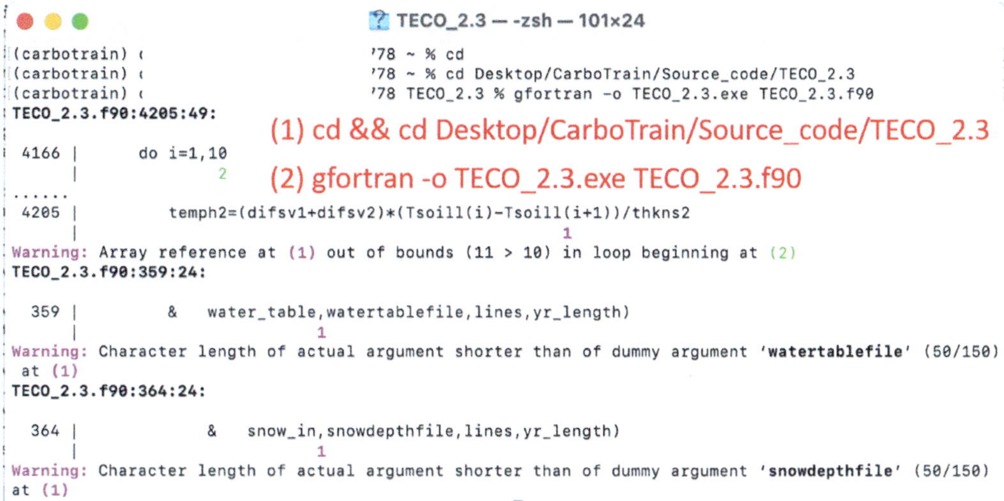

FIGURE A3.19 Steps to compile the TECO model.

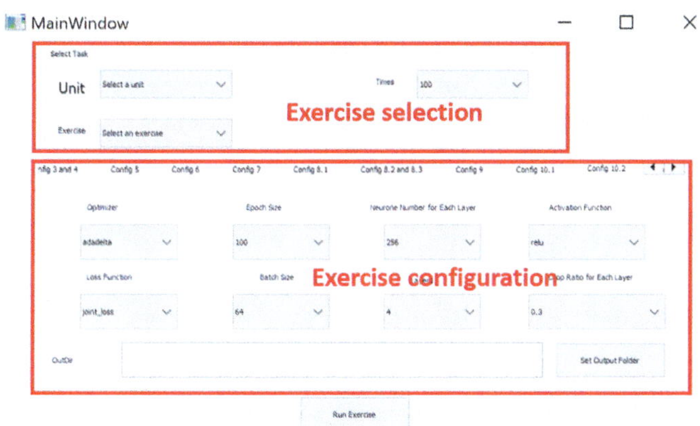

FIGURE A3.20 The *CarboTrain* GUI showing the two parts.

You now have all the required software installed and are ready for the practices in the training course or doing practices in these practice chapters of this book.

Use of CarboTrain

CarboTrain will be used for the practices for Units 2 to 10. Unit 2 to 4 make use of the matrix version of the TECO ecosystem model and Unit 5 is about traceability. In Unit 6, you will practice data assimilation (DA) with a simple version of TECO. In Units 7 and 8, you will use an intermediately complex version of the TECO model to perform data assimilation and ecological forecasting. Units 9 and 10 practices cover deep learning. The instructions for each practice are described in detail in the corresponding unit. Here we describe the general steps for using the software.

To use *CarboTrain*, you first need to launch it. In Windows, copy the path of *CarboTrain*, locate to the path you copied, and run the software in a CMD window as below:

1. cd path_you_copied
2. python main.py

In macOS, use the following commands to run the software:

1. cd && cd Desktop/CarboTrain
2. Python main.py

Once you have launched the software, you will see a graphical user interface (GUI) similar to Figure A3.1. The GUI of the software consists of two parts: exercise selection, and exercise configuration, as shown in Figure A3.20. Each tab shown in the "Exercise configuration" part contains one or a set of exercises, as shown in TABLE A3.1.

The general steps for each practice are: (1) select a unit, (2) select an exercise, (3) configure the exercise you just selected, and (4) run the exercise. For example, to run Exercise 1 in Unit 10, you need to go through the steps shown in Figure A3.21. The first two steps are for selecting an exercise and step 3 is for configuring the exercise. The configuration required may be different depending on the details of the exercise. For this example, configuration involves selecting the output folder. In some exercises, you need to modify the source code, change the settings or customize the exercise according to your own questions. Once you have finished exercise selection and configuration, it is ready for step 4: clicking "Run Exercise" to start the exercise.

When you have clicked the "Run Exercise" button in each exercise, a pop-up window will show up with the message "Task submitted!" and you need to click "OK" before you can run the exercise. Once an exercise is done, a pop-up window with the message "Finished!" will inform you that the exercise has been completed with no error. If any errors occur, please double-check the steps you went through and then run the exercise again following the correct procedure. If you still cannot figure out the cause of the error, please ask instructors.

TABLE A3.1
Exercises in different tabs.

Config	Exercise(s)
2	Exercise 2 of Unit 2
3	Exercise of Unit 3
4	Exercise of Unit 4
5	Exercises of Unit 5
6	Exercises of Unit 6
7	Exercises of Unit 7
8-DA	Exercises 2 and 3 of Unit 8
8-Forecast	Exercises 4 and 5 of Unit 8
9	Exercises of Unit 9
10.1	Exercise 1 of Unit 10
10.2	Exercise 2 of Unit 10
10.3	Exercise 3 of Unit 10

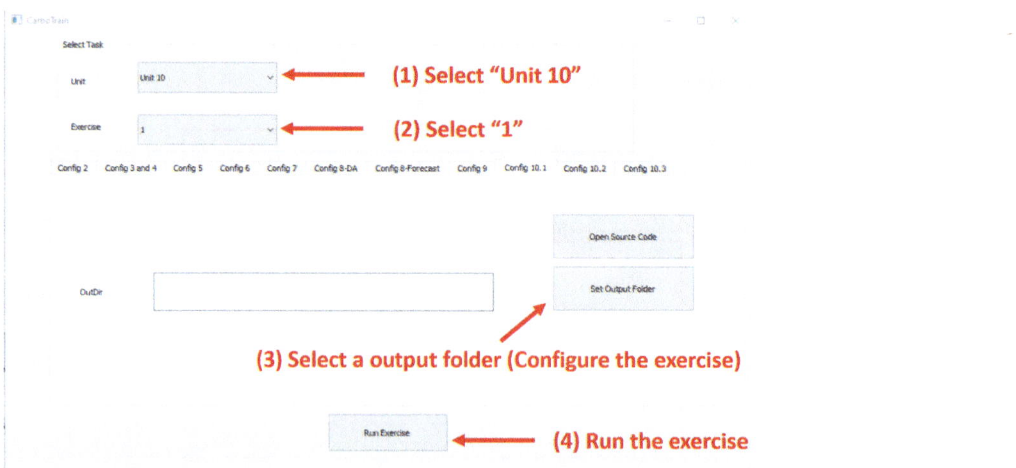

FIGURE A3.21 Steps to run Exercise 1 in Unit 10.

Bibliography

Ahlström, A., Schurgers, G., & Smith, B. (2017). The large influence of climate model bias on terrestrial carbon cycle simulations. *Environmental research letters, 12*(1), 014004.

Ahlström, A., Smith, B., Lindström, J., Rummukainen, M., & Uvo, C. B. (2013). GCM characteristics explain the majority of uncertainty in projected 21st century terrestrial ecosystem carbon balance. *Biogeosciences, 10*(3), 1517–1528. doi:10.5194/bg-10-1517-2013

Ahlström, A., Xia, J., Arneth, A., Luo, Y., & Smith, B. (2015). Importance of vegetation dynamics for future terrestrial carbon cycling. *Environmental research letters, 10*(5), 054019.

Allaire, J. J., & Chollet, F. (2020). keras: R Interface to 'Keras'. R package version 2.3. 0.0. Computer software]. https://CRAN.R-project.org/package=keras

Anderson, D. H. (2013). *Compartmental modeling and tracer kinetics* (Vol. 50): Springer Science & Business Media.

Anderson, D. H., & Roller, T. (1991). Equilibrium points for nonlinear compartmental models. *Mathematical biosciences, 103*(2), 159–201.

Andren, O., & Paustian, K. (1987). Barley straw decomposition in the field: a comparison of models. *Ecology, 68*(5), 1190–1200.

Arora, V. K., Katavouta, A., Williams, R. G., Jones, C. D., Brovkin, V., Friedlingstein, P., …Cadule, P. (2020). Carbon–concentration and carbon–climate feedbacks in CMIP6 models and their comparison to CMIP5 models. *Biogeosciences, 17*(16), 4173–4222.

Ball, J. T., Woodrow, I. E., & Berry, J. A. (1987). *A model predicting stomatal conductance and its contribution to the control of photosynthesis under different environmental conditions.* Paper presented at the Progress in photosynthesis research: volume 4 proceedings of the VIIth international congress on photosynthesis providence, Rhode Island, USA, August 10–15, 1986.

Bastin, G., & Guffens, V. (2006). Congestion control in compartmental network systems. *Systems & control letters, 55*(8), 689–696.

Bauer, P., Thorpe, A., & Brunet, G. (2015). The quiet revolution of numerical weather prediction. *Nature, 525*(7567), 47–55.

Bell, J. (2020). *Machine learning: Hands-on for developers and technical professionals*: John Wiley & Sons.

Bishop, C. M. (2006). Pattern recognition. *Machine learning, 128*(9).

Blackard, J. A., Finco, M. V., Helmer, E. H., Holden, G. R., Hoppus, M. L., Jacobs, D. M., …Tymcio, R. P. (2008). Mapping U.S. forest biomass using nationwide forest inventory data and moderate resolution information. *Remote sensing of environment, 112*(4), 1658–1677. doi:https://doi.org/10.1016/j.rse.2007.08.021

Blagodatsky, S., Blagodatskaya, E., Yuyukina, T., & Kuzyakov, Y. (2010). Model of apparent and real priming effects: linking microbial activity with soil organic matter decomposition. *Soil biology and biochemistry, 42*(8), 1275–1283.

Bodesheim, P., Jung, M., Gans, F., Mahecha, M. D., & Reichstein, M. (2018). Upscaled diurnal cycles of land–atmosphere fluxes: a new global half-hourly data product. *Earth syst. sci. data, 10*(3), 1327–1365. doi:10.5194/essd-10-1327-2018

Bolin, B. (1981). Steady state and response characteristics of a simple model of the carbon cycle. *Carbon cycle modelling*, 315–331.

Bolin, B., & Rodhe, H. (1973). A note on the concepts of age distribution and transit time in natural reservoirs. *Tellus, 25*(1), 58–62.

Bolker, B. M., Pacala, S. W., & Parton Jr, W. J. (1998). Linear analysis of soil decomposition: insights from the century model. *Ecological applications, 8*(2), 425–439.

Bonan, G. B., & Doney, S. C. (2018). Climate, ecosystems, and planetary futures: The challenge to predict life in Earth system models. *Science, 359*(6375), eaam8328.

Box, G. E. (1979). Robustness in the strategy of scientific model building. In *Robustness in statistics* (pp. 201–236): Elsevier.

Burrows, S., Maltrud, M., Yang, X., Zhu, Q., Jeffery, N., Shi, X., …Tang, J. (2020). The DOE E3SM v1. 1 biogeochemistry configuration: Description and simulated ecosystem-climate responses to historical changes in forcing. *Journal of advances in modeling earth systems, 12*(9), e2019MS001766.

Byrne, M. P., & O'gorman, P. A. (2016). Understanding decreases in land relative humidity with global warming: Conceptual model and GCM simulations. *Journal of climate, 29*(24), 9045–9061.

Cai, A., Liang, G., Zhang, X., Zhang, W., Li, L., Rui, Y., …Luo, Y. (2018). Long-term straw decomposition in agro-ecosystems described by a unified three-exponentiation equation with thermal time. *Science of the total environment, 636*, 699–708.

Canham, D. C. (2003). *Models in ecosystem science*: Princeton University Press.

Carvalho, A., Langa, J. A., & Robinson, J. (2012). *Attractors for infinite-dimensional non-autonomous dynamical systems* (Vol. 182): Springer Science & Business Media.

Cheng, X., Luo, Y., Su, B., Zhou, X., Niu, S., Sherry, R., …Zhang, Q. (2010). Experimental warming and clipping altered litter carbon and nitrogen dynamics in a tallgrass prairie. *Agriculture, ecosystems & environment, 138*(3), 206–213. doi:https://doi.org/10.1016/j.agee.2010.04.019

Ciais, P., Sabine, C., Bala, G., Bopp, L., Brovkin, V., Canadell, J., …Wania, R. (2014). Carbon and other biogeochemical cycles. In *Climate change 2013: The physical science basis. Contribution of working group I to the fifth assessment report of the intergovernmental panel on climate change* (pp. 465–570): Cambridge University Press.

Collier, N., Hoffman, F. M., Lawrence, D. M., Keppel-Aleks, G., Koven, C. D., Riley, W. J., …Randerson, J. T. (2018). The International Land Model Benchmarking (ILAMB) system: design, theory, and implementation. *Journal of advances in modeling earth systems, 10*(11), 2731–2754.

Craiu, R. V., & Rosenthal, J. S. (2014). Bayesian computation via markov chain monte carlo. *Annual review of statistics and its application, 1*, 179–201.

Cui, E., Huang, K., Arain, M. A., Fisher, J. B., Huntzinger, D. N., Ito, A., …Michalak, A. M. (2019). Vegetation functional properties determine uncertainty of simulated ecosystem productivity: A traceability analysis in the East Asian monsoon region. *Global biogeochemical cycles, 33*(6), 668–689.

Davidson, E. A., Janssens, I. A., & Luo, Y. (2006). On the variability of respiration in terrestrial ecosystems: moving beyond Q10. *Global change biology, 12*(2), 154–164.

De Kauwe, M. G., Disney, M. I., Quaife, T., Lewis, P., & Williams, M. (2011). An assessment of the MODIS collection 5 leaf area index product for a region of mixed coniferous forest. *Remote sensing of environment, 115*(2), 767–780. doi:https://doi.org/10.1016/j.rse.2010.11.004

Dietze, M. C. (2017). Prediction in ecology: A first-principles framework. *Ecological Applications, 27*(7), 2048–2060.

Dietze, M. C., Lebauer, D. S., & Kooper, R. (2013). On improving the communication between models and data. *Plant, cell & environment, 36*(9), 1575–1585.

Dormaar, J. (1979). Organic matter characteristics of undisturbed and cultivated chernozemic and solonetzic A horizons. *Canadian journal of soil science, 59*(4), 349–356.

Du, Z., Weng, E., Jiang, L., Luo, Y., Xia, J., & Zhou, X. (2018). Carbon–nitrogen coupling under three schemes of model representation: a traceability analysis. *geoscientific model development, 11*(11), 4399–4416.

ElGhawi, R., Kraft, B., Reimers, C., Reichstein, M., Körner, M., Gentine, P., & Winkler, A. J. (2023). Hybrid modeling of evapotranspiration: inferring stomatal and aerodynamic resistances using combined physics-based and machine learning. *Environmental research letters, 18*(3), 034039.

Elman, J. (1990). Finding Structure in Time. *Cognitive Science*, 14 (2), 179–211.

Fer, I., Gardella, A. K., Shiklomanov, A. N., Campbell, E. E., Cowdery, E. M., De Kauwe, M. G., …Dietze, M. C. (2021). Beyond ecosystem modeling: A roadmap to community cyberinfrastructure for ecological data-model integration. *Global change biology, 27*(1), 13–26. doi:https://doi.org/10.1111/gcb.15409

Fisher, R. A., Koven, C. D., Anderegg, W. R., Christoffersen, B. O., Dietze, M. C., Farrior, C. E., …Lawrence, P. J. (2018). Vegetation demographics in Earth System Models: A review of progress and priorities. *Global change biology, 24*(1), 35–54.

Forrester, J. W. (1961). *Industrial dynamics*. Cambridge, Massachusetts: MIT Press.

Forster, P., Ramaswamy, V., Artaxo, P., Berntsen, T., Betts, R., Fahey, D. W., …Myhre, G. (2007). Changes in atmospheric constituents and in radiative forcing. Chapter 2. In *Climate change 2007. The physical science basis.*

Friedlingstein, P., Cox, P., Betts, R., Bopp, L., von Bloh, W., Brovkin, V., …Fung, I. (2006). Climate–carbon cycle feedback analysis: results from the C4MIP model intercomparison. *Journal of climate, 19*(14), 3337–3353.

Friedlingstein, P., Jones, M. W., O'Sullivan, M., Andrew, R. M., Bakker, D. C. E., Hauck, J., …Zeng, J. (2022). Global Carbon Budget 2021. *Earth syst. sci. Data, 14*(4), 1917–2005. doi:10.5194/essd-14-1917-2022

Friedlingstein, P., Meinshausen, M., Arora, V. K., Jones, C. D., Anav, A., Liddicoat, S. K., & Knutti, R. (2014). Uncertainties in CMIP5 climate projections due to carbon cycle feedbacks. *Journal of climate, 27*(2), 511–526.

Gelman, A., Carlin, J. B., Stern, H. S., Dunson, D. B., Vehtari, A., & Rubin, D. B. (2014). *Bayesian data analysis* (Vol. 2): CRC press Boca Raton, FL.

Gelman, A., Roberts, G. O., & Gilks, W. R. (1996). Efficient Metropolis jumping rules. *Bayesian statistics, 5*(599-608), 42.

Georgiou, K., Jackson, R. B., Vindušková, O., Abramoff, R. Z., Ahlström, A., Feng, W., …Torn, M. S. (2022). Global stocks and capacity of mineral-associated soil organic carbon. *Nature communications, 13*(1), 3797. doi:10.1038/s41467-022-31540-9

Giglio, L., Randerson, J. T., van der Werf, G. R., Kasibhatla, P. S., Collatz, G. J., Morton, D. C., & DeFries, R. S. (2010). Assessing variability and long-term trends in burned area by merging multiple satellite fire products. *Biogeosciences, 7*(3), 1171–1186. doi:10.5194/bg-7-1171-2010

Gill, A. L., Giasson, M. A., Yu, R., & Finzi, A. C. (2017). Deep peat warming increases surface methane and carbon dioxide emissions in a black spruce-dominated ombrotrophic bog. *Global change biology, 23*(12), 5398–5411.

Goodfellow, I., Bengio, Y., & Courville, A. (2016). *Deep learning*: MIT press.

Green, J., Berry, J., Ciais, P., Zhang, Y., & Gentine, P. (2020). Amazon rainforest photosynthesis increases in response to atmospheric dryness. *Science advances, 6*(47), eabb7232.

Green, J. K., Ballantyne, A., Abramoff, R., Gentine, P., Makowski, D., & Ciais, P. (2022). Surface temperatures reveal the patterns of vegetation water stress and their environmental drivers across the tropical Americas. *Global change biology, 28*(9), 2940–2955.

Greve, P., Orlowsky, B., Mueller, B., Sheffield, J., Reichstein, M., & Seneviratne, S. I. (2014). Global assessment of trends in wetting and drying over land. *Nature geoscience, 7*(10), 716–721. doi:10.1038/ngeo2247

Guanter, L., Frankenberg, C., Dudhia, A., Lewis, P. E., Gómez-Dans, J., Kuze, A., …Grainger, R. G. (2012). Retrieval and global assessment of terrestrial chlorophyll fluorescence from GOSAT space measurements. *Remote sensing of environment, 121*, 236–251. doi:https://doi.org/10.1016/j.rse.2012.02.006

Guckenheimer, J., & Holmes, P. (2013). *Nonlinear oscillations, dynamical systems, and bifurcations of vector fields* (Vol. 42): Springer Science & Business Media.

Guo, F., Yost, R., Hue, N., Evensen, C., & Silva, J. (2000). Changes in phosphorus fractions in soils under intensive plant growth. *Soil science society of America journal, 64*(5), 1681–1689.

Haario, H., Saksman, E., & Tamminen, J. (2001). An adaptive Metropolis algorithm. *Bernoulli, 7*(2), 223–242.

Haddix, M., Plante, A., Conant, R., Six, J., Steinweg, J. M., Magrini-Blair, K., …Paul, E. (2011). The role of soil characteristics on temperature sensitivity of soil organic matter. *Soil science society of America journal, 75*(1), 56–68.

Hanson, P., Gill, A., Xu, X., Phillips, J., Weston, D., Kolka, R., …Hook, L. (2016). Intermediate-scale community-level flux of CO 2 and CH 4 in a Minnesota peatland: putting the SPRUCE project in a global context. *Biogeochemistry, 129*(3), 255–272.

Hanson, P.J., J.R. Phillips, J.S. Riggs, W.R. Nettles, and D.E. Todd. 2014. SPRUCE Large-Collar In Situ CO2 and CH4 Flux Data for the SPRUCE Experimental Plots. Carbon Dioxide Information Analysis Center, Oak Ridge National Laboratory, U.S. Department of Energy, Oak Ridge, Tennessee, U.S.A. http://dx.doi.org/10.3334/CDIAC/spruce.006

Hanson, P. J., Griffiths, N. A., Iversen, C. M., Norby, R. J., Sebestyen, S. D., Phillips, J. R., …Ricciuto, D. M. (2020). Rapid Net Carbon Loss From a Whole-Ecosystem Warmed Peatland. *AGU advances, 1*(3), e2020AV000163. doi:https://doi.org/10.1029/2020AV000163

Hanson, P. J., Riggs, J. S., Nettles, W. R., Phillips, J. R., Krassovski, M. B., Hook, L. A., …Ricciuto, D. M. (2017). Attaining whole-ecosystem warming using air and deep-soil heating methods with an elevated CO 2 atmosphere. *Biogeosciences, 14*(4), 861–883.

Hararuk, O., Smith, M. J., & Luo, Y. (2015). Microbial models with data-driven parameters predict stronger soil carbon responses to climate change. *Global change biology, 21*(6), 2439–2453.

Hararuk, O., Xia, J., & Luo, Y. (2014). Evaluation and improvement of a global land model against soil carbon data using a Bayesian Markov chain Monte Carlo method. *Journal of geophysical research: Biogeosciences, 119*(3), 403–417.

Harrell, J., Frank E, & Harrell, F. E. (2015). Multivariable modeling strategies. *Regression modeling strategies: With applications to linear models, logistic and ordinal regression, and survival analysis*, 63–102.

Hart, S. (1989). Shapley value. In *Game theory* (pp. 210–216): Springer.

Hastie, T., & Tibshirani, R. (2009). & Friedman, J.(2008). The elements of statistical learning; data mining, inference and prediction. In: Springer, New York.

Hengl, T., de Jesus, J. M., Heuvelink, G. B., Gonzalez, M. R., Kilibarda, M., Blagotić, A., …Bauer-Marschallinger, B. (2017). SoilGrids250m: Global gridded soil information based on machine learning. *PLoS One, 12*(2), e0169748.

Hochreiter, S., & Schmidhuber, J. (1997). Long short-term memory. *Neural computation, 9*(8), 1735–1780.

Horgan, J. (2016). Bayes Theorem: What's the big deal. *Scientific American.*

Hou, E., Lu, X., Jiang, L., Wen, D., & Luo, Y. (2019). Quantifying soil phosphorus dynamics: a data assimilation approach. *Journal of Geophysical Research: Biogeosciences, 124*(7), 2159–2173.

Hou, E., Ma, S., Huang, Y., Zhou, Y., Kim, H.-S., López-Blanco, E., …Luo, Y. (2023). Across-model spread and shrinking in predicting peatland carbon dynamics under global change. *Global change biology, 29*(10), 2759–2775. doi:https://doi.org/10.1111/gcb.16643

Hou, E., Tan, X., Heenan, M., & Wen, D. (2018). A global dataset of plant available and unavailable phosphorus in natural soils derived by Hedley method. *Scientific data, 5*(1), 1–13.

Hou, E. L., Xingjie; Jiang, Lifen; Wen, Dazhi; Luo, Yiqi. (2019). *A soil phosphorus dynamics (SPD) model.* Retrieved from: https://doi.org/10.6084/m9.figshare.8273816.v4

Huang, X., Lu, D., Ricciuto, D. M., Hanson, P. J., Richardson, A. D., Lu, X., …Luo, Y. (2021). A model-independent data assimilation (MIDA) module and its applications in ecology. *Geosci. model dev., 14*(8), 5217–5238. doi:10.5194/gmd-14-5217-2021

Huang, Y., Jiang, J., Ma, S., Ricciuto, D., Hanson, P. J., & Luo, Y. (2017). Soil thermal dynamics, snow cover, and frozen depth under five temperature treatments in an ombrotrophic bog: Constrained forecast with data assimilation. *Journal of geophysical research: Biogeosciences, 122*(8), 2046–2063.

Huang, Y., Lu, X., Shi, Z., Lawrence, D., Koven, C. D., Xia, J., …Luo, Y. (2018). Matrix approach to land carbon cycle modeling: A case study with the Community Land Model. *Global change biology, 24*(3), 1394–1404.

Huang, Y., Stacy, M., Jiang, J., Sundi, N., Ma, S., Saruta, V., …Hanson, P. J. (2019). Realized ecological forecast through an interactive Ecological Platform for Assimilating Data (EcoPAD, v1. 0) into models. *Geoscientific model development, 12*(3), 1119–1137.

Huang, Y. Y., Zhu, D., Ciais, P., Guenet, B., Huang, Y., Goll, D. S., …Luo, Y. Q. (2018). Matrix-Based Sensitivity Assessment of Soil Organic Carbon Storage: A Case Study from the ORCHIDEE-MICT Model. *Journal of advances in modeling earth systems, 10*(8), 1790–1808. doi:10.1029/2017ms001237

Hugelius, G., Bockheim, J. G., Camill, P., Elberling, B., Grosse, G., Harden, J. W., …Yu, Z. (2013). A new data set for estimating organic carbon storage to 3 m depth in soils of the northern circumpolar permafrost region. *Earth syst. sci. data, 5*(2), 393–402. doi:10.5194/essd-5-393-2013

Jacquez, J. A., & Simon, C. P. (1993). Qualitative theory of compartmental systems. *Siam review, 35*(1), 43–79.

Jiang, J., Huang, Y., Ma, S., Stacy, M., Shi, Z., Ricciuto, D. M., …Luo, Y. (2018). Forecasting responses of a northern peatland carbon cycle to elevated CO2 and a gradient of experimental warming. *Journal of geophysical research: Biogeosciences, 123*(3), 1057–1071.

Jiang, L., Shi, Z., Xia, J., Liang, J., Lu, X., Wang, Y., & Luo, Y. (2017). Transient Traceability Analysis of Land Carbon Storage Dynamics: Procedures and Its Application to Two Forest Ecosystems. *Journal of advances in modeling earth systems, 9*(8), 2822–2835.

Jones, C., Robertson, E., Arora, V., Friedlingstein, P., Shevliakova, E., Bopp, L., …Kawamiya, M. (2013). Twenty-first-century compatible CO2 emissions and airborne fraction simulated by CMIP5 earth system models under four representative concentration pathways. *Journal of climate, 26*(13), 4398–4413.

Jones, T. R., Carpenter, A. E., Lamprecht, M. R., Moffat, J., Silver, S. J., Grenier, J. K., …Golland, P. (2009). Scoring diverse cellular morphologies in image-based screens with iterative feedback and machine learning. *Proceedings of the national academy of sciences, 106*(6), 1826–1831.

Jung, M., Reichstein, M., & Bondeau, A. (2009). Towards global empirical upscaling of FLUXNET eddy covariance observations: validation of a model tree ensemble approach using a biosphere model. *Biogeosciences, 6*(10), 2001–2013.

Jung, M., Schwalm, C., Migliavacca, M., Walther, S., Camps-Valls, G., Koirala, S., …Reichstein, M. (2020). Scaling carbon fluxes from eddy covariance sites to globe: synthesis and evaluation of the FLUXCOM approach. *Biogeosciences, 17*(5), 1343–1365. doi:10.5194/bg-17-1343-2020

Kloeden, P. E., & Rasmussen, M. (2011). *Nonautonomous dynamical systems*: American Mathematical Soc.

Koven, C., Riley, W., Subin, Z., Tang, J., Torn, M., Collins, W., …Swenson, S. (2013). The effect of vertically resolved soil biogeochemistry and alternate soil C and N models on C dynamics of CLM4. *Biogeosciences, 10*(11), 7109.

Kraft, B., Jung, M., Körner, M., Requena Mesa, C., Cortés, J., & Reichstein, M. (2019). Identifying dynamic memory effects on vegetation state using recurrent neural networks. *Frontiers in big data, 2*, 31.

Kumar, S. V., Peters-Lidard, C. D., Tian, Y., Houser, P. R., Geiger, J., Olden, S., …Sheffield, J. (2006). Land information system: An interoperable framework for high resolution land surface modeling. *Environmental modelling & software, 21*(10), 1402–1415. doi:https://doi.org/10.1016/j.envsoft.2005.07.004

Kuzyakov, Y. (2010). Priming effects: interactions between living and dead organic matter. *Soil biology and biochemistry, 42*(9), 1363–1371.

Lardy, R., Bellocchi, G., & Soussana, J.-F. (2011). A new method to determine soil organic carbon equilibrium. *Environmental modelling & software, 26*(12), 1759–1763.

Lawrence, D. M., Fisher, R. A., Koven, C. D., Oleson, K. W., Swenson, S. C., Bonan, G., …Kennedy, D. (2019). The Community Land

Model version 5: Description of new features, benchmarking, and impact of forcing uncertainty. *Journal of advances in modeling earth systems, 11*(12), 4245–4287.

LeBauer, D. S., Wang, D., Richter, K. T., Davidson, C. C., & Dietze, M. C. (2013). Facilitating feedbacks between field measurements and ecosystem models. *Ecological monographs, 83*(2), 133–154.

LeCun, Y., Bengio, Y., & Hinton, G. (2015). Deep learning. *Nature, 521*(7553), 436.

LeCun, Y., Bottou, L., Bengio, Y., & Haffner, P. (1998). Gradient-based learning applied to document recognition. *Proceedings of the IEEE, 86*(11), 2278–2324.

Lee, H., Calvin, K., Dasgupta, D., Krinner, G., Mukherji, A., Thorne, P., …Barrett, K. (2023). *Climate Change 2023: Synthesis Report.* Contribution of Working Groups I, II and III to the Sixth Assessment Report of the Intergovernmental Panel on Climate Change.

Lee, H.-y. (2016). Deep Learning Tutorial. Retrieved from https://speech.ee.ntu.edu.tw/~tlkagk/slide/Tutorial_HYLee_Deep.pdf

LeVeque, R. J. (2007). *Finite difference methods for ordinary and partial differential equations: steady-state and time-dependent problems*: SIAM.

Liang, J., Li, D., Shi, Z., Tiedje, J. M., Zhou, J., Schuur, E. A., …Luo, Y. (2015). Methods for estimating temperature sensitivity of soil organic matter based on incubation data: A comparative evaluation. *Soil biology and biochemistry, 80*, 127–135.

Liang, J., Xia, J., Shi, Z., Jiang, L., Ma, S., Lu, X., …Penton, C. R. (2018). Biotic responses buffer warming-induced soil organic carbon loss in Arctic tundra. *Global change biology, 24*(10), 4946–4959.

Liang, J., Zhou, Z., Huo, C., Shi, Z., Cole, J. R., Huang, L., …Luo, Z. (2018). More replenishment than priming loss of soil organic carbon with additional carbon input. *Nature communications, 9*(1), 1–9.

Liao, C., Lu, X., Huang, Y., Tao, F., Lawrence, D. M., Koven, C. D., …Luo, Y. (2023). Matrix Approach to Accelerate Spin-Up of CLM5. *Journal of advances in modeling earth systems, 15*(8), e2023MS003625. doi:https://doi.org/10.1029/2023MS003625

Liu, W., He, H., Wu, X., Ren, X., Zhang, L., shi, L., …Lv, Y. (2023). Importance of the memory effect for assessing interannual variation in net ecosystem exchange. *Agricultural and forest meteorology, 341*, 109691. doi:https://doi.org/10.1016/j.agrformet.2023.109691

Lu, D., Ricciuto, D., Stoyanov, M., & Gu, L. (2018). Calibration of the E3SM land model using surrogate-based global optimization. *Journal of advances in modeling earth systems, 10*(6), 1337–1356.

Lu, X., Du, Z., Huang, Y., Lawrence, D., Kluzek, E., Collier, N., …Luo, Y. (2020). Full Implementation of Matrix Approach to Biogeochemistry Module of CLM5. *Journal of advances in modeling earth systems, 12*(11), e2020MS002105.

Lu, X., Wang, Y.-P., Luo, Y., & Jiang, L. (2018). Ecosystem carbon transit versus turnover times in response to climate warming and rising atmospheric CO_2 concentration. *Biogeosciences, 15*(21), 6559–6572.

Luo, Y., Ahlström, A., Allison, S. D., Batjes, N. H., Brovkin, V., Carvalhais, N., …Finzi, A. (2016). Toward more realistic projections of soil carbon dynamics by Earth system models. *Global biogeochemical cycles, 30*(1), 40–56.

Luo, Y., Keenan, T. F., & Smith, M. (2015). Predictability of the terrestrial carbon cycle. *Global change biology, 21*(5), 1737–1751.

Luo, Y., Ogle, K., Tucker, C., Fei, S., Gao, C., LaDeau, S., …Schimel, D. S. (2011). Ecological forecasting and data assimilation in a data-rich era. *Ecological applications, 21*(5), 1429–1442.

Luo, Y. Q., Randerson, J. T., Abramowitz, G., Bacour, C., Blyth, E., Carvalhais, N., ... & Zhou, X. H. (2012). A framework for benchmarking land models. *Biogeosciences, 9*(10), 3857–3874.

Luo, Y., & Reynolds, J. F. (1999). Validity of extrapolating field CO2 experiments to predict carbon sequestration in natural ecosystems. *Ecology, 80*(5), 1568–1583.

Luo, Y., & Schuur, E. A. (2020). Model parameterization to represent processes at unresolved scales and changing properties of evolving systems. *Global change biology, 26*(3), 1109–1117.

Luo, Y., Shi, Z., Lu, X., Xia, J., Liang, J., Jiang, J., …Ahlström, A. (2017). Transient dynamics of terrestrial carbon storage: mathematical foundation and its applications. *Biogeosciences, 14*(1), 145.

Luo, Y., Su, B., Currie, W. S., Dukes, J. S., Finzi, A., Hartwig, U., …Parton, W. J. (2004). Progressive nitrogen limitation of ecosystem responses to rising atmospheric carbon dioxide. *Bioscience, 54*(8), 731–739.

Luo, Y., & Weng, E. (2011). Dynamic disequilibrium of the terrestrial carbon cycle under global change. *Trends in ecology & evolution, 26*(2), 96–104.

Luo, Y., Weng, E., Wu, X., Gao, C., Zhou, X., & Zhang, L. (2009). Parameter identifiability, constraint, and equifinality in data assimilation with ecosystem models. *Ecological applications, 19*(3), 571–574.

Luo, Y., White, L. W., Canadell, J. G., DeLucia, E. H., Ellsworth, D. S., Finzi, A., …Schlesinger, W. H. (2003). Sustainability of terrestrial carbon sequestration: A case study in Duke Forest with inversion approach. *Global biogeochemical cycles,* 17, 1021, doi:10.1029/2002GB001923, 1

Luo, Y., Wu, L., Andrews, J. A., White, L., Matamala, R., Schäfer, K. V., & Schlesinger, W. H. (2001). Elevated CO2 differentiates ecosystem carbon processes: deconvolution analysis of Duke Forest FACE data. *Ecological monographs, 71*(3), 357–376.

Ma, S., Jiang, J., Huang, Y., Shi, Z., Wilson, R. M., Ricciuto, D., …Luo, Y. (2017). Data-constrained projections of methane fluxes in a northern Minnesota peatland in response to elevated CO2 and warming. *Journal of geophysical research: Biogeosciences, 122*(11), 2841–2861.

Maechler, M. (2018). Cluster: cluster analysis basics and extensions. *R package version 2.0.* 7–1.

Malhotra, A., Brice, D. J., Childs, J., Graham, J. D., Hobbie, E. A., Vander Stel, H., …Iversen, C. M. (2020). Peatland warming strongly increases fine-root growth. *Proceedings of the national academy of sciences, 117*(30), 17627–17634.

Masson-Delmotte, V., Zhai, P., Pirani, A., Connors, S. L., Pean, C., Berger, S., …Zhou, B. (2021). *IPCC, 2021: Climate Change 2021: The Physical Science Basis. Contribution of Working Group I to the Sixth Assessment Report of the Intergovernmental Panel on Climate Change.* Retrieved from Cambridge, UK: https://oceanrep.geomar.de/id/eprint/58039/

Matis, J. H., Patten, B. C., & White, G. C. (1979). *Compartmental analysis of ecosystem models* (Vol. 10): International Cooperative Publishing House.

Medlyn, B. E., Zaehle, S., De Kauwe, M. G., Walker, A. P., Dietze, M. C., Hanson, P. J., …Parton, W. (2015). Using ecosystem experiments to improve vegetation models. *Nature climate change, 5*(6), 528–534.

Meng, L., Hess, P. G. M., Mahowald, N. M., Yavitt, J. B., Riley, W. J., Subin, Z. M., …Fuka, D. R. (2012). Sensitivity of wetland methane emissions to model assumptions: application and model testing against site observations. *Biogeosciences, 9*(7), 2793–2819. doi:10.5194/bg-9-2793-2012

Metzler, H., Müller, M., & Sierra, C. A. (2018). Transit-time and age distributions for nonlinear time-dependent compartmental systems. *Proceedings of the national academy of sciences, 115*(6), 1150–1155.

Metzler, H., & Sierra, C. A. (2018). Linear autonomous compartmental models as continuous-time Markov chains: Transit-time and age distributions. *Mathematical geosciences, 50*(1), 1–34.

Molnar, C. (2020). *Interpretable machine learning*: Lulu. com.

Mulholland, R. J., & Keener, M. S. (1974). Analysis of linear compartment models for ecosystems. *Journal of theoretical biology, 44*(1), 105–116.

Müller, M., & Sierra, C. A. (2017). Application of input to state stability to reservoir models. *Theoretical ecology, 10*(4), 451–475.

Muñoz-Sabater, J., Dutra, E., Agustí-Panareda, A., Albergel, C., Arduini, G., Balsamo, G., …Thépaut, J. N. (2021). ERA5-Land: a state-of-the-art global reanalysis dataset for land applications. *Earth syst. sci. data, 13*(9), 4349–4383. doi:10.5194/essd-13-4349-2021

Murphy, K. P. (2012). *Machine learning: a probabilistic perspective*: MIT press.

Myhre, G., Shindell, D., & Pongratz, J. (2014). Anthropogenic and natural radiative forcing. In T. Stocker (Ed.), *Climate change 2013: the physical science basis; Working Group I contribution to the fifth assessment report of the Intergovernmental Panel on Climate Change* (pp. 659–740). Cambridge: Cambridge University Press.

Nakagawa, R., Chau, M., Calzaretta, J., Keenan, T., Vahabi, P., Todeschini, A., …Kang, Y. (2023). Upscaling global hourly GPP with Temporal Fusion Transformer (TFT). *arXiv preprint arXiv:2306.13815*.

Norby, R. J., Childs, J., Hanson, P. J., & Warren, J. M. (2019). Rapid loss of an ecosystem engineer: Sphagnum decline in an experimentally warmed bog. *Ecology and evolution, 9*(22), 12571–12585.

Novick, K., Williams, C., Runkle, B., Anderegg, W., Hollinger, D., Litvak, M., …Anderson, C. (2022). *The science needed for robust, scalable, and credible nature-based climate solutions in the United States: Full Report*.

Odum, E. P., & Barrett, G. W. (1971). *Fundamentals of ecology* (Vol. 3): Saunders Philadelphia.

Odum, H. T. (1971). Environment, power, and society: John Wiley & Sons Inc.

Oreskes, N. (2003). The role of quantitative models in science Naomi Oreskes. In e, Canham, CD, Cole, JJ, and Lauenroth, WK (Eds.), *Models in ecosystem scienc*, 13–31.

Oreskes, N., Shrader-Frechette, K., & Belitz, K. (1994). Verification, validation, and confirmation of numerical models in the earth sciences. *Science, 263*(5147), 641–646.

Papale, D., & Valentini, R. (2003). A new assessment of European forests carbon exchanges by eddy fluxes and artificial neural network spatialization. *Global change biology, 9*(4), 525–535.

Parton, W. J., Stewart, J. W. B., & Cole, C. V. (1988). Dynamics of C, N, P and S in grassland soils - a model. *Biogeochemistry, 5*(1), 109–131. doi:10.1007/bf02180320

Pastorello, G., Trotta, C., Canfora, E., Chu, H., Christianson, D., Cheah, Y.-W., …Papale, D. (2020). The FLUXNET2015 dataset and the ONEFlux processing pipeline for eddy covariance data. *Scientific data, 7*(1), 225. doi:10.1038/s41597-020-0534-3

Peters-Lidard, C. D., Houser, P. R., Tian, Y., Kumar, S. V., Geiger, J., Olden, S., …Sheffield, J. (2007). High-performance Earth system modeling with NASA/GSFC's Land Information System. *Innovations in systems and software engineering, 3*(3), 157–165. doi:10.1007/s11334-007-0028-x

Pinnington, E., Quaife, T., Lawless, A., Williams, K., Arkebauer, T., & Scoby, D. (2020). The Land Variational Ensemble Data Assimilation Framework: LAVENDAR v1.0.0. *Geosci. model dev., 13*(1), 55–69. doi:10.5194/gmd-13-55-2020

Ploton, P., Mortier, F., Réjou-Méchain, M., Barbier, N., Picard, N., Rossi, V., …Bayol, N. (2020). Spatial validation reveals poor predictive performance of large-scale ecological mapping models. *Nature communications, 11*(1), 1–11.

Poplin, R., Varadarajan, A., Blumer, K., Liu, Y., McConnell, M., Corrado, G., …Webster, D. Predicting cardiovascular risk factors from retinal fundus photographs using deep learning. arXiv 2017. *arXiv preprint arXiv:1708.09843*.

Qu, Y., Maksyutov, S., & Zhuang, Q. (2018). An efficient method for accelerating the spin-up process for process-based biogeochemistry models. *Biogeosciences, 15*(13), 3967–3973.

Rafique, R., Xia, J., Hararuk, O., Asrar, G. R., Leng, G., Wang, Y., & Luo, Y. (2016). Divergent predictions of carbon storage between two global land models: Attribution of the causes through traceability analysis. *Earth System Dynamics, 7*(3), 649–658.

Rafique, R., Xia, J., Hararuk, O., Leng, G., Asrar, G., & Luo, Y. (2017). Comparing the performance of three land models in global C cycle simulations: A detailed structural analysis. *Land degradation development, 28*(2), 524–533.

Rasmussen, M. (2007). *Attractivity and bifurcation for nonautonomous dynamical systems*: Springer.

Rasmussen, M., Hastings, A., Smith, M. J., Agusto, F. B., Chen-Charpentier, B. M., Hoffman, F. M., …Wang, Y.-P. (2016). Transit times and mean ages for nonautonomous and autonomous compartmental systems. *Journal of mathematical biology, 73*(6-7), 1379–1398.

Reichstein, M., Camps-Valls, G., Stevens, B., Jung, M., Denzler, J., Carvalhais, N., & Prabhat. (2019). Deep learning and process understanding for data-driven Earth system science. *Nature, 566*(7743), 195–204. doi:10.1038/s41586-019-0912-1

Rey, A., & Jarvis, P. (2006). Modelling the effect of temperature on carbon mineralization rates across a network of European forest sites (FORCAST). *Global change biology, 12*(10), 1894–1908.

Ricciuto, D., Sargsyan, K., & Thornton, P. (2018). The impact of parametric uncertainties on biogeochemistry in the E3SM land model. *Journal of advances in modeling earth systems, 10*(2), 297–319.

Richardson, A. D., Hufkens, K., Milliman, T., Aubrecht, D. M., Furze, M. E., Seyednasrollah, B., …Heiderman, R. R. (2018). Ecosystem warming extends vegetation activity but heightens vulnerability to cold temperatures. *Nature, 560*(7718), 368–371.

Riley, W., Subin, Z., Lawrence, D., Swenson, S., Torn, M., Meng, L., …Hess, P. (2011). Barriers to predicting changes in global terrestrial methane fluxes: analyses using CLM4Me, a methane biogeochemistry model integrated in CESM. *Biogeosciences, 8*(7), 1925–1953.

Roe, S., Streck, C., Beach, R., Busch, J., Chapman, M., Daioglou, V., …Lawrence, D. (2021). Land-based measures to mitigate climate change: Potential and feasibility by country. *Global change biology, 27*(23), 6025–6058. doi:https://doi.org/10.1111/gcb.15873

Rolnick, D., Donti, P. L., Kaack, L. H., Kochanski, K., Lacoste, A., Sankaran, K., …Bengio, Y. (2022). Tackling Climate Change with Machine Learning. *ACM comput. surv., 55*(2), Article 42. doi:10.1145/3485128

Russell, B. C., Torralba, A., Murphy, K. P., & Freeman, W. T. (2008). LabelMe: a database and web-based tool for image annotation. *International journal of computer vision, 77*(1-3), 157–173.

Rykiel Jr, E. J. (1996). Testing ecological models: the meaning of validation. *Ecological modelling, 90*(3), 229–244.

Saatchi, S. S., Harris, N. L., Brown, S., Lefsky, M., Mitchard, E. T. A., Salas, W., …Morel, A. (2011). Benchmark map of forest carbon stocks in tropical regions across three continents. *Proceedings of the national academy of sciences, 108*(24), 9899–9904. doi:doi:10.1073/pnas.1019576108

Saunois, M., Stavert, A. R., Poulter, B., Bousquet, P., Canadell, J. G., Jackson, R. B., …Patra, P. K. (2020). The global methane budget 2000–2017. *Earth system science data, 12*(3), 1561–1623.

Schädel, C., Luo, Y., Evans, R. D., Fei, S., & Schaeffer, S. M. (2013). Separating soil CO 2 efflux into C-pool-specific decay rates via inverse analysis of soil incubation data. *Oecologia, 171*(3), 721–732.

Schädel, C., Schuur, E. A., Bracho, R., Elberling, B., Knoblauch, C., Lee, H., …Turetsky, M. R. (2014). Circumpolar assessment of permafrost C quality and its vulnerability over time using long-term incubation data. *Global change biology, 20*(2), 641–652.

Scheffer, M., Carpenter, S., Foley, J. A., Folke, C., & Walker, B. (2001). Catastrophic shifts in ecosystems. *Nature, 413*(6856), 591–596.

Segers, R. (1998). Methane production and methane consumption: a review of processes underlying wetland methane fluxes. *Biogeochemistry, 41*, 23–51.

Sellers, P., Randall, D., Collatz, G., Berry, J., Field, C., Dazlich, D., …Bounoua, L. (1996). A revised land surface parameterization (SiB2) for atmospheric GCMs. Part I: Model formulation. *Journal of climate, 9*(4), 676–705.

Shen, C., Appling, A. P., Gentine, P., Bandai, T., Gupta, H., Tartakovsky, A., …Lawson, K. (2023). Differentiable modelling to unify machine learning and physical models for geosciences. *Nature reviews earth & environment, 4*(8), 552–567. doi:10.1038/s43017-023-00450-9

Shi, X., Ricciuto, D. M., Thornton, P. E., Xu, X., Yuan, F., Norby, R. J., …Hanson, P. J. (2021). Extending a land-surface model with Sphagnum moss to simulate responses of a northern temperate bog to whole ecosystem warming and elevated CO 2. *Biogeosciences, 18*(2), 467–486.

Shi, X., Thornton, P. E., Ricciuto, D. M., Hanson, P. J., Mao, J., Sebestyen, S. D., …Bisht, G. (2015). Representing northern peatland microtopography and hydrology within the Community Land Model. *Biogeosciences, 12*(21), 6463–6477.

Shi, Z., Crowell, S., Luo, Y., & Moore, B. (2018). Model structures amplify uncertainty in predicted soil carbon responses to climate change. *Nature communications, 9*(1), 2171.

Shi, Z., Yang, Y., Zhou, X., Weng, E., Finzi, A. C., & Luo, Y. (2016). Inverse analysis of coupled carbon–nitrogen cycles against multiple datasets at ambient and elevated CO2. *Journal of plant ecology, 9*(3), 285–295.

Shukla, P., Skea, J., Calvo Buendia, E., Masson-Delmotte, V., Pörtner, H., Roberts, D., …Van Diemen, R. (2019). IPCC, 2019: Climate Change and Land: an IPCC special report on climate change, desertification, land degradation, sustainable land management, food security, and greenhouse gas fluxes in terrestrial ecosystems.

Sierra, C. A., Ceballos-Núñez, V., Metzler, H., & Müller, M. (2018). Representing and understanding the carbon cycle using the theory of compartmental dynamical systems. *Journal of advances in modeling earth systems, 10*(8), 1729–1734.

Sierra, C. A., & Müller, M. (2015). A general mathematical framework for representing soil organic matter dynamics. *Ecological monographs, 85*(4), 505–524.

Sierra, C. A., Müller, M., Metzler, H., Manzoni, S., & Trumbore, S. E. (2017). The muddle of ages, turnover, transit, and residence times in the carbon cycle. *Global change biology, 23*(5), 1763–1773.

Sitch, S., Friedlingstein, P., Gruber, N., Jones, S. D., Murray-Tortarolo, G., Ahlström, A., …Huntingford, C. (2015). Recent trends and drivers of regional sources and sinks of carbon dioxide. *Biogeosciences, 12*(3), 653–679.

Sontag, E. D. (2013). *Mathematical control theory: Deterministic finite dimensional systems* (Vol. 6): Springer Science & Business Media.

Stanford, G., & Smith, S. (1972). Nitrogen mineralization potentials of soils. *Soil science society of America journal, 36*(3), 465–472.

Strogatz, S. H. (1994). Nonlinear dynamics and chaos: with applications to physics. *Biology, chemistry and engineering*, 1.

Sun, Y., Goll, D. S., Chang, J. F., Ciais, P., Guenet, B., Helfenstein, J., …Zhang, H. C. (2021). Global evaluation of the nutrient-enabled version of the land surface model ORCHIDEE-CNP v1.2 (r5986). *Geoscientific model development, 14*(4), 1987–2010. doi:10.5194/gmd-14-1987-2021

Sun, Y., Goll, D. S., Huang, Y., Ciais, P., Wang, Y. p., Bastrikov, V., & Wang, Y. (2023). Machine learning for accelerating process-based computation of land biogeochemical cycles. *Global change biology, 29*(11), 3221–3234.

Tang, J., & Riley, W. (2013). A total quasi-steady-state formulation of substrate uptake kinetics in complex networks and an example application to microbial litter decomposition. *Biogeosciences, 10*(12), 8329-8351.

Tang, J., & Riley, W. J. (2015). Weaker soil carbon–climate feedbacks resulting from microbial and abiotic interactions. *Nature climate change, 5*(1), 56–60.

Tao, F., Huang, Y., Hungate, B. A., Manzoni, S., Frey, S. D., Schmidt, M. W. I., …Luo, Y. (2023). Microbial carbon use efficiency promotes global soil carbon storage. *Nature, 618*(7967), 981–985. doi:10.1038/s41586-023-06042-3

Tao, F., & Luo, Y. (2022). PROcess-guided deep learning and DAta-driven modelling (PRODA). In Y. Luo & B. Smith (Eds.), *Land carbon cycle modeling: matrix approach, data assimilation, and ecological forecasting*: Taylor and Francis.

Tao, F., Zhou, Z., Huang, Y., Li, Q., Lu, X., Ma, S., …Luo, Y. (2020). Deep Learning Optimizes Data-Driven Representation of Soil Organic Carbon in Earth System Model Over the

Conterminous United States. *Frontiers in big data, 3*(17). doi:10.3389/fdata.2020.00017

Taussky, O. (1949). A recurring theorem on determinants. *The american mathematical monthly, 56*(10P1), 672–676.

Thornton, P. E., Law, B. E., Gholz, H. L., Clark, K. L., Falge, E., Ellsworth, D. S., …Falk, M. (2002). Modeling and measuring the effects of disturbance history and climate on carbon and water budgets in evergreen needleleaf forests. *Agricultural and forest meteorology, 113*(1-4), 185–222.

Thornton, P. E., & Rosenbloom, N. A. (2005). Ecosystem model spin-up: Estimating steady state conditions in a coupled terrestrial carbon and nitrogen cycle model. *Ecological modelling, 189*(1-2), 25–48.

Todd-Brown, K., Randerson, J., Post, W., Hoffman, F., Tarnocai, C., Schuur, E., & Allison, S. (2013). Causes of variation in soil carbon simulations from CMIP5 Earth system models and comparison with observations. *Biogeosciences, 10*(3), 1717–1736.

Torn, M. S., Trumbore, S. E., Chadwick, O. A., Vitousek, P. M., & Hendricks, D. M. (1997). Mineral control of soil organic carbon storage and turnover. *Nature, 389*(6647), 170.

Tramontana, G., Jung, M., Schwalm, C. R., Ichii, K., Camps-Valls, G., Ráduly, B., …Papale, D. (2016). Predicting carbon dioxide and energy fluxes across global FLUXNET sites with regression algorithms. *Biogeosciences, 13*(14), 4291–4313. doi:10.5194/bg-13-4291-2016

Troy Baisden, W., & Amundson, R. (2003). An analytical approach to ecosystem biogeochemistry modeling. *Ecological Applications, 13*(3), 649–663.

Tuomi, M., Thum, T., Järvinen, H., Fronzek, S., Berg, B., Harmon, M., …Liski, J. (2009). Leaf litter decomposition—Estimates of global variability based on Yasso07 model. *Ecological modelling, 220*(23), 3362–3371.

Turner, M. G. (2018). Yellowstone Rebounded from an Epic 1988 Fire—That May Be Harder in Future. *Scientific America.*

Ueyama, M., Ichii, K., Iwata, H., Euskirchen, E. S., Zona, D., Rocha, A. V., …Oechel, W. C. (2013). Upscaling terrestrial carbon dioxide fluxes in Alaska with satellite remote sensing and support vector regression. *Journal of geophysical research: Biogeosciences, 118*(3), 1266–1281. doi:https://doi.org/10.1002/jgrg.20095

VanderJagt, D. J., Morales, M., Thacher, T. D., Diaz, M., & Glew, R. H. (2001). Bioelectrical Impedance Analysis of the Body Composition of Nigerian Children with Calcium-deficiency Rickets. *Journal of tropical pediatrics, 47*(2), 92–97.

Venables, W. N., & Ripley, B. D. (2013). *Modern applied statistics with S-PLUS*: Springer Science & Business Media.

Walter, B. (1998). *Development of a process-based model to derive methane emissions from natural wetlands for climate studies* (Doctoral dissertation, University of Hamburg Hamburg).

Walter, B. P., & Heimann, M. (2000). A process-based, climate-sensitive model to derive methane emissions from natural wetlands: Application to five wetland sites, sensitivity to model parameters, and climate. *Global biogeochemical cycles, 14*(3), 745–765.

Wang, H., Yan, S., Ciais, P., Wigneron, J.-P., Liu, L., Li, Y., …Li, S. (2022). Exploring complex water stress–gross primary production relationships: Impact of climatic drivers, main effects, and interactive effects. *Global change biology, 28*(13), 4110–4123. doi:https://doi.org/10.1111/gcb.16201

Wang, Y. P., Chen, B. C., Wieder, W. R., Leite, M., Medlyn, B. E., Rasmussen, M., ... & Luo, Y. Q. (2014). Oscillatory behavior of two nonlinear microbial models of soil carbon decomposition. *Biogeosciences, 11*(7), 1817–1831.

Wang, Y. P., Jiang, J., Chen-Charpentier, B., Agusto, F. B., Hastings, A., Hoffman, F., ... & Luo, Y. Q. (2016). Responses of two nonlinear microbial models to warming and increased carbon input. *Biogeosciences, 13*(4), 887–902.

Wang, Y. P., Kowalczyk, E., Leuning, R., Abramowitz, G., Raupach, M. R., Pak, B., …Luhar, A. (2011). Diagnosing errors in a land surface model (CABLE) in the time and frequency domains. *Journal of geophysical research: Biogeosciences, 116*(G1).

Wania, R., Ross, I., & Prentice, I. (2010). Implementation and evaluation of a new methane model within a dynamic global vegetation model: LPJ-WHyMe v1. 3.1. *Geoscientific model development, 3*(2), 565–584.

Warner, D. L., Bond-Lamberty, B., Jian, J., Stell, E., & Vargas, R. (2019). Spatial Predictions and Associated Uncertainty of Annual Soil Respiration at the Global Scale. *Global biogeochemical cycles, 33*(12), 1733–1745. doi:https://doi.org/10.1029/2019GB006264

Wasserman, L. (2004). *All of statistics: a concise course in statistical inference* (Vol. 26): Springer.

Weng, E., & Luo, Y. (2008). Soil hydrological properties regulate grassland ecosystem responses to multifactor global change: A modeling analysis. *Journal of geophysical research: Biogeosciences, 113*(G3).

Weng, E., & Luo, Y. (2011). Relative information contributions of model vs. data to short-and long-term forecasts of forest carbon dynamics. *Ecological Applications, 21*(5), 1490–1505.

White, J. T., Hunt, R. J., Fienen, M. N., & Doherty, J. E. (2020). *Approaches to highly parameterized inversion: PEST++ Version 5, a software suite for parameter estimation, uncertainty analysis*, management optimization and sensitivity analysis (No. 7-C26). US Geological Survey.

Wilson, R., Hopple, A., Tfaily, M., Sebestyen, S., Schadt, C., Pfeifer-Meister, L., …Kolton, M. (2016). Stability of peatland carbon to rising temperatures. *Nature communications, 7*(1), 1–10.

Winkler, K., Yang, H., Ganzenmüller, R., Fuchs, R., Ceccherini, G., Duveiller, G., …Ciais, P. (2023). Changes in land use and management led to a decline in Eastern Europe's terrestrial carbon sink. *Communications earth & environment, 4*(1), 237. doi:10.1038/s43247-023-00893-4

Wu, J. (2012). *Advances in K-means clustering: a data mining thinking*: Springer Science & Business Media.

Wu, Z., Hugelius, G., Luo, Y., Smith, B., Xia, J., Fensholt, R., …Ahlström, A. (2019). Approaching the potential of model-data comparisons of global land carbon storage. *Scientific reports, 9*(1), 3367.

Wutzler, T., & Reichstein, M. (2008). Colimitation of decomposition by substrate and decomposers–a comparison of model formulations. *Biogeosciences, 5*(3), 749–759.

Xia, J., Luo, Y., Wang, Y.-P., Weng, E., & Hararuk, O. (2012). A semi-analytical solution to accelerate spin-up of a coupled carbon and nitrogen land model to steady state. *Geoscientific model development, 5*(5), 1259–1271.

Xia, J., Luo, Y., Wang, Y. P., & Hararuk, O. (2013). Traceable components of terrestrial carbon storage capacity in biogeochemical models. *Global change biology, 19*(7), 2104–2116.

Xiao, H., Rasul, K., & Vollgraf, R. (2017). Fashion-mnist: a novel image dataset for benchmarking machine learning algorithms. *arXiv preprint arXiv:1708.07747.*

Xu, T., White, L., Hui, D., & Luo, Y. (2006). Probabilistic inversion of a terrestrial ecosystem model: Analysis of uncertainty in parameter estimation and model prediction. *Global biogeochemical cycles, 20*(2).

Xu, X., Shi, Z., Li, D., Rey, A., Ruan, H., Craine, J. M., ...Luo, Y. (2016). Soil properties control decomposition of soil organic carbon: Results from data-assimilation analysis. *Geoderma, 262*, 235–242.

Yang, H., Ciais, P., Wigneron, J.-P., Chave, J., Cartus, O., Chen, X., ...Yao, Y. (2022). Climatic and biotic factors influencing regional declines and recovery of tropical forest biomass from the 2015/16 El Niño. *Proceedings of the national academy of sciences, 119*(26), e2101388119. doi:doi:10.1073/pnas.2101388119

Yang, Y., Luo, Y., & Finzi, A. C. (2011). Carbon and nitrogen dynamics during forest stand development: a global synthesis. *New phytologist, 190*(4), 977–989.

Zeng, N., & Hausmann, H. (2022). Wood Vault: remove atmospheric CO2 with trees, store wood for carbon sequestration for now and as biomass, bioenergy and carbon reserve for the future. *Carbon balance and management, 17*(1), 2.

Zhang, D., Hui, D., Luo, Y., & Zhou, G. (2008). Rates of litter decomposition in terrestrial ecosystems: global patterns and controlling factors. *Journal of plant ecology, 1*(2), 85–93.

Zhang, Y., Li, C., Trettin, C. C., Li, H., & Sun, G. (2002). An integrated model of soil, hydrology, and vegetation for carbon dynamics in wetland ecosystems. *Global biogeochemical cycles, 16*(4), 9-1-9-17.

Zhang, Y., Xiao, X., Wu, X., Zhou, S., Zhang, G., Qin, Y., & Dong, J. (2017). A global moderate resolution dataset of gross primary production of vegetation for 2000–2016. *Scientific data, 4*, 170165.

Zhou, J., Xia, J., Wei, N., Liu, Y., Bian, C., Bai, Y., & Luo, Y. (2021). A traceability analysis system for model evaluation on land carbon dynamics: design and applications. *Ecological Processes, 10*(1), 1–14.

Zhou, S., Liang, J., Lu, X., Li, Q., Jiang, L., Zhang, Y., ...Sitch, S. (2018). Sources of uncertainty in modeled land carbon storage within and across three MIPs: Diagnosis with three new techniques. *Journal of climate, 31*(7), 2833–2851.

Zhu, Q., Liu, J., Peng, C., Chen, H., Fang, X., Jiang, H., ...Zhou, X. (2014). Modelling methane emissions from natural wetlands by development and application of the TRIPLEX-GHG model. *Geosci. model dev., 7*(3), 981–999. doi:10.5194/gmd-7-981-2014

Zhuang, Q., Melillo, J. M., Kicklighter, D. W., Prinn, R. G., McGuire, A. D., Steudler, P. A., ...Hu, S. (2004). Methane fluxes between terrestrial ecosystems and the atmosphere at northern high latitudes during the past century: A retrospective analysis with a process-based biogeochemistry model. *Global biogeochemical cycles, 18*(3).

Index

Note: Locators in **bold** refer to tables and those in *italics* to figures.

1-3-5 scheme of diagnosis 57–62
3D space, describing model outputs 57, 59–60

A

accelerated decomposition approach (AD) 90
afforestation 72
  see also carbon sequestration
age distributions 94–96
Agriculture, Forestry and other Land Uses (AFOLU) mitigation 72
air dryness 224–228
air temperature 7
  DALEC model 176
  data assimilation 133
  machine learning 215, 225, *228*, 254
  meteorological data 220
  soil carbon decomposition 76
  SPRUCE experiment 163, 196
  TECO model 114, 205
algebra for matrices 263–267
Amazon rainforest 224, 225, 227
Application Programming Interface (API) 202
artificial intelligence (AI) modeling 253–257
  see also machine learning
artificial neural networks (ANNs) *212*, 212–213
  advantages and disadvantages 225–228
  interpretability 227–228
  perturbation analysis 226–227
Attractor
  Moving attractor 3, 188
  Global attractor 85–87
autonomous compartmental systems
  compared to nonautonomous 46
  linear 47, 48
  nonlinear 47–49
  time characteristics 94–95
autotrophic respiration 18, 176, 182–184

B

balance equations *see* carbon balance equations
Bayes' theorem 140–141, 143
  CARDAMOM approach 176–177
  see also Markov Chain Monte Carlo
Bayesian statistics 140
Beijing Climate Center (BCC) model 59
Benchmark analysis
benchmark 111, 117, 121–125, 202, 203, 228, 238
bias, geographical 247–248, 251
bias in data 122–123, 175
reference data 121–124, 248, 250
scoring system 121, *122*, 125
bioenergy with carbon capture and storage (BECCS) 71
biogeochemical cycles 89

C

CABLE *see* Community Atmosphere-Biosphere-Land Exchange model
Canada (CAN) model 59
carbon balance equations 18–23
  deriving the matrix equation 30–33
  exercises 25–26, 51–54
  matrix version 29–30
  TECO model 21–23
carbon cycle data assimilation system (CCDAS) 180
carbon dioxide (CO_2)
  coupled carbon-nitrogen matrix models 34–44
  flow diagrams 18–23, *19*
  free-air CO_2-enrichment (FACE) experiment 59–61
  as a greenhouse gas compared to methane 163
  principles underlying removal from the atmosphere 69
  soil incubation experiments 146
carbon dioxide removal (CDR) 69–71
  current warming and the purpose of *70*
  land-based strategies 71–72
  principles for evaluating and designing strategies 72–73
carbon flow 18
carbon flow diagrams 18–23, *19*, *20*
  exercises 25–26
carbon input
  3D space 59, *60*
  balance equations 26, 29
  and carbon storage capacity 76–77
  CDR techniques 72
  contribution to storage capacity 76–77
  deriving the matrix equation 30–33
  flow diagrams 21–23
  mathematical foundation 74–76
  model uncertainty 57, 74
  nitrogen influence 109
  SASU approach 89, 98–102
  SOC profiles 246, 251
  traceability analysis 60
carbon pools
  data assimilation 152–159, 175–177
  soil incubation experiments 146–147
  traceability analysis 105–110
  transient traceability analysis 112–114
carbon residence time
  3D space 59, *59*
  across different ecosystems 107
  in carbon cycle modeling 74–75
  and carbon storage capacity 76–77
  model uncertainty *61*, *62*
  nitrogen's influence 109
  potential to increase **72**, 72–73
  SASU approach 89, 90–91, 98–102
  traceability framework 105–106
  transient traceability analysis 117, *118–119*
carbon sequestration 34
  by the Amazon rainforest 224
  data assimilation 133, *134*
  forest restoration 71–72
  uncertainty 57
carbon stocks 18
  see also carbon pools
carbon storage capacity 75–77
carbon storage potential 75–76
carbon use efficiency (CUE)
  SOC relationship 251–252
  uncertainty analysis 61
CarboTrain
  carbon balance equations 51–54
  data assimilation 157–159
  SASU approach 98–102
  SPRUCE experiment 182–184, 205
  user guide 275–284
CARDAMOM approach 175–180
Celery *194*, 195
CENTURY model
  deriving the matrix model from carbon balance equations 51–54
  flow diagram for *25*, 25
climate change
  future states of ecosystems 121
  global climate treaties 180
  nationally determined contributions (NDCs) 69, *70*, 72
  net global emissions *70*
  predictability of carbon cycle 9
  recent warming 69
climate mitigation 72
  see also carbon dioxide removal (CDR)
coding, introduction to programming in Python 268–274
Community Atmosphere-Biosphere-Land Exchange (CABLE) model
  Semi-Analytic Spin-Up *91*, 91–93, 98–102
  traceability analysis 106–108, 126–130
  uncertainty analysis 58
community cyberinfrastructures 199–203
Community Earth System Model Biogeochemical module (CESM GBC) 59
Community Earth System Model (CESM) 165
Community Land Model version 4 (CLM4), comparison to other models 123, *124*
Community Land Model version 4.5 (CLM4.5)
  benchmark analysis 122
  C-N coupling 109
  carbon transfer 7
  comparison to other models 123, *124*
  data assimilation 246
  flow diagram 23
  matrix equation 30
  PRODA approach 245, 246
  soil carbon module 245
  traceability analysis 114
Community Land Model version 5 (CLM5)
  C-N coupling 36–44, *43*
  CLM-CASA model 108
  comparison to other models 123, *124*
  PRODA approach 245–252, 261–262
  soil carbon module 245
  uncertainty analysis 57, 58
compartmental dynamical systems

classification of 46–47
definition 45–46
properties and long-term behavior
autonomous 47–49
nonautonomous 49–50
time characteristics 94–96
compartmentalization of carbon processes 5
complexity, modeling 173–174
coupled carbon-nitrogen matrix models 34–44
CLM5 36–44
Semi-Analytic Spin-Up (SASU) 90–93
TECO model 34–36
traceability analysis *110*, 112, 117
Coupled Model Intercomparison Project Phase 5 (CMIP5)
data assimilation 135
transient traceability analysis 112
uncertainty 57
Coupled Model Intercomparison Project Phase 6 (CMIP6)
transient traceability analysis 112
uncertainty 105
cyberinfrastructures 199–203

D

DALEC *see* Data Assimilation Linked Ecosystem Carbon model
DART *see* Data Assimilation Research Testbed
data assimilation 133
CARDAMOM approach 175–180
community cyberinfrastructures 199–203
ecological forecasting 189
EcoPAD 133, 189–190, 192–198
key challenges and opportunities 180
Markov Chain Monte Carlo 140–144
matrix phosphorus model 64, 66–67
methane modeling 167–172
need for 133–135
PRODA approach 246–247, 261–262
scientific values 137–138
seven-step procedure 135–137, 152–159
soil incubation experiments 147–151
SPRUCE experiment 163–166
TECO model 184
Data Assimilation Linked Ecosystem Carbon (DALEC) model 175–177
Data Assimilation Research Testbed (DART) 200
data ingestion 202
data-model integration 174–175
decision trees 211–212, *212*
deep learning 244–252, 258–262
digital elevation models (DEMs) 213
direct air capture (DAC) 71
Disequilibrium 3, 9, 59, **188**
disturbance events 188, 189
donor pool-dominated carbon transfer *4*, 4–5
dry season 224, 227
Dynamic disequilibrium 3, 9, 59
dynamical systems *see* compartmental dynamical systems

E

earth observation (EO) 173, 175, *177*, *178*, 179
earth system models (ESMs) 14
complex carbon cycle interactions 224–228
ecological forecasting 188–189
ELM (land component) 165
hybrid modeling 253–257
land models included in 121–125
neural networks 242–243
optimizing parameterization 244–252
traceability analysis 105–110
uncertainty 57
ecological and dynamics constraints (EDCs) 177
ecological forecasting 187
community cyberinfrastructures 199–203
data availability 189
EcoPAD 133, 189–190, 192–198
models and predictability of the terrestrial carbon cycle 187–189
at the SPRUCE site 204–206, *207*
workflow system 189–190
Ecological Platform for Assimilating Data (EcoPAD)
application in SPRUCE experience 196–197
as cyberinfrastructure 199, 200–201
data
data assimilation 133, *134*, 189–190, 192–198
preparation of 204–205
structure and types of 193–195
need for 192–193
ecosystems models
compartmental approach 50
future states 121
matrix representation 45
soil P dynamics 67–68
terrestrial carbon cycle variations 107
Terrestrial Ecosystem Model (TEM) 90
Terrestrial Ecosystem (TECO) model
carbon balance equations 6
coupled C-N version 34–36, 109, *110*
data assimilation 136–137, 152–159, 184, 194
datasets 204–205
efficiency and convergence of SASU 98–102
methane modeling 167–172
SPRUCE experiment 165
traceability analysis 109, *110*
transient traceability analysis 113–114, *114*
ELM land-surface model 165
energy flow models 13, *14*
evergreen broadleaf forests 107

F

Findable, Accessible, Interoperable, and Reusable (FAIR) 200
fire 7, *178*
first-order kinetics of carbon transfer 5
flow diagrams
exercises 25–26
methane (CH4) emissions *168*
terrestrial carbon processes *20*, 21–23
uses of 18, *19*
forecasting *see* ecological forecasting; weather forecasting
forest biomes
afforestation and carbon sequestration 71–72
Amazon rainforest 224, 225, 227
evergreen broadleaf forests 107
transient traceability analysis 113–116
tundra 107
forward modeling
CARDAMOM approach 176
data assimilation 133–135, 194
ecological forecasting 204
fractional release coefficients 46
free-air CO_2-enrichment (FACE) experiment 59–61

G

general circulation models (GCMs) 108, 203
geographical bias 247–248, 251
global asymptotically stable (GAS) 49
global attractors 85–87
Graphical User Interface (GUI) 199
greenhouse gas (GHG) emissions 69
machine learning applications 213–216, **214**
methane compared to carbon dioxide 163
see also carbon dioxide
gross primary production (GPP)
complex carbon cycle interactions 224–228
EcoPAD 196–197, *197*
Long Short-Term Memory (LSTM) 217–223
reference data sets 122
studies of 59

H

heterotrophic respiration 18, 165, 168, 176
historical simulations 89, 171, 220, 244
history of modeling 13
homogenized world soil carbon database (HWSD) 135
hybrid modeling 253–257

I

image classification, machine learning 241–242
industrial revolution 89
input-to-state stability (ISS) 50
instantaneous steady-state 85
interannual variability 122
Intergovernmental Panel on Climate Change (IPCC) 105
International Land Model Benchmarking (ILAMB) project 121–125

J

Jacobian matrix 48–49

L

land carbon cycle
dynamic disequilibrium 9–10
ecosystems convergence 3
predictability 9
Land Information System (LIS) framework 201–202
Land surface Data Toolkit (LDT) 201
land surface models (LSMs) 14
Land surface Verification Toolkit (LVT) 202
land use change
carbon storage capacity 76–77
industrial revolution 89
predictability of the land carbon cycle 9
Land Variational ENsemble Data Assimilation Framework (LAVENDAR) 201
leaf area index
CARDAMOM approach *179*, 179
earth observation 175
process rates 173–174
reference data sets **123**
linear compartmental systems 47, 48–49

LIS *see* Land Information System framework
litterfall
  in carbon flow 18
  role in the land carbon cycle 4–5
Long Short-Term Memory (LSTM) 217–223
LPJ-GUESS model 108, *109*

M

machine learning (ML) 211, 237
  applications of 237
    carbon cycle research 213–216
  complex carbon cycle interactions 224–228
  hybrid modeling 253–257
  LSTM model 217, 220
  model performance 217
  neural networks 237–243
  random forest concept 229–233
  types of ML algorithms 211–213
Markov Chain Monte Carlo (MCMC)
  Bayes' theorem 140–141, 143
  CARDAMOM approach 176–177
  convergence of results 144
  deep learning 258
  as method 141–144
  PRODA approach 246
  sampling 136
  soil incubation experiments 149–150
  SPRUCE experiment 166
mass-balanced systems
  compartmental dynamical systems 45–50
  matrix approach to model representation 45
matrix algebra 263–267
matrix approach to model representation 5–8
  coupled carbon-nitrogen matrix models 35–44
  diagnostic variables 74–75
  exercises for carbon balance equations and coding 51–54
  literature overview 45
  paradox of the matrix equation 8–9
  purpose of 45
matrix phosphorus model 63
  construction of the P matrix model 64–66
  data selection and description 64
  new knowledge emerging 66–68
  soil P dynamics 63–64
  validation and data assimilation 66
matrix version of the carbon balance equation 29–30
  deriving the matrix equation 30–33
maximum likelihood estimates (MLEs) 150
meteorological data 7, 220
  DALEC model 176
  machine learning 215, 225, *228*, 254
  soil carbon decomposition 76
  SPRUCE experiment 163, 196
  TECO model 114, 205
methane (CH4) emissions
  flow diagram *168*
  as a greenhouse gas 163
  major parameters **169**
  modeling 167–172
Metropolis-Hasting criterion 136, 155
Michaelis-Menten equations
  data assimilation 137
  role in the land carbon cycle 5
microbial processes 5, 251
model-data fusion (MDF) analysis *178*
Model-Independent Data Assimilation module (MIDA) 201
model intercomparison projects (MIPs)
  3D space 60
  spin-up 89
  traceability analysis exercises 126–130
  transient traceability analysis 112, 117, *120*
  uncertainty 57, 62, 105
modeling
  benchmark analysis 121–125
  challenges for 173
  complexity 173–174
  data assimilation as part of 133–135
  data-model integration 174–175
  definition 11
  errors 174
  matrix approach 5–8
  in research 11–12
  system dynamics 12–13
  types of land carbon cycle models 13–14, **15**
  ways of using 12
  workflow 15–17
Monte Carlo method 141
  see also Markov Chain Monte Carlo (MCMC)
Mother Nature 3, 9
Multiscale Synthesis and Terrestrial Model Intercomparison Project (MsTMIP) protocol 89

N

n-pool model 83–84
  global attractor of 87
National Center for Atmospheric Research (NCAR) 200
National Ecological Observatory Network (NEON) 192
nationally determined contributions (NDCs) 69, *70*, 72
native dynamics spin-up (ND) 89–90, 98–102
net primary production (NPP)
  across different ecosystems 107
  CARDAMOM approach *180*
  free-air CO_2-enrichment (FACE) experiment 59–61
  matrix approach to model representation 6
  reference data sets 122, **123**
  traceability analysis 106–107, 126–130
  transient traceability analysis 114–116, *115*, 117, *118–120*
  uncertainty 57
neural networks
  artificial *212*, 212–213
    advantages and disadvantages 225–228
    interpretability 227–228
    perturbation analysis 226–227
  deep learning 258–262
  image classification 241–242
  in machine learning 237–243
  under/overfitting 240–241
nitrogen (N)
  coupled carbon-nitrogen matrix models 34–44, 90–93
  global change 121
  influence on carbon input and residence time 109
  soil incubation experiments 146
  soil nitrogen content 215
  spin-up 90, 91, 92–93
  traceability analysis *110*, 112, 117
nitrogen cycle 7
nonautonomous compartmental systems
  compared to autonomous 46
  linear 49
  nonlinear 49–50
  time characteristics 95–96
nonautonomous ordinary differential equations (ODEs) 81–85, 87
nonautonomous systems 8–9
nonlinear compartmental systems 47–48, 49–50
nonlinearity, spin-up 100, 101
numerical weather prediction (NWP) models 12

O

one-pool model 82
ORCHIDEE-MICT model
  flow diagram 23
  matrix equation 29–33, *30*, *33*

P

PacAN *see* Predictive Ecosystem Analyzer
peatland bog ecosystems
  methane modeling 167–172
  SPRUCE experiment 163–166
peatland restoration 72
perturbation analysis 226–227
phase shift 122
phosphorus (P)
  matrix phosphorus model 63
    matrix approach and data assimilation 64–66
    new knowledge emerging 66–68
  role of 63
photosynthesis
  carbon flow diagrams 18, *19*
  data assimilation 133, *134*
  process rates 173–174
  role in the land carbon cycle 4
  Semi-Analytic Spin-Up (SASU) 90
  traceability analysis 105–110
photosynthetic sensitivity 224–228
photosynthetically active radiation (PAR)
  machine learning 215, 217, 225–226
  predicting GPP 217
  TECO model 196, 205
plant biomass
  reference data sets 122, **123**
  Semi-Analytic Spin-Up (SASU) 90
precipitation
  data assimilation 133
  future states 121
  and leaf area index *124*
  machine learning 215, 217, 224–228, *226*, *227*
  meteorological data 220
  SPRUCE experiment 163, 189–190, 196
  TECO model 114, 205
predictability, land carbon cycle 9
Predictive Ecosystem Analyzer (PacAn) 201
primary production *see* gross primary production; net primary production
priming effect 137–138
probability, Bayes' theorem 140–141
probability density functions (PDFs) 148
PROcess-Guided Deep Learning and DAta-Driven Modeling (PRODA) 244–252, 261
pullback attracting solution 49

pullback attractor 49
punctuated nitrogen addition approach (PN) 90
Python
  data assimilation 152–159
  downloading and installing 276, *277*
  introduction to programming in 268–274

Q

quantities of interest (QoIs) 165–166

R

random forest concept 229–233
Random Forest (RF) 211–212, *212*
recurrent neural network (RNN) 217
regression analysis, seven-step procedure 135–136
relative humidity
  machine learning 225
  SPRUCE experiment 164
  TECO model 114, 196, 205
Representational State Transfer (RESTful) 195
residence time *see* carbon residence time
ReSOM model, carbon flow diagram *25*, 25
respiration 18
  carbon flow diagrams 18–19, 22
  CARDAMOM approach 174–175
  data assimilation 133, *134*, 146
  land models 121
  reference data sets 122, **123**
  SPRUCE experiment 165, 168, 182–184
root-mean-square-error (RMSE) 122–123

S

Semi-Analytic Spin-Up (SASU) 90–93
  community cyberinfrastructures 203
  efficiency and convergence in TECO 98–102
Sevilleta Long-Term Ecological Research (LTER) site 189
Shapley values 215–216, 227–228
simulation modeling
  Bayes' theorem 140, 141
  data assimilation 133–135, *134*
  Markov Chain Monte Carlo 142
  SOC distribution 250
soil incubation experiments 146
  based on soil carbon models 146–147
  data assimilation 147–151
soil moisture 224–228
soil nitrogen content 215
soil organic carbon (SOC)
  data assimilation 137, 150–151
  machine learning 229–233, **231**
  PRODA approach 244–252, 261–262
  role in the land carbon cycle 5
soil organic matter (SOM)
  coupled carbon-nitrogen matrix models 34–36, *35*
  data assimilation 152, 176
  matrix representation 45
  modeling 16
  Semi-Analytic Spin-Up (SASU) 90
soil P dynamics 63–64
  example of applying a matrix model and data assimilation 64–66
  matrix approach 64
  new knowledge emerging 66–68
Soil-Plant-Atmosphere (SPA) model 175–176
Soil-Vegetation-Atmosphere Transfer (SVAT) schemes 14
spatial distributions 122
Sphagnum moss, SPRUCE experiment 163–166
spin-up
  definition 89
  development of 89
  Semi-Analytic Spin-Up (SASU) 90–93, 98–102
Spruce and Peatland Responses Under Changing Environments (SPRUCE) 163–166, 167, 182–184
  ecological forecasting 189, 204–206, *207*
  use of EcoPAD 196–197
stability analysis
  compartmental dynamical systems 48–49
  general stability statements 87–88
  global attractor 85–87
  steady state 85
steady state
  instantaneous 85
  spin-up 89
  traceability analysis 105–110
  transient traceability analysis 112–120
system dynamics modeling 12–13

T

terrestrial biosphere models (TBMs) 14
Terrestrial Ecosystem Model (TEM) 90
Terrestrial Ecosystem (TECO) model
  carbon balance equations 6
  coupled C-N version 34–36, 109, *110*
  data assimilation 136–137, 152–159, 184, 194
  datasets 204–205
  efficiency and convergence of SASU 98–102
  methane modeling 167–172
  SPRUCE experiment 165
  traceability analysis 109, *110*
  transient traceability analysis 113–114, *114*
three-pool model 84–85
time characteristics of compartmental systems 94–96
traceability analysis 60–61, *61*
  design and key components 105–110
  diagnostic tools 74
  exercises 126–130
  identification of uncertainty sources 105
  transient framework 112–120
TraceME system 110
transient traceability analysis 112–120
transit time distributions 94–96
tundra biomes 107

U

uncertainty
  community cyberinfrastructures 203
  data assimilation 137, 157
  diagnostic tools 74
  identification of sources 105
  in land carbon cycle modeling 57
  methane modeling 167, 171
  spin-up 89
  SPRUCE experiment 165–166, 205–206
uncertainty analysis
  shrinking model uncertainty to zero 61–62
  unified diagnostic system 57–61
unified diagnostic system
  uncertainty analysis 57–61
  use of one formula 57–58
United States Forest Service (USFS) 163

V

validation of models
  benchmark analysis 121
  CARDAMOM approach 173, 174–175
  data assimilation 66
  distinction from verification 16–17
  machine learning 213, 220
  uncertainty quantification 165
    see also verification
vapor pressure deficit (VPD)
  machine learning 224–228
  meteorological data 220
  model-data fusion (MDF) analysis *178*
  SPRUCE experiment 164
  TECO model 114, 205
vector auto-regression (VAR) 205
vegetation-atmosphere energy flux 14, **15**
vegetation carbon dynamics 7
  benchmark analysis 121
  C-N coupling 36–37, 41, 42, *43*
  data assimilation 189
  ecological forecasting 187
    community cyberinfrastructures 199–203
    data availability 189
    EcoPAD 133, 189–190, 192–198
    models and predictability of the terrestrial carbon cycle 187–189
    SPRUCE site 204–206, *207*
    workflow system 189–190
  external forcings 108
  matrix equation 75
  SPRUCE experiment 163, 164, 166
  uncertainty 57–58, 108
verification
  distinction from validation 16
  global climate treaties 180
  Land surface Verification Toolkit (LVT) 202
vertical mixing
  coupled carbon-nitrogen matrix models 41, 42
  spin-up approaches 90
  TECO model 168
  uncertainty analysis 61

W

weather forecasting 187
wind speed
  meteorological data 220
  TECO model 196, 205

For Product Safety Concerns and Information please contact our EU
representative GPSR@taylorandfrancis.com Taylor & Francis Verlag GmbH,
Kaufingerstraße 24, 80331 München, Germany

Printed and bound by CPI Group (UK) Ltd, Croydon, CR0 4YY
27/03/2025
01837554-0001